# MULTI-STOREY
# PRECAST CONCRETE
# FRAMED STRUCTURES

**Other books of interest**

*Structural Details in Concrete*
M. Y. H. Bangash
0 632 02853 X

*Structural Masonry Designers' Manual*
Second Edition (Revised)
W. G. Curtin, G. Shaw, J. K. Beck and W. A. Bray
0 632 03899 3

*Composite Structures of Steel and Concrete*
*Volume 1: Beams, Slabs, Columns, and Frames for Buildings*
Second Edition
R. P. Johnson
0 632 02507 7

*Glass Fibre Reinforced Cement*
A. J. Majumdar and V. Laws
0 632 02904 8

*Reinforced Concrete*
S. S. Ray
0 632 03724 5

# MULTI-STOREY PRECAST CONCRETE FRAMED STRUCTURES

**Kim S. Elliott**

BTech, PhD, CEng, MICE

*b*

**Blackwell
Science**

© 1996 by
Blackwell Science Ltd
Editorial Offices:
Osney Mead, Oxford OX2 0EL
25 John Street, London WC1N 2BL
23 Ainslie Place, Edinburgh EH3 6AJ
350 Main Street, Malden
  MA 02148 5018, USA
54 University Street, Carlton
  Victoria 3053, Australia
10, rue Casimir Delavigne
  75006 Paris, France

Other Editorial Offices:

Blackwell Wissenschafts-Verlag GmbH
Kurfürstendamm 57
10707 Berlin, Germany

Blackwell Science KK
MG Kodenmacho Building
7–10 Kodenmacho Nihombashi
Chuo-ku, Tokyo 104, Japan

First published 1996
Reissued in paperback 2000

Set in 10pt Times
by Setrite Typesetters Ltd, Hong Kong
Printed and bound in Great Britain by
MPG Books Ltd, Bodmin, Cornwall

The Blackwell Science logo is a
trade mark of Blackwell Science Ltd,
registered at the United Kingdom
Trade Marks Registry

DISTRIBUTORS

Marston Book Services Ltd
PO Box 269
Abingdon
Oxon OX14 4YN
(*Orders:* Tel: 01235 465500
          Fax: 01235 465555)

USA
Blackwell Science, Inc.
Commerce Place
350 Main Street
Malden, MA 02148 5018
(*Orders:* Tel: 800 759 6102
             781 388 8250
        Fax: 781 388 8255)

Canada
Login Brothers Book Company
324 Saulteaux Crescent
Winnipeg, Manitoba R3J 3T2
(*Orders:* Tel: 204 837-2987
        Fax: 204 837-3116)

Australia
Blackwell Science Pty Ltd
54 University Street
Carlton, Victoria 3053
(*Orders:* Tel: 03 9347 0300
        Fax: 03 9347 5001)

A catalogue record for this title
is available from the British Library

ISBN 0-632-05529-4

Library of Congress
Cataloging-in-Publication Data

Elliott, Kim S.
    Multi-storey precast concrete framed
    structures/Kim S. Elliott.
        p.    cm.
    Includes index.
    ISBN 0-632-03415-7 (alk. paper)
    1. Precast concrete construction.   2. Tall
    buildings – Design and construction.
    I. Title
    TH1498.E43   1995
    693′.522—dc20                        95-30820
                                             CIP

For further information on
Blackwell Science, visit our website:
www.blackwell-science.com

# Contents

*Preface*                                                                                                      xiii

*Notation*                                                                                                      xvi

**Chapter 1  Precast Concepts, History and Design Philosophy**                                                    1
    1.1  An historical note on the development of precast frames                                1
    1.2  The scope for prefabricated buildings                                                  7
    1.3  Current attitudes towards precast concrete structures                                15
    1.4  Recent trends in design, and a new definition for precast concrete                    18
    1.5  Precast superstructure simply explained                                              20
        1.5.1  Differences in precast and cast insitu concrete structures      20
        1.5.2  Structural stability                                           26
        1.5.3  Floor plate action                                             28
        1.5.4  Connections and joints                                         30
        1.5.5  Foundations                                                    30
    1.6  Precast design concepts                                                              33
        1.6.1  Devising a precast solution                                    33
        1.6.2  Construction methods                                           36

**Chapter 2  Procurement and Documentation**                                                                    43
    2.1  Initial considerations for the design team                                           43
    2.2  Design procurement                                                                   44
        2.2.1  Definitions                                                    44
        2.2.2  Responsibilities                                               45
        2.2.3  Routes to procurement                                          46
        2.2.4  Design office practice                                         46
        2.2.5  Project design stages                                          48
        2.2.6  Structural design calculations                                 50
        2.2.7  Layout drawings                                                50
        2.2.8  Component schedules and the engineer's instructions to factory and
              site                                                      55
    2.3  Construction matters                                                                 55
        2.3.1  Design implications                                            55
    2.4  Codes of practice, design manuals, text books and technical literature               61
        2.4.1  Codes and building regulations                                 61
        2.4.2  Non-mandatory design documents                                 63

|  |  | 2.4.3 Other literature on precast structures | 64 |
|---|---|---|---|
|  | 2.5 | Definitions | 65 |
|  |  | 2.5.1 General structural definitions | 65 |
|  |  | 2.5.2 Components | 66 |
|  |  | 2.5.3 Connections and jointing materials | 67 |

**Chapter 3  Architectural and Framing Considerations**                     68

    3.1   Frame and component selection                                      68
    3.2   Component selection                                                72
          3.2.1   General principles                                        72
          3.2.2   Roof and floor slabs                                      72
                  3.2.2.1   Hollow core units                              73
                  3.2.2.2   Double-tee slabs                               81
                  3.2.2.3   Composite plank floor                         84
          3.2.3   Staircases                                                84
          3.2.4   Roof and floor beams                                    91
          3.2.5   Beam-to-column connections                              95
          3.2.6   Columns                                                   97
          3.2.7   Bracing walls                                           100
    3.3   Special features                                                 103
          3.3.1   Hybrid construction                                     103
          3.3.2   Precast–insitu concrete structures                     105
          3.3.3   Precast concrete–steelwork or timber structures        108
          3.3.4   Precast concrete–masonry structures                    110
    3.4   Balconies                                                        111

**Chapter 4  Design of Skeletal Structures**                               120

    4.1   Basis for the design                                            120
    4.2   Materials                                                        122
          4.2.1   Concrete                                                123
          4.2.2   Concrete admixtures                                     126
          4.2.3   Reinforcement                                           126
          4.2.4   Prestressing steel                                      126
          4.2.5   Structural steel and bolts                              127
          4.2.6   Non-cementitious materials                             127
    4.3   Structural design                                               127
          4.3.1   Terminology                                             127
          4.3.2   Design methods                                          128
          4.3.3   Design of beams                                         130
          4.3.4   Non-composite reinforced concrete beams                131
          4.3.5   Beam boot design                                        136
          4.3.6   Upstand design                                          141
          4.3.7   Non-composite prestressed beams                        150
                  4.3.7.1   Flexural design                              150
                  4.3.7.2   Boot reinforcement                           156
                  4.3.7.3   Shear design                                 156
          4.3.8   Beam end shear design                                   160
          4.3.9   Recessed beam ends                                      161

# Chapter vii

4.3.10  Design methods for end shear                                   165
4.3.11  Hanging shear cages for wide beams                             173
4.3.12  Prefabricated shear boxes                                      179
    4.3.12.1   Wide box design                    180
    4.3.12.2   Narrow plate design                 185
4.4  Columns subjected to gravity loads                                188
  4.4.1   General design                                     188
    4.4.1.1   Design rules in BS 8110 for columns in precast structures   189
  4.4.2   Column in braced structures                        192
  4.4.3   Columns in unbraced structures                     192
  4.4.4   Columns in partially braced structures             193
4.5  Staircases                                                        198
  4.5.1   Reinforced concrete staircases                     198
  4.5.2   Prestressed concrete staircases                    200
  4.5.3   Staircase and landing end reinforcement            201

**Chapter 5   Design of Precast Floors used in Precast Frames**       206
5.1  Flooring options                                                  206
5.2.  Hollow core slabs                                                207
  5.2.1   General                                            207
  5.2.2   Design                                             214
  5.2.3   Design of cross-section                            215
  5.2.4   Web thickness                                      216
  5.2.5   Edge profiles                                      218
  5.2.6   Reinforcement                                      218
  5.2.7   Lateral load distribution                          220
  5.2.8   Flexural capacity                                  224
    5.2.8.1   Serviceability limit state of flexure   225
    5.2.8.2   Ultimate limit state of flexure      228
    5.2.8.3   Strength sensitivity exercise        228
  5.2.9   Precamber and deflections                          229
  5.2.10  Shear capacity                                     229
    5.2.10.1   Shear capacity in the uncracked region, $V_{co}$   230
    5.2.10.2   Shear capacity in the region cracked in flexure, $V_{cr}$   232
  5.2.11  Anchorage and bond development lengths             233
  5.2.12  Cantilever design using hollow core slabs          235
  5.2.13  Wet cast reinforced hollow core flooring           235
5.3  Double-tee slabs                                                  240
  5.3.1   General                                            240
  5.3.2   Design                                             243
  5.3.3   Flexural and shear capacity, precamber and deflections   246
  5.3.4   Special design situations                          247
5.4  Composite plank floor                                             247
  5.4.1   General                                            247
  5.4.2   Design                                             248
  5.4.3   Voided composite slab                              251
5.5  Precast beam and plank flooring                                   254
  5.5.1   General                                            254

|  |  |  |  |
|---|---|---|---|
|  | 5.5.2 | Design of prestressed beams in the beam and plank flooring system | 255 |
| 5.6 | Design calculations |  | 255 |
|  | 5.6.1 | Hollow core flooring | 255 |

**Chapter 6   Composite Construction** — 258
| 6.1 | Introduction |  | 258 |
| 6.2 | Texture of precast concrete surfaces |  | 259 |
|  | 6.2.1 | Classification of surface textures | 259 |
|  | 6.2.2 | Surface treatment and roughness | 261 |
| 6.3 | Calculation of stresses at the interface |  | 262 |
| 6.4 | Losses and differential shrinkage effects |  | 264 |
|  | 6.4.1 | Losses in prestressed composite sections | 264 |
|  | 6.4.2 | Design method for differential shrinkage | 266 |
|  | 6.4.3 | Cracking in the precast and insitu concrete | 270 |
| 6.5 | Composite floors |  | 271 |
|  | 6.5.1 | General considerations | 271 |
|  | 6.5.2 | Flexural analysis for prestressed concrete elements | 273 |
|  |  | 6.5.2.1   At serviceability state | 273 |
|  |  | 6.5.2.2   At ultimate limit state | 275 |
|  | 6.5.3 | Propping | 276 |
|  | 6.5.4 | Design calculations | 277 |
| 6.6 | Economic comparison of composite and non-composite hollow core floors |  | 281 |
| 6.7 | Composite beams |  | 283 |
|  | 6.7.1 | Flexural design | 283 |
|  | 6.7.2 | Propping | 287 |
|  | 6.7.3 | Horizontal interface shear | 288 |
|  | 6.7.4 | Shear check | 288 |
|  | 6.7.5 | Deflections | 289 |

**Chapter 7   Design of Connections and Joints** — 293
| 7.1 | Development of connections |  | 293 |
| 7.2 | Design brief |  | 296 |
| 7.3 | Joints and connections |  | 301 |
| 7.4 | Criteria for joints and connections |  | 302 |
|  | 7.4.1 | Design criteria | 302 |
|  |  | 7.4.1.1   Strength | 303 |
|  |  | 7.4.1.2   Influence of volume changes | 303 |
|  |  | 7.4.1.3   Ductility | 303 |
|  |  | 7.4.1.4   Durability | 303 |
| 7.5 | Types of joints |  | 304 |
|  | 7.5.1 | Compression joints | 304 |
|  | 7.5.2 | Tensile joints | 311 |
|  | 7.5.3 | Shear joints | 313 |
|  | 7.5.4 | Flexural and torsional joints | 319 |
| 7.6 | Bearings and bearing stresses |  | 320 |
|  | 7.6.1 | Average bearing stresses | 320 |
|  | 7.6.2 | Localized bearing stresses | 324 |
| 7.7 | Connections |  | 325 |

|  | 7.7.1 | Pinned connections | 325 |
|  | 7.7.2 | Moment resisting connections | 325 |
| 7.8 | Design of specific connections in skeletal frames | | 326 |
|  | 7.8.1 | Floor slab-to-beam connections | 326 |
|  | 7.8.2 | Connections at supports | 326 |
|  | 7.8.3 | Connections at longitudinal joints | 331 |
|  | 7.8.4 | Floor connections at load bearing walls – load bearing components | 333 |
| 7.9 | Beam-to-column, and beam-to-wall connections | | 336 |
|  | 7.9.1 | Definitions for different assemblies | 336 |
|  | 7.9.2 | Connections to continuous columns using hidden steel inserts | 337 |
|  | 7.9.3 | Beam-to-column inserts at C | 337 |
| 7.10 | Column insert design | | 339 |
|  | 7.10.1 | General considerations | 339 |
|  | 7.10.2 | Single-sided insert connections | 344 |
|  | 7.10.3 | Additional welded reinforcement to inserts | 352 |
|  | 7.10.4 | Double-sided billet | 356 |
|  | 7.10.5 | Three- and four-way connections | 360 |
|  | 7.10.6 | Narrow plate column inserts | 365 |
|  | 7.10.7 | Cast-in sockets | 366 |
|  | 7.10.8 | Bolts in sleeves | 368 |
| 7.11 | Connections to columns on concrete ledges | | 369 |
|  | 7.11.1 | Corbels | 369 |
|  |  | 7.11.1.1  Shallow corbels | 372 |
|  |  | 7.11.1.2  Deep corbels | 376 |
|  | 7.11.2 | Haunched columns | 379 |
|  | 7.11.3 | Connections to the tops of columns | 384 |
| 7.12 | Beam-to-beam connections | | 387 |
| 7.13 | Column splices | | 396 |
|  | 7.13.1 | Types of splices | 396 |
|  | 7.13.2 | Column-to-column splices | 396 |
|  | 7.13.3 | Coupled joint splice | 397 |
|  | 7.13.4 | Welded plate splice | 400 |
|  | 7.13.5 | Grouted sleeve splice | 402 |
|  | 7.13.6 | Grouted sleeve coupler splice | 403 |
|  | 7.13.7 | Steel shoe splices | 403 |
|  | 7.13.8 | Columns spliced on to beams or other precast components | 408 |
| 7.14 | Column base connections | | 409 |
|  | 7.14.1 | Columns in pockets | 409 |
|  | 7.14.2 | Columns on base plates | 419 |
|  | 7.14.3 | Columns on grouted sleeves | 428 |

**Chapter 8  Designing for Horizontal Loading** | | | **430** |
| 8.1 | Introduction | | 430 |
| 8.2 | Distribution of horizontal loading | | 431 |
| 8.3 | Horizontal diaphragm action in precast concrete floors without structural screeds | | 441 |
|  | 8.3.1 | Background | 441 |
|  | 8.3.2 | Details | 443 |

| | | | |
|---|---|---|---|
| | 8.3.3 | Structural models for diaphragm action | 444 |
| | 8.3.4 | Diaphragm reinforcement | 451 |
| | 8.3.5 | Design by testing | 454 |
| | 8.3.6 | Finite element analysis of the floor plate | 458 |
| 8.4 | Diaphragm action in composite floors with structural screeds | | 460 |
| 8.5 | Horizontal forces due to volumetric changes in precast concrete | | 462 |
| 8.6 | Vertical load transfer | | 465 |
| | 8.6.1 | Introduction | 465 |
| | 8.6.2 | Unbraced structures | 466 |
| | 8.6.3 | Deep spandrel beams in unbraced structures | 469 |
| | 8.6.4 | Braced structures | 471 |
| | 8.6.5 | Uni-directionally braced structures | 474 |
| | 8.6.6 | Partially braced structures | 474 |
| 8.7 | Methods of bracing structures | | 478 |
| | 8.7.1 | Infill shear walls | 478 |
| | 8.7.2 | Design methods for infill concrete walls | 483 |
| | 8.7.3 | Design method for brickwork infill panels | 487 |
| | 8.7.4 | Infill walls without beam-framing elements | 489 |
| | 8.7.5 | Use of slip-formed hollow core walls as infill walls | 490 |
| | 8.7.6 | Cantilever shear walls and shear boxes | 497 |
| | 8.7.7 | Hollow core cantilever shear walls | 500 |
| | 8.7.8 | Solid cantilever shear walls | 503 |

**Chapter 9  Structural Integrity and the Design for Accidental Loading** — 510

| | | | |
|---|---|---|---|
| 9.1 | Precast frame integrity – the vital issue | | 510 |
| 9.2 | Ductile frame design | | 510 |
| 9.3 | Background to the present requirements | | 515 |
| 9.4 | The fully tied solution | | 521 |
| | 9.4.1 | Horizontal ties | 521 |
| | 9.4.2 | Calculation of tie forces | 526 |
| | 9.4.3 | Horizontal ties to columns | 530 |
| | 9.4.4 | Ties at balconies | 534 |
| | 9.4.5 | Vertical ties | 534 |
| 9.5 | Catenary systems in precast construction | | 537 |

**Chapter 10  Site Practice and Temporary Stability** — 541

| | | | |
|---|---|---|---|
| 10.1 | The effects of construction techniques on design | | 541 |
| 10.2 | Designing for pitching and lifting | | 544 |
| | 10.2.1 | Early lifting strengths | 544 |
| | 10.2.2 | Lifting points | 546 |
| | 10.2.3 | Handling | 556 |
| | 10.2.4 | Cracks | 556 |
| 10.3 | Temporary frame stability | | 561 |
| | 10.3.1 | Propping | 561 |
| | 10.3.2 | The effect of erection sequence | 562 |
| | 10.3.3 | Special consideration for braced frames | 564 |
| | 10.3.4 | Special considerations for unbraced frames | 565 |
| | 10.3.5 | Temporary loads | 567 |

| | |
|---|---|
| 10.4  On-site connections | 568 |
| 10.4.1  Effect of fixing types | 568 |
| 10.4.2  Strength and maturity of connections | 569 |
| 10.5  Erection procedure | 570 |
| 10.5.1  Site preparation | 570 |
| 10.5.2  Erection of precast superstructure | 570 |
| 10.5.2.1   Columns | 571 |
| 10.5.2.2   Beams | 575 |
| 10.5.2.3   Stability walls | 576 |
| 10.5.2.4   Two-dimensional units, including floor slabs | 577 |
| 10.6  Insitu concrete | 579 |
| 10.6.1  General specification | 579 |
| 10.6.2  Concrete screeds and joint infill in floors | 580 |
| 10.6.3  Grouting | 581 |
| 10.7  Hand-over | 581 |
| *References* | 583 |
| *Index* | 595 |

# Preface

Of all the major forms of multi-storey building construction, structural precast concrete is perhaps the least understood. The most common 'black spot' in the building profession is associated with the design and construction of large or small multi-storey precast and prestressed concrete frames. This is due mainly to two factors:

- the notion of using a modular form of construction, such as precast concrete, is not encouraged at undergraduate level because it is thought of as being too narrow in the wider application of theory and design instruction
- precast concrete design is usually carried out in house by engineers employed by the precast manufacturing companies.

As a result, the trainee structural designer is not exposed to the virtues, good or indifferent, of using concrete in this way. Even where precast concrete is accepted as a viable alternative form of construction to, say steelwork for medium- to high-rise structures, or to insitu concrete for some of the more complex shape buildings, and even to masonry for low-rise work, the opportunities to study the basic concepts adopted in the design, manufacturing and site erection stages have not been made available to the vast majority of trainees. Thus, precast frame design has long been considered as having a secondary role to the main structural work, and rarely have intelligent challenges been made to precast designers questioning the fundamental principles they are using. In a conversation with the author an eminent structural engineer in London referred to precast frame design as a 'black box'.

To meet with the increased building specifications of the past 15 to 20 years, precast manufacturing companies have refined the design of their product. They have formed highly effective product associations dealing with not only the marketing and manufacturing of the 'product', but technical matters relating to common design problems, research initiatives, education, unified design approaches, responses to proposals for Eurocodes and, most importantly, the encouragement of a wider appreciation of precast structures in the professional design office. Even so, the structural and architectural complexity of some of the more recent precast frames has deepened the gap between precast designers and the rest of the profession. The latter have nowhere to turn for guidance on how the former are working. Satisfying codes of practice and the building regulations plays only a minor role in the total package. Many gaps still need to be filled.

Despite this, precast reinforced and prestressed concrete for multi-storey framed buildings is widely regarded as an economic, structurally sound and architecturally versatile building medium. Design concepts have evolved over the last 30 years to satisfy a wide range of commercial and industrial building needs. 'Precast concrete frames' is a term which is now synonymous with high quality, strength, stability, durability and robustness. Design is carried out to the highest standard of exactness unsurpassed in the concrete industry and yet the knowledge, for reasons given above, remains essentially within the precast industry itself.

Precast concrete buildings do not behave in the same way as cast insitu ones. The components which make up the completed precast structure are subjected to different forces and movements from the concrete in the monolithic structure. It is necessary to understand where these physical effects come from, where they go to, and how they are transferred through the structure.

This book is aimed at making the first step towards the widespread understanding of the procedures involved in precast structural design, from drawing office practice to explaining the reasons for some of the more intricate operations performed by precast contractors on site. The author has chosen to concentrate on skeletal type structures. These are defined as frameworks consisting of beams, columns, slabs and a small number of shear walls. There is no doubt that skeletal frames are the most demanding of all precast structures, from the structural and architectural viewpoints. They contain the smallest quantity of structural concrete per unit volume. The precast components can be co-ordinated into the architectural façade, both internally and externally to meet the social, economic and ecological demands that the twenty-first century will inevitably bring. There is no doubt that the construction industry, which is seeking greater challenges in total building design, will turn to high specification prefabricated concrete for its advancement. There is new terminology being used for this, it is 'factory engineered concrete'.

The chapters in this book have been arranged so that different parts of the design process can either be isolated, for example in the cases of precast flooring or connections without necessarily referring to the overall frame design, or read sequentially to realize the entire design. Similarly, it is also possible to read the subject on a superficial, i.e. non-technical, level in Chapters 1 to 3. Chapters 4 to 9 describe, in detail, the design procedures as would be carried out in a precast manufacturing company's design office. Chapter 10 describes the relevant site construction methods. Numerous examples have been used to demonstrate the application of design rules, many of which are not code dependent.

There are many aspects to the design of precast skeletal frames that are appearing in print for the first time, e.g. designs which, before today, have evolved through the natural development of precast frame design since the 1950s and yet have never been committed in print. This information has not been easy to find. For example, the information leading to the calculation of the lever arm in a precast floor diaphragm is found in a paper presented at a Dutch seminar in 1990; the validation of the assumed behaviour of a precast column connector appeared in an American journal in 1980. To add to this the precast concrete industry has kept closely guarded secrets about design techniques, in particular connector design, and has been criticized in the past by developers and consultants for doing so. Although this information is now more freely available since the expiry of many patentable ideas, it has left many unchartered design procedures. One of the main purposes of this book is to bring together in a coherent manner the wide and varied design methods used in the industry for the benefit of everyone. This First Edition has therefore to be put to the test.

Precast concrete designs are not entirely code dependent, but the parts which comply are in accordance with the British code BS 8110 despite the fact that the publication of the book coincides with the release of the Eurocode EC2. There are two good reasons for this. First, this book 'breaks a lot of new ground' and many of the designs are interpolations of existing practices. A safe approach has therefore been adopted in using established design rules. Second, the part of EC2 dedicated to prefabricated concrete, Part 1.3, is not complete. Any attempt to design this code would have involved speculation on the final outcome of some of the clauses.

The author is indebted to the following individuals and companies for their personal assistance and corporate help in the preparation of this book: Belcon Structures (Hoddesdon, UK), Birchwood Ltd (Somercoates, UK), Bison Floors Ltd (Litchfield, UK), Bison Holdings Ltd (Slough, UK), Blatcon Ltd (Wells, UK), Breton Precast (Glasgow, UK), Brick Development Association (Windsor, UK), British Cement Association (Crowthorne, UK), British Precast Concrete Federation (Leicester, UK), B + FT (Germany), Composite Structures (Eastleigh, UK), Costain Building Products (formerly DowMac) (Stamford, UK), Crendon Structures (Aylesbury, UK), Andrew T. Curd & Partners (USA), Deecrete (Washington, Tyne & Wear, UK), F.C. Ltd (Derby, UK), Hume Industries Berhard (Malaysia), Richard Lees Ltd (Derby, UK), New Zealand Concrete Society, Partek Ltd (Belgium), Span Deck (USA), Tarmac Precast (UK) and Trent Concrete Ltd (Nottingham, UK) for allowing the publication of technical literature, and photographs of their works and/or construction sites. Grateful acknowledgement is made to Mr Roy Wright (formerly Trent Concrete Ltd) for his reference, and to research colleagues Dr Gwynne Davies, Dr Wahid Omar, Mr Reza Adlparvar, Mr Halil Gorgun, Dr Ali Mahdi and Mr Kamel Bensalem. Thanks also go to my wife for enduring the hundreds of hours of family life lost during the preparation of this book.

Kim S. Elliott

# Notation

| | |
|---|---|
| $a$ | lever arm distance; distance to wall from shear centre |
| $a'$ | distance from compression face to point where crack width is calculated |
| $a_b$ | clear distance between bars; cover to inside face of bars |
| $a_{cr}$ | distance from nearest bar to point where crack width is calculated |
| $a_{eff}$ | effective bearing length at corbel |
| $a_f$ | tangent of coefficient of friction, i.e. $\mu = \tan a_f$ |
| $a_u$ | sway deflection |
| $a_v$ | lever arm distance to shear force |
| $b$ | breadth of section |
| $b_e$, $b_{eff}$ | effective breadth |
| $b_1$ | length of bearing |
| $b_p$ | breadth of bearing |
| $b_t$ | breadth of section at centroid of steel in tension |
| $b_v$ | breadth of shear section or shear web |
| $b_w$ | breadth of web |
| $c$ | cover distance; distance to centre of bar |
| $c_{min}$ | minimum distance to bar in tension |
| $c_w$ | crack width |
| $d$ | effective depth of section to tension steel; depth of web in steel sections |
| $d'$ | effective depth to compression steel |
| $d''$ | effective depth to tension steel in boot of beam |
| $d_f$ | effective depth to edge of foundation pocket |
| $d_h$ | effective depth of half joint |
| $d_n$ | depth to centroid of compression zone |
| $e$ | eccentricity |
| $e'$ | effective eccentricity; lack of verticality |
| $e_{add}$ | second order eccentricity (infill wall) |
| $e_{net}$ | net eccentricity |
| $e_x$ | minimum eccentricity (infill wall) |
| $f_b$ | ultimate bearing stress; limiting flexural compressive stress in concrete |
| $f_{bc}$ | bottom fibre stress due to prestress after losses |
| $f_{bci}$ | bottom fibre stress due to prestress at transfer |
| $f_c$ | compressive strength of bearing material |
| $f_c'$ | effective compressive strength of precast insitu concrete joint |
| $f_{cc}$ | prestress at centroid of tendons after losses |
| $f_{cci}$ | prestress at centroid of tendons at transfer |
| $f_{ci}$ | characteristic compressive cube strength of concrete at transfer |

| | |
|---|---|
| $f_{cp}$ | prestress at centroidal axis (taken as positive) |
| $f_{cpx}$ | prestress at centroidal axis at distance $x$ from end of section |
| $f_{ct}$ | limiting flexural tensile stress in concrete |
| $f_{cu}$ | characteristic compressive cube strength of concrete |
| $f_{cu}'$ | characteristic compressive cube strength of infill concrete, ditto at lifting |
| $f_{cyl}$ | characteristic compressive cylinder strength of concrete |
| $f_k$ | characteristic compressive strength of brickwork |
| $f_{pb}$ | design tensile stress in tendons/wires |
| $f_{pe}$ | final prestress in tendons/wires after losses |
| $f_{pu}$ | characteristic strength of prestressing tendons/wires |
| $f_s$ | stress in reinforcing steel bars |
| $f_{tc}$ | top fibre stress due to prestress after losses |
| $f_{ti}$ | top fibre stress due to prestress at transfer |
| $f_t$ | limiting direct (splitting) tensile stress in concrete |
| $f_v$ | characteristic shear strength of brickwork |
| $f_y$ | characteristic strength of reinforcing steel bars |
| $f_{yb}$ | characteristic strength of bolts |
| $f_{yv}$ | characteristic strength of reinforcing steel links/stirrups |
| $g_k$ | characteristic uniformly distributed dead load |
| $h$ | floor-to-floor height; depth of section |
| $h'$ | clear floor-to-floor height; reduced depth at half joint |
| $h_{agg}$ | nominal size of aggregate |
| $h_s$ | depth of slab in composite construction |
| $k$ | core distance $= I/0.5\,hA$ |
| $k_s$ | shear stiffness in joints |
| $k_b$ | flexural stiffness in joints |
| $k_j$ | rotational stiffness in joints |
| $l$ | distance between column-to-column centres |
| $l_b$ | bearing length |
| $l_e$ | effective length |
| $l_o$ | clear height between restraints |
| $l_p$ | prestress development length |
| $l_r$ | distance between columns or walls (stability ties) |
| $l_t$ | prestress transmission length |
| $l_w$ | length of weldment |
| $l_x$ | penetration of starter bar into hole |
| $l_z$ | distance between positions of zero bending moment |
| $m$ | distance from centre of starter bar to holding down bolt (base plate) |
| $m$ | modular ratio |
| $m_s$ | modular ratio (of elastic modulii) at service |
| $m_u$ | modular ratio (of strength) at ultimate |
| $n$ | number of columns in one plane frame, number of bars in tension zone of wall; number of storeys |
| $p_b$ | bearing strength of steel plate |
| $p_{weld}$ | strength of weld material |
| $p_y$ | strength of steel plate |
| $q$ | pressure (key elements) |
| $q_k$ | characteristic uniformly distributed live load |

| | |
|---|---|
| $r$ | radius of gyration; bend radius of reinforcing bar |
| $s$ | spacing of reinforcing bars |
| $t$ | thickness of section; temperature range |
| $t_{ef}$ | effective thickness |
| $u$ | perimeter distance |
| $v$ | thickness of insitu infill; ultimate shear stress |
| $v_{ave}$ | average interface shear stress |
| $v_c$ | design concrete shear stress |
| $v_h$ | design interface shear stress |
| $v_t$ | design torsion stress |
| $w$ | length of steel plate; uniformly distributed load; breadth of compressive strut |
| $w'$ | diagonal length of infill shear wall |
| $w_k$ | characteristic uniformly distributed wind pressure |
| $x$ | dimension of stirrup in boot of beam |
| $x$ | distance to centroid of stabilizing system |
| $y_{po}$ | half bearing breadth $(b_p/2)$ |
| $y_o$ | half section breadth $(b/2)$ |
| $z$ | lever arm |
| | |
| $A$ | area; cross-section area |
| $A_b$ | area of bolts |
| $A_{bst}$ | area of bursting reinforcement |
| $A_c$ | cross-section area of concrete |
| $A_c'$ | gross cross-section area of hollow core slabs |
| $A_{c' (net)}$ | net cross-section area of hollow core slabs |
| $A_d$ | area of diagonal reinforcement |
| $A_f$ | cross-section area of flange |
| $A_{hd}$ | area of diaphragm reinforcement |
| $A_k$ | contact area in castellated joint |
| $A_{ps}$ | area of prestressing reinforcement |
| $A_s$ | area of tension reinforcement |
| $A_s'$ | area of compression reinforcement |
| $A_s''$ | area of longitudinal reinforcement in top of boot of beam |
| $A_{sc}$ | total area of reinforcement in column |
| $A_{sh}$ | area of horizontal reinforcement |
| $A_{shv}$ | area of horizontal punching shear reinforcement |
| $A_{sv}$ | area of shear reinforcement |
| $A_{sw}$ | area of reinforcing bars welded to plate |
| $B$ | breadth of void in slab; breadth of building; breadth of foundation |
| $C$ | compressive force |
| $D$ | depth of pocket in foundation; depth of floor diaphragm; depth of hcu |
| $E$ | Young's modulus of elasticity |
| $E'$ | equivalent Young's modulus in precast insitu joint |
| $E_c$ | Young's modulus of concrete |
| $E_c'$ | Young's modulus of insitu concrete |
| $E_{ci}$ | Young's modulus of concrete at transfer |
| $E_i$ | Young's modulus of infill concrete |

| | |
|---|---|
| $E_s$ | Young's modulus of steel |
| $F$ | force |
| $F_c$ | compressive force in concrete |
| $F_{cR}$ | ultimate compressive resistance force |
| $F_d$ | tensile force in diagonal reinforcing bars |
| $F_h$ | tensile force in horizontal reinforcing bars |
| $F_s$ | tensile force in reinforcing bars |
| $F_t$ | notional tensile force in stability ties |
| $F_t'$ | tensile force in stability ties |
| $F_{tR}$ | ultimate tensile resistance force |
| $F_v$ | shear force to one side of interface (composite construction) |
| $G$ | shear modulus |
| $G_k$ | characteristic dead load |
| $H$ | total height of building; length of foundation; horizontal force |
| $H_{bst}$ | bursting force |
| $H_v$ | horizontal resistance of infill wall |
| $I$ | second moment of area of section |
| $I_c$ | transformed second moment of area of composite section |
| $K$ | stress factor $M/f_{cu}\,bd^2$; shrinkage factor |
| $K_b$ | flexural stiffness of connections between members |
| $K_s$ | shear stiffness of connections between members |
| $K_t$ | bond length parameter |
| $L$ | span; length of void in slab; length of building |
| $L$ | base plate overhang distance |
| $L'$ | clear opening between columns; length of wall |
| $L_e$ | effective length |
| $L_s$ | clamping length |
| $L_{sb}$ | bond length |
| $L_1$ | distance from face of column to load |
| $L_2, L_3$ | length of stress block (insert design) |
| $L_4$ | length of embedment of insert |
| $M$ | bending moment |
| $M'$ | moment of resistance based on strength of concrete $= 0.156\,f_{cu}bd^2$ |
| $M_{add}$ | additional bending moment due to deflection $= Na_u$ |
| $M_b$ | bending moment due to shrinkage |
| $M_c$ | balancing bending moment due to shrinkage |
| $M_h$ | bending moment in floor diaphragm |
| $M_{max}\ M_{min}$ | maximum and minimum bending moment |
| $M_{net}$ | $M_{max} - M_{min}$ |
| $M_o$ | decompression bending moment |
| $M_R$ | moment of resistance |
| $M_{sr}$ | serviceability moment of resistance |
| $M_{ur}$ | ultimate moment of resistance |
| $M_1, M_2, M_3$ | strength of connections in frames |
| $N$ | ultimate axial force; number of strands/wires/bolts |
| $P$ | prestressing force |
| $P_f$ | prestressing force after all losses |
| $P_i$ | initial prestressing force |

| | |
|---|---|
| $P_t$ | prestressing force at transfer |
| $Q_k$ | characteristic live load |
| $R$ | prop force |
| $R_a$ | roughness factor |
| $R_v$ | diagonal resistance of infill wall |
| $S$ | first moment of area of section; plastic section modulus |
| $S_c$ | first moment of area to one side of interface |
| $T$ | tension force; torque |
| $V$ | reaction force; shear force |
| $V_{co}$ | shear resistance in flexurally uncracked prestressed section |
| $V_{cr}$ | shear resistance in flexurally cracked prestressed section |
| $V_d$ | shear force in dowel |
| $V_h$ | horizontal shear force |
| $V_r$ | increased shear capacity due to additional reinforcement (insert design) |
| $V_{uh}$ | ultimate horizontal shear force |
| $W$ | self weight of column |
| $W_k$ | characteristic wind load |
| $W_u$ | ultimate wind load |
| $X$ | distance to neutral axis; stress block depth factor |
| $Z$ | elastic section modulus |
| $Z_b, Z_t$ | elastic section modulus at extreme bottom and top fibres |
| $Z_c$ | elastic cracked concrete section modulus |
| $Z_u$ | elastic uncracked concrete section modulus |
| | |
| $\alpha$ | angle; ratio $Z_t/Z_b$; coefficient of thermal expansion; characteristic contact length in infill wall |
| $\alpha_c$ | ratio of sum of column stiffness to beam stiffness |
| $\alpha_{cmin}$ | minimum value of $\alpha_c$ |
| $\beta$ | ratio of propped to unpropped moment; column effective length factor; angle |
| $\delta$ | deflection; shear slip |
| $\delta_s$ | shrinkage induced deflection |
| $\varepsilon$ | strain |
| $\varepsilon_b$ | free shrinkage strain in precast beam or slab |
| $\varepsilon_{bb}$ | free shrinkage strain in bottom of precast beam or slab |
| $\varepsilon_{bt}$ | free shrinkage strain in top of precast beam or slab |
| $\varepsilon_c$ | concrete strain |
| $\varepsilon_f$ | free shrinkage strain in insitu concrete flange or topping |
| $\varepsilon_m$ | average strain at level where crackwidth is calculated |
| $\varepsilon_s$ | steel strain; relative shrinkage strain $\varepsilon_f - \varepsilon_{bt}$ |
| $\varepsilon_{sh}$ | shrinkage strain |
| $\varepsilon_1$ | strain at the level where crackwidth is calculated |
| $\phi$ | rotation |
| $\gamma$ | partial factor of safety |
| $\eta$ | total losses in prestressing force; force reduction factors |
| $\lambda$ | relative stiffness parameter; joint deformability |
| $\mu$ | coefficient of friction; degree of prestress force $= P_i/P$ |
| $\theta$ | angle; slope of infill wall |

| | |
|---|---|
| $\rho$ | reinforcement ratio $= A_s/bd$ |
| $\sigma$ | stress |
| $\sigma_{sp}$ | spalling stress |
| $\tau$ | shear stress |
| $\zeta$ | bursting coefficient; prestress loss due to elastic shortening |
| $\Delta$ | second order deflection; deformation; construction tolerance distance |
| $\Phi$ | reinforcing bar or dowel diameter; ductility factor |

# Chapter 1

# Precast Concepts, History and Design Philosophy

*The background to the relevance of precast concrete as a modern construction method for multi-storey buildings is described. The design method is summarized.*

## 1.1 AN HISTORICAL NOTE ON THE DEVELOPMENT OF PRECAST FRAMES

Precast concrete is not a new idea. William H. Lascelles (1832–85) of Exeter, England devised a system of precasting concrete wall panels, 3 ft × 2 ft × 1 in. thick strengthened by forged square iron bars of $\frac{1}{8}$ in. sides. The cost was 3d (£0.01) per ft$^2$. Afterwards the notion of 'pre-casting' concrete for major structural purposes began in the late nineteenth century when its most obvious application – to span over areas with difficult access – began with the use of flooring joists. François Hennebique (1842–1921) first introduced precast concrete into a cast insitu flour mill in France, where the self weight of the prefabricated units was limited to the lifting capacity of two strong men! White [1.1] and Morris [1.2] give good historical accounts of these early developments.

The first precast and reinforced concrete (rc) frame in Britain was Weaver's Mill in Swansea. In referring to the photograph of the building, shown in Fig. 1.1, an historical note states

'… the large building was part of the flour mill complex of Weaver and Co. The firm established themselves at the North Dock basin in 1895–6, and caused the large ferro-concrete mill to be built in 1897–8. It was constructed on the system devised by a Frenchman, F. Hennebique, the local architect being H. C. Portsmouth …'.

The structure was a beam-column skeletal frame, generally of seven storeys in height, with floor and beam spans of about 20 ft. The building has since been demolished because of changes in land utilization, but as a major precast and rc construction it predates the majority of early precast frames by about 40 years.

During the First World War store houses for various military purposes were prefabricated using rc walls and shells. Later, the 1930s saw expansions by companies such as Bison, Trent Concrete and Girling, with establishments positioned close to aggregate reserves in the Thames and Trent Valley basins. The reason why precast concrete came into being in the first place varies from country to country. One of the main reasons was that structural timber became more limited. Some countries, notably the former Soviet Union, Scandinavia and Northern Continental Europe, who together possess more than one-third of the world's timber resources, but experience long and cold winters, regarded its development as a major part of their indigenous national economy. Outside of the US structural steelwork was not a major competitor at the time.

During the next 25 years developments in precast frame systems, prestressed concrete long span rafters (up to 70 ft) and precast cladding increased the precasters' market share to around 15 per cent in the industrial, commercial and domestic sectors. Influential

**Fig. 1.1** Weaver's Mill, Swansea. The first precast concrete skeletal frame in the UK, constructed in 1897–98 (courtesy of Swansea City Archives).

articles in such journals as the 'Engineering News Record' encouraged some companies to begin producing prestressed floor slabs and, in order to provide a comprehensive service by which to market the floors, diversified into frames. In 1960 the number of precast companies manufacturing structural components in Britain was around 30. Today it is about eight.

Early structural systems were rather cumbersome compared with the slimline components used in modern construction. Structural zones of up to 36 in., giving rise to span–depth ratios of less than 9, were used in favour of more optimized precasting techniques and designs. This could have been called the 'heavy' period, as shown in C. Glover's now classic handbook 'Structural Precast Concrete' [1.3]. Some of the concepts shown by Glover are still practised today and one cannot resist the thought that the new generation of precast concrete designers should take heed of books such as this. It is also difficult to avoid making comparisons with the 'lighter' precast period that was to follow in the 1980s where the saving on total building height could, in some instances, be as much as 100 mm to 150 mm per floor.

Attempts to standardize precast building systems in Britain led to the development of the National Building Frame (NBF) and later the Public Building Frame (PBF). The real initiative in developing these systems was entrenched more in central policy from the then Ministry of Public Building and Works than by the precasting engineers' devotion to the building industry. The NBF was designed to provide

'... a flexible and economical system of standardised concrete framing for buildings up to six storeys in height. It comprises a small number of different precast components produced from a few standard moulds.' [1.4].

The consumer for the PBF was the Department of Environment for use within the public sector's expanding building programme of the 1960s. Unlike the NBF, which was controlled by licence, the PBF was available to any designer without patent restrictions. The structural models were simple and economical: simply supported long span prestressed concrete slabs up to 20 in. deep were half recessed into beams of equal depth. By controlling the main variables, such as loading ($3 + 1$ kN/m$^2$ superimposed was used throughout), concrete strength and reinforcement quantities, limiting spans were computed against structural floor depths. Figures 1.2 and 1.3 show some of the details of these frames. Diamant [1.5] records the international development of industrialized buildings between the early 1950s and 1964. During this period the authoritative Eastern European work by Mokk [1.6] was translated into English, and with it the documentation of precast concrete had begun.

Unfortunately, façade architecture was reflected in the modular design philosophy. The results were predictable, exemplified in the photograph of Highbury Technical College in Portsmouth (now a part of the University of Portsmouth) shown in Fig. 1.4, that is a legacy the precast industry is finding difficult to dispel.

Following the demise of the NBF and PBF, precast frame design evolved towards more of a client-based concept. Standard frame systems gave way to the incorporation of standardized components into bespoke solutions. The result, shown in Fig. 1.5(a)–(c) of Western House (Swindon 1983), Surrey Docks (London 1990) and Merchant's House (London 1991), respectively, established the route to the versatile precast concrete concepts of the present day.

In the mid-1980s, the enormous demands on the British construction industry led developers to look elsewhere for building products as the demands on the British precast industry far exceeded its capacity. Individual frame and cladding companies (with annual turnovers of between £1million and £3million) were being asked to tender for projects that were singularly equal or greater in value to their annual turnovers. Programmes were unreasonably tight and it seemed, at one point, that the lessons learned of mass, market lead production techniques of the 1960s were unheeded. One solution was to turn to Northern Europe where the larger structural concrete prefabricators were able to cope with these demands. Concrete prefabricated in Belgium was duly transported to London's Docklands project, as shown for example in Fig. 1.6.

In making a comparison of developments in Europe and North America, Nilson [1.8] states:

'Over the past 30 years, developments of prestressed concrete in Europe and in the United States has taken place along quite different lines. In Europe, where the ratio of labor cost to material cost has been relatively low, innovative one-of-a-kind projects were economically feasible. ... In the U.S. the demand for skilled on-site building labor often exceeded the supply, economic conditions favoured the greatest possible standardisation of construction ...'

North America's production capabilities are an order of magnitude greater than in Europe. Figure 1.7 shows the construction of a 30 storey, 5000 room hotel and leisure complex in Las Vegas. The conditions to which Nilson refers are changing. The gap between labour and material costs is closing to such an extent that it will soon be feasible for high quality precast components, e.g. façades, to be manufactured in the US destined for European building sites. The process may have begun already.

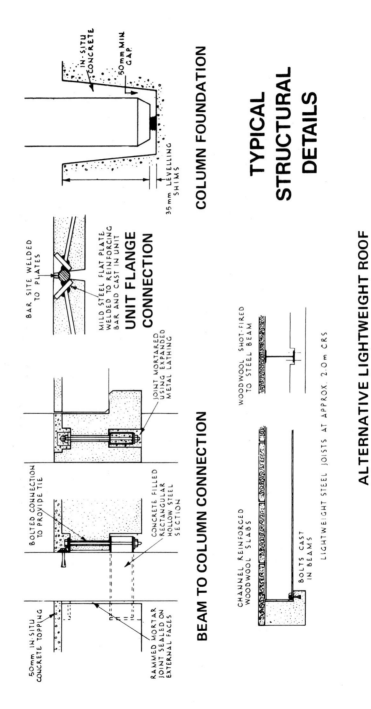

**Fig. 1.2** Typical structural details for the National Building Frame [1.4].

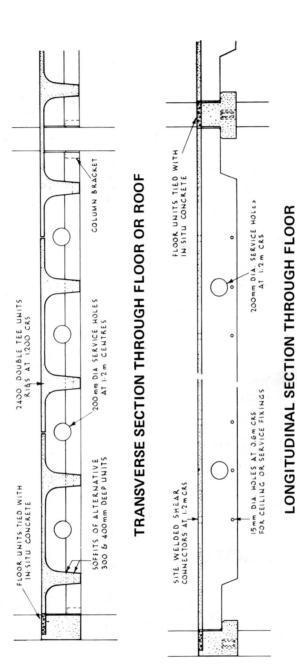

**TRANSVERSE SECTION THROUGH FLOOR OR ROOF**

FLOOR UNITS TIED WITH IN SITU CONCRETE

SOFFITS OF ALTERNATIVE 300 & 400mm DEEP UNITS

2400 DOUBLE TEE UNITS RIBS AT 1200 CRS.

200 mm DIA SERVICE HOLES AT 1·2 m CENTRES

COLUMN BRACKET

**LONGITUDINAL SECTION THROUGH FLOOR**

SITE WELDED SHEAR CONNECTORS AT 1·2 m CRS.

15mm DIA HOLES AT 0·6m CRS FOR CEILING OR SERVICE FIXINGS

200mm DIA SERVICE HOLES AT 1·2 m CRS

FLOOR UNITS TIED WITH IN SITU CONCRETE

**Fig. 1.3** Floors used in the National Building Frame [1.4].

**Fig. 1.4** Precast construction of the 1960s using the National Building Frame. The building is Highbury College, now part of the University of Portsmouth (courtesy of Costain Building Products).

(a)

**Fig. 1.5** Examples of precast construction of the 1980s. (a) Western House, Swindon (courtesy Trent Concrete Ltd).

**Fig. 1.5** (*continued*) (b) Surrey Docks, London (courtesy Crendon Structures); (c) Merchant House, London [1.7].

## 1.2   THE SCOPE FOR PREFABRICATED BUILDINGS

The precast industry is still labouring under the misconceptions of modular precast concrete buildings. This is not surprising as many texts, old and new, refer to

'... the design of a precast concrete structure on a modular grid. The grid should preferably have a basic module of 0.6 m ...' [1.9].

The Continental Europeans used the phrase 'modular co-ordination' which meant the interdependent arrangement of dimensions, based on a primary value accepted as a module [1.10]. This dimension was 30 cm horizontally and 10 cm vertically. Moreover, the storey height in precast concrete apartment buildings was fixed at 280 cm with the horizontal

**Fig. 1.6** Precast concrete imported in to the UK from Belgium. Canary Wharf, London.

**Fig. 1.7** MGM Hotel and Casino at Las Vegas, US constructed in 1992 (courtesy of A. T. Curd).

grid dimension on a 30 cm incremental scale between 270 cm and 540 cm. Strict observance of these rules facilitated the optimum assembly of prefabricated structures. In other words all prefabricated buildings looked the same.

There is a clear distinction between 'modular co-ordination' and 'standardization'. The precast industry deplores the former and encourages the latter. What is the difference and how can this be?

Modulation offers zero flexibility off of the modular grid. The end product is evident in the comparison of the two buildings adjacent to Vauxhall Bridge in London, and shown in Fig. 1.8. Interior architectural freedom is possible only in the adoption of module quantities and configuration, and one cannot escape the geometrical dominance and lack of individuality of the older building on the left of the photograph. Exterior façades may of course be varied indefinitely, as in the 'twin façade' system shown in Figs 1.9(a) and (b), but this requires a full precast perimeter wall. As far as skeletal frames are concerned, one need go no further than the standardization of families of precast concrete components to obtain the optimum solution for any building, within reasonable limits.

Industrial modularized buildings were introduced in Europe in the 1950s during the mass construction period following the Second World War. The problems in the architectural and social environment brought a re-emergence of traditional methods, and the closer control on design and factory production. This has inevitably led to a new philosophy in what is called the 'modulated hierarchical building systems' [1.11] which aims at the subdivision of a building into:

- functional systems, i.e. space utilization both vertically and horizontally, personel co-ordination, adaptability to changing needs
- technical systems, i.e. the structural design, the façade, mechanical and electrical, waste disposal and air conditioning systems.

Precast frame manufacturers have been able to synthesize these requirements through

**Fig. 1.8** Examples of past and present use of precast concrete. Vauxhall Bridge, London.

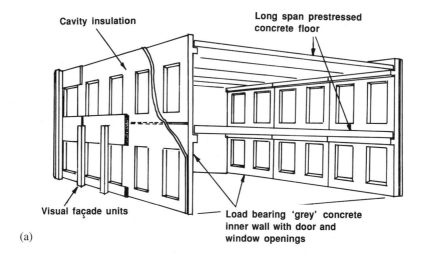

Cavity insulation

Long span prestressed concrete floor

Visual façade units

Load bearing 'grey' concrete inner wall with door and window openings

(a)

(b)

**Fig. 1.9** Principle of the twin facade system. (a) diagrammatic representation and (b) offices, Brussels.

continuous development of improved products and creative use of 'standardized' precast concrete products.

Standardization is quite different from modulation. It refers to the manner in which a set of predetermined components are used and connected. Most of the buildings shown in Figs 1.10 to 1.12 were constructed using more or less the same family of standardized components; the Reinforced Concrete Council has published case studies on precast frames, e.g. [1.7], where this may also be appreciated. By adjusting beam depths, column lengths, wall positions, etc. the same components in any of these buildings could have been used to make a completely different structure. This is not possible with modular systems, and any reference to it should be solemnly abandoned alongside the architectural visions of the 1960s.

The three basic types of precast concrete structures are:

(1) The wall frame, Fig. 1.13, consisting of solid or voided vertical wall and horizontal slab units only, and used extensively for multi-storey hotels, retail units, hospitals, housing and offices. Here the structural walls serve also as acoustic and thermal partitioning.

**Fig. 1.10** Structural 'grey' precast frame at Nottingham, UK (1995).

**Fig. 1.11** Commercial office development (courtesy of Crendon Structures, 1988).

**Fig. 1.12** Visual structural units with a polished concrete finish (courtesy of Trent Concrete Ltd, 1992).

    (2) The portal frame, Fig. 1.14, consisting of columns and roof rafters or beams, and used for single-storey retail warehousing and industrial manufacturing facilities.

    (3) The skeletal structure, Fig. 1.15, consisting of columns, beams and slabs for low- to medium-rise buildings, with a small number of walls for high rise. Skeletal frames are used chiefly for commercial offices and car parks.

**Fig. 1.13** Wall frames used in hotel construction.

This book is concerned only with skeletal frames. These are the most architecturally and structurally demanding because, in both disciplines, designers feel that they have a free rein to exploit the structural system by creating large uninterrupted spans whilst reducing structural depths and the extent of the bracing elements. This results in a large proportion of the connections being highly stressed and difficult to analyse.

Non-structural or structural load-bearing panels may also be incorporated in the frame, usually in the perimeter, and may be used as decorative cladding with a very wide range of finishes, textures, etc., thermal and sound insulation, fixings and other provisions. As decorative cladding, they are beyond the scope of this book and are documented elsewhere [1.12–1.14].

The skeletal structure is distinguished from other types because imposed gravity loads are carried to the foundations by beams and columns, and horizontal loads by columns and/or walls. The structural efficiency of a building is related, to a certain degree although not absolutely, to the mass of the structure. The volume of structural concrete in a skeletal precast structure is in the order of 1 to 2 per cent of the volume of the building (for typical 3 m to 3.5 m storey heights and 6 m to 8 m beam spans). For a cross-wall frame the figure is closer to 4 per cent. Flooring is the most significant factor with prestressed

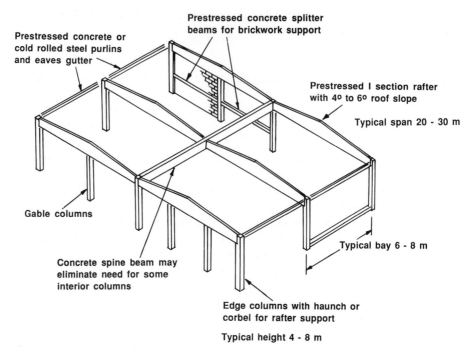

**Fig.1.14** Typical arrangement of portal frames.

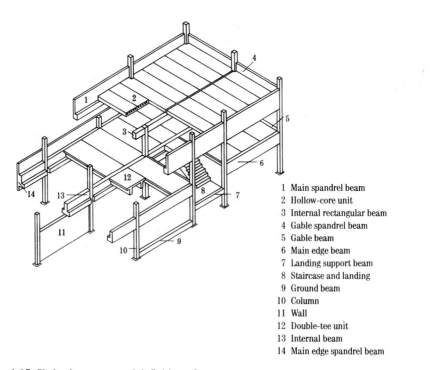

1 Main spandrel beam
2 Hollow-core unit
3 Internal rectangular beam
4 Gable spandrel beam
5 Gable beam
6 Main edge beam
7 Landing support beam
8 Staircase and landing
9 Ground beam
10 Column
11 Wall
12 Double-tee unit
13 Internal beam
14 Main edge spandrel beam

**Fig. 1.15** Skeletal structure and definition of components.

concrete (psc) voided slabs adding a further 4 per cent, compared with 7 to 8 per cent for solid concrete slabs.

## 1.3  CURRENT ATTITUDES TOWARDS PRECAST CONCRETE STRUCTURES

The latest generation of multi-storey precast concrete frames has evolved over the last 20 years. In this time the market share of commercial buildings, offices, schools, hotels, hospitals and car parks, etc. has fluctuated between as little as 5 per cent and to no more than about 15 per cent. Compared with Continental Europe, Scandinavia and North America, the UK's precast concrete frames, flooring and cladding market place is small; in 1988 the turnover in the UK was, respectively, approximately £41million, £108million and £40million [1.15]. The market share data given in Fig. 1.16(a)–(c) are fairly typical of the situation in both periods of growth and recession.

To place this into context, Table 1.1 shows the approximate quantities of precast concrete flooring used in Europe and Scandinavia in 1990 [1.16].

Speaking in relative terms, the figures for precast buildings are of the same order. This discrepancy is not apparent on design and production matters where concrete technology and production engineering is found to be broadly similar [1.15].

Unlike the attitude in Continental Europe, North America and New Zealand towards medium-rise buildings (five to ten storeys), precast concrete is still thought of as an alternative means of construction to insitu concrete and structural steel. The possible reasons for this are:

(1)  There is a widespread lack of knowledge on the structural mechanics of precast construction because the topic is not taught at university and is not covered by post-graduate training in professional design practices. Precast has a low profile in academia, unlike the situation in Continental Europe (where Chairs in Precast Concrete have been created), because very little fundamental research has been sponsored by industry or the Government. Few training courses have been organized and little in the way of general, non-specific guidance has been made available. As a result, precast concrete is perceived as difficult to specify.

(2)  Structural design and erection is nearly always (95 per cent of cases) carried out by the nominated precast manufacturer, thus depriving the consulting engineer of direct control over the frame design.

(3)  There is a lack of manufacturing capacity. Large scale precast production demands capital investment. In a volatile economy, as has been seen in the UK since 1970, manufacturing companies have been forced to ride the crests and troughs with equal acumen if production capacities are to be maintained. Whereas structural steel can be stockpiled ready for cutting, and insitu concrete be poured at short notice, precast products must be manufactured to order.

In some countries precast concrete is viewed as a 'value added product' and therefore carries a VAT surcharge over cast insitu concrete.

The survey [1.15] suggested that by the year 2000 the precast industry will need to double its design and production capabilities to meet the predicted growth. For this to happen the key attitudes of designers to modern building form will involve conceptual and environmental design, durability, serviceability, detailing and quality control. These

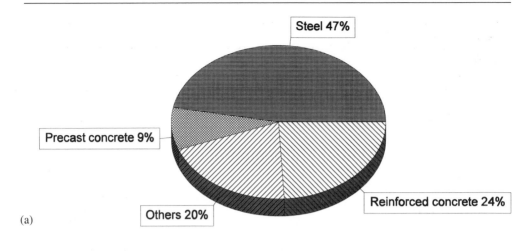

(a)

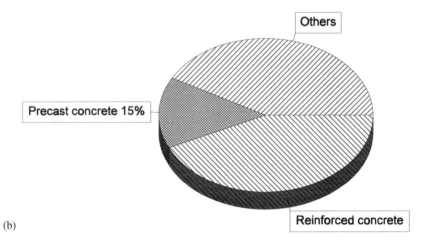

(b)

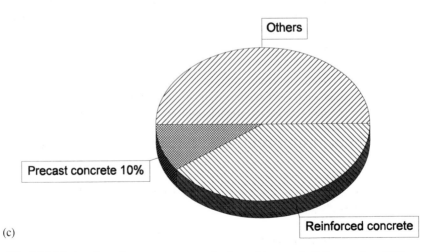

(c)

**Fig. 1.16** Market share for precast concrete in the multi-storey building market, 1988 [1.15]. (a) Market share, 1988; (b) non-housing and (c) housing.

**Table 1.1** Approximate quantities of precast concrete flooring used in Europe and Scandinavia in 1990.

| Country | Production of hollow cored flooring (millions $m^2$) | $m^2$ per capita |
|---|---|---|
| Germany | 0.6 | 0.01 |
| France | 1.6 | 0.03 |
| UK | 2.8 | 0.05 |
| Italy | 3.4 | 0.06 |
| Belgium | 0.8 | 0.08 |
| Denmark | 0.7 | 0.15 |
| Norway | 0.9 | 0.22 |
| Sweden | 2.5 | 0.30 |
| Netherlands | 5.1 | 0.36 |
| Finland | 3.0 | 0.65 |

*Note*: On average 1 $m^2$ equates to about 0.12 $m^3$ of concrete, weighing 290 kg.

demands on designers and manufacturers has already resulted in a wide range of high quality buildings.

On a worldwide scale the scope for high specification precast concrete in buildings is still very large, with the exception of South America, which has been stagnant since 1975, and perhaps in Italy where the market for precast has focused mainly on flooring. The trend seems to be that in places where the contract is controlled by the contractor precast is used mainly for the horizontal components, such as prestressed floors. It is well known that precast manufacturers make less money from making vertical components than horizontal ones.

Furthermore, in certain regions, e.g. in the Middle East and Far East, the demand is growing at a rate greater than the increase in available technology. The major restraint for precast in the Far East is the very competitive cost for cast insitu concrete, where site labour costs less than $\frac{1}{20}$ per capita than in Europe. Building development is moving at such a swift pace that the frame geometry is fixed before the precast manufacturer has a chance to submit a more economical layout.

Precast can be grouped into three markets:

- residential (mainly 2–4 storeys)     5–15%
- commercial (offices, hotels, retail supermarkets, factories)     50–70%
- services (car parks, hospitals, colleges, sports stadiums     20–40%.

Product break down based on volume of concrete produced can be grouped into four markets:

- architectural façades, structural and non-structural     20–30%
- walls (non-decorative)     5–30%*
- floor slabs     50–60%
- beam and columns     10–15%.

* The production of walls for wall frames varies widely in different countries.

Note that volume of concrete produced is possibly not the best indicator with which to assess the industry. For example, the cost and the effort expended in manufacturing floor slabs is considerably less than an architectural panel of similar weight, by a factor of 10 in some cases at least.

There is no doubt that the key to a successful precast concrete business is size. Output capacity has had a significant influence on the survival of precast companies in building in the last 20 years. Using the American market as a role model (although a similar but scaled down trend was seen in the UK and Europe), the profit margins for US companies were as shown in Table 1.2 [1.17].

Over this same period the approximate volume of precast reinforced and prestressed concrete produced per capita increased from 0.015 $m^3$ to 0.025 $m^3$ by the mid-1980s, but has fallen back to about 0.015 $m^3$ today.

## 1.4   RECENT TRENDS IN DESIGN, AND A NEW DEFINITION FOR PRECAST CONCRETE

Responding to the reduced market of the 1990s the industry saw a new movement in the design of precast structures, and a new definition to the word 'design'. It no longer meant '$wL^2/8$'. In a shrinking market the design of structural concrete frames for prestigious commercial buildings came under closer scrutiny as architects, designer engineers and contractors strived to find optimum economy, speed of erection and the highest specification for their projects. The construction industry required a multiple choice in the selection of building components, and it was likely that the increasing demand on the performance of these components would overtake the existing technology used in their manufacture and utilization.

Building design therefore became a multi-functional process where the optimum use of all the components forming the building were maximized. Attention was directed towards the structural frame, which could no longer be considered as serving only a structural function, but must be harmonized with the requirements of the building in total. Architectural-structural precast concrete components were therefore used on an increasing number of prestigious commercial buildings as 'specifiers' became more aware of the high quality finishes possible in prefabricated units. However, changes had to be made to the way that traditional precast concrete structures were conceived and designed. Despite the cutbacks in staffing levels the precast concrete industry was ideally placed to accommodate these higher demands by using experienced design teams and skilled labour

**Table 1.2** Profit margins for US companies.

| Decade | Annual turnover ($million) | Annual profit (approx.) (%) |
|---|---|---|
| 1970s | <3.0 | 1.1 |
| | >3.0 | 4.6 |
| 1980s | <7.5 | 1.8 |
| | >7.5 | 8.2 |
| Early 1990s | <15.0 | 1.2 |
| | >15.0 | 6.6 |

in a quality controlled environment to produce high specification components, which served the structural, architectural and services functions simultaneously.

An excellent example of the use of architectural-structural components is shown in Fig. 1.12. White marble was originally chosen as separate cladding material to the external façade of a plain precast concrete frame, including the expressed external columns. During the conceptual design stage it was realized that a white marble façade would have poor weathering characteristics. This was replaced with a white polished concrete manufactured using 10 mm size Spanish Dolomite aggregate and white Portland cement, wet ground and polished to a 'marble' type of finish. The external façade was cast simultaneously with the structural components to reduce the overall cross-section and eliminate unsightly joints to the front of the columns and beams. A paper co-written by the author describes this and other similar  projects in greater detail [1.18].

A further example of structural finished units is shown in Fig. 1.17. Structural and architectural variety was incorporated into the scheme at several levels. Straight blocks formed in 'L' shapes and quadrangles were joined with towers of glass blocks which

**Fig. 1.17** Prefabricated brick piers and concrete floor units.

both accommodated the stairs and became as stack-effect flues to help drive the natural ventilation without the need for air conditioning. Energy-saving features included in the superstructure were windows shaded by projecting piers which, together with the mass of the concrete floors, limited peak summer temperatures. The structure has the capacity for absorbing heat by night-time pre-cooling. The design of the buildings includes load-bearing brick piers supporting precast concrete units, expressed as a rippling concrete band of vaulted units at each floor level (Fig. 1.18). The components were manufactured in a single piece with dummy joints to give the appearance of a number of individual arched units. The alternative design would have been to support vaulted panel units on longitudinal beams spanning between the piers. The advantage in using a single element with dummy joint lines is therefore clear.

The advantages in using single piece visual concrete components are also evident from the photograph in Fig. 1.19 of a building where cantilever columns support a light-weight steel roof. Concrete strengths of 50 MPa were achieved using white Portland cement and 10 mm size limestone aggregates. The structural capacity of this type of unit is no different from a column made from ordinary grey concrete.

## 1.5 PRECAST SUPERSTRUCTURE SIMPLY EXPLAINED

### 1.5.1 Differences in precast and cast insitu concrete structures

Cast insitu concrete structures behave as three-dimensional (3-D) frameworks. Continuity of displacements and equilibrium of bending and torsional moments, and shear and axial forces is achieved by reinforcing the joints so that they have equal strength as the members. However for design purposes the frames are designed in 2-D plane stress, although the presence of the floor slab in the third dimension will affect the manner in which the

**Fig. 1.18** Vaulted precast concrete floor units span 13 m between precast brick piers (courtesy of Trent Concrete Ltd, 1994).

**Fig. 1.19** Profiled white concrete columns (courtesy of Crendon Structures Ltd).

plane frame behaves. This is particularly important in the end bay where the beams and columns in the 2-D idealization may be subject to torsional stresses and biaxial bending, respectively.

Figure 1.20(a) shows the approximate deflected shapes and bending moment distributions for a two-storey continuous frame subjected to gravity and horizontal loads. The foundations are assumed to be fully encastré in this example which we may call frame F1. The relative magnitudes of the moments in the beams and columns depend on the relative stiffness ($EI/L$) of the columns and beams meeting at a joint. The joints depend on having equal strength, $M_1$, to the members.

If the strength of the joints in either of the beams in this frame were deliberately weakened to $M_2$ (e.g. by omitting reinforcement) the behaviour of the frame (called F2) would be equal to that of frame F1 up to the point where the moment in the joint reached $M_2$. Upon further loading, frame F2 would therefore develop plastic hinges at the joints. The difference between the bending moments in the column above and below the joint would be equal to the moment in the beam, $M_2$.

Taking this to the limit, if the joints were weakened to $M_3 = 0$, a pinned connection would result and the deflections and moments would be as shown in Fig. 1.20(b). Note how the column moments have increased to allow for the fact that they can no longer be distributed into the beams. The difference between the bending moment in the upper and lower column at the joint is now zero. This is how a 'skeletal' precast structure behaves because the beam–column connections are 'pinned' and the structure can be readily precasted with the individual components being bolted, dowelled or welded together on site. If there are no other stabilizing elements in the structure the ends of the columns at the foundations must be encastré.

A general synopsis of the design of skeletal structures was given by the author in a paper published in the PCI Journal in 1992 [1.19].

The term 'pinned-jointed' refers essentially to the manner in which connections between columns, beams and floor slabs are made. The form of construction does not lend itself

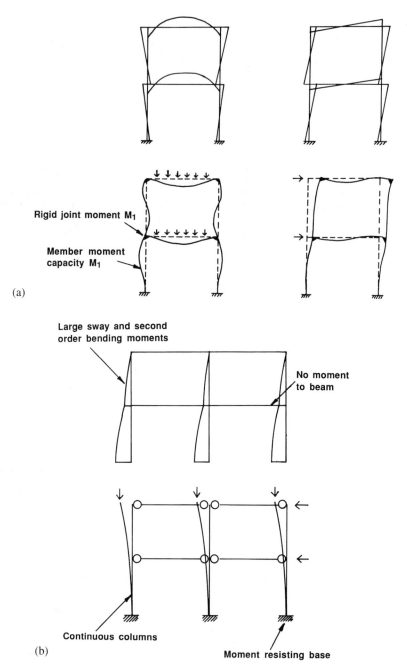

(a)

Rigid joint moment M₁

Member moment capacity M₁

Large sway and second order bending moments

No moment to beam

Continuous columns

(b)

Moment resisting base

**Fig. 1.20** Moments and deflections. (a) In continuous frameworks and (b) in an unbraced structure.

to seismic design, and conversely in seismic zones fully-rigid frame connections do not lend themselves to the safest and most economical use of precast concrete. Between these two extremes are semi-rigid connections, but here design methods are not sufficiently

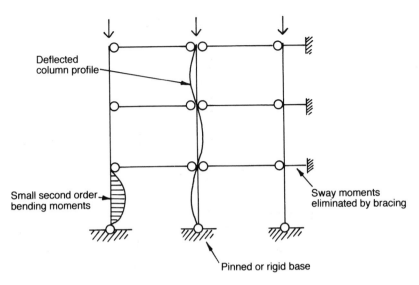

**Fig.1.21** Moments and deflections in a braced structure.

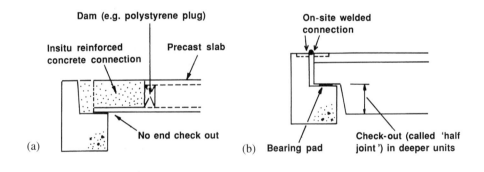

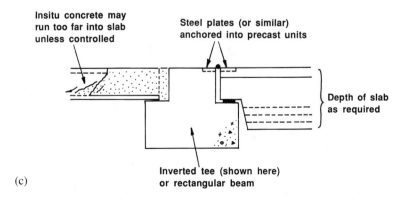

**Fig. 1.22** Detail of slab to beam bearing. (a) Edge beam to hollow core slab; (b) edge beam to double-tee slab and (c) internal beam to slab connections.

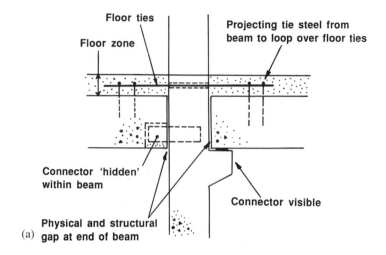

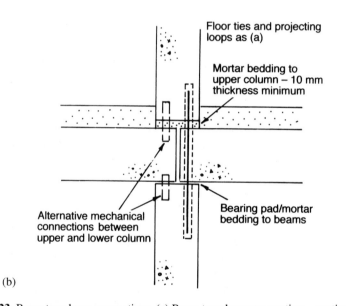

**Fig. 1.23** Beam to column connections. (a) Beam to column connection – continuous column; (b) beam to column connection – discontinuous column and beam and (c) beam to column connection – continuous beam.

understood in the post-elastic regime to be considered at present for the ultimate limit state design of columns in precast structures.

In buildings of more than about three storeys the horizontal sway deflections may become excessive so that additional bracing must be used. Thus stability walls, cores or other forms of bracing are used. The usual practice is to place the stabilizing units around lift shafts or stair wells so that the open plan office space is not interrupted. The thickness

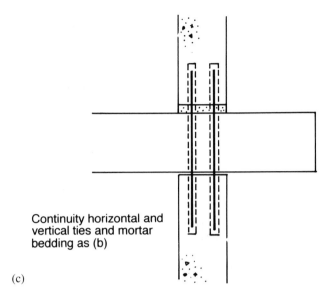

Continuity horizontal and
vertical ties and mortar
bedding as (b)

(c)

**Fig. 1.23** (*continued*)

of the walls varies from 125 mm to about 250 mm, depending on the size of the building. It is possible to cast door or window openings in walls providing that the force paths are maintained.

The structure is now classed as 'braced', Fig. 1.21, and the foundations may now be pinned. This simplifies foundation construction considerably and means that braced precast structures can be erected off insitu concrete retaining walls, beams, etc. and poor ground. The stabilizing elements are so massive that the stiffness of the frame elements and connections is not important. Bending moments due to sway are small and columns can only deflect between floors as pin ended struts.

In all cases, braced or not, horizontal wind loads are transmitted through the precast floor, sometimes using unscreeded slabs, as though the floor were a deep beam. One-way spanning prestressed (or reinforced) precast floor slabs are seated on or are recessed into beams. The structural zone is sometimes less than in an equivalent fire protected steel frame (excluding slim floor), often showing a saving of 100 mm per floor [1.20, 1.21]. The slab bearing, shown in Figs 1.22(a), (b) and (c), is designed as a simple pinned joint despite the presence of the reinforced concrete insitu infill which penetrates into the floor slab and obviously provides some moment restraint. Beam design is to ordinary rc or psc principles, and composite construction with the floor slab is occasionally appropriate. Precast staircases and landings are designed as inclined solid rc or psc units, and are omitted from the floor plate action.

Beams are connected to columns and walls using connectors which are mainly de-signed as pinned joints (Fig. 1.23(a)). An eccentric loading is applied to the column, which is continuous at the connection, and the bending moment is distributed in the column according to simplified 2-D frame analysis. Alternatively, the beam may be supported on the tops of columns (Fig. 1.23(b)), or made continuous across the top of single-storey columns (Fig 1.23(c)), and connected to its neighbour away from the highly stressed connection. Columns are considered continuous at floor joints, even though

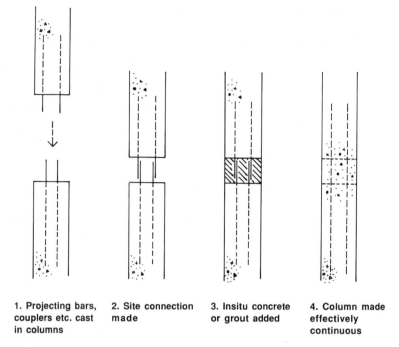

1. Projecting bars,    2. Site connection    3. Insitu concrete    4. Column made
   couplers etc. cast       made                   or grout added         effectively
   in columns                                                             continuous

**Fig. 1.24** Principle of column splices. The substructure member could also be a wall or beam.

mechanical connections, called 'splices', are often made at this level, as shown in Fig. 1.24.

In this book architectural aspects will be dealt with in Chapter 3, with the frame analysis and design methods in Chapter 4. Component design will be discussed in Chapters 5 and 6, the design of all the connections mentioned above in Chapter 7, and horizontal stability in Chapter 8.

### 1.5.2  Structural stability

Structural stability is the most crucial issue in precast concrete design because it involves both the design of the precast concrete components and of the connections between them.

The first design features are stability and robustness. Precast systems are scrutinized by checking authorities more for structural stability, integrity, resistance against abnormal loading and robustness, than for the design of individual precast components (slabs, beams, etc.) which usually have adequate factors of safety. In general, 'stability' means adequate resistance against side sway; 'integrity and robustness' means correct joint design, attention to details and the prevention of progressive collapse. Chapter 9 will address the problem in detail.

The problem is to ensure adequate ultimate strength and stiffness, but more importantly to ensure that  the failure mode is ductile. Large factors of safety are less relevant to the overall performance criteria if brittle or sudden failures involving the release of large amounts of energy result.

Two design stages are considered:

- **Temporary stability during frame erection.** This has certain implications on design, e.g. the axial load capacity of temporary column splices must be greater than (at least) the self weight of the upper column before the insitu infill grout, which renders the splice permanent, has hardened. Chapter 10 will discuss this topic.
- **Permanent stability**. This may be subdivided into four further stages:
  - horizontal diaphragm action in the precast floor slab
  - transfer of horizontal loading from the floor slab and into the vertical bracing elements and the foundations
  - component design
  - connection and joint design.

The contribution to the lateral strength and stiffness of the structure from the insitu reinforced concrete infill strips is paramount. These strips provide the necessary tie forces which eliminate relative displacements between the various parts of the frame and ensure interaction between the components. The general idea is shown in Fig. 1.25. These positions are not always easy to define, and no experimental work has addressed this problem directly. Engineers are cautious not to allow service openings or novel connections in these highly sensitive areas.

Stability may be achieved in several ways, but in the vast majority of cases it is usually based on either:

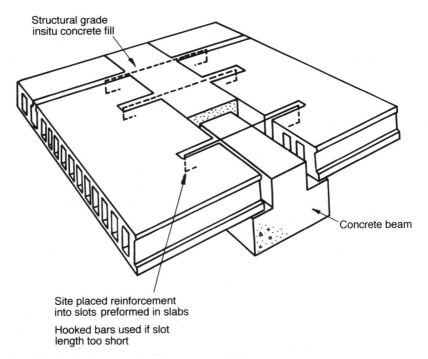

Structural grade
insitu concrete fill

Concrete beam

Site placed reinforcement
into slots preformed in slabs

Hooked bars used if slot
length too short

**Fig. 1.25** Continuity ties in the insitu concrete infill.

- an 'unbraced' (or 'sway') structure (Fig. 1.20), where stability is provided by the skeletal structure, by cantilever action of the columns
- a 'braced' structure (Fig. 1.21), where resistance against horizontal loading is not provided by the skeleton of the beams, columns and slabs; in other terminology, e.g. using insitu concrete, this would be called a 'no-sway' frame.

Combinations of the above are possible, which is known as a 'partially braced' structure (Fig. 1.26). It is perfectly reasonable for the frame to be braced in one plane, and rely on column action in the other plane, particularly if the building is long and narrow. The design of these structural systems will be dealt with in Chapter 8, but it is important that sufficient bracing elements, e.g. walls, boxes, columns, etc. are provided at the very outset of design, not at the end.

The positioning of shear walls is often a contentious issue. Conflicts with architects as to the number of walls and their positions is usually the biggest problem. In general, it is necessary to 'balance' the flexural resistances (i.e. summation of stiffness) of the walls in the structure (in two mutually perpendicular directions) to avoid torsional effects as far as is reasonably practical. A design method for the different types of walls used in precast concrete frames is given in Chapter 8.

### 1.5.3   Floor plate action

Floor plate action means the transfer of horizontal loading across a building. It is sometimes called 'diaphragm action' because the floor is a relatively thin membrane. There has been a considerable debate on whether a precast concrete floor, consisting of a large number of discrete elements, tied together only around their edges, as shown in Fig. 1.27(a), can in fact be used as a floor diaphragm. Various structural models have been proposed for the shear transfer mechanism, which relies on the development of

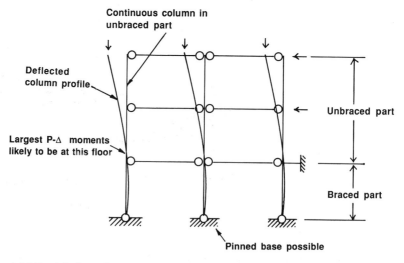

**Fig. 1.26** Partially braced structure.

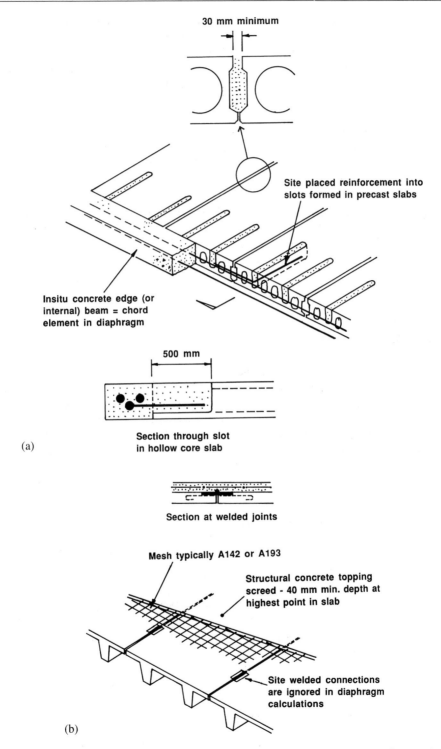

**Fig. 1.27** Basic principles of the precast floor diaphragm. (a) Precast concrete floors only and (b) precast floors with an insitu structural topping.

clamping forces generated in the tie bars placed in the insitu concrete strips at the ends of the slabs. Shear forces parallel with the span of the flooring units are transmitted to shear cores or other stabilizing facilities by deep beam action. Shear forces perpendicular to the span of the floor slab are carried to successive bays of flooring by the shear friction reinforcements in the insitu concrete strips in the beam to slab connections.

If this cannot be achieved, for example in precast floor units with discontinuous or thin flanges, a structural topping screed is used as shown in Fig. 1.27(b).

### 1.5.4   Connections and joints

The term 'connection' refers to major structural connections between precast components, whereas the term 'joint' is used to describe a more simplified jointing between components. For example, the junction between beam and column is called a connection, whereas the halving detail between landing and stairflight is classed as a joint.

Connections between components are the most important factors influencing the design, construction and in-service behaviour of precast structures. The many different types of connections used by designers makes it difficult to generalize on rules and guidance notes, particularly because different methods of making the same type of connection are practised by engineers. The major connections are between:

- column or wall to the foundation
- column to column
- beam to column or wall
- beam to beam
- slab and/or staircase to beam or wall
- slab to slab
- structural steelwork, insitu concrete, timber and masonry to precast concrete components.

Joints are required mainly (but not exclusively) to transmit compression, tension and/or shear stresses. Bending and torsion moments can usually be resolved into these three components. Typical forms of joint construction involves grouting, dry pack grout, adhesives, neoprene pads, welds, bolted cleats, dowels and screwed rods. Examples of joints are:

- bearings between slabs and beams or walls
- interfaces between precast and insitu concrete, e.g. structural screed
- scarf or halving joints between stair components
- trimmer angles forming holes in floor decks
- site erection aids (temporary or permanent).

### 1.5.5   Foundations

Loads and moments in columns are transferred to insitu concrete foundations through deep pockets, base plates or grouted sleeves. Wall loads and moments are usually spread over such a large area that the connections are simple compression and tension joints.

The various options for foundations are shown in Figs 1.28(a), (b) and (c) and (d). The available depth of foundation may be the deciding factor in the type of stabilizing system used in the structure, e.g. a shallow footing requires that a braced structure is used. The main criteria for foundations are simplicity in design and erection of the precast superstructure.

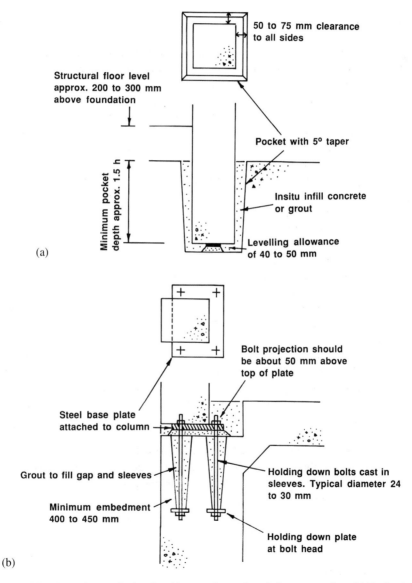

**Fig. 1.28** Alternative methods of making a column–foundation connection. (a) Pocket in pad footing; (b) base plate to insitu wall.

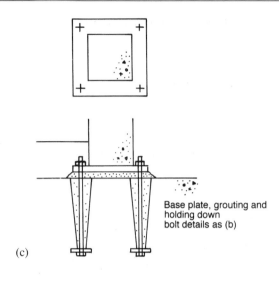

Base plate, grouting and
holding down
bolt details as (b)

(c)

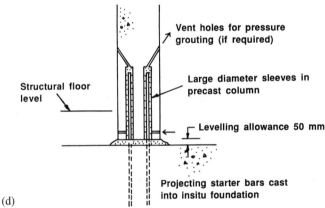

Structural floor
level

Vent holes for pressure
grouting (if required)

Large diameter sleeves in
precast column

Levelling allowance 50 mm

Projecting starter bars cast
into insitu foundation

(d)

**Fig. 1.28** (*continued*) Alternative methods of making a column-foundation connection.
(c) Base plate to pad footing and (d) grouted projecting bars in pad footing or basement
wall.

Foundation design is much simplified in braced frames where the only loading is
vertical axial with small bending moments due to carry-over moments resulting from
eccentric loads at the first floor beam connection. Simple pad footings are used in most
instances, although the columns in a braced frame may be spliced onto the tops of insitu
concrete retaining walls.

Moment resisting foundations are required in unbraced frames. These may be very
expensive in poor ground, and because of this it is often advisable to liaise with other
bodies on the overall economics of unbraced structures. Structures exceeding three storeys
or about 12 m in height should certainly be designed as braced. Precast box foundations

have been tried, but the preparation of level ground for them outweighs most of the advantages.

## 1.6   PRECAST DESIGN CONCEPTS

### 1.6.1   Devising a precast solution

In their most recent book Bruggeling and Huyghe [1.22] state:

'Prefabrication does not mean to 'cut' an already designed concrete structure into manageable pieces …'.

The correct philosophy behind the design of precast concrete multi-storey structures is to consider the frame as a total entity, not an arbitrary set of elements each connected in a way that ensures interaction between no more than the two elements being joined. Thus it is clear that all the aspects of component design and structural stability are dealt with simultaneously in the designer's mind. The main aspects at the preliminary stage include:

- structural form
- frame stability and robustness
- component selection
- connection design.

These items cannot be dealt with in isolation. For example, the nature of the column–beam connection dictates the arrangement and function of the reinforcement in the ends of beams, and the manner in which floor slabs are connected to edge beams influences the torsional behaviour of the beam. Frame design is therefore an integrated process in which many of the iterative steps are not so obvious because they are now hidden within the natural evolution of the design, detailing and site erection procedures. It must also be remembered that two additional procedures, namely manufacture and site erection, are also directly influential in making design decisions.

The correct order for dealing with the framing plan is:

- positions of construction joints
- positions of lift shafts and stairwells which provide the locations for shear walls and/or cores
- positions of shear walls (if more than three storeys, or if on poor ground)
- positions of columns
- structural floor zones
- cantilevers
- direction of span and length of span of beams and slabs, at all floors and roof if different in plan
- use of non-preferred structural components
- use of structural precast concrete cladding
- positions of service areas and major holes in floor slabs
- staircases and precast liftshaft boxes.

The main building function is assessed with respect to:

- architectural and aesthetic requirements
- loadings, i.e. dead, imposed dead, live, wind, seismic (occasionally), lateral load due to temperature, creep and shrinkage, and miscellaneous loads such as vibrating or mobile machinery
- fire and durability
- building regulations and codes of practice
- design guides and manuals.

The first task is therefore to establish an economical plan layout for the optimization of the minimum number of the least cost components versus overall building requirements. The optimum is usually found in a rectangular grid where the beams span in a direction parallel with the greater dimension of the frame.

Figure 1.29 shows three options, (a), (b) and (c), each with a weighted function reflecting design, detailing, manufacture and construction effort and relative costs. There is a delicate balance between the architect's wishes and the precaster's pragmatism. Precast component selection will follow logically and simply once the optimum layout has been achieved. So too will service routes, as the most favourable option will allow easy highways for pipes and cables.

The optimum solution in terms of building efficiency and cost is not always obvious. Consider the three different floor layouts shown in Fig. 1.29. Scheme A may appear to give the best solution in terms of number of building components, and this is certainly the most favoured solution. However, as the data in Fig. 1.30 show, Scheme C offers the lowest floor zone to bay area ratio, with a saving of about 100 mm to 150 mm per floor.

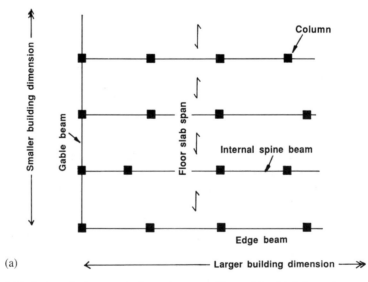

**Fig. 1.29** Options for layout of primary beams and floor slabs. (a) Scheme A: one-way spanning system with beams spanning in the larger building dimension; (b) Scheme B: one-way spanning system with beams spanning in the smaller building dimension and (c) Scheme C: two-way spanning system.

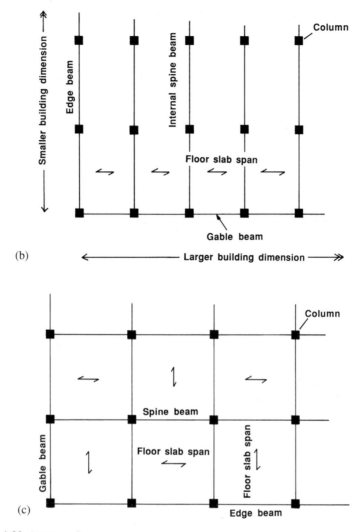

(b)

(c)

**Fig. 1.29** (*continued*)

If, in this case, the cost of the peripheral cladding was such that the saving in surface area exceeded the additional expense of the precast components, Scheme C would be the cheapest. This simple example must be treated with scepticism as there are many more variables in the final equation.

The rules for determining whether a structure is to be braced or unbraced are fairly clear cut, and they affect both prefabricated steel and concrete frames alike. Table 1.3 gives guidance.

The robustness of the structure must also be considered at the conceptual stage, particularly if a structural floor screed is not being used. Making sure that all the necessary peripheral and internal ties can indeed be placed and be continuous and fully anchored at their ends is a design exercise in its own right. Dangerous congestion of ties can be avoided if the layout of beams is thought through at this early stage.

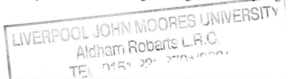

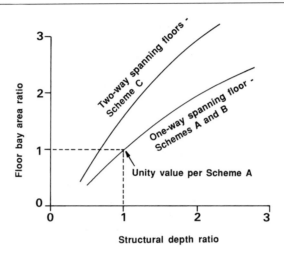

**Fig. 1.30** Structural zones versus bay area for the structural layouts shown in Fig. 1.29.

**Table 1.3** Storey heights of precast concrete frames and type of bracing [1.23].

| Approximate economical range for number of storeys | Type of frame | Bracing element(s) |
|---|---|---|
| 2 | Unbraced | Cantilevered columns |
| Up to 3 | | Cant. columns (roof load small) |
| 3 or 4 to 6 | | Precast wind posts (deep columns) |
| Up to 4 | Braced | Steel cross bracing |
| Up to 5 or 6 | | Brick or block infill walls |
| 3 to 5 or 6 | | Precast hollow core infill walls |
| 3 to 10 | | Precast solid infill walls |
| 3 to 10 | | Precast solid cantilever walls |
| 3 to 12 | | Precast hollow core shear walls |
| 10 to 15 | | Precast concrete shear box(es) |
| 15 to 20 | | Insitu concrete shear core(s) |
| Up to 5 | Partially braced * | Brick or block infill walls |
| 5 to 10 | | Precast infill walls |
| 5 to 12 | | Precast hollow core, shear walls |

\* Uppermost one or two storeys unbraced.

## 1.6.2   Construction methods

Precast frames can greatly improve buildability because many of the sensitive site operations are moved to the protective environment of the factory. Seasonal variations are less critical to site progress, and totally nullified at the factory. Depending on the circumstances of the design, size, and complexity of the building and the conditions at the construction site (i.e. access), prefabrication has the following approximate savings over cast insitu construction at the site:

- scaffolding material and labour to erect scaffold                80–90%
- shuttering and formwork                                          90–95%
- delivery and pouring wet concrete*                              75–95%
- delivery and fixing of loose reinforcement                     90–95%
- time of construction of superstructure (above foundations)     25–50%
- total construction time                                         10–30%
- site labour on superstructure                                   75–90%
- total site labour                                               50–75%.

* Lower value where structural floor screeds are used.

One of the key issues is programming the deliveries to site so that the construction team is not under pressure to construct hastily, nor fix any of the components out of sequence. This could impair the temporary stability of the structure as the height to the centre of the mass of the concrete above the level at which the frame is stable could be prohibitive to further progress. The rule is that components should not be fixed more than two storeys ahead of the last floor to be fully tied in to the stabilizing system. This allows time for the insitu concrete at the lower levels to mature. Theoretically it is possible to erect precast frames of up to about seven or eight storeys in height before the components are permanently structurally tied together. However, there is evidence that this is folly and it leaves no room for error.

On-site construction methods have a significant influence on design. Most of these concern connection details, jointing materials and temporary stability. In many instances the construction sequence will dictate the design of the frame. Often, the positions of shear walls, sizes of beams, spans, etc. can only be finalized when the construction programme is finalized, and the type and capacity of the crane is agreed. Significant economies can be achieved if the designer takes into account all the benefits available on site. In this respect construction sequences are self-defined; with columns, walls, beams, slabs and staircases being the obvious progression. However, the main decision to be made at the design stage is a logistical one. The two main options are:

(1) completion of frame floor by floor, Fig. 1.31
(2) completion of frame to roof block by block, Fig. 1.32.

The following points should be considered:

- positions of shear walls and maturity of connections may dictate site progress
- possibility of temporary stability and the need to design some of the key components to eliminate the need for it if foundation fixity is available
- availability and/or positioning of equipment to transport and erect every component
- size and weight of components
- safety and speed of construction
- tolerances for economical construction.

The vast majority of precast structures are erected by fixing gangs with many years of experience in handling precast concrete. Many precasting companies employ their own fixing teams, and this is obviously beneficial as far as feedback to the design office and factory is concerned. Information about tolerances is essential to the smooth running of a site. The UK's Precast Concrete Industry Training Association [1.24] and Precast

**Fig. 1.31** Construction sequence floor by floor (courtesy of Trent Concrete Ltd).

Flooring Federation [1.25] have published guides to the safe erection of precast concrete frames and flooring.

Programming can increase the overall speed of construction by allowing parts of the building to be released to following trades whilst work continues on erecting the rest of the precast structure. To construct bays to the full height whilst backing out of the building enables the structure to be released bay by bay. The alternative method to construct floor by floor releases lower floors whilst work continues on the erection of the floors above. Access can often be gained within two or three weeks of starting precast construction.

The construction cost element, which includes transportation costs, varies with building size, number of storeys and the structural grid. In general it is about 8 to 12 per cent of the cost of the precast structure (not the finished building). The information may sometimes deceive as the cheapest site cost may have hidden extras, e.g. the cheapest beams are the longer, which means greater depth and hence a greater structural zone.

Speed of construction is a major consideration in most building projects and it is here that the design of precast structures should be carefully considered. This advantage is maximized if the layout and details are not too complex. Designing for maximum repetition will make manufacture of the precast units easier and construction faster, but precast concrete can also be used in complex and irregular structures, although it may not then provide the same efficiency of construction as a rationalized design. The fixing rates shown in Table 1.4 vary depending on the shape and size of the structure, i.e. the number of linear components such as beams and columns per unit area of the building. The data refer to typical site progress using a fixing gang of no more than five persons and one crane. A precast solution can stretch to a considerable number of non-standard details, say about 30 to 40 per cent of the total, before the fixing rates reduce dramatically.

**Fig. 1.32** Construction sequence block by block (courtesy of Blatcon Ltd).

As an example, a recently completed three-storey building shown in Fig.1.33 [1.26] using a 6 m × 6 m grid layout achieved the following fixing rates:

- columns (14.5 m long)          30 per week
- beams (6 m long generally)          45 per week
- floor slabs (6 m to 17 m span)      2335 m$^2$ per week
- façade panels          750 m$^2$ per week.

**Table 1.4** Site fixing rates for generalized frame layouts.

| Frame layout and example bay area* (m$^2$) | Site fixing rate per week (m$^2$) |
| --- | --- |
| Rectangular grid, 100 | 1000–1500 |
| Rectangular grid, 40–50 | 600–800 |
| Non-rectangular grid, 100 | 900–1200 |
| Non-rectangular, staggered columns, 40–50 | 300–400 |

* A 'bay' is defined by the area enclosed between four columns.

**Fig. 1.33** Skeletal frame (courtesy of B + FT Wisenbaden).

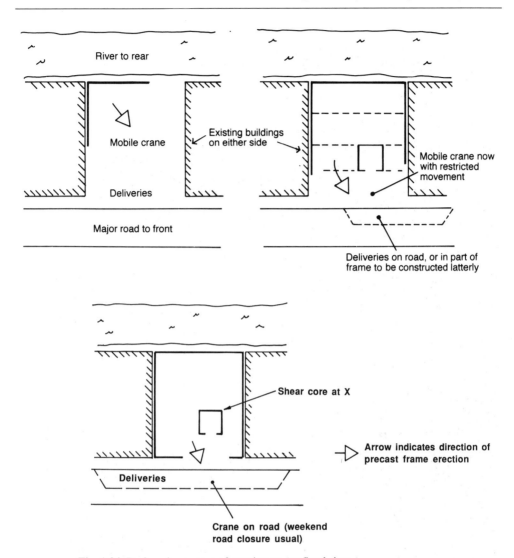

**Fig. 1.34** Preferred sequence of erection on confined sites.

The structural form of a precast structure may be influenced by the manner in which it is built, the site access for long vehicles, and the capacity of the craneage. One of the most important decisions pertaining to temporary stability is the availability of a stiff shear core, or shear walls (or other forms of bracing) in two orthogonal directions at all times during construction. This may not seem to be a problem, but consider this structure shown in Fig. 1.34 which would be constructed to its full height, bay by bay. If the shear core were available only up to point X, the previous bays would be unstable. In low-rise buildings of less than four storeys the columns can be stabilized by propping at the ground level (assuming a sound footing is available, which is not always the case). However, once a column splice is included, permanent stability is required.

In all cases the precast frame is correctly aligned and levelled as each floor is completed. Following trades may take up residence at each floor level immediately after the floor level above is completed. Precast frames may lend themselves to partial completion and a phased handover. This depends very much on the form of the structure, and the divisions are dictated by the positions of fully stabilized sub-frames.

Finally, the construction programme may be dictated by the sequence in which components may be fixed. If the plan area is small, say less than 300 m$^2$, it is possible that the fixing rate of progress is faster than the maturity rate of the connections. It may therefore be advantageous to design a connection which has no maturity restriction, e.g. a bolted or welded connection, rather than specify a cast insitu one.

# Chapter 2

# Procurement and Documentation

*This chapter deals with the role of the precast frame designer as part of the project design team, and describes the design methodology. The literature is reviewed.*

## 2.1  INITIAL CONSIDERATIONS FOR THE DESIGN TEAM

The Fédération Internationale de la Précontrainte (FIP) is an international organization devoted to the cause of precast concrete in every form. Its publications are well respected throughout the world and many of the recommendations are used in design offices, and several have been adopted into national codes. The FIP Commission on Prefabrication has produced a planning and design handbook [2.1] in which the following advice for a design organization is given:

'Every construction system has its own characteristics which to a greater or lesser extent influence the layout, storey height, stability, statical system of the building. For best results the initial design should respect the specific and particular demands of the intended structure. The potential for profit seen at the signing of the contract documents by the client and the contractor often disappears during the construction period … It is very important to realise that the best design for a precast concrete structure is arrived at if the structure is conceived as a precast structure from the very outset and is not merely adapted from the traditional cast insitu or masonry methods.'

This is a pertinent statement in the modern world and particularly important with regard to structural stability and the integrity of a precast concrete structure, where special details have evolved through the natural sequence of a 'design – manufacture – erection – design' loop used in the precasting industry.

The major beneficial advantages of a precast concrete solution will accrue at the conceptual design stage. Figure 2.1 is a schematic representation of the 'scope for project potential', whatever that term may imply, as a function of the four major decisions taken prior to the construction of a precast building, when it is practically too late to implement substantial changes. Note that additional scope for potential savings is always present at the beginning of each of the five phases. Unfortunately the conceptual part of the design process is often carried out in the absence of the precast concrete engineer. It may be necessary to adjust the structural function of the building when the means of achieving maximum potential with the precast structure are realized. The initial design should be carried out at least with the assistance of the precast engineer who will contribute to reducing the construction time and cost and ensure better quality by in-house expert knowledge.

The FIP handbook [2.1] concludes that

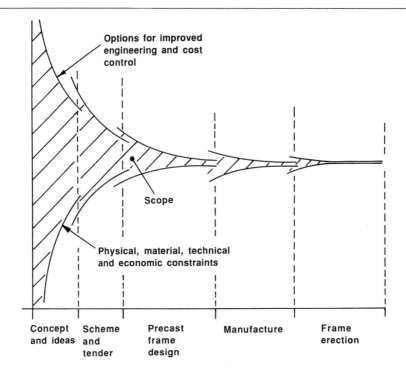

**Phases of Project**

**Fig. 2.1** Scope for optimum project potential.

'... designers should therefore consider the possibilities, restrictions, advantages and disadvantages of precast concrete, its detailing, manufacture, transport and erection and serviceability stages before completing a design in precast concrete. Precast concrete organisations will usually make available design and production information to the client, architect, consulting engineer, services engineers and all other disciplines to give unified guidance to the entire design team. This will ensure that all parties are aware of the particular methods adopted in all phases of the project, leading to maximum efficiency and benefit. This is particularly true with the manufacturing and erection stages as many consulting engineers may not be familiar with some of the methods used.'

The FIP is conscientious of the fact that precast concrete buildings are often adaptations of the more traditional and well understood forms of construction, such as insitu concrete, structural steelwork or masonry.

## 2.2   DESIGN PROCUREMENT

### 2.2.1   Definitions

The terminology used in this book to describe the functions of designers and contractors is as follows:

Precast company:           organization responsible for design, detailing, manufacture and erection.

In-house designer:         designer employed by the company.

Specialist designer:       private organization or individuals offering a dedicated service to the manufacturer and occasionally the company.

Consultant designer:       private consultancy or government-based design office offering a service to the manufacturer and occasionally the company, but not necessarily a dedicated specialist to precast.

Precast manufacturer:      manufacturer of components made to details supplied by the designer, specialist or occasionally the company.

Precast contractor:        frame erector specializing in this field.

### 2.2.2  Responsibilities

The chart in Fig. 2.2 shows the responsibilities within the precasting organization. It is apparent that the involvement of the principal design engineer is wider than the structural design itself. It is important that this engineer is in close contact with the architect, engineer and services personnel, as well as the factory co-ordinator, estimator and the contracting staff. The need for excellent lines of communication between the design office, the factory and the site cannot be stressed too highly.

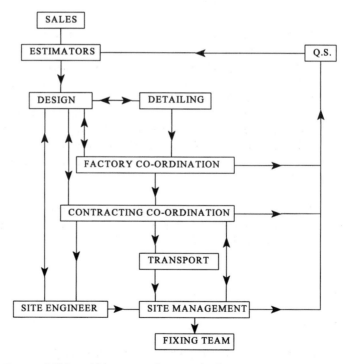

**Fig. 2.2** Responsibilities within a precasting organization.

### 2.2.3 Routes to procurement

The design manual [2.2] includes a comprehensive chapter on precast project procurement (which was written by Andrew Dyson, Director of Advantage Precast, Edinburgh, UK). The essential issues raised in this text were on the selection of the precasting company, and how the decisions leading to the appointment of the company were taken, e.g. the checks on quality control, qualifications of personnel, etc. Section 2.2.4 takes this a step further and describes the activities that follow when the precaster is invited to tender and after the precaster has been appointed, i.e. design procurement.

### 2.2.4 Design office practice

Design and drawing office practice is different to those experienced in a consulting engineers' practice because the design procedures are highly dependent on the manufacturing and construction methods adopted by the particular precasting company. In some instances, design decisions are influenced more by ease of manufacture, repetition, handling and erection constraints, than by economy of materials or form. Some prefabricators have excellent in-house mould making facilities and are therefore able to

(a)

**Fig. 2.3** Manufacture of visual concrete units. (a) Timber moulds for prefabrication (courtesy of Trent Concrete Ltd) and (b) the final product (courtesy of Trent Concrete Ltd).

(b)

**Fig. 2.3** (*continued*)

offer structural units with intricate architectural features. The moulds and resulting units shown in Figs 2.3 (a) and (b) respectively are examples of this skill.

Design office practice is inextricably linked to the various shop floor operations; bar bending, cage assembly, mould making, cage and connector positioning, casting, curing, lifting and handling are operations which must all be satisfied simultaneously in design to ensure that an economical and structurally sound product results. The thought process in the design of multi-storey buildings is essentially in two parts:

- the conceptual use of precast technology
- detailed design work with a knowledge of load paths and prestressed and rc stress analysis.

The designer attempts to utilize, as far as possible, a range of predetermined structural components in the skeletal structure whilst satisfying, again as far as possible, all the architectural and functional requirements of the building. Although the emphasis throughout is on the maximum use of factory cast prefabricated concrete, there may be many economic, aesthetic or structural constraints which prevent the designer from using a totally precast solution. The reasons for these are numerous and the subject will be discussed in the appropriate chapter. However, the norm is to prefabricate, without question, a stable structure (columns, walls, beams and slabs) and then to consider the options of precasting other items such as staircases, liftshafts, balconies, arches, etc. to obtain the most economical solution for the completed building.

### 2.2.5   Project design stages

In the office precast design is accomplished in four steps:

(1) preparation of the schematic building layout and quotation
(2) design calculations
(3) layout drawings
(4) component scheduling.

Step (1) is often termed 'schemes' or 'quotes', whereas Steps (2) to (4) are carried out upon receipt of an order and are termed 'projects'. Schemes and projects engineers may be one and the same group, or may be affiliated to different parts of the company, or other companies with the relevant expertise. Whichever is the case, it is important that the schemes engineer is fully conversant with project design practice.

The preparation of the schematic building layout drawings and the quotation forms the basis of the conceptual stage in which the capability of the precast system is assessed and structural components are notionally designed for estimating purposes. The non-precast areas are identified at this stage and shown on the tender drawing as work by the general contractor, such as foundations or other ground work (culverts, piling, ground beams, retaining walls). The precaster may choose to carry out a small amount of insitu work in order to secure a contract, for example it may be more economical and not too disruptive to the construction programme if unusually shaped staircases are cast insitu by a subcontractor employed by the precast company. In this case the area would be labelled 'insitu staircase by the precaster'.

The tender drawing carries sufficient information for the client to judge the merits of the structure. The drawing(s) usually contain:

(1) a plan of every roof and floor level (or small variations to a single plan are noted if the floor layouts are similar)

(2) a cross-section of the structure (in total or in part if the floor level details are repeated)

(3) a selection of structural components in cross-section, e.g. hollow cored flooring, spine and edge beams, spandrel beams and non-rectangular columns

(4) details to other insitu work, e.g. column pockets and/or base plate, or steelwork connections.

Items (3) and (4) are often considered optional but are included at this early stage to assist other trades in assessing the viability of the precast structure. Points to look out for in this stage are:

- structural zones shown on the architect's drawing
- particularly large voids, service holes, etc. in the floor slab
- service routes above or beneath the floor (these may affect the direction in which beams and floors will span)
- heavy loading in one particular area, e.g. plant rooms, where beam spans selected for other lightly loaded areas are too great for the plant room area
- the possibility of using a structural steel or timber roof, e.g. truss or 'Mansard' type roof
- the possibility of using load bearing or infill structural masonry, either prefabricated or built insitu
- obstructions to column positions, and the continuity of columns from roof to foundation
- possible positions of structural shear walls
- balconies, overhangs, etc. which must be structural in form, i.e. cannot be completed by the following trades
- splays in beams, chases in columns, holes in walls, etc., all of which may prove embarrassing at the project design stage.

At the completion of this exercise the designer should be confident that temporary stability is guaranteed no matter how the site erector may choose to build, and that the areas of insitu concrete may be allowed to cure and develop strength before further construction proceeds. To this end the scheme drawing(s) is annotated for internal use during the billing stage with information to assist the estimator to assess craneage and site costs in addition to the relatively simple task of pricing the precast concrete components.

A typical schedule for costing would include:

- beam cross-section, either in dimensions or unit reference coding
- beam length
- reinforcement, usually in cage reference coding for both flexure and shear, or individually referred if no code exists
- beam end connector capacity
- any voids (size and position), chases, cut out, splays, etc.
- special requirements for surface finishes, exposed aggregates, etc.

For other components such as exposed aggregate spandrels, the texture, relief, colour, jointing, etc. is given on the drawing. Omission of this information may have very serious implications far in excess of omitting the exposed aggregate materials. An obvious example which comes to mind is the cost for a new mould in which to manufacture the unit.

Consideration must also be given to specifying a small number of very heavy units in situations where a larger number of smaller units would suffice. Although the golden rule to precast efficiency is an optimum number of components of approximately all the same weight, there are exceptions where the cost of hiring and using a large capacity crane outweighs the benefit of a small number of components.

Preparing a final scheme and quotation is usually an iterative procedure: seldom are original layouts accepted after the first attempt. The points of contention are usually positions of shear walls, depths of beams and voids for service routes. Time spent on preparing schematic layouts may even exceed the time spent on final design, and unsuccessful tenders should not be considered as totally abortive; the experience bodes well for the next one.

### 2.2.6   Structural design calculations

Design calculations commence after discussions with the architect (or developer) and centre around the concurrence of details shown on the architect's current drawing with the precaster's latest scheme drawing. Loads are finalized and structural design commences in the manner described in Chapters 4–8. An example calculation for the design of hollow core floors is shown in Fig. 2.4. The terminology used here will be explained in Chapter 5. Draughtspersons are often keen to begin structural drawings early, and because of the increased use of computer-aided design or draughting (CAD) facilities this is not the problem it was as recently as the late 1980s. Recent advances in computer controlled detailing and scheduling of 2-D units, such as wall panels or floor slabs, has resulted in the kind of output shown in Fig. 2.5, for example.

It is not always possible to delay tackling those areas marked 'items in abeyance', returning to them at the end of the project. A case in point is the size of a void in the floor, where beyond a certain size (depending on spans and loading), additional measures may have to be taken to cater for the increased loads on adjacent units and the transfer of these loads to beams and columns.

Throughout the structural design process it is important not to lose sight of the overall behaviour. The sizes and positions of stability ties, often left until the last, must be catered for in the design of the floor and/or beams. As with insitu concrete or structural steelwork the design of a precast structure may be subdivided into a team effort, but there should be one engineer in charge to co-ordinate the input and output of information.

The design of most precast components can be committed to in-house or commercial software packages resulting in predetermined design and detailing information. (In-house packages are merely an extension of the tabulated/graphical design data and standardized drawings for standard frames of yesteryear.) The main distinguishing feature of precast design software is that design output may be in coded form to be compatible with the chosen method for scheduling components.

### 2.2.7   Layout drawings

Until very recently (c. 1990) these were all prepared manually, but more than 95% of

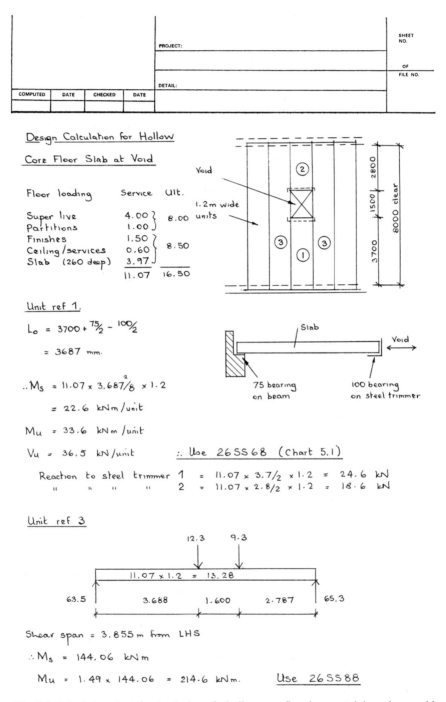

| COMPUTED | DATE | CHECKED | DATE |
|---|---|---|---|

PROJECT:

DETAIL:

SHEET NO.

OF

FILE NO.

Design Calculation for Hollow
Core Floor Slab at Void

| Floor loading | Service | Ult. |
|---|---|---|
| Super live | 4.00 ⎫ | 8.00 units |
| Partitions | 1.00 ⎭ | |
| Finishes | 1.50 ⎫ | |
| Ceiling/services | 0.60 ⎬ | 8.50 |
| Slab (260 deep) | 3.97 ⎭ | |
| | 11.07 | 16.50 |

Void

1.2 m wide

Unit ref 1.

$$L_o = 3700 + \frac{75}{2} - \frac{100}{2}$$

$$= 3687 \text{ mm.}$$

$$\therefore M_s = 11.07 \times \frac{3.687^2}{8} \times 1.2$$

$$= 22.6 \text{ kNm/unit}$$

$$M_u = 33.6 \text{ kNm/unit}$$

$$V_u = 36.5 \text{ kN/unit} \qquad \therefore \text{Use } 265568 \text{ (Chart 5.1)}$$

Slab

Void

75 bearing
on beam

100 bearing
on steel trimmer

Reaction to steel trimmer 1 $= 11.07 \times \frac{3.7}{2} \times 1.2 = 24.6 \text{ kN}$
"           "      "        "     2 $= 11.07 \times \frac{2.8}{2} \times 1.2 = 18.6 \text{ kN}$

Unit ref 3

12.3    9.3

11.07 × 1.2 = 13.28

63.5     3.688    1.600    2.787    65.3

Shear span = 3.855 m from LHS

$$\therefore M_s = 144.06 \text{ kNm}$$

$$M_u = 1.49 \times 144.06 = 214.6 \text{ kNm.} \qquad \text{Use } 265588$$

**Fig. 2.4** Calculation sheet for the design of a hollow core floor bay containing a large void.

drawings are now made using CAD methods. Those which do not use CAD facilities claim that because of the one-off nature of most commercial buildings today, the system is not an economic use of drafting skills. The aims of the layout drawings, e.g. Fig. 2.6,

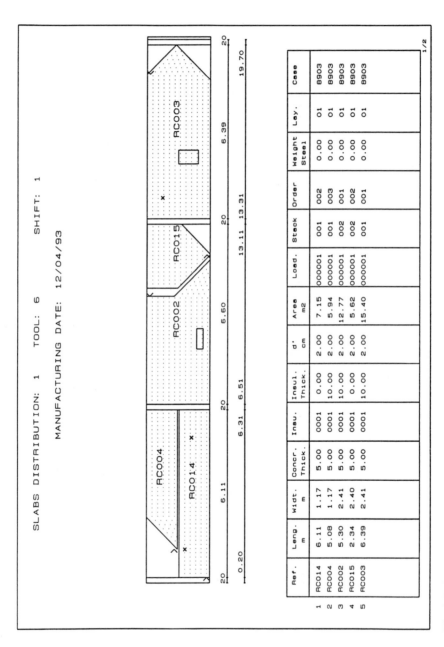

**Fig. 2.5** Computer-aided drawing and scheduling of precast components.

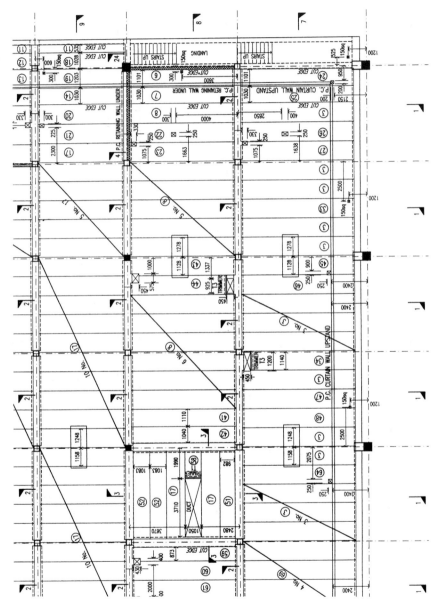

**Fig. 2.6** Example of a precast layout drawing (courtesy of Trent Concrete Ltd).

is (i) to show the architect, engineer and other trades the details of the structure and its connections, and (ii) to transmit to site information necessary to erect the structure.

To this end, precast concrete frame drawings tend to resemble structural steelwork layouts more than drawings for insitu concrete. Components are referred to individually and a reference number, which corresponds with its unit schedule number, is prominent on the drawing. Individual details, cross-referenced on the plan, are shown to an appropriate scale in Fig. 2.7. Beams, walls, columns, staircases and cladding components are detailed on separate drawings.

The project drawings are not sufficient to describe fully the details of the precast components. They are to be read in conjunction with a set of 'project product drawings'. Because these drawings are generated within CAD software it is possible to store hundreds of details, recalling them when the specific requirements of a project are known. These drawings are used to show the exact details of the cast-in connections, reinforcement joint details, connections, splices, ties, etc. assumed in the preparation of the project design and drawings. An example of a project product drawing is shown in Fig. 2.8.

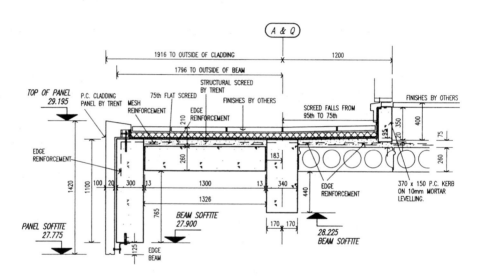

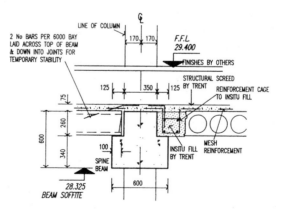

**Fig. 2.7** Examples of cross-sections (courtesy of Trent Concrete Ltd).

## 2.2.8   Component schedules and the engineer's instructions to factory and site

Unit schedules are prepared when the layout drawings are finalized or well advanced. CAD systems have usurped manual scheduling for the same reasons as those given in Section 2.2.7. An example of a beam schedule (made using 'Autocad') is shown in Fig. 2.9 and this too must be read in conjunction with the project product drawings (Fig. 2.8) *and* project layout drawings (Fig. 2.6) to be meaningful. Overall dimensions and tolerances are given in the schedules. The design procedure ends with the submission of component schedules to the casting factory. The casting works will make the necessary adjustments to cast-in inserts and moulds without violating the overall dimensions.

Sometimes instructions to the factory are not contained on product drawings and must be relayed by memorandum. These include:

- special requests for additional quality control, i.e. cube testing, cover distances, curing, handling or storage
- special surface finishes, including exposed aggregate, retarded and scabbled faces and smooth finishes
- use of special mixes, e.g. using lightweight aggregates, selected aggregates and grading, white cement, sulphate resisting cement, pigments, additives, etc.

Instructions to site are usually contained, in a manner of speaking, in a 'code of practice' which is prepared by the precasting organization at the tendering stage. Other instructions to site are self-evident from the drawings. The procedures adopted on site have grown out of many years' experience, and safe and structurally competent standard practices have evolved. Many site fixing details are contained in the product drawings, e.g. the details for column-to-column splices are fully described, although the exact procedural methods are not. Only where special or unusual components or joint details are being used are specific instructions corresponded to site. These might include:

- the sequence of erection, with particular reference to temporary stability (usually this is the decision of the contracting division, but special circumstances may prevail)
- temporary propping of components
- extra care, or specific instructions in the handling and fixing of certain components
- changes to the specification or use of expanding agents, cementitious grouts, mortars, epoxy-based materials, adhesives, bolting and welding
- additional degree of quality control, e.g. cube sampling, curing protection, compaction
- extra details for the placement of important stability ties
- additional requirements on construction tolerances.

## 2.3   CONSTRUCTION MATTERS

### 2.3.1   Design implications

The precast superstructure is designed and detailed to enable the safe handling, transportation and fixing, without damage to the component and framework. The effect of small hairline cracks which may occur in certain components are assessed on the

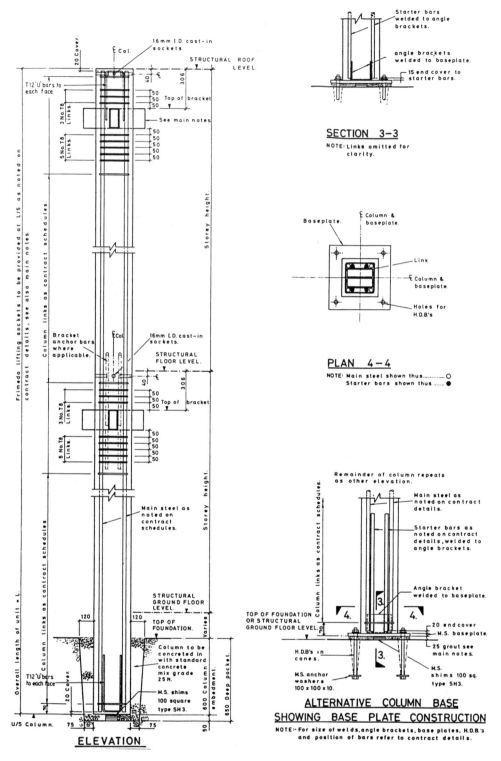

**Fig. 2.8** Standard detail for a column foot-to-steel base plate (courtesy of Crendon Structures Ltd).

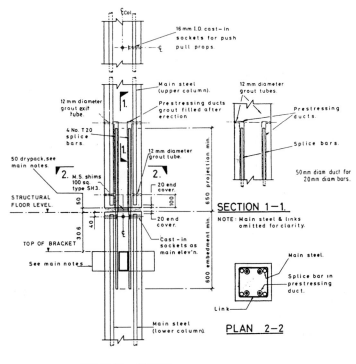

16 mm I.D. cast–in
sockets for push
pull props.

Main steel
(upper column).

12 mm diameter
grout tubes.

12 mm diameter
grout exit
tube.

Prestressing ducts
grout filled after
erection

Prestressing
ducts.

4 No. T 20
splice
bars.

12 mm diameter
grout tube.

Splice bars.

50 drypack, see
main notes.

M.S. shims
100 sq.
type SH3.

20 end
cover.

50mm diam duct for
20mm diam bars.

### SECTION 1–1.

STRUCTURAL
FLOOR LEVEL.

20 end
cover.

NOTE: Main steel & links
omitted for clarity.

TOP OF BRACKET

Cast–in
sockets as
main elev'n.

See main notes

650 projection min.

600 embedment min.

Main steel.

Splice bar in
prestressing
duct.

Link

Main steel
(lower column).

### PLAN 2–2

### COLUMN SPLICE

NOTE: All links omitted
for clarity.

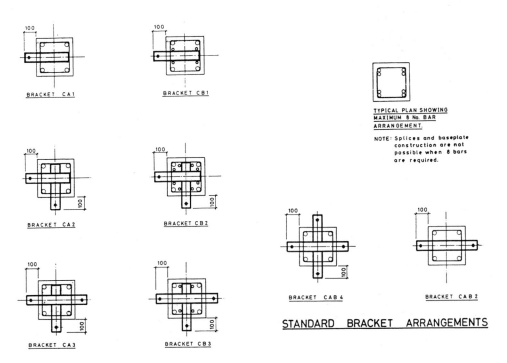

BRACKET CA1

BRACKET CB1

**TYPICAL PLAN SHOWING
MAXIMUM 8 No. BAR
ARRANGEMENT.**

NOTE: Splices and baseplate
construction are not
possible when 8 bars
are required.

BRACKET CA2

BRACKET CB2

BRACKET CAB 4

BRACKET CAB 2

BRACKET CA3

BRACKET CB3

### STANDARD BRACKET ARRANGEMENTS

**Fig. 2.8** (*continued*)

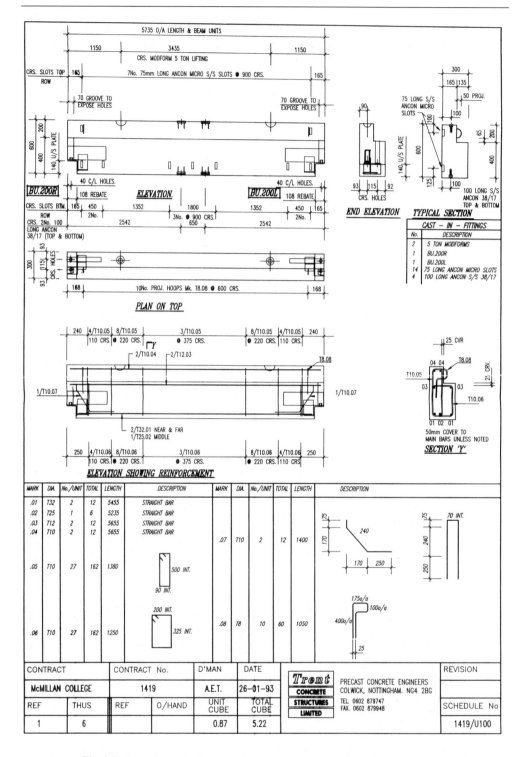

**Fig. 2.9** Examples of beam and column component schedules (courtesy of Trent Concrete Ltd).

basis of the nature of the component and its function within the structure. Due consideration should also be given to the location of the cracks, particularly in highly stressed regions close to connectors. In the event of structural cracks appearing in any component, reference is made to the designer prior to completion of the fixing.

The designer ensures that all lifting devices, whether bespoke or proprietary, are of sufficient load capacity to deal with all possible load magnitudes and directions. The lifting devices are located at the points given in the project product drawings. Detailed instructions and special provisions, etc. are given on working drawings regarding the safe handling, lifting and storage of components, and any specific features regarding temporary and/or permanent stability. The sequence in which the structure is erected and the temporary measures taken to ensure the corect transfer of load to the ground are available in writing for the fixing contractor to follow. Information relating to the point at which an entire framework, or part thereof (whether vertically or horizontally partitioned), is considered fully stable and may be allowed to stand free of any external restraint is stated.

Loading arrangements on delivery vehicles are such that the components may not be subjected to forces and stress not catered for in the design. The components are loaded evenly so that the distribution of weight is uniform on each component, particularly when the vehicle is cornering, breaking or inclined (Fig. 2.10). The type of lifting equipment (crane, hoist, etc.) is compatible with the geometry of the structure, weight and size of components, the nature of ground and site access. Crawler cranes of 40 to 100 tonne capacity, and smaller mobile cranes are frequently used where all round access is available. Otherwise 5 to 10 tonne capacity tower cranes are used. Lifting beams are used in situations where the inclination of the lifting chains will either cause cracking of the component, or will exceed the safe working load of the chains.

**Fig. 2.10** Correct transportation of precast components.

Components which are to be inclined by pitching are first placed on firm ground, and then pitched in a separate operation (Fig. 2.11). A special (steel, concrete or timber) shoe may be required when pitching heavy and thin units.

Temporary imposed loads due to the combined storage of materials, storage of precast components and construction traffic and plant are assessed, and maximum allowances are given to the various parts of the framework. Where a precast component has been designed to act compositely, consideration is given to the non-composite strength and stiffness of the component. Working drawings carry this information in the form of allowable loads per unit area, or per individual component, irrespective of the location of the imposed load.

The total dead load present (fixed or stored) above the highest level in the stabilized structure are known to the contractor, who will keep a record of events during construction. The contractor will record dimensional inaccuracies, such as component dimensions (length, cross-section, position of connectors, etc.), fixing tolerances (clear gaps, squareness, verticality), and report any major deviations from permitted tolerances to the designer.

There is a wide range of connections in a precast structure, many of which are used to perform a unique function. The contractor is made aware of the importance, or not, of

**Fig. 2.11** Pitching columns on site.

each fixing type, and of the permitted deviations. The designer should ensure that the fixing(s) perform their intended function, either by strict compliance with the manufacturer's details, or full scale testing. In the former, the influence of the position of the fixing in the component, and the measures taken to ensure the full safe working capacity of the fixing are observed. Construction tolerances are stated on the drawings.

The strength of bolted or welded connections are assessed for both the temporary and permanent condition such that no element is overstressed at any time. The effect of torsion and shear stresses induced by non-symmetrical loading (e.g. slabs onto one side of a beam) which would not normally occur in service are catered for.

The contribution to strength and stiffness from insitu concreted connections are assessed with respect to the time taken for the concrete to mature. Although adequate strength may be achieved in compression the bond resistance between rebar/dowel and the concrete may take longer to fully develop. The connections are protected from rapid drying or rain penetration.

## 2.4   CODES OF PRACTICE, DESIGN MANUALS, TEXT BOOKS AND TECHNICAL LITERATURE

### 2.4.1   Codes and building regulations

Past and present UK design practice has used a combination of statutory and non-statutory documents to satisfy the general requirements of the building regulations with respect to strength, safety, reliability and stability. In effect this means the design is carried out in accordance with BS 8110 [2.3] which allows the design of structures or structural components to be based on testing and permits the engineer to design in this manner which:

'... may be deemed satisfactory on the basis of results from an appropriate model test coupled with the use of model analysis to predict the behaviour of the actual structure ...' (Part 1, Clause 2.6.1),

or:

'... if the analytical or empirical basis of the design has been justified by development testing of prototype units ...' (Part 1, Clause 2.6.1).

For connections the code allows:

'Any other type of connection that can be shown to be capable of carrying the ultimate loads acting on it may be used' (Part 1, Clause 5.3.5.3).

In other words, full scale testing of structural components and connections is permitted.

Precast concrete designers have seized the opportunity to use the results of ultimate loads tests and fire tests to demonstrate the adequacy of connections and components. Full scale ultimate load testing has been carried out by independent authorities to determine factors of safety for a range of connector types and capacities (Fig. 2.12). Beam-to-column shear connectors have been the subject of considerable load testing, but most of

**Fig. 2.12** Beam end shear testing at Nottingham University (courtesy of Trent Concrete Ltd).

the data are confidential to the manufacturer. Certificates of validation for the connectors are issued to checking authorities as part of the design submission. In some instances it has been possible to supplement test results for connectors with a rudimentary structural design (Section 7.10), but in general the behaviour is far too complex for simple analysis and ultimate load testing is the sensible approach towards acceptance.

A similar approach has been adopted for fire resistance, particularly with hollow cored flooring units where the hollow core renders fire engineering calculations susceptible to over simplification. There has never been a fully co-ordinated approach to fire testing in Europe and it is proving to be a difficult task to disseminate the available data. One stumbling block is that the results of fire tests are frequently confidential to the manufacturer, but in many cases resistances greater than those obtained by calculation with respect to BS 8110 have been measured.

Although precast concrete design is carried out according to BS 8110, not all sections of the codes are used. The following is a summary of the sections used:

(1) In Part 1:

- structural analysis and design, loading, materials and robustness; Section 2
- ultimate limit state for the design of reinforced precast concrete components in flexure, shear and compression; Sections 3.1 to 3.4, 3.8 to 3.10, and 3.12
- serviceability and ultimate limit state for the design of prestressed concrete components in flexure and shear; Sections 4.1, 4.3 to 4.10, and 4.12
- precast construction; Sections 5.1. to 5.3
- composite precast insitu construction; Section 5.4
- concrete and tolerances; Section 6.

(2) In Part 2:

- reinforced concrete components in torsion: Section 2.4
- effective column height: Section 2.5
- robustness: Section 2.6
- serviceability limit state for reinforced concrete: Section 3
- fire resistance: Section 4
- volumetric changes in concrete: Section 7
- movement joints: Section 8.

### 2.4.2   Non-mandatory design documents

*2.4.2.1   Structural Joints in Precast Concrete.* The Institution of Structural Engineers [2.4].
This document has found wide acceptance in the UK and many design methods are used in conjunction with the ultimate design stresses given in BS 8110. The only criticism of the design methods is that many of the finer points are dealt with only in a qualitative manner, leaving the designer to propose numerical solutions to fit the specific problem. For example, column-to-column splices are described and well illustrated, but no design method to determine ultimate load or moment capacity is offered. The other drawback with this manual is that some of the connections shown are highly impractical and have rarely been used in the UK (e.g. Figs 4 and 18 in the manual). The danger here is that the inexperienced designer may be tempted to specify these types of connections without fully understanding precast construction.

*2.4.2.2   PCI Design Handbook.* The Prestressed Concrete Institute [2.5].
Although many of the design methods presented are not appropriate to limit state design, this manual is often used for detailing purposes. Some of these are used widely in Europe and successful designs have been converted from the American model. Many of the designs and details presented in this handbook are the product of the experimental and analytical data generated in over three decades of research work. (Most of this work originates in the PCI Journal.) However, because many of the design principles are based on empiricism, they do not satisfy European legislation without national certification. Because of this, alternative methods have been sought by European designers and PCI rules are largely ignored. Many design equations are dimensional and care must be taken in using them.

*2.4.2.3   Design of Multi-Storey Precast Concrete Structures.* Fédération Internationale de la Précontrainte, FIP Commission on Prefabrication [2.6].
This document focuses attention on the stability of medium to high-rise structures and presents broad principles on structural analysis with very little quantitative design data. The document is therefore useful only in the overall assessment of the integrity of precast structures, but is of little use in detailed design. The illustrations show only the principles involved in the design of joints and of structural stability, and as many appear to be outdated the document has not found favour in some countries. A more comprehensive edition has been published, see Section 2.4.2.4.

**2.4.2.4** *Planning and Design of Precast Concrete Structures.* Fédération Internationale de la Précontrainte, FIP Commission on Prefabrication [2.1].

This book is primarily meant for beginners. It illustrates the wider issues of precast in many contexts – buildings, housing, stadiums, cladding – and takes the reader through the basic philosophy behind precast construction before focusing on specific details regarded as good practice within the Commission. The information is collected from more than 15 countries worldwide, showing the enormous scope in precast. There are no specific design rules.

**2.4.2.5** *Precast Prestressed Hollow Core Floors.* Fédération Internationale de la Précontrainte, FIP Commission on Prefabrication [2.7].

This document has been used as a guide to both the design and detailing of hollow core floors and their connections to precast structural components, and to other hollow core units. The information has been used to design a structurally robust hollow core floor unit, capable of resisting flexural, shear, bond and bearing stresses in an economical manner. Connection details between adjacent slabs and at the ends of slabs, utilizing the minimum volume of insitu concrete and expedient placement of site reinforcement, have been widely adopted in all European countries.

At the present time (1996), the FIP is near to the publication of a more comprehensive second edition.

**2.4.2.6** *Guidelines for the Use of Structural Precast Concrete in Buildings.* Study Group of the New Zealand Concrete Society, and National Society of Earthquake Engineering [2.8].

This document is of primary importance to seismic resistant structures, although some of the details and recommendations are relevant to non-seismic conditions. Special attention is given to support conditions and continuity in floors and frame connections, and to grouted and embedded connections. A positive aspect is the manner in which the document shows the relationship between the way in which connections are made and their influence on the behaviour of the structure as a whole. There is a very good section on tolerances.

**2.4.2.7** *Precast Concrete Frame Buildings – A Design Guide.* British Cement Association [2.2].

This publication is very much a forerunner to this book. The guide establishes the potential use of precast concrete in multi-storey building and although the basic design principles, both in the permanent and temporary stages, are comprehensively covered, no computational data are presented. Its place in the design process is therefore to harmonize design principles in a qualitative manner, and establish some basic rules for the design of precast buildings, e.g. stability methods, floor diaphragm action, connections, etc. There are only references to specific design rules.

### 2.4.3   Other literature on precast structures

Precast design methods are not well documented because structural design is carried out mainly in-house by the precasting companies and there has been no real requirement to publish. Nilson [2.9] covers precast building design, but most of the details and methods are based on PCI and ACI rules and therefore have limited applicability in Europe. A new edition of Sheppard & Phillips' extensive book [2.10] also focuses on American practice and covers such items as the thermal, acoustic and dynamic properties of precast products. Bruggeling and Huyghe [2.11] gives good coverage to most aspects of precast design, particularly in the design of joints, but as with Lewicki [2.12] and Haas [2.13] it leans towards the European ideal. Bljuger [2.14] deals with precast in a very analytical manner, and provides a firm basis for many design aspects particularly connections and wall frames. Levitt's book [2.15] is useful for production purposes, and as such has a place in the designer's library, but there is no structural design and the text is becoming rapidly outdated, particularly in Scandinavia and northern Europe where nearly automated factories exist [e.g. 2.16]. Richardson has been a pioneer in the line of quality through manufacture and his books are particularly instructive [2.17, 2.18].

The PCI and FIP have large conventions every two or four years, in which approximately 10 per cent is devoted to prefabricated building. Recent events of note have been held in Washington DC (1993 and 1994), Tokyo (1993), and Budapest (1992). The Secretariat for these organizations should be contacted for further details [2.19, 2.20].

In addition to the above there have also been a number of significant seminars and conferences dealing with precast frames design and construction, namely the DoE & CIRIA Seminar on The Stability of Precast Concrete Structures [2.21], Delft University Seminar on Prefabrication of Concrete Structures [2.22], Noteworthy Developments in Prestressed and Precast Concrete, Singapore [2.23] and The Institute of Engineering Malaysia, Seminar on Trends, Innovation and Performance of Prestressed and Precast Concrete [2.24].

Many papers may be found in academic and professional journals including (in approximate order of numbers of publications) the PCI Journal (US), Nordic Concrete Research (Scandinavia), Concrete Precasting Plant and Technology, otherwise known as BF+T (Germany), ACI Structural Journal (US), ASCE Structural Division (US), The Structural Engineer (UK), Proceedings of the Institution of Civil Engineers (UK), Journal of Structural Engineering (US), Magazine for Concrete Research (UK), Heron (Netherlands), and the Canadian Journal of Civil Engineering.


## 2.5   DEFINITIONS

It is useful to define the main terms to be used in this book. The notation is given on page xvi.


### 2.5.1   General structural definitions

*Frame*: a framework comprising columns, beams and slabs in which there is some degree of flexural (or torsional) continuity between the members, and part of the resistance to sway loads is carried by the horizontal members.

*Skeletal structure*: a structure comprising columns, walls, beams, slabs and staircases, in which the connections are all designed as pin-jointed, and the resistance to sway loads is carried by the vertical members only.

*Precast concrete structure*: the total structure including any precast structural cladding, insitu floor screeds.

*Structural system*: the load bearing components for both gravity and horizontal loads and the way in which these components are assumed to behave.

*Structural cladding*: externally load bearing exposed panels serving both structural and architectural requirements simultaneously. Usually eliminates use for columns, beams and walls in external part of the structure.

*Unbraced structure*: in which resistance to horizontal loading is provided by columns in both orthogonal directions.

*Braced structure*: in which resistance to horizontal loading is provided by bracing in both orthogonal directions.

*Partially braced structure*: in which part of the structure, usually the lower levels, is braced and the remainder is unbraced in both orthogonal directions.

*Uniaxially braced structure*: in which the structure is braced in one direction and unbraced in the other.

*Diaphragm, or floor plate, action*: mechanism in which horizontal loads are transferred in the floor slab to the vertical bracing.

*Chord elements*: compression and tension beam members in the floor diaphragm.

*Stability ties*: horizontal tie bars insitu concreted into spaces between precast concrete components.

*Components*: the discrete precast concrete units from which the structure is assembled.

*Members*: two or more components working interactively with one another.

## 2.5.2   Components

*Column*: vertical load and bending moment resisting linear component.

*Wind post*: deep columns in the direction of horizontal loading.

*Shear wall*: general term given to vertical bracing walls.

*Cantilever wall*: a structurally continuous shear wall designed as a deep cantilever.

*Hollow core wall*: a precast wall containing full height vertical holes into which reinforced insitu concrete is placed to form a continuous cantilever wall.

*Infill wall*: a discontinuous solid precast shear wall positioned between adjacent columns and beams.

*Hollow core infill wall*: a factory cast wall containing unfilled vertical or horizontal voids positioned between adjacent columns and beams.

*Wall panel*: a solid precast shear wall positioned between columns only and connected vertically to one another.

*Beams*: the primary load carrying horizontal linear component.

*Spandrel*: a beam with a deep upstand web, and which is flush with the front edge of the column.

*Outstand spandrel*: one which the upstand projects beyond the front edge of the column.

*Slabs*: primary 2-D floor and roof components.

*Composite slabs*: combined slab comprising precast concrete slab (also providing permanent shuttering) and insitu concrete topping (or screed).

*Hollow core slab*: rectangular section slab with full length longitudinal circular or oval (or similar) shape voids.

*Double-tee slab*: T-shaped solid units comprising two deep webs and a full width top flange.

*Flat plank*: rectangular solid section, used either alone, or more usually as permanent shuttering in a composite floor.

### 2.5.3    Connections and jointing materials

*Base plate*:  steel plate anchored to end of column for foundation connection.

*Splice*:  connection between successive columns.

*Connector*:  mechanical device for the transfer of forces between precast components.

*Shear box*:  mechanical device for the sole purpose of transmitting end shear forces into beams.

*Column insert*:  mechanical device for the sole purpose of transmitting beam end shear forces in to columns.

*Corbel*:  bearing surface to a column or wall made by a reinforced concrete projection of finite depth.

*Haunch*:  bearing surface to a column made by increasing the cross-section of the column below the bearing level.

*Scarf (or halving) joint*:  made between the recessed ends of two components, one bearing on top of the other.

*Simple bearing*:  a direct dry bearing between precast components.

*Extended bearing*:  a reinforced insitu bearing made by concreting reinforcement left projecting from the adjoining components.

*Dry bearing*:  no additional material is used between the precast components.

*Wet-bedding bearing*:  mortar is used in the bearing surface before erection.

*Dry-packed bearing*:  mortar is packed in the gap between precast components after erection.

*Elastomeric bearing*:  neoprene (or similar) pads are used.

*Simple connection*:  where only shear and axial forces are transmitted.

*Moment connection*:  where bending and torsional moments, and shear and axial forces are transmitted.

*Trimmed hole*:  a void made in a floor slab by supporting the curtailed members in line with the void (usually) on steel angles, which in turn are supported on adjacent components.

*Cast-in-sockets, channels, etc.*:  threaded sockets, channels, etc. fully anchored into concrete.

# Chapter 3

# Architectural and Framing Considerations

*This chapter describes the main features in the preliminary design stages which influence the interaction between the architectural concepts, the building's function and the precast concrete solution.*

## 3.1 FRAME AND COMPONENT SELECTION

The criteria for the choice of internal and external layout are different in most buildings. The interior is governed by the location, size and orientation of lift shafts, stairwells, mezzanine floors, major partitions etc., whilst for the exterior the precast structure can be designed for a wide range of architectural features. It is easy for the precast frame designer to respond to both of these requirements simultaneously by specifying an external structure that is, in the main, totally different from the internal arrangement. There are three main tasks in this selection process:

- selection of the structure
- recognition of the functional requirement of the building
- preparation of the framing plan by making optimum use of precast components.

These are discussed below.

(1) Selection of the structure.

The possibilities for the structure are:

- an external structural envelope comprising either visual concrete spandrel beams, or structural precast concrete cladding, Fig. 3.1
- non-structural precast concrete cladding, insitu masonry, steel sheeting, or glass curtain walling supported on a plain concrete beam–column structure, Fig. 3.2.

The external structure can be physically massive, being overclad in some form, and may therefore be used as a primary horizontal stabilizing structure. This means that the internal structure may be relatively lighter in construction and need not necessarily be established on the same dimensional grid as the exterior. The options are:

- a totally precast internal structure comprising a stable beam–column framework, floors, walls, staircases and lift shaft walls, Fig. 3.3
- a hybrid internal structure (see also Section 3.3.1) comprising a precast skeleton of beams, columns and slabs designed collectively with one or more of the following:

  ○ insitu concrete basement or ground floor podium

**Fig. 3.1** Architectural precast spandrels (courtesy of Bellcon Ltd).

**Fig. 3.2** Natural and reconstituted stone precast into concrete panels at Northgate House, Halifax (courtesy of Trent Concrete Ltd).

- ○ structural steel or timber roof (Mansard or Warren girder type) atrium or gallery, Fig. 3.4
- ○ structural steel portal frame (e.g. warehouse).

**Fig. 3.3** Precast skeletal structure under construction.

**Fig. 3.4** Timber roof trusses seated on long span precast roof edge beams.

(2) Recognition of the functional requirement of the building.

Different areas of the structure may require a higher building specification than other areas or floor levels, for example with respect to fire resistance, thermal, acoustic or vibration characteristics. Other criteria may include the requirements for open spaces, natural light, the absence of non-demountable walls, unlimited freedom for vertical and horizontal service routes, future vertical and horizontal extensions, and changes in ownership. The design engineer must address all the aspects, and more, listed above. This cannot be stressed too highly.

(3) Preparation of the framing plan by making optimum use of precast components. The example building plan shown in Fig. 3.5 is subdivided into component rectangles and triangles. The flooring should span on to the beams or walls making an angle of intersection of less than 45°.

An orthogonal grid layout is specified as far as possible with the columns and shear walls placed on beam lines. Primary columns are located at the strategic points (corners, changes in floor level, around stairwells and lift shafts) and secondary columns are introduced to satisfy architectural requirements, such as window bay widths, internal partitioning, or to obtain structural economy by using the minimum number of components giving acceptable structural zones. The distances between beams are equalized as far as possible in order to divide the building into equal bay widths. Beams should be placed beneath any major partitions, say 215 mm thick brick or block walls, rather than to support them on the floor slabs.

There are general rules regarding the provision and location of holes and cast-in fixings in precast components. Their availability varies with the different methods of manufacture. For example, some units are cast in steel moulds and it may be less expensive and time consuming to drill and fix on site. Alternatively if *ad hoc* timber or glass fibre moulds are used then it is relatively easy to fix fittings to the inside of the mould. The best advice is to contact the precast manufacturer early in the project to ascertain which of these may be incorporated in the normal course of manufacture and which require additional attention. Some general guidelines are given in Sections 3.2–3.4.

The surface finish to all ex-mould faces is Type A to BS 8110, Part 1, clause 6.10.3. All non-contact surfaces have either a steel trowelled finish, or are rough tamped for reasons of wishing to develop a shear key with other concrete.

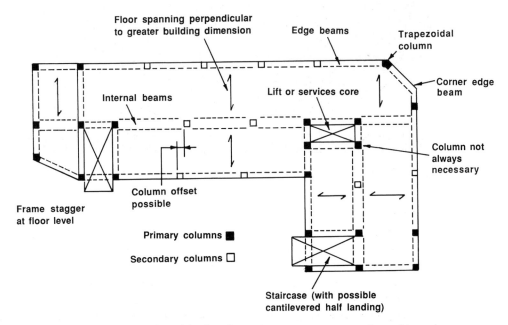

**Fig. 3.5** Subdivision of the floor layout into component rectangles and trapeziums.

## 3.2 COMPONENT SELECTION

### 3.2.1 General principles

Precast manufacturers have standardized their components by adopting a range of 'preferred' cross-sections for each type of component. In fact it is the moulds which are standardized and therefore non-preferred shapes will incur cost penalties because of the alterations to the mould. If a non-preferred cross-section is to be used there should be a sufficient number giving a piece-to-mark ratio (units per mould alteration) of ten, at least, to justify these alterations, The framing plan may be adjusted according to this philosophy.

For the purpose of this study the components will be described in the design sequence:

- roof and floor slabs
- staircases
- roof and floor beams
- columns
- bracing walls.

### 3.2.2 Roof and floor slabs

The wide range of precast concrete floors used in precast skeletal structures has, in recent years, reduced to five main types (Fig. 3.6 (a)–(e)):

- prestressed  hollow core floor
- reinforced and prestressed double-tee floor

which may or may not be used with a structural topping screed, and:

- composite prestressed plank-floor
- composite beam and plank
- beam (including wide beam) and block floors

which must always be used with a structural screed.

Table 3.1 gives a performance guide to the weights and maximum spans for some of these units.

In most cases the floor construction is voided, thus producing a lighter floor having the same structural performance as a continuous insitu slab. The percentage void (volume of voids to total volume of a solid slab of equal depth) for hollow core slabs is between 30 per cent and 50 per cent (deeper units latter). For double-tee and beam/plank flooring the voids occupy about 75 per cent of an equivalent solid section.

Reinforced concrete versions of the above are equally valid, usually in short lengths for standard applications, e.g. housing. However, as explained in Chapter 1, unless the prefabricator is wanting to increase the turnover for the project and manufacture the units as reinforced, the advantages gained from maximum performance often outweighs the additional costs.

Beam and block flooring, shown in Fig. 3.6(d), is restricted mainly to low-rise domestic or commercial work, whilst other forms of solid rc slabs or solid prestressed joists cannot compete structurally or economically with hollow core or double-tee flooring systems. Comprehensive design and detailing information is given in Chapter 5. The following is a brief introduction to these units.

Table 3.2 compares the relative structural performance and costs of a selection of hollow core and double-tee units. The data are for the cost of units supplied and fixed on site, but exclude the cost of pouring insitu infill and laying structural screeds. The higher costs for the larger double-tee units reflect the additional craneage and haulage costs associated with these units. The data were collected in the UK in 1993 and were based on the design of floor units for 5 kN/m$^2$ superimposed loading over a 6 m span. The haulage distance used in the exercise was 100 km.

### 3.2.2.1  *Hollow core units*

Hollow core units (hcus) are now the most widely used type of precast flooring; in Europe annual production is about 20 million m$^2$ representing 40 to 60 per cent of the precast flooring market. This success is largely due to the highly efficient design and production methods, choice of unit depth and capacity, surface finish and (last but not least) structural efficiency. In this context only the machine cast units will be discussed, although it is equally possible for hcus to be wet cast into moulds.

The design of dry cast hcus originated in the US in the late 1940s following the development of high strength 3-wire and 7-wire helical strand in 1951 that could be reliably pretensioned over distances of 100 m to 150 m. This coincided with advancements in zero slump (hence the term 'dry') concrete production which inevitably led to factory engineered units. Concrete extrusion techniques were first introduced in the US following their development in Germany in the 1930s. Slip forming followed later. The first units were 2 ft wide, and later 1.2 m wide, as is the case today. Attempts to produce units 3.6 m wide in the factory have not been successful, although very wide units (11 ft) have been produced in the open air for specific projects in the US. During the 1960s the industry experimented with different methods of forming the cores, using flexible pneumatic tubes (called 'flexicore') or tubes loosely filled with stones, but production rate could not compete with the extrusion or slip-forming production line methods.

For many years the maximum depth of hcus was 300 mm; the restriction in depth was due mainly to instability in the narrow web. However, recent developments in Northern Europe have led to a 500 mm deep unit, shown in Fig. 3.7, which became available in 1994. This slab has a 53 per cent void content and requires a special sturdy saw to cut it to length.

Hcus are manufactured using the long line extrusion or slip-forming process in which the degree of prestress and depth of unit are the two main design parameters. Steel beds, of very high accuracy, are used in lengths of up to 150 m. Many plants in warm and dry climates are uncovered where there is less handling as the stock pile can be placed alongside the bed. Nearly all European plants are under cover, many in heated environments. Dimensional deviations are less than ±5 mm in depth and width, and ±10 mm in length. Cross-section, concrete strength and surface finish are standard to each system of manufacture. Small variations include increased fire resistance by raising the level of the centroid of the tendons, provisions for vertical service holes, opening of cores for special fixings, cut-outs at columns, etc. Openings and cut-outs are easily formed

(a)

(b)

**Fig. 3.6** Most common types of precast floors used in skeletal structures. (a) Hollow core units and double-tee slabs used in the same structure; (b) double-tee slabs used in parking structures (courtesy of Frank Graham Consulting Engineers), (c) composite plank floors (courtesy of MWE Precast Concrete Bhd.); (d) beam and block floor (courtesy of Bison Floors) (e) wide beam and block floor.

whilst the concrete is 'green', i.e. less than 12 h old, as shown in Fig. 3.8, but afterwards the operation is more expensive.

The width of units are based on a nominal 1200 mm grid. More than 95 per cent of units produced are 1200 mm wide. The actual width of 1197 mm allows for constructional tolerances and prevents overrunning of the floor layout due to cumulative errors. To

(c)

(d)

**Fig. 3.6** (*continued*)

maintain repetition in the flooring layout, and hence in the detailing of slabs and beams, the most economical framing plan is columns centred on a 1.20 m modular grid. Recently, outdoor production of 2.4 m and 3.35 m wide units was achieved for a specified project in the US where a casting yard was established close to the site.

The slabs are cut to length using a circular saw. Although the cut gives a square end as standard, skew or cranked ends, which are necessary in a non-rectangular framing plan, may be specified. Longitudinal cutting is possible, as shown in Fig. 3.9. This operation

(e)

**Fig. 3.6** (*continued*)

**Table 3.1** Basic properties and performance characteristic of precast flooring.

| Type | Widths (mm) | Typical depths (mm) | Approx. self weight (kN/m$^2$) | Approx. max. span[#] (m) |
|---|---|---|---|---|
| Hollow core | 1200, 600, | 110 | 1.90 | 7.0 |
| | 400, 333 | 150 | 2.25 | 8.5 |
| | | 200 | 3.00 | 10.5 |
| | | 250 | 3.35 | 11.0 |
| | | 300 | 3.60 | 14.5 |
| | | 400 | 4.80 | 18.5 |
| | | 500 | 5.50 | 22.5 |
| Double-tee | 2400, 3000 | 400 | 3.60 | 14.5 |
| | | 500 | 4.10 | 17.0 |
| | | 600 | 4.50 | 19.5 |
| | | 700 | 4.95 | 21.0 |
| | | 840 | 5.50 | 24.0 |
| Composite plank | 2400 max. | 65/75* | 3.36 | 4.50 |
| | | 75/75* | 3.60 | 5.25 |
| | | 100/100* | 4.80 | 7.25 |
| Composite beam and plank | 900–2400 beam centres | 455/115** | 4.2–6.6 | 17.8 |
| | | 550/115** | 4.5–7.4 | 20.9 |

[#] Span for superimposed loading of 1.5 kN/m$^2$ plus finishes of 1.5 kN/m$^2$
\* Depth of precast plank/depth of insitu topping
\*\* Depth of precast beam/depth of plank and topping.

**Table 3.2**  Economic and structural comparison of hollow core and double-tee units.

| Flooring type | Depth (mm) | Structural performance index* | Mean cost index* |
|---|---|---|---|
| Hollow core | 150 | 0.6 to 0.7 | 0.94 |
| | 200 | 1.00 | 1.00 |
| | 250 | 1.5 to 1.6 | 1.17 |
| Double-tee | 400 | 2.0 to 2.8 | 1.7 |
| | 500 | 3.0 to 4.0 | 2.0 |
| | 700 | 5.5 to 6.5 | 2.6 |

* Based on 200 mm deep hollow core unit.

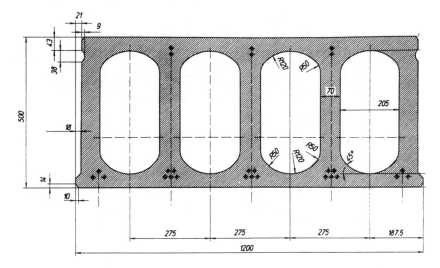

**Fig. 3.7**  Cross-section of 500 mm deep hollow core floor unit (courtesy of B + FT).

shows how insitu concrete, which was placed into the hollow core through a slot formed during the manufacturing process, penetrated the hollow core for a distance of about 600 mm along the slot. It also shows the degree of compaction possible when concrete is placed into hollow core slabs; information which is particularly useful where a solid section is specified.

Hollow cores may be used as warm air heating ducts; see Bruggeling and Huyghe [3.1] or even for air conditioning cooling systems. The UK's Building Research Establishent has been studying the cleanliness of conditioned air through the voids, and found them to be satisfactory [3.2]. Hollow core slabs are also being used to cool the structure in summer and heat it in winter [3.3]. The webs of the slabs are punctured at various positions to create one continuous 'tube' snaking along the floor slab. The proprietary system 'Termodeck' was developed in Scandinavia for this purpose [3.4]. There are limitations to this method of air transport, particularly across beam lines where the cores are filled with insitu concrete, but alternative routes may be found at such locations.

Most of the fire tests and associated research work on hollow core slabs was carried out between 1960 and 1980. The work covered a wide range of slab cross-sections together

**Fig. 3.8** Cut-outs for services or check-outs around obstructions are made whilst the concrete is young.

with variations in reinforcement quantities and position, cover to soffit and void, etc. Standard designs are based on a requirement for a 1 or 2 h fire resistance in accordance with the relevant legislation. This offers three options, namely calculation, fire testing or tabulated data, the latter being commonly adopted by the industry. There is a slight penalty to pay in the flexural strength (about 10 per cent) of the units with increased fire resistance. Greater fire ratings of up to 4 h can be achieved by the application of soffit finishes.

Holes in the floor are dealt with in one of two ways depending on size. Small holes of less than about 600 mm in size may be formed in the precast unit (except in 600 mm wide hollow core units where the maximum size is 300 mm) during the manufacturing stage and before the concrete has hardened. The cost is small and the result is as shown in Fig. 3.8. The rough edges are beneficial to the shear key with insitu concrete infill. The voids are formed manually, although there are plans to automate this within the machine's operations.

The maximum size of hole which may be formed in the units depends on the size of the voids in the slab and how much reinforcement may be removed without jeopardizing the strength of the unit. For example, in a 200 mm deep extruded unit (e.g. Spiroll) the

**Fig. 3.9** Longitudinal cutting of hollow core slab is expensive, but often unavoidable.

diameter of the void is 150 mm and therefore a hole formed through the void may be of equal size. In a 200 mm deep slip-formed unit (e.g. Roth) the diameter of a hole is limited to the width of the void which is only 100 mm. General FIP information on restrictions on hole size and location is given in Fig. 3.10(a) [3.5]. Figure 3.10(b) gives further information from product literature for a range of 1200 mm wide units produced by the Partek Company [3.6]. Referring to Fig. 3.10(b), the limits shown in Table 3.3 are imposed. Small holes, up to about 150 mm diameter, can be core drilled on site. The holes should pass through the hollow core, and the designers *must* be consulted if any of the reinforcing bars are accidentally removed. This can impose quite severe restrictions on the M&E services, but a way of increasing freedom is to provide additional reinforcement in the slab and to specify permitted 'zones' in which a limited number of holes at a given spacing and size may be site drilled.

Larger voids which are wider than the width of the precast units are 'trimmed' using transverse supports such as steel angles (i.e. perpendicular to the span of the floor) as shown in Fig. 3.11(a) or concrete beams for larger openings shown in Fig. 3.11(b). A typical minimum section size for the steel trimmer is $100 \times 100 \times 8$ mm equal leg rolled mild steel angle. The angles carry point loads to the edges of the adjacent units. The maximum practical size of hole is about 2.4 m $\times$ 2.4 m for normal office loading, but is less for superimposed floor loads greater than about 7.5 kN/m$^2$. There is a certain margin of safety in the design of trimmer angles because the grouted longitudinal joints between the slabs will transfer some of the superimposed loading to adjacent slabs. The steel angles must be properly fire protected either by precasting concrete to the soffit and sides of the angle or spraying with a fire protective coating.

It is not possible to cast sockets or other fixings into the soffits or sides of prestressed floor slabs. These must be formed on site using proprietary anchor or toggle bolts. Shot fired fixings are not recommended for prestressed units. Ceiling hangers are used

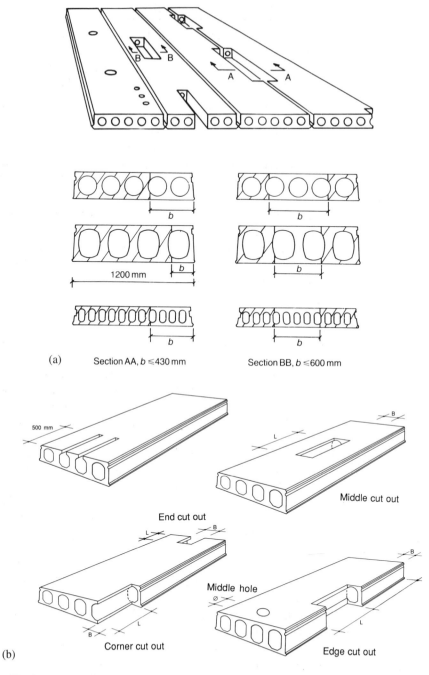

(a)  Section AA, $b \leqslant 430$ mm   Section BB, $b \leqslant 600$ mm

(b)

**Fig. 3.10** Rules for the permitted sizes and location of voids in hollow core slabs. (a) Data given by FIP [3.5] and (b) additional guidance given by manufacturers.

extensively in commercial buildings where ceilings and extensive electrical and mechanical services are suspended. Information sheets giving comprehensive details of these and other provisions are available from the flooring manufacturers.

**Table 3.3** Limits imposed (refer to Fig 3.10(b)).

| Depth of slab (mm) | Corner cut-out L × B (mm) | Edge cut-out L × B (mm) | End cut-out L × B (mm) | Middle cut-out L × B (mm) | Middle hole diameter $\phi$ (mm) |
|---|---|---|---|---|---|
| 150 | 600 × 400 | 600 × 400 | 1000 × 400 | 1000 × 400 | 80 |
| 200 | 600 × 380 | 600 × 400 | 1000 × 380 | 1000 × 400 | 130 |
| 265 | 600 × 260 | 600 × 400 | 1000 × 260 | 1000 × 400 | 130 |
| 320 | 600 × 260 | 600 × 400 | 1000 × 260 | 1000 × 400 | 170 |
| 400 | 600 × 260 | 600 × 400 | 1000 × 260 | 1000 × 400 | 170 |

One note of caution when using hollow core slabs in wet and cold climates is that water may penetrate into the hollow cores through the open ends of the cores, or even through small drag cracks in the top flange. If this water is allowed to gather and expand by freezing there is a possibility that the bottom flange will explode off the unit. A simple remedy is to drill weep holes in the bottom of the slabs, usually during manufacture.

### 3.2.2.2   *Double-tee slabs*

These units originate from the US where they have been used extensively since the 1940s for what are termed 'medium spans', i.e. up to 25 m simply supported span. Double-tees formed the principal slab units in the NBF (Fig. 1.3), and they enjoyed the leading market share in precast flooring until the late 1960s when hollow core became the preferred slab. Single-tee units were found to be less economical in terms of quantity of concrete per unit area. In the early days of development the units were reinforced; prestressing followed later.

Typical cross-sections through prestressed precast double-tee units are shown in Fig. 3.12. Various attempts were made to hide the longitudinal joint over the centres of the webs in an F configuration, shown in Fig. 3.13. This failed due to non-symmetrical prestressing requirements and differential stiffness problems where the flange of one unit was bearing over the stiff web of its neighbour. After exhaustive research the double-tee section is still used. The main advantages in using these type of units (over the hcus) are:

- load carrying capacity, although this advantage has been eroded by the introduction of 400 mm and 500 mm deep hollow core slabs
- the ends of the units can be notched to one-third of overall depth to form a halving joint to reduce the overall structural depth
- the units are manufactured as standard up to 2400 mm wide (actually 2390 mm) or 2500 mm wide – thus reducing the number of units to be fixed on site. However, there is no reason not to manufacture other widths, e.g. 3.0 m, to meet the requirements of a particular project.

However, due to the shallow flange depth an insitu reinforced concrete structural screed is required to ensure vertical shear transfer between adjacent units and horizontal diaphragm action in the floor plate.

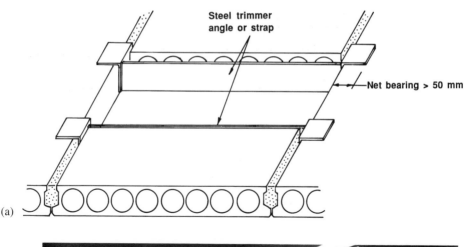

**Fig. 3.11** Trimmer details around floor voids: (a) steel trimmer angles used to support floor units around a small void; (b) concrete trimmer beams used to support floor units around a large void (greater than about 2.4 m side).

The standard end profile is square, although contoured ends may be specified by shaping the flanges to suit the structure. The ends of ribs are always square. If vertical service voids are required adjacent to supporting beams forked ends may be formed by cutting back the flanges over the full width of the unit as shown in Fig. 3.14. The ribs must be maintained at full length to facilitate the welded connections at the supporting beam.

Double-tee units have a minimum fire rating of 2 h providing the correct depth of screed is used. Other units with wider ribs and thicker flanges achieve a 4 h rating. Greater fire rating can be achieved on all units by the application of soffit finishes. The slabs are

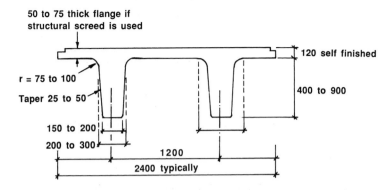

**Fig. 3.12** Shapes of double-tee floor units.

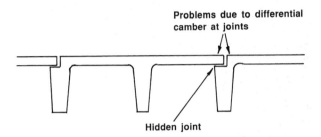

**Fig. 3.13** Historical development of shapes of double-tee slabs.

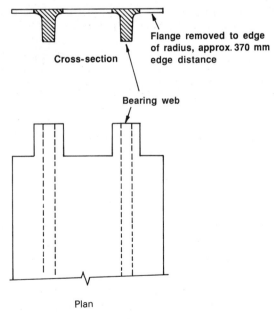

**Fig. 3.14** Forked ends in double-tee floor units.

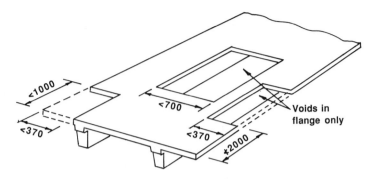

**Fig. 3.15** Guidelines for the permitted sizes and location of voids in double-tee floor units.

manufactured in steel moulds with a high degree of dimensional stability and excellent surface finish. The soffit is Type A to BS 8110.

Holes may be formed in the positions shown in Fig. 3.15. In no circumstances should vertical holes be formed through the webs of double-tee units. The circular holes through the webs are sometimes provided as standard at 1200 mm centres so that they provide a passage for services where the double-tee units cross at right angles to each other.

The positions and sizes of additional horizontal holes need to be planned in advance because they may affect the shear capacity of the slab, or alter the position of the prestressing reinforcement. Holes and notches in the flanges are best formed in the units immediately after manufacture and before the concrete has hardened, although site cutting is permitted with the manufacturer's consent and within its limitations.

### 3.2.2.3 *Composite plank floor*

The principle is quite simple: rectangular precast concrete beams or slabs are laid between supports and used as permanent formwork for an insitu concrete topping. Steel bars, wires or tendons placed in the precast units act as the flexural sagging reinforcement, and a light steel mesh (e.g. A142) in the insitu concrete acts as hogging reinforcement. The floor slab may therefore be designed as continuous.

Figures 3.16 and 3.17 show the construction of the composite floor using beams and slabs, and slabs alone, respectively. In terms of robustness this floor is considered equal to that of cast insitu construction. This floor slab has advantages in using a prefabricated soffit unit (smooth finish, no formwork, up to 2.4 m wide units, rapid fixing), but carries performance penalities on span and self-weight; see Table 3.1. This slab may be made lighter by the use of lightweight block (or even polystyrene) void formers, or by using a lightweight aggregate concrete topping. Splayed end, or other special shapes may be formed, usually at an extra charge due to making alterations to the mould. Where the floor slabs bear on to beams (Fig. 3.16) maximum spans between the primary beams in the structure is in the order of 18 m.

Service openings up to 2.4 m wide may be formed by increasing the strength of the precast unit and placing additional steel around the hole in the insitu topping. Cantilevers up to 1.5 m span may be formed by placing additional top steel in the insitu topping. The usual fire resistance is 2 h. Thicker precast planks (up to 125 mm) are used for more onerous fire, durability and loading conditions.

### 3.2.3 Staircases

Precast concrete staircases are a viable alternative to cast insitu staircases providing

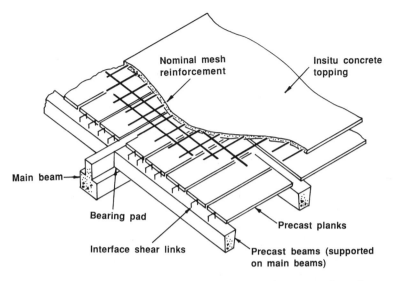

**Fig. 3.16** Composite beam and plank construction (the beam forms part of a tertiary system).

**Fig. 3.17** Composite plank construction.

there are sufficient numbers and a reasonable amount of repetition. There were twelve identical stair towers on the six storey Grand Island Project in Manchester, UK enabling the circular staircase shown in Fig. 3.18 to be designed economically. Precast staircases with high quality finishes from steel moulds may be used in other types of structures, such as the insitu concrete frame shown in Fig. 3.19.

Consistency in tread height and going are the two main factors affecting repetition.

The method of manufacture enables the depth of the waist, number of treads and the width of the flight to be varied readily. The most important factor in the use of the precast stairs is plan configuration and compatibility with the structure. This implies making optimum use of the structure to avoid introducing additional components to cater for the staircase. Alternative layouts are shown in Figs 3.20(a), (b) and (c), and 3.21.

Figure 3.20(a) makes optimum use of the structure – only two short beams are required for support. It is the least satisfactory solution from structural (greater span hence depth) and manufacturing points of view. Differential levels at floor and half landings are difficult to avoid and either a finishing screed or an insitu levelling piece is necessary.

This would not be the case in Fig. 3.20(b) where optimum use of individual precast flights and landings is made possible by the use of scarf joints, or support brackets, intended to eliminate construction errors. However, Fig. 3.22 only emphasizes the need for greater dimensional tolerances where close fitting elements are to be positioned between two fixed points. The major disadvantage in Fig. 3.20(b) is that beams or stub walls parallel with the flight (typically 4 m to 5 m long) are required both at floor and half landing levels. Insitu masonry may be used as half landing support, but this enforces a strict programme both on the main contractor and the precast fixing team. Efforts should be made to make the floor and half landings identical.

Figure 3.20(c) makes optimum use of the supporting beams and precast staircase units by casting the flight and half landing in one piece. This solution enables cantiliver half

**Fig. 3.18** Circular staircase tower, later to be overclad in steel sheeting.

**Fig. 3.19** Precast concrete staircase positioned in an insitu concrete frame (courtesy of Breton Precast, UK).

landings to be expressed beyond the edge of the building, as shown in Fig. 3.23. These may be rectangular, semi-circular or chamfered in plan, and the external cladding may be carried by the precast members. A minimum waist of 150 mm is, in all cases, required for handling purposes and fire resistance.

In the three flight staircase shown in Fig. 3.21, various options are available to the designer but the importance of minimizing the number of precast components whilst maintaining simplicity of manufacture cannot be stressed too highly. Cantilevers may be specified providing consideration is given to an adequate tie back and the design of the support beam which must be profiled to suit the different landing levels.

Fixings for balustrades, lighting, security, etc. may be incorporated in precast staircase and landing units, but it is often more economical and accurate to drill and fix on site. The edge distance to holes should be 50 mm minimum. Stair nosings, granolithic or non-slip surfaces are more easily added on site although rebates or rounded arrises at the nose can be precasted.

The finishes to stair units depend on the method of manufacture. Units cast with the treads face down will have a surface finish (Type A to BS 8110) from timber moulds, with a steel trowelled finish to the soffit and sides. Units cast in adjustable angle steel moulds with the treads face up will have all Type A finishes except for the treads which will be trowelled. Information sheets supplied by the manufacturer of precast stairs provide

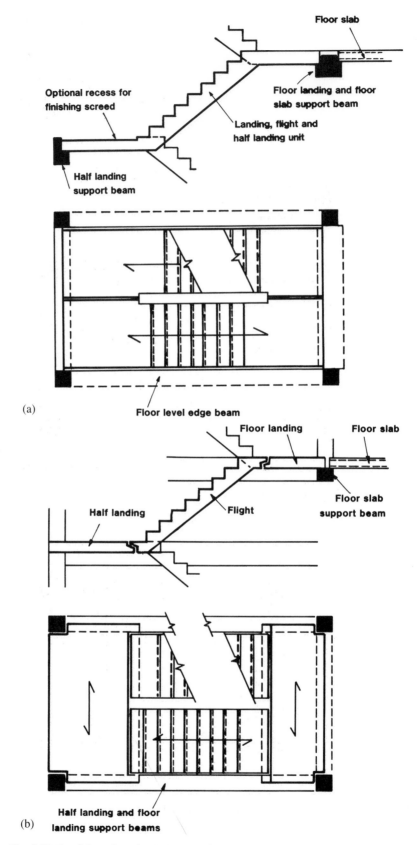

**Fig. 3.20** Possible options for the layout of a two flight staircase. (a) Option 1; (b) Option 2 and (c) Option 3 [3.7].

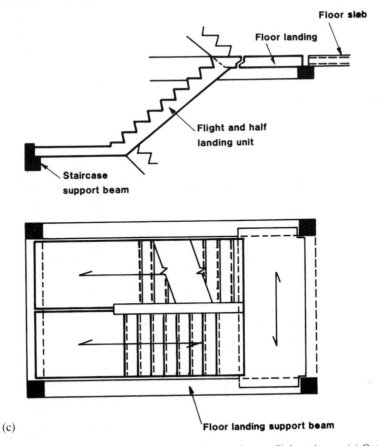

(c)

**Fig. 3.20** (*continued*)  Possible option for the layout of a two flight staircase. (c) Option 3 [3.7].

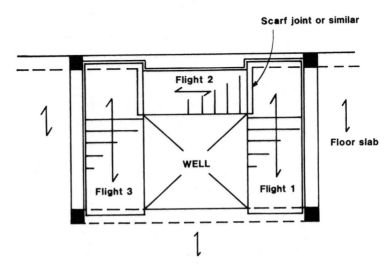

**Fig. 3.21**  Possible layout of a three flight staircase [3.7].

**Fig. 3.22** Problems encountered with lack of tolerances in fitting a stairflight between two fixed point landings.

**Fig. 3.23** Cantilevered half landing with semi-circular balcony.

further details on this. Inserts and holes are formed and located with exceptional accuracy, ±3 mm in position and less than 1° in alignment in most cases. Staircase and landing units can be designed with a fire rating equal to, or exceeding, that for the rest of the structure.

### 3.2.4 Roof and floor beams

A wide range of beams are available in preferred cross-sections, mainly composed of rectangles, and an infinite range in non-preferred sections. What distinguishes a preferred section is simply a matter of available moulds in manufacturers' works. Beams may be classified either as:

- 'internal', where floor loading is approximately symmetrical
- 'external', where floor loading is predominantly non-symmetrical.

External beams may be manufactured using visual concrete incorporating external texture, colour, and weather protection, or in plain grey concrete. A net structural section is used in the design of decorative components, allowing a thickness of about 40 to 50 mm for the finishes.

Internal beams may be psc or rc. The most common, shown in section in Fig. 3.24(a), are called 'inverted tee' or 'double boot' where part of the structural section falls within the floor zone, thus reducing overall structural depth. Typical span to depth ratios for these beams are about 10 to 15, the former representing the worst possible situation using a reinforced beam rather than a prestressed beam.

The boot width is governed by two factors: (i) an adequate floor slab bearing distance of 100 mm minimum, i.e. 75 mm bearing length plus 25 mm tolerance, and (ii) the physical size of the connection unit cast in to the beam. Rectangular beams are less efficient structurally because a part of the floor slab is not concealed within the depth of the beam, but they may be more economical if the structural zones is not a limiting factor. The centre line of internal beams should preferably correspond with the centre of the column to avoid bi-axial bending effects in the column. Although the effects may be catered for in designing the connection, the maximum eccentricity off the centre of the column should not be greater than $b/3$. The exception to this rule may be a lift shaft or staircase where a flush edge at the beam–column interface is architecturally required. In this case the beam will be lightly loaded. The outstand boot of the beam is not continuous past the faces of the column.

Changes in floor level may be accommodated by using either an L beam (or single boot beam), or building up one side of an inverted-tee beam as shown in Fig. 3.24(a) to form what is known as a 'dropped boot beam'. If the change of floor level exceeds about 750 mm an improved solution is to use two L beams, back to back and separated by a small gap (say 100 mm), to unhinder site fixing.

External beams are usually reinforced because of their unsymmetrical L-shape cross-section, and as the depth of edge beams is not usually a restrictive dimension there is not a lot to be gained by attempting eccentric prestressing techniques. Preferred sections vary with different manufacturers but a typical range is shown in Fig. 3.24(b). Minimum dimensions are again dictated by the size of the beam connector making the connection with the column. The upstand serves two functions by contributing to the strength of the beam and providing a permanent shuttering to the insitu infill concrete used with the floor slab. The depth of the upstand may be varied indefinitely. L beams with large upstands are known as 'spandrel' beams (Fig. 3.25). Structurally efficient they form an immediate dry environment to the structure by providing the following trades with a simple means of weather protection. Parapet spandrel beams may contain specially shaped crenellations for weather proofing.

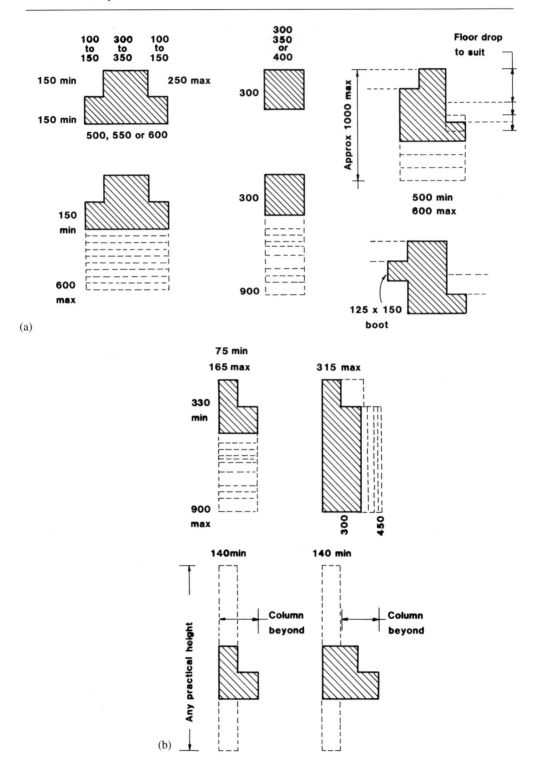

**Fig. 3.24** Shapes of standardized beams. (a) Internal beams and (b) edge and spandrel beams.

**Fig. 3.25** Example of long span roof spandrel beams (note the cast-in plate top connection).

The width of L shape and spandrel beams may be confined within the width of the column, or may project forward of the column to form an outstand spandrel. Realistically, the gap between the face of the column and inside edge of the upstand cannot be greater than about 250 mm. Architecturally they provide the structure with a continuous perimeter façade free of columns or other obstructions. Vertical joints between outstand spandrel beams may be pointed or sealed using a cold mastic. Alternatively, the column may be revealed in front of the spandrel, as shown in Fig. 3.26.

Because of their non-standard nature, these units may be fitted with window, brickwork and other service fixings at minimal additional cost. The range of decorative finishes is as wide as those available for cladding itself. Horizontal service holes may be provided transversely to the span of the beam. They should be located near to the neutral axis of the beam and/or near to the ends of the beam. Consultation with individual manufacturers with regard to maximum size and position of the holes should be made early in the design appraisal.

Table 3.4 is a guide to the different sizes and uses of the main types of preferred sections for beams. Provisions for holes and fixings in beams and spandrels depend largely on the nature of the mould in which they are cast. In general, units having a

**Fig. 3.26** Polished concrete columns and beams offset by up to 500 mm (courtesy of Trent Concrete Ltd, UK).

preferred cross-sectional profile (as defined in Section 3.1) are manufactured in steel moulds, whilst the remainder are cast in timber moulds. Fixings in concrete which are in contact with the sides and soffits of steel moulds are battened in the mould as shown in Fig. 3.27. In plain concrete units the rebate is not usually structurally or architecturally embarrassing. In some cases manufacturers prefer to drill holes in the mould and repair these later by plugging and grinding the inside face of the mould smooth.

The manufacturer can provide the following built-in fixings and fittings:

- cast-in-sockets (see also Table 3.5)
- galvanized steel dovetailed slots for masonry and other types of cladding
- dovetailed timber battens for window fixings, etc.

**Table. 3.4** Guide to sizes and uses of beams.

| Location | Type | Range of typical sizes (mm) | | | |
|---|---|---|---|---|---|
| | | Overall breadth | Upstand breadth | Overall depth | Depth below slab |
| Floor and roof main and gable edge beams | L beam | 250 to 450 | 75 to 315 | 330 to 900 | 150 to 750 |
| | Spandrel | 250 to 450 | 140 to 300 | No limit* | – |
| | Out spandrel | 400 min. | 140 min. | No limit* | – |
| Internal main spine beams | Rectangular | 300 | – | 300 to 900 | 300 to 900 |
| | Inverted-tee | 500 to 750 | 300 to 450 | 300 to 800 | 150 to 600 |
| Int. secondary | Rectangular | 300 | – | 300 to 500 | 300 to 500 |
| Staircase, lift shaft beams | L beam (mainly) | 300 | 75 to 165 | 330 to 600 | 150 to 400 |

* Within practical and architectural limits.

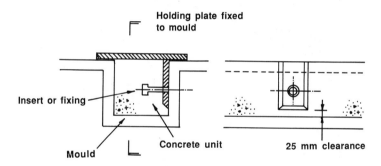

**Fig. 3.27** Fixing inserts inside a steel mould.

- crenellations for weather proofing
- insitu brickwork support in either a bolted-on steel angle or a cast-on concrete nib with drip
- horizontal service holes.

### 3.2.5 Beam-to-column connections

One of the most important connections in multi-storey precast construction is the beam-to-column connection. The design, which will be dealt with in detail in Chapter 7, is based on a pinned joint, i.e. the beam is simply supported and no continuity moment transfer between the beam and column is considered. A bending moment is induced in the column but this derives from the eccentric load reaction from the end of the beam, acting at about 50 mm from the face of the columns.

The main architectural decision to be taken is whether the connection is to be confined within the depth of the beam, or may be permitted to project below the soffit of the beam,

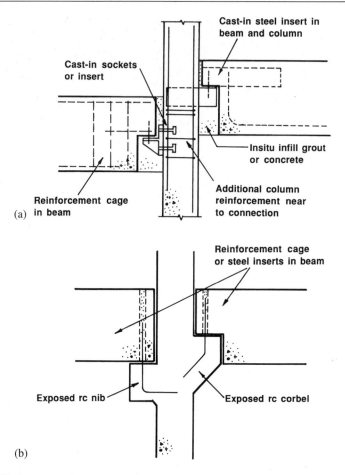

**Fig. 3.28** Beam-to-column connections: (a) enclosed; (b) expressed.

as in Fig. 3.28. Concealed connections shown in (a) are frequently required at internal connections because of the restrictions on floor zones. Because of the reduced depth of beam and the highly localized forces in the column, steel inserts are cast into both beam and column components to ensure that a positive connection is made. All steel inserts are protected against fire and exposure using an insitu grout or concrete. Corbelled connections shown in (b) may be used in the perimeter of the building where external cladding will hide the connection from view. However, it is unlikely that both types of connections will be used in the same building because of the different requirements on moulds.

In assessing the suitability of a precast concrete structure at certain intersections, due consideration must be made for the specialized nature of this connection. For example, it is not always possible to terminate beams on the column grid. In these circumstances a direct connection between the end of a secondary beam and a primary beam is made as shown in Fig. 3.29. This is usually a non-standard detail which requires special attention particularly in the primary beam. As the shear force capacity of this type of connection is restricted, the use of columns in the framing plan should be optimized accordingly.

**Fig. 3.29** Beam-to-beam connections (courtesy of Trent Concrete Ltd).

The beam-to-floor connection is also based on a simple support despite the presence of reinforced insitu concrete strips, or welded connections, as shown in Fig. 1.22. Flooring is usually laid directly onto the shelf provided by the boot of the beam, but neoprene bearing pads or wet bedding onto grout is also used in certain circumstances.

The fire resistance of all beams is based on the recommendations of BS 8110, Part 2. Typical ratings for rc beams using a 'standardized' cage (i.e. a predetermined arrangement of bars most suited to the preferred range of beams ) is 2 h, although 4 h may be achieved without additional external protection. Fire ratings for most standardized prestressed beams is 1.5 to 2 h, although again with the use of a fine mesh and increased cover to the prestressing tendons 4 h may be achieved. Lightweight aggregates are not used for beams of structural importance because of restrictions in strength, particularly at the connections. However, lightweight aggregates have been successfully incorporated into the upstand of deep spandrel beams, i.e. away from the structural connection, to reduce weight.

### 3.2.6  Columns

Column cross-sections, shown in Fig. 3.30(a) and (b), are based on a minimum of 300 mm × 300 mm plan dimension because of the nature of the column-to-beam connection. Preferred increments in size are 50 mm or 75 mm, in one or both faces. Non-preferred shapes (e.g. trapezoidal, hexagonal, round) are based around a column core size of 300 × 300 mm to accommodate structural connections, Fig. 3.31. Columns may be erected in one piece, i.e. without mechanical splicing, up to 20 m in length, as in the five storey building shown in Fig. 3.32 where an unbroken finish in an exposed aggregate was required. However, the usual practical and more economic limit is 12 to 13 m.

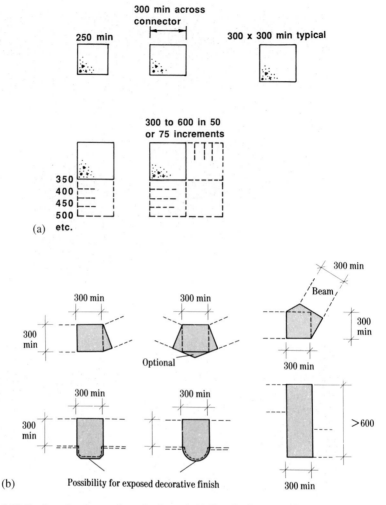

**Fig. 3.30** Preferred cross-section of columns. (a) Standard rectangular columns (in mm) and (b) non-standard columns (in mm) [3.7].

Column sizes in braced structures may be approximated by assuming that the cross-sectional area $(mm^2)$ is equal to the maximum ultimate axial load (N) divided by 28. In using this simple approximation, check that the slenderness ratio is not greater than 15. In unbraced structures the quotient depends on the magnitudes of both axial force and overturning moment, which are due to many parameters, but in the broadest possible sense column sizes for three storey buildings are typically 350 mm square, and 300 mm square for two storey buildings. These values take into account the effects of slenderness.

Column finishes may either be plain grey or incorporate a decorative finish on any or all of the faces, as shown in Fig. 1.12 where polished concrete, using white Portland cement and dolomitic limestone, was specified. In the latter case, conforming to standard shape or size does not carry cost penalties because of the special nature of the decorative finish. The design of the column makes due allowance for the finishes in computing a net

**Fig. 3.31** Non-preferred shapes of columns.

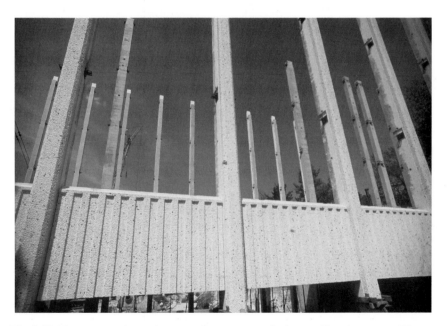

**Fig. 3.32** Five storey columns in exposed aggregate at Orpington, Kent (courtesy of Trent Concrete Ltd).

cross-section. Unless the concrete retains its gross section, e.g. by polishing or acid etching, a thickness of 40 to 50 mm is subtracted for the finishes.

High quality visual concrete in columns has only recently been fully exploited. Although the increase in cost in using visual concrete compared with plain concrete is between 50 and 250 per cent, depending on the complexity of the finish, a net saving is often made.

Consider the following example for an external column in a prestigious four storey office building:

*Option*                                                                                                   *Cost index*

(1) Visual finish polished concrete column, grade C40 concrete,
    13 m  long ×350 mm × 350 mm  section,  reinforced with 4
    no. T25 bars vertically and R10 links @ 300 c/c.          *Total Option 1 =*    *1.60*

(2) Plain  concrete  column,  grade  C50,  13 m  long × 300 mm ×
    300 mm, reinforced with 4 no. T32 bars vertically and R10
    links @ 300 c/c.                                                                       1.00
    Polished concrete U shape cladding bolted on three sides
    of column, grade C30, four pieces totalling 13 m long ×
    75 mm thick, reinforced with 5 no. T10 bars vertically and
    R8 @ 200 c/c horizontally.                                                    1.40
    Total plan dimensions in Option 2 = 470 mm × 385 mm.  *Total Option 2 =*    *2.40*

Table 3.5 lists the additional provisions that may be incorporated in columns, and the sizes and restrictions in using each facility. Columns may rise to the full height of the building, or may be terminated and stepped back into the structure to suit architectural features. As in any form of construction, it is desirable that the internal grid layout is not varied too extensively except in terminating columns in places where the floor (or roof) construction can adequately span over the row of columns omitted beneath. Figure 3.33 shows examples of permissible practice.

### 3.2.7   Bracing walls

Precast skeletal structures of more than three storeys in height are usually braced. The need for permanent walls or cross-bracing is implicit in the design of a braced structure,

**Table 3.5** Additional provisions in columns.

| Provision | Typical sizes | Restrictions |
| --- | --- | --- |
| Cast-in sockets and dovetailed channels for precast concrete, metal or other cladding | 6 to 24 mm diameter Length >30 mm | Centres >2 × diameter Edge distance >50 mm |
| Internal rain water pipes* | 100 mm diameter | PVC (plastic) only, no steel May increase column size Not passable through certain splices |
| Protective cloaking for impacts and damage resistance | 50 × 50 mm bent plate × 1 m long | Galvanized or stainless steel only |
| Chases for electrical fittings | 50 mm wide, full height if required | 20 mm depth Edge distance >30 mm Centres >100 mm |
| Holes for horizontal services | Width <$b$ – 200 mm Length <200 mm* | May increase column size Not possible in beam zone |
| Ties for brickwork | 100 long @ 150 c/c | |

*Note*:  Consultation with precast concrete manufacturers is strongly advised before any of these
          provisions is considered, particularly those marked*.

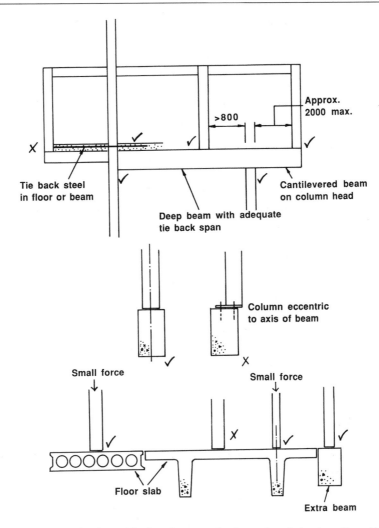

**Fig. 3.33** Guidance for the positioning of staggered columns founded on to walls and beams.

and the design team must appreciate the need for these obstructions to clear open-plan floor areas, which may not always be necessary in a cast insitu frame. The number and positions of walls in a precast structure is often a contentious issue, one that has to be accepted within the limitations of any prefabricated building system.

Precast concrete walls serve to provide stability and as surrounding walls or boxes for staircases and lift shafts. Walls may be classified as 'infill' or 'cantilever'. Users of infill walls attempt to design the wall as a single unit, but if the opening between beams and columns is particularly large, say 8 m × 4 m, it may be necessary to 'stitch' two separate units using a vertical reinforced insitu concrete joint. The effects of slenderness (= length/thickness <30) are included in design. It is shown in Chapter 8 that the usual limiting strength for infill walls is given when the ultimate horizontal interface shear stress between the precast wall and the foundation is taken as 0.45 N/mm$^2$.

The thickness of the wall units varies between (the practical minimum thickness of) 150 mm and 300 mm, and is governed either by lifting in the factory or the ultimate shear capacity in service. Because the design considers the wall acting as a compressive diagonal strut it is not acceptable to provide large voids for windows, doors, etc. in these units. Service holes, up to about 500 mm in diameter, may be accommodated in the walls providing they do not interrupt the structural continuity by being less than approximately 250 mm from the edge of the wall. The wall openings shown in Fig. 3.34 give an indication of the possible sizes in this respect.

Cantilever walls are designed as 'deep beams' with adequate tension anchorage to the foundation and between successive storey height units, Fig. 3.35. In the case of isolated walls, the wall is braced laterally by other walls in the structure. No frame elements are used in this type of construction. Therefore, seating corbels for flooring or stair landing and inserts for connecting beams or spandrels are required.

Precast shear cores or boxes, shown in Fig. 3.36, are an extension of the single cantilever wall. These units can contain large openings for doors and windows. They may be formed on site by 'stitching' two or more units, or manufactured as a complete storey (or part-storey) height box (with door and window openings as required) in the factory. The twin lift shaft shown in Fig. 3.37 was manufactured as two separate E-shaped units and stitched toe-to-toe on site. Lift motor rooms and basement pit boxes may be used to complete the prefabricated lift shaft. Lift motor slabs may also be precasted but tend to be disproportionately expensive (per m$^2$) due to their individual nature. Precast staircase boxes may be used similarly to provide access and/or architectural features to basements, roofs, plant rooms, etc.

The location and distribution of walls should be so their centre of resistance coincides

**Fig. 3.34** Holes in solid shear walls.

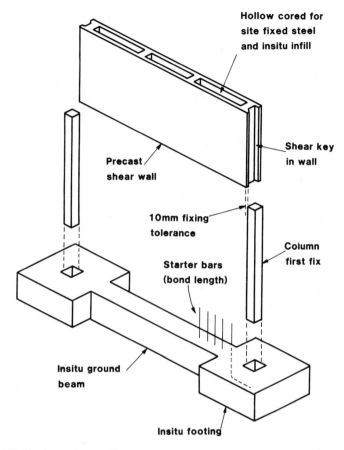

**Fig. 3.35** Hollow core shear walls.

closely with the centre of mass and the geometric centroid of the completed building. More precise design guidance will be given in Chapter 8. The most obvious location is around stair wells or lift shafts, although other internal or external walls may have to be used in order to 'balance' the structure.

## 3.3  SPECIAL FEATURES

### 3.3.1  Hybrid construction

The term 'hybrid' is used to describe mixed construction where precast concrete is used in combination with other building media, such as cast insitu concrete, steelwork, masonry and timber. The term must not be confused with 'composite' construction, which also uses both precast and another material, but where the structural performance relies on the interaction between the two. Reference should be made to the publication by the UK Reinforced Concrete Council [3.8] which deals extensively with this topic.

Figures 3.38 and 3.39 show the combined use of steelwork and precast concrete in various situations. Figures 3.38(a) and (b) show a steel Universal beam providing flexural strength with the concrete being used for the fire and corrosion protection only. Figure 3.39 shows separate precast concrete and steelwork members in hybrid construction.

**Fig. 3.36** Shear core box complete with lift door opening and holes for electrical fittings (courtesy of Costain Building Products).

**Fig. 3.37** Twin lift shaft manufactured from two precast pieces (courtesy of Blatcon Ltd).

In this latter case a conscious decision has been made to use precast in one part of the structure and a different material in another part. Other examples include:

- insitu concrete basement or lower storey podium

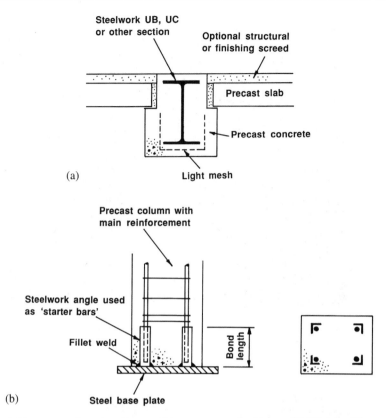

**Fig. 3.38** Structural steelwork encased in concrete at the factory. (a) UBs cased in concrete for fire protection and bearing purposes and (b) steel sections used for starter bars where the area of rebar would be too great in normal practical situations.

- structural steel roof atrium, gallery, or portal frame
- timber roof or balcony
- masonry load bearing walls or shear walls.

In a hybrid structure, structural precast concrete members are replaced with equivalent steel, timber, cast insitu, or masonry members because there is a cost or performance benefit, or because the architect has called for this specific type of structure. A good example of the use of steel and precast concrete is shown in Fig. 3.40 where the more heavily loaded parts of the structure were manufactured using precast rc or psc, and the lightly loaded roof members were steel. This was a truly hybrid structure because the concrete foundations were cast insitu, and the diagonal bracing was in steel.

### 3.3.2   Precast–insitu concrete structures

The integration of precast and insitu superstructures is not widely accepted, especially in multi-storey work. It is common for insitu concrete to be used solely for the substructure (e.g. underground car park, access ramps, retaining walls and foundations) to a precast

**Fig. 3.39** Precast concrete structural steelwork in a hybrid building.

**Fig. 3.40** Structural steelwork used in the roof of this precast structure.

structure. Structural compatibility is not a problem if the differential shrinkage rates are allowed for in the analysis. Providing the differences between continuous rigid insitu frame and pin-jointed precast structure behaviour can be accommodated in the stability analysis, the main problems are in the joints between two structural media.

Precast–insitu connections rely on accuracy in on-site work because the tolerances in precast work are small. Typical beam connections are shown in Fig. 3.41(a)–(c). Insitu–precast connections may be less tolerant towards accuracy, but rely on adequate bond developing between the two concretes, and checking that the precast element can resist

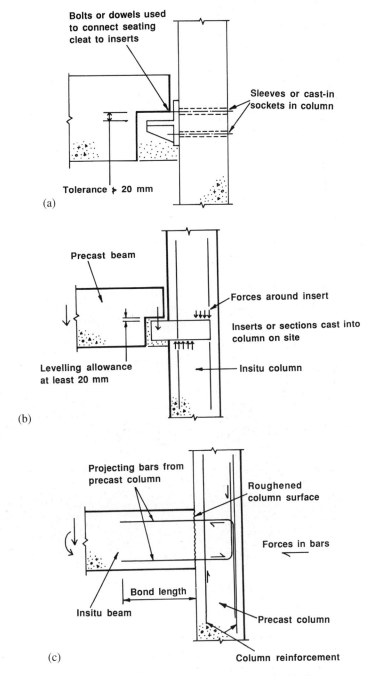

**Fig. 3.41** Beam–column connection details between precast and insitu concrete.

the new set of forces resulting from the changes in detail. Figure 3.41 illustrates this point. The effects of shrinkage must be considered particularly if the connection is of major structural importance, as in the case of a moment resisting joint.

Precast columns have been used in insitu concrete buildings for the dual purpose of speed and quality of finish. The main benefit is in multi-storey construction where immediate load carrying capacity is provided. The most common division of precast and insitu concrete is to construct the precast superstructure over the top of a basement or podium, which for reasons of structural form, strength or function cannot be part of the precast structure. Figure 3.42 shows insitu concrete being used in columns with U-shaped precast beams.

### 3.3.3   Precast concrete–steelwork or timber structures

Structural steelwork and structural timber are used mainly for lightly loaded roofs, either in the form of portals or trusses. In all cases the precast concrete elements provide the substructure in the marriage. Simple connections are made to the precast components as shown in Fig. 3.43(a)–(c).

**Fig. 3.42** Precast superstructure founded on to insitu concrete basement (courtesy of New Zealand Concrete Society).

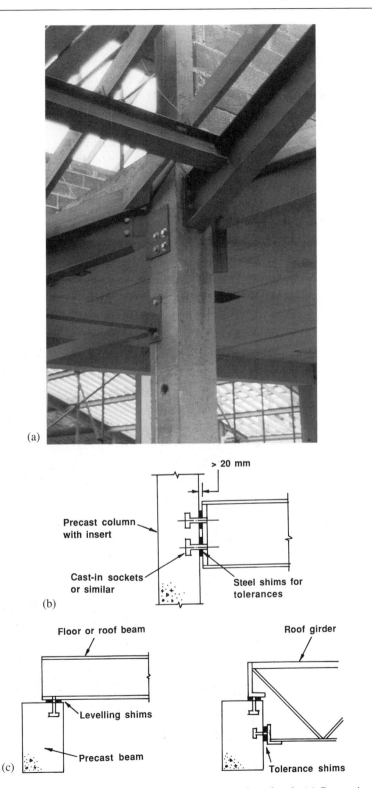

**Fig. 3.43** Connection details between precast and structural steelwork. (a) Connection details; (b) between precast column or wall and (c) to precast beam.

Fixings for structural steelwork can easily be accommodated in precast concrete column, beam and wall elements. This is because most precast components utilize steel inserts or cast-in sockets in their connections to other precast members anyway. Providing the stability of the structure is not impaired, the substitution of a steel girder, truss or portal frame for a precast component is taken care of in the connection detail. Frame selection can therefore proceed to satisfy the economic and architectural aspects. The main problems occur if connecting members are not coincident with the framing grid and connections are required in the floor slab. These are made as shown in Fig. 3.44(a). An alternative method, but which is less satisfactory in terms of steelwork efficiency, shown in Fig. 3.44(b), may achieve the same objective.

### 3.3.4   Precast concrete–masonry structures

Infill masonry is used extensively in shear walls in buildings of up to seven storeys.

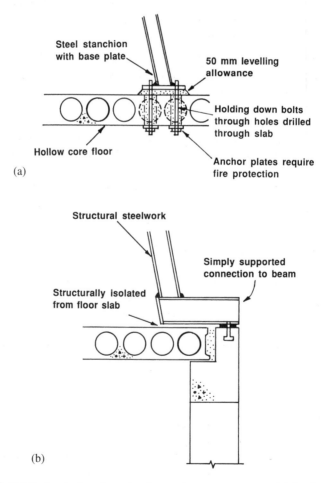

**Fig. 3.44** Details for steel roof construction on to precast structure. (a) Steel sections to precast components and (b) off-grid steel Mansard roof connection.

There is ambiguity over whether this is 'composite construction' because the precast structure relies on composite action with the wall, or whether it is 'hybrid construction'. The answer is not very important because the methodology is proven in either case.

Load bearing masonry walls are not used to support precast concrete elements because of on-site fixing sequences; the speed of erection of a precast structure is too rapid for the strength development of the load bearing masonry. Occasionally masonry walls are built to support a special item, such as a lift motor room floor which is not on the critical path of the frame erection.

## 3.4  BALCONIES

Cantilevers may be formed in a number of ways depending on the span of the balcony and its position relative to the main part of the structure:

- Edge beams adopted to provide a small projection up to about 700 to 800 mm overhang, Fig. 3.45(a) and (b). The beam is a special casting, but if the quantity is large the cost implications of this are fairly small. Insitu concrete tie backs may be necessary as shown in the figure.

- Beams projecting over columns as shown in Fig. 3.45(c) provide cantilevers of up to about 2.5 m. This distance may be varied to produce a splayed effect. The disadvantages are:

  - column splicing at every floor level
  - a greater number of beams are required than in the usual solution
  - an additional edge beam is required at the end of the cantilever.

- Certain flooring, e.g. the double-tee, used to cantilever up to about 1.5 m directly over edge beams. The overall structural zone may be large because it is not possible to use halving joints at the supporting beam. Hollow core slabs are not recommended for direct cantilever action.
- Tee-columns – this solution of casting cantilever beams integral with the columns has not been used extensively in the UK. The shape of the unit (sometimes referred to as 'Christmas tree' columns) and the connection detail are shown in Fig. 3.45(d). The limiting span of the cantilever, which is governed by the moment transfer capacity in the column, is about 2.0 to 2.5 m for typical floor loadings and external cladding arrangements. Manufacturing costs are likely to be high, particularly for a small number of units, due to the individual nature of the units and mould costs.

## EXERCISE 3.1

Devise a precast concrete solution to the five storey building shown in Fig. 3.46 using a braced skeletal structure. The total depth of the floor zone may not exceed 600 mm and no structural screed is to be used. The floor finish is to be raised timber, and the ceiling suspended. The external façade may be taken as 50 per cent brickwork and 50 per cent glazing, except around the curved façade which is to be aluminium curtain walling. The only internal blockwork (100 mm thick $\times$ 1500 kg/m$^3$ density) is around the service

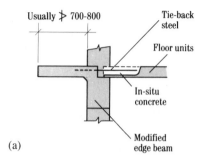

Usually $\not> 700\text{-}800$

Tie-back steel

Floor units

In-situ concrete

(a)

Modified edge beam

(b)

**Fig. 3.45** Options for forming cantilever balconies. (a) Cross-section of overhang beam [3.7]; (b) use of overhang beam, (c) cantilever beam [3.7] and (d) cantilever tee-columns [3.7].

areas adjacent to the lift shaft. The side walls to the main staircases are to be full height glazing.

Take the characteristic superimposed live and dead loads as 5 kN/m$^2$ and 2 kN/m$^2$ for the floors and 1.5 kN/m$^2$ and 2 kN/m$^2$ for the roof, respectively, and the characteristic horizontal wind loading as 1 kN/m$^2$. Ignore local wind pressure coefficients for this exercise.

**Solution**

The building is symmetrical about the $y$-axis. Therefore consider one half only. Assume that the top of the roof parapet beam is 1.0 m above the roof level. Thus the total height of the building $H = 17$ m.

**Step 1   Positions of stabilizing walls, possibly at any or all of lift shafts, main staircase and emergency staircase**

Total ultimate horizontal force is the greater of the wind force, $W_u = \gamma_f\, w_k\, L\, H$, or 1.5

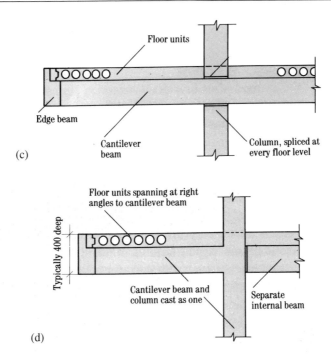

(c)

Floor units

Edge beam

Cantilever beam

Column, spliced at every floor level

(d)

Floor units spanning at right angles to cantilever beam

Typically 400 deep

Cantilever beam and column cast as one

Separate internal beam

**Fig. 3.45** (*continued*)

per cent $G_k$. In the y-direction the wind force is $W_{uy} = 1.4 \times 1.0 \times 72 \times 17 = 1713$ kN, and $W_{ux} = 952$ kN in the x-direction. The corresponding values for 1.5 per cent $G_k$ is 930 kN < wind force (see Step 4). Choose to use infill concrete walls. (Brickwork would be equally suitable but adopt a total precast solution here if possible.)

Assume column sizes are 400 mm square at the ground level and 300 mm square at higher levels. Thus the length of wall at the ground = grid distance – 400 mm.

In the y-direction, the total length of wall available is $3 \times 2.6$ m = 7.8 m at the lift shaft, plus $2 \times 2.6$ m = 5.2 m at the main staircase plus $2 \times 2.2$ m = 4.4 m at the escape staircases. Hence $L = 17.4$ m.

If the maximum horizontal shear stress = 0.45 N/mm$^2$ (Section 3.2.7), the thickness of shear wall = $1713/17.4 \times 0.45 = 218$ mm. Use $t = 225$ mm thick precast walls. The positioning of the walls is symmetrical and well balanced throughout the structure.

In the x-direction, $L = (4 \times 5.6 + 2 \times 2.6) = 27.6$ m. Hence $t = 952/27.6 \times 0.45 = 77$ mm. This is less than the minimum recommended thickness of 150 mm. It is likely that the thickness of the walls at the escape staircase will be governed by slenderness, i.e. $5600/30 = 187$ mm. Therefore use 225 mm thickness as per the walls in the y-direction.

### Step 2  Positions and shapes of primary columns
These are shown hatched solid in Fig. 3.46. They are located at all the intersections identified so far, i.e. corners, lift shafts and staircases. Assume the columns are all rectangular or square in cross-section. The special shaped columns in the curved part will be dealt with in Step 7 (1).

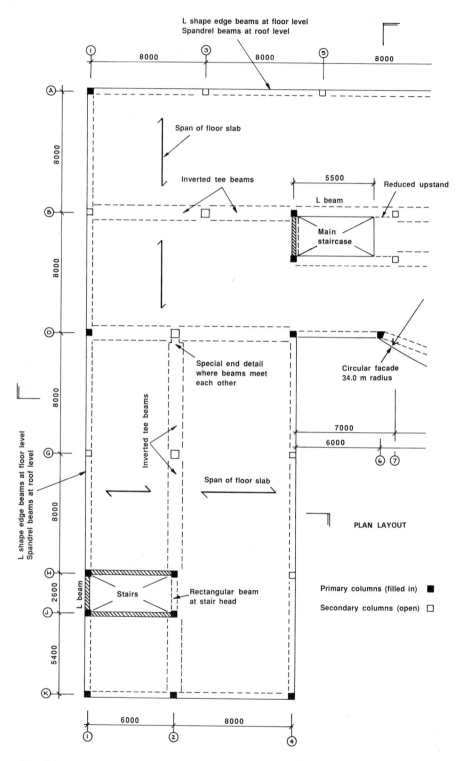

**Fig. 3.46** Schematic precast concrete structure used in Exercise 3.1.

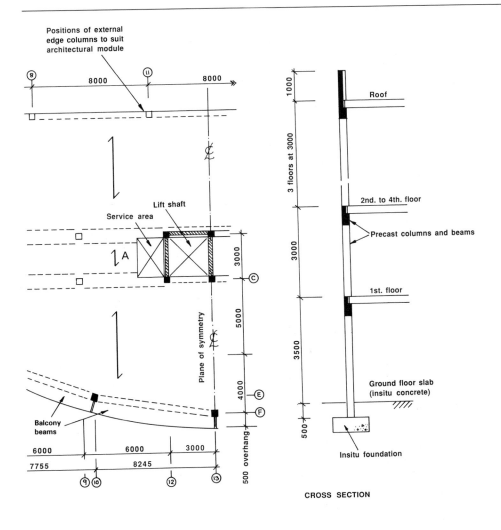

Positions of external edge columns to suit architectural module

8000 | 8000

Lift shaft

Service area

Balcony beams

6000 | 6000 | 3000

7755 | 8245

Plane of symmetry

3000

5000

4000

500 overhang

1000

3 floors at 3000

3000

3500

500

Roof

2nd. to 4th. floor

Precast columns and beams

1st. floor

Ground floor slab (insitu concrete)

Insitu foundation

CROSS SECTION

**Fig. 3.46**  (*continued*)

### Step 3   Positions and shapes of secondary columns

The positioning of these columns is not always straightforward. It is often dictated by architectural requirements, such as window bay sizes. Assume that the columns are all square in cross-section. Split the task into two parts.

**(1) Internal secondary columns.** A suggested layout of columns is shown as open squares in Fig. 3.46. The internal columns on grid line 2/A–K have been equispaced at 8 m centre-to-centre. The distance between the main stair (grid 7) and lift shaft (grid 12) is 12.0 m. This distance could be divided in many ways depending on the architecture, but choose 6 + 6 m for this exercise. The distance between beams on grid lines B–C/9–12 has been made 3.0 m to enable the floor slab to be used to support the walls around the service void (see Step 7 (2)).

The maximum clear beam span (at the higher floor levels where the column size is 300 mm) is therefore 7.7 m, giving a span/allowable depth ratio = 7700/600 = 12.8, which according to Section 3.2.4 is within range. Therefore, internal beam depth = 600 mm.

**(2) External secondary columns.** There is no requirement that where flooring is spanning perpendicular to a building façade, the centres of external and internal columns should be equal. Thus on grid line A column centres may be equidistant = 8.0 m. Similarly in the curved part, columns are placed at 8 m centre-to-centre (x-axis), and on grid lines 1 and 4 also at 8 m centre-to-centre.

Where the flooring is spanning parallel with the edge of the building the internal and external column centres should preferably be equal to avoid cranked beams and splayed floor slabs. Thus, the maximum external beam span = 7.7 m, giving span/depth = 12.8, again within range. Therefore, external beam depth = 600 mm.

### Step 4   Floor depths

The maximum floor span is 9 m centre-to-centre grid 13/C–F, giving a clear span of 8.5 m (assuming the beams are 500 mm wide). For this type of building the choice of slab is realistically restricted to hollow core or double-tee slabs, although with the majority of spans being 5 to 8 m, hollow core flooring will be preferred from Table 3.1, depth of hollow core slab = 250 mm at the floors and 200 mm at the roof.

Thus maximum slab loading is $(kN/m^2)$:

|        |                                      | Dead | Live |
|--------|--------------------------------------|------|------|
| Floors | = 3.0 self wt + 2.0 finishes = 5.0   |      | 5.0* |
| Roof   | = 2.5 self wt + 2.0 finishes = 4.5   |      | 1.5  |

\* With the provision to reduce the live load by 20 per cent for the design of the five storey column.

The total dead load $G_k$ in the building may be estimated from the above floor loads plus allowances of 0.6 $kN/m^2$ for the beams and columns, 0.3 $kN/m^2$ for the shear walls, and 1 $kN/m^2$ for the external brickwork/glazing. Thus, total equivalent floor $udl$ = 6.9 $kN/m^2$, and roof $udl$ = 5.1 $kN/m^2$. Floor area = 1900 $m^2$. Hence total $G_k$ = 62 000 kN.

### Step 5   Column sizes

Some knowledge of column design is required to make a good estimate of column sizes, but referring to Section 3.2.6, use gross area = axial load/28.

### (1)  Internal columns

Maximum axial load occurs at grid D/2 where the floor area supported by that column is $8 \times [(6 + 8)/2] = 56$ m$^2$, and the length of beam $= 3.0 + 4.0 + 4.0 = 11.0$ m. Allowing 5 kN/m and 3 kN/m for the self weight of the beams and column, respectively, the maximum axial force (kN) at the foundation is:

|  | *Dead* | *Live* |
|---|---|---|
| Roof slab = 56 m$^2$ | 252 | 84 |
| Roof beam = 11 m | 55 | 0 |
| Floor slab = 4 × 56 = 224 m$^2$ | 1120 | 896* |
| Floor beams = 4 × 11 = 44m | 220 | 0 |
| Column = 17 m | 51 | 0 |
|  | 1698 | 980 |

\* Reduced live load.

Ultimate force $N = 3945$ kN, say 4000 kN.
Hence $b = h = \sqrt{4000 \times 10^3/28} = 377$ mm. Use 400 mm at the ground to third floor. The slenderness ratio $= 4000/400 = 10 < 15$, hence the column is not slender. This makes the length of the first and second lift columns 10 m and 7 m, respectively. A column splice can be made and concealed within the depth of the third floor slab (where 250 mm depth is available in which to make the connection).
   The axial force (kN) in the column above the third floor is:

|  | *Dead* | *Live* |
|---|---|---|
| Roof slab = 56 m$^2$ | 252 | 84 |
| Roof and floor beam = 22 m | 110 | 0 |
| Floor slab = 56 m$^2$ | 280 | 280 |
| Column = 7 m | 21 | 0 |
|  | 663 | 364 |

Ultimate force $N = 1510$ kN.
Hence $b = h = \sqrt{1510 \times 10^3/28} = 232$ mm. Use 300 mm at the roof to third floor. The slenderness ratio $= 3000/300 = 10 < 15$, hence the column is not slender.

### (2)  External columns

External columns are usually smaller in cross-section than internal columns. The maximum load in an external column occurs on grid line A (neglecting for the moment the columns in the curved façade which are non-rectangular and will be dealt with in Step 7), where the area of floor slab supported by the column is $8 \times 8/2 = 32$ m$^2$. The external façade *udl* is assessed to be 6.6 kN/m for 1.5 m high double skin brickwork, plus 0.5 kN/m glazing. Thus repeating the above procedures, at the foundation $N = 2750$ kN, $b = h = 315$ mm. Use 350 mm square ground to second floor, 300 mm square thereafter. The external column splice has been placed at the second floor in order to stagger the level of the splices. Note that changes in the external column section may cause severe problems in modular bay width for window fixings, brickwork, etc. and should be discussed with the architect.

### Step 6   Staircases

Referring to Fig. 3.20, (c) would be used for all staircases, and L-shaped edge beams would be used at the half landing level. The stair side beams on grid lines B, C, H and J would be a rectangular shape with a cut-out to support the floor landing.

The depth of the flight unit would be approximately (5500/25) + 30 cover = 250 mm, i.e. equal to the floor depth. The landing could be either a 250 mm deep hollow core slab or a solid rc unit.

### Step 7   Special details

#### (1)  Columns and edge beams to curved façade.

The details in Fig. 3.47 show the proposed shapes of these units. Because the angle offset at each of the three interior columns (grids 10, 13) is equal, these columns may be similar in cross-section. The beam span = 8.31 m.

The column profile between floor levels would be circular, and in the region of the beam connection would be a rhombus. The maximum area of floor slab supported by these columns (grid F/13) is (4.5 + 0.5 overhang) × 8.245 = 41.23 m².

The maximum axial load on these columns is approximately 3000 kN. Thus the diameter of the column $b$ = 369 mm. Use 400 mm diameter over the full height. A special mould would be required to cast these columns. In order to reduce the cost of this mould, the column should be spliced at the third floor level. (Here it is better for the upper column to be the shorter length.)

The 4 no. curved edge beams would all be equal length and identical in shape, requiring only 1 no. additional mould. The ends of the beams would be square to fit against the rhombus shape column.

Careful attention would be required when scheduling the lengths and splayed ends to the hollow core floor slabs in this area.

#### (2)  Framing around large floor voids

It is very important to appreciate the effect of a major floor void. It may involve additional components rather than merely strengthening existing components.

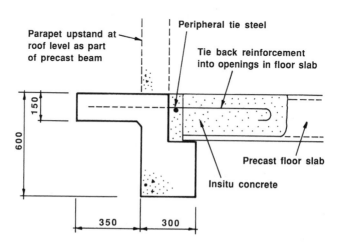

**Fig. 3.47** Edge details for the special beams used in the precast structure studied in Exercise 3.1.

In this case the major services void occurs adjacent to the lift shaft at grid B/12. Referring to the detailed plan in Fig. 3.46, the walls around the void (2 m × 2 m) may be supported on the floor slabs marked A. The total ultimate load of the 100 mm blockwork (plus plaster) around each side of the void = 14 kN. Thus the additional load on the slab marked A = 14 kN. This is easily catered for using 250 mm deep units.

### Step 8   Stability ties

Although the calculation for tie forces is carried out at the detailed design, it is necessary to ensure at this stage that the ties can be positioned and made structurally continuous throughout the building. In floors without structural screeds the ties will be collected over the tops of beams and pass either through sleeves in the columns or alongside the column. This poses no problems except at staircases and lift shafts where the ties cannot be continued straight through. Threaded bars are screwed into cast-in threaded sockets at these positions.

The perimeter ties approaching the external corners (e.g. grid A/1, K/4, etc.) are allowed to pass in an arc around the corner. However, the ties approaching the re-entrant corner (grid D/4) are not (because the resultant force pulls towards the outside of the structure). Here the ties should pass straight on and be fully anchored in the floor slab beyond. The solution is shown in Fig. 9.21.

### Step 9   Construction matters

At this point it is worth considering the cranage requirements, the construction sequence and the self weights of the units obtained in the preliminary design. It may be assumed that the site is a 'green field' with no obstructions to crane or lorry access. For the purpose of this exercise it may be assumed that the nearest approach the crane may make is to a line drawn at 5 m from the edge of the structure, and that the crane may be positioned at the most convenient point on this loci. It is necessary to determine the maximum moment (weight of unit × reach) for the crane as shown in Table 3.6.

**Table 3.6** Determining the maximum moment for the crane.

| Unit | Grid location | Weight (tonne) | Maximum reach (m) | Moment (t m) |
|---|---|---|---|---|
| 250 mm deep hollow core floor | C–F/13 | 3.3 | 10.0 | 33 |
| Edge beams | A/1–13 | 4.0 | 5.0 | 20 |
| Edge beams in curved façade | D–F/6–13 | 4.8 | 6.0 | 29 |
| Internal beams | B/2–4 | 4.0 | 13.0 | 52 |
| Lift shaft walls at ground level | B–C/13 | 5.6 | 14.5 | 81 |
| Main stair wall at ground level | B–C/4 | 5.6 | 14.5 | 81 |
| Escape stair wall at ground level | H–J/1–2 | 9.5* | 8.5 | 81 |
| Escape stair wall at upper level | H–J/1–2 | 6.6 | 8.5 | 56 |
| Main stairflight | B–C/4–7 | 6.0 | 12.3 | 73 |
| Escape stairflight | H–J/1–2 | 5.7 | 8.5 | 48 |
| Internal column | D/2 | 3.9 | 13.0 | 51 |
| External circular column | F/13 | 3.0 | 5.0 | 15 |

\* This unit may be split into two.

# Chapter 4

# Design of Skeletal Structures

*Detailed design methods, with worked examples, are given for the design of beams, staircases and columns subjected to gravity loads. Material specifications are given.*

## 4.1  BASIS FOR THE DESIGN

The correct philosophy for the design of precast concrete multi-storey structures is to consider it as a total entity, not an arbitrary set of components. The BCA's Frame Guide [4.1] states:

> 'Precast concrete frames are often, but mistakenly considered to consist of individual components which need to be designed and connected together in a manner that ensures adequate strength and interaction between them. ... An assessment should be made of the requirements for each component and its contribution to the building as a whole'.

Precast structural frame design is an interactive procedure, and several iterations within the design routine are carried out. The result is that the subtleties of the process have been refined by the precasters to such a degree that it is difficult for the uninitiated to gain access to the thought process. Final solutions are borne out of many years of trial and error in the design office, testing laboratory and on site. Once the general (Chapter 1) and architectural (Chapter 3) aspects have been resolved, the order in which the design proceeds is:

(1) precast concrete components, i.e. roof and floor slabs, staircases, beams, columns and walls
(2) connections between components
(3) stability, i.e. floor diaphragm action and walls, columns, i.e. temporary safety
(4) structural integrity, i.e. robustness, ties and progressive collapse.

The components of the structure are classified by considering their functions as either:

(a) slabs, simply supported between beams and providing the horizontal floor diaphragm
(b) beams, supported between columns and providing the chords to the horizontal diaphragm, or
(c) columns and walls, carrying gravity forces to the foundations and providing the stabilizing system.

In this context staircases do not contribute to the structural system. The design considers individual precast components, the connections between them and their behaviour in the global structure at both the serviceability and ultimate limiting states. In the former the components are proportioned so that crack and deflection criteria are satisfied and the effects of thermal, creep and shrinkage movements are not adverse. At ultimate, the components and their connections are checked for sufficient strength, stability and ductility.

Simplified 2-D substructuring of frames is used to determine the internal forces and moments in the members and connections, Fig. 4.1. An elastic analysis of the structure is carried out and plastic methods are used for the design of the components. No redistribution of bending moments is considered. The effects of imperfections, e.g. lack of verticality, positions of shear walls, creep, thermal movement and shrinkage are catered for in the global structural frame analysis. The latter items are not considered in buildings where the smallest dimension is less than 30 m. Equilibrium is checked on the basis of an undeformed structure, using first order deflections to compute force eccentricities. Second order deflections are used in the design of individual components, e.g. columns and walls. Occasionally numerical finite element methods or full scale experimental strain analysis of localized areas, e.g. supports, concentrated load points, anchorage zones and abrupt changes in geometric section, are used to validate design assumptions.

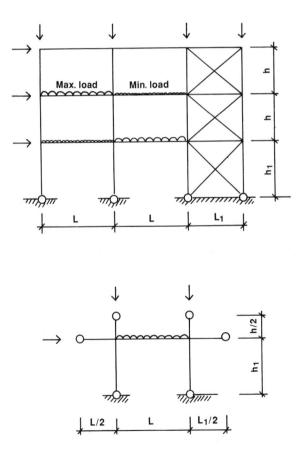

**Fig. 4.1** Substructuring methods for precast frame design.

In the main, precast concrete buildings are analysed as pin-jointed braced structures, Fig. 4.2(a). The beams are simply supported between columns or walls, and the columns are required to resist axial loads and small moments produced by eccentric forces at the ends of beams. The foundations are pinned, and subjected only to axial compression. Although the design of the stabilizing system is separate from the design of the components, the two are carried out simultaneously.

In certain cases, for example in buildings of up to three storeys, the structure is considered unbraced, relying for stability on cantilever action in the columns and moment-fixity at the foundation. As an alternative to every column resisting moments, a small number of large columns acting as deep beams may be used as wind posts at intervals along the structure, Fig. 4.2(b). In both these cases the stability design is an integral part of the component design.

Partially braced structures, Fig. 4.2(c), offer a compromise between fully braced and fully unbraced structures. Here the component design is directly influenced by the stabilizing system, and vice versa.

However, it is generally accepted that the most economic solution, in terms of manufacture and erection, is to use the pin-jointed braced version. The main advantages are:

- minimization of precast components, e.g. columns may be multi-storey and beams need only be provided primarily in one direction, Fig. 3.5
- pin-jointed connections are made on site without the necessity of a seven day (or longer) curing period required for insitu concrete joints
- the structure is stable floor by floor
- foundations are optimized because the column base is pin-jointed.

The main disadvantage is that shear walls, etc. are often sought in places other than staircases and liftshafts in order to avoid torsional effects, see Section 8.2. Conflicting architectural and structural requirements occasionally lead to additional framing to satisfy both parties.

## 4.2   MATERIALS

Prefabrication of rc and psc components has much greater potential for economy, structural performance and durability than in cast insitu concrete. Most precasting works use computer controlled batching and mixing equipment, leading to a reduction in the standard deviation in the appropriate performance index, i.e. workability, strength gain, ultimate strength and uniformity. Typical values for the s.d. of compressive cube strengths are in the order of 3 to 4 $N/mm^2$ for mean design strengths of 50 to 60 $N/mm^2$.

The working platform for casting concrete is adjacent to the mould and is therefore conducive to geometrical accuracy and good concrete compaction. The result is that the grade of concrete used can be exactly suited to the requirements of each type of component in order to expedite the use of the more expensive and exhaustible materials. Enhanced concrete stresses, i.e. up to $0.8 f_{cu}$, may be used in certain situations. Tests on bespoke precast components and connections have been used to substantiate this assumption, which would not be possible in cast insitu work.

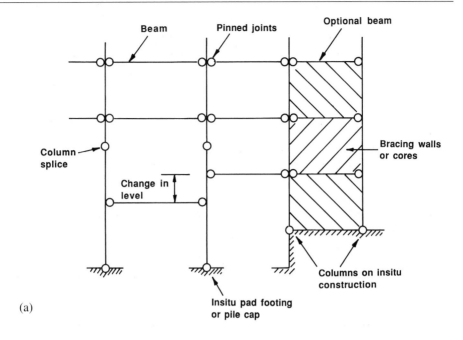

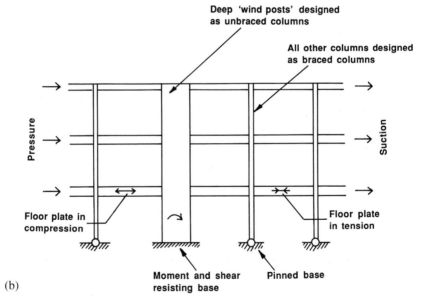

**Fig. 4.2** Principles of braced structures. (a) Principle of a pin-jointed braced structure; (b) principle of a bracing using deep columns.

### 4.2.1   Concrete

The 28 day characteristic and design strengths and short-term elastic modulii (N/mm$^2$ units) for the range of standard mixes used are given in Table 4.1. The 28 day cube

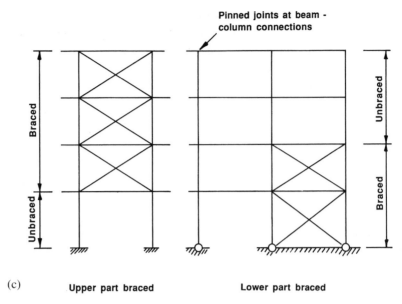

(c)          **Upper part braced**                    **Lower part braced**

**Fig. 4.2** (*continued*) Principles of braced structures. (c) Principle of a partially braced structure.

**Table 4.1** Design strengths and moduli used in precast component [4.2].

| Component | Type | Grade | $f_{cu}$ | $0.45 f_{cu}$ | $f_{ct}$ | $E_c$ |
|---|---|---|---|---|---|---|
| Beams, staircases, floors, shear walls | rc | C40 | 40.0 | 18.0 | — | 28 000 |
| Columns Load-bearing walls | rc | C50 | 50.0 | 22.5 | — | 30 000 |
| Beams, floors Staircases | psc | C60 | 60.0 | 27.0 | 3.50 | 32 000 |

*Note*: $E_c$ is the 28 day value. Long-term values are derived from the creep coefficients in BS 8110, Part 2, clause 7.3 [4.2].

crushing strengths $f_{cu}$ refer to cubes stored in water at 20°C (not air cured), and to strength grade C.

The design of concrete mixes is in accordance with national design data, e.g. in the UK this is 'Design of Normal Concrete Mixes' [4.3]. Producers may modify the mix proportions to obtain the greatest benefit from local materials, superior in quality than in the national standard. One of the main mix design parameters is the strength of concrete at an age of 24 h for beam, column, wall and staircase units, and between 12 and 24 h for prestressed hollow core slabs. This is the optimum age when the units are removed from the mould, and the strength at this point in time is termed the lifting strength. For prestressed units it coincides with the de-tensioning strength. It is important that the cubes used for de-tensioning purposes are cured in the identical manner to the actual concrete. This may present a problem where electrical methods, such as induction in the tendons, is used. Accelerated cured cubes, possibly no more than 24 h old, are frequently used to predict the 28 day strengths. These strengths may be as high as 85 to 90 per cent of the 28 day value, and excellent correlation (within 5 per cent) is found.

Notional indicative values for concrete lifting strengths and design cube strengths (N/mm$^2$), and mix proportions (kg/m$^3$) are given in Table 4.2. The sand is medium zone, and the aggregates are normal weight gravels, crushed limestone or crushed granite.

Cast insitu concrete is used to complete many types of connections. The strength of the insitu concrete is only required to be the same as the precast concrete where the load carrying capacity must be equal to that in the parent precast member, e.g. column splices, but should not be less than grade C25. A special mix may be specified for pumped concrete to structural screeds. Notional values for infill insitu concrete strengths (N/mm$^2$) and mix proportions (kg/m$^3$) for the site placed concrete are given in Table 4.3.

**Table 4.2** Indicative strength gain and mix proportions in standard mixes. (Actual values used depend on quality of local materials and the lifting requirements of the precast producer.)

| Grade | Lifting strength | 28 day $f_{cu}$ | Mix proportions | | | | | |
|---|---|---|---|---|---|---|---|---|
| | | | RHPC | Water | Sand | Agg. 14 mm | Agg. 10 mm | Admixture[1] |
| C40 | 21 | 40 | 375 | 170 | 575 | 850 | 430 | 0 |
| C50 | 25 | 50 | 400 | 150 | 525 | 890 | 445 | 0 |
| C60 | 30 | 60 | 410 | 150 | 500 | 895 | 445 | 0 |
| C60[2] | 35–38 | 60 | 410 | 100 | 685 | 775 | 480 | 0–60[3] |

*Note:* [1]See Section 3.3.3; [2]refers to hollow core production and [3]includes 60 kg/m$^3$ Pozzolana used in certain hollow core slab mixes; however, problems with blow holes in the soffits have detained the widespread use of these additives.

There is a marked difference in the mix proportions of the C60 and C60$^2$ grades, which is due to the latter being used for machine production rather than casting.

**Table 4.3** Indicative infill insitu concrete strengths and mix proportions. Actual values used depend on quality of local materials and the specific requirements of the contractor.

| Location | Grade | 28 day $f_{cu}$ | Mix proportions (kg/m$^3$) | | | | | | |
|---|---|---|---|---|---|---|---|---|---|
| | | | OPC | RHPC | Water | Sand | Agg.[1] 20 mm | Agg. 6 mm | Admixture |
| Floor infill, structural screeds, beam end connections[2] | C25 | 25.0 | 325 | — | 180 | 660 | 1235 | — | 0 |
| Beam end connections, wall infill | C40 | 40.0 | — | 425 | 250 | 1700 | — | 0 | 0–4 |
| | | | | 425 | 250 | 850 | | 850 | 0–4 |
| Column splices and column-to-foundation connections | C50 | 50.0 | — | 500 | 300 | 1575 | — | 0 | 0–5 |
| | | | | 500 | 300 | 785 | | 785 | 0–5 |

*Note:* [1]Graded through 20–5 mm and [2] some end connections are large enough for concrete, rather than grout, to be used (see Section 7.9).

Expanding agents are occasionally used as an admixture. The data given here should not be used in favour of manufacturer's instructions which must be followed to give the correct dosage with respect to the type of cement used and the intended purpose of the mix.

Creep and shrinkage strains are determined according to BS 8110, Part 2, Figs 7.1 and 7.2 [4.2]. The age at loading is taken as 28 days. Relative humidity is taken as 45 per cent in internal situations, and 85 per cent externally.

### 4.2.2   Concrete admixtures

The main types of admixtures sometimes used in factory cast concrete are:

- Pozzolana, used as a partial cement replacement in certain prestressed units.
- retarding agents, used to retard the hardening of concrete surfaces which are to be hammered to reveal the aggregate matrix. The retarding agent is applied only to the surface of the mould, and is not added to the mix.
- colouring pigments, for visual concrete.

### 4.2.3   Reinforcement

High tensile ribbed bar is used in the majority of cases, including shear links, where the increased tensile strength and ductility outweigh the additional costs. This is particularly relevant in the highly stressed areas around connections and supports where smaller volumes of reinforcement give greater access for vibrating pokers and the like. Welded fabric (or mesh), produced from reduced plain high tensile bar or indented wire, is used in 2-D units such as slabs and walls. It is false economy to specify too many different bar types and sizes in precast production.

Stainless steel ribbed reinforcement is used in very exceptional circumstances where the thickness of a section is small with respect to the conditions of exposure, e.g. 75 mm thickness if the exposure category is severe or worse. Epoxy fusion bonded reinforcement is not usually used.

The bar diameters commonly used are 6 and 8 mm for column stirrups, 10 and 12 mm for beam links and distribution (crack control) bars, and 16, 25, 32 and 40 mm for the main flexural bars. Popular mesh sizes are A142 (6 mm bars at 200 centres) and A193 (7 mm bars at 200 centres) for flat precast units such as plank floor slabs and walls, and structural screeds. C283 mesh (6 mm bars at 100 centres in one direction and 5 mm bars at 400 centres in the other) is used in the flanges of double-tee slabs. The 5 per cent characteristic strengths ($N/mm^2$) for hot rolled high yield steel is taken as 460 $N/mm^2$ and for mild steel at 250 $N/mm^2$. The modulus of elasticity is taken as 200 $kN/mm^2$.

### 4.2.4   Prestressing steel

Two main types of prestressing steel tendons are used:

- 5 or 7 mm diameter indented (or crimped) wire or plain wire

- 9.3, 10.9, 12.5 or 15.2 mm diameter 7-wire helical strand.

Smaller diameters are possible but less frequently used. A popular type of helical strand, called 'Dyform', is drawn through a dye to partly flatten the six outer wires. Both wire and strand is classified as 'class 2:5 per cent low relaxation' meaning the final stress in the wire after elastic relaxation over 1000 h will be 95 per cent of the original prestress. The 5 per cent characteristic strengths (N/mm$^2$) is taken as $f_{pu}$ = 1750 to 1770 N/mm$^2$. Super-stress strand is also available where $f_{pu}$ = 1860 N/mm$^2$, and the relaxation is as low as 2.5 per cent.

The modulus of elasticity is taken as 195 kN/mm$^2$ for helical strand, and 205 kN/mm$^2$ for wire.

### 4.2.5   Structural steel and bolts

Rolled steel sections and bent or flat steel plate are welded to form steel connectors in many highly stressed support situations where direct contact between concrete surfaces is to be avoided. The steel used most is grade 43 or grade 50. Welded electrodes are of minimum grade E43. The most common types of rolled sections used in beam shear boxes and/or column inserts are rectangular hollow sections (typically 200 × 100 mm or 150 × 100 mm cross-section), square hollow sections (typically 150 mm or 100 mm size), channels and angles. Solid steel billets are also used in column inserts.

Hot dipped galvanized steel is used for exposed connections, usually of secondary structural significance, such as in channels for brick ties. The basic plate is grade 43 steel, and grade 50 is used in the more highly stressed plates.

Black bolts grade 4:6 and 8:8 are used in many connections. High strength friction grip bolts are used in special circumstances where the integrity and safety (both temporary and permanent) of connections made with ordinary bolts in clearance holes cannot be guaranteed.

### 4.2.6   Non-cementitious materials

Polymers are used exclusively in site work. Epoxy-based mortars are used to make, either partially or completely, connections where a rapid gain in strength is required, e.g. up to 40 N/mm$^2$ in 2 to 3 h. Care is taken to ensure that these materials have not exceeded their shelf life, are being used correctly and for the right application. Neoprene, rubbers and mastics are for soft bearings, backing strips, etc. The PCI Manual on Architectural Precast Concrete [4.4] gives extensive guidance to the use of these materials. Although they are not used extensively in precast structures, a typical range of applications is given in Table 4.4. See also the Institution of Structural Engineers, Section 3 [4.5].

## 4.3   STRUCTURAL DESIGN

### 4.3.1   Terminology

The term 'component' is used to describe the single prefabricated artefact, e.g. a beam.

**Table 4.4** Use of non-cementitious materials in precast construction.

| Material | Application | Data (at 20°C) |
|---|---|---|
| Elastomeric bearings, e.g. neoprene, rubber | Bearing pads | Comp. strength = 7 N/mm$^2$ <br> Shear strength = 1 N/mm$^2$ <br> Compressive strain = 15% |
| Bitumen impregnated sealing strip | Backing strip to concrete joints | Compressibility = 85% |
| Polysulphide sealants | Expansion joints | Elongation strain < 50% |
| Epoxy resin mortars | Compression, shear or tension joints | Comp. strength = 55–110 N/mm$^2$ <br> Tensile strength = 9–20 N/mm$^2$ <br> Elongation strain < 15% |
| Polyester resin mortars | Compression, shear or tension joints | Comp strength = 55–110 N/mm$^2$ <br> Tensile strength = 6–15 N/mm$^2$ <br> Elongation strain < 2% |
| Polystyrene | Filler, back-up blocks | |

Occasionally the term 'unit' is used to describe single floor components, such as hollow core units.

Precast components are divided into the following groups:

- Linear components: beams and columns
- 2-D components: floors, walls, stairs and landings
- 3-D components: towers and cores
- Others: foundations, cladding panels, plant room roofs, etc.

The term 'members' is used to describe a situation where the construction is completed by the use of two or more components, e.g. composite members consist of precast floor components and insitu concrete. 'Non-isolated members' are those which interact with other members and would sustain loads in the event of failure to their supporting member. For example, hollow core floor units with fully grouted joints would sustain load if the supporting beam failed.

## 4.3.2  Design methods

Horizontal components are designed as simply supported. The effective span is calculated either in accordance with BS 8110, Part 1, Clause 3.4.1 [4.2] or by determining the point of effective rotation between the members (Fig. 4.3). Maximum flexural and torsional moments and shear forces are computed for the fundamental serviceability and ultimate limit states, the accidental load condition, lifting at the factory and on site.

Although minimum section sizes are nearly always governed by influences other than design strength, the minimum dimensions for components are 50 mm for the flanges of solid slabs, 75 mm for solid slabs, and 140 to 150 mm for horizontally cast beams, columns and shear walls.

Dimensional deviations from the nominal cross-section are according to BS 8110, Part 1, Clause 6.11.3. (Section 3 of the BCA Manual [4.1] gives further guidance on

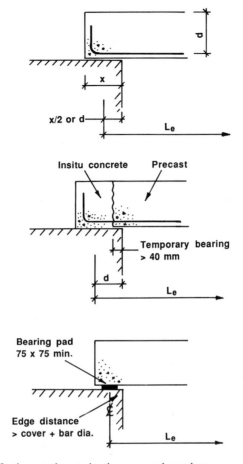

**Fig. 4.3** Point of effective rotation at simply supported members.

tolerances.) The design assumes that the effects of deviations are included as part of the relevant partial safety factors and the designer should not design using the reduced sections. The results of a sensitivity analysis on 300 mm wide × 300 mm and 600 mm deep rectangular psc beams given in Table 4.5 show the effects that deviations of both the position of the centroid of the tendons and the dimension of the beam have on strength (see also Fig. 4.4).

The results for the ultimate condition were broadly similar, but with slightly smaller differences for the larger deviations. Table 4.5 shows that the effects of negative beam and positive tendon deviations, and vice versa, cancel each other out irrespective of the depth of the beam, e.g. the relative strength of a beam 305 mm deep with the tendons raised by 5 mm = 1.00. Where the 5 mm deviations are cumulative the reductions or increase are in the order of 8 percentage points for the 300 mm deep beam and 3 percentage points for the 600 mm deep beam.

Minimum and maximum volumes of reinforcement can be optimized to give the most benefit in prefabricated concrete. Maximum volume ratios in the order of 10 per cent are possible in flexural components, particularly where the higher strength concretes

**Table 4.5** Effects of dimensional deviation on flexural serviceability strength of prestressed beams.

| Nominal cross-section $h \times b$ (mm) | Dimension deviation (mm) | Relative flexural strength for deviation of positions of tendons* | | |
|---|---|---|---|---|
| | | −5 mm | 0 | +5 mm |
| | −5 | 0.92 | 0.96 | 1.00 |
| 300 × 300 | 0 | 0.95 | 1.00 | 1.04 |
| | +5 | 1.00 | 1.04 | 1.07 |
| | −5 | 0.96 | 0.98 | 1.00 |
| 600 × 300 | 0 | 0.98 | 1.00 | 1.02 |
| | +5 | 1.01 | 1.02 | 1.04 |

*Note:* *−ve value means nearer to the neutral axis of the section.
All prestressing parameters are constant for each size of beam.

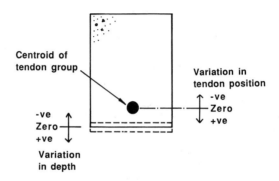

**Fig. 4.4** Definitions of dimensions used in Table 4.5.

(grade > C60) are used, in order to minimize depth. Because factory cast units are cured in a controlled environment, it is possible for minimum quantities of reinforcement to be less than insitu concrete. All major structural components are reinforced according to BS 8110, Part 1, clause 3.12.5 for cracking, and in BS 8110, Part 1, Table 3.27 generally [4.2]. This is summarized in Table 4.6.

The spacing of reinforcement will be described in the individual component design in Sections 4.3.3 to 4.3.5. The cover to all reinforcement (including links) is as indicated in Fig. 4.5. The exposure class, according to BS 8110, Part 1, Table 3.2 [4.2] is mild for internal use. The dimensional deviation to the reinforcement cover is taken as 5 mm.

### 4.3.3 Design of beams

The design of beams is based on ordinary rc or psc principles for specified loads and support conditions. Complications to design are associated with beam-to-beam connections, large asymmetrical loadings and the provision of service holes near to the ends of beams where special shear cages or shear boxes are provided to transmit shear forces to the support. A predetermined set of standardized beam sections is selected according

**Table 4.6** Maximum and minimum reinforcement quantities [4.2].

| Component | Steel | Primary | | Transverse* | |
|---|---|---|---|---|---|
| | | Min. (%) | Max. (%) | Min. (%) | Max. (%) |
| Beam | HT | 0.13 | 4.0 | 0.10 | 4.0 |
| | MS | 0.24 | 4.0 | 0.18 | 4.0 |
| Staircases | HT | 0.13 | 4.0 | >20% primary | >20% primary |
| | MS | 0.24 | 4.0 | >20% primary | >20% primary |
| Columns | HT | 0.40 | 8 or 10 at laps | clause 3.12.7.1 | ** |
| Walls | HT | 0.40 | 4.0 | 0.25 or clause 3.12.7.4 | 4.0 or clause 3.12.7.5 |

*Note*: HT = high tensile rebar $f_y$ = 460 N/mm$^2$; MS = mild steel plain bar $f_y$ = 250 N/mm$^2$.
* Transverse bars at 90° to primary bars.
** No requirement, but in practice limited to between 1.5 and 2.0 per cent.

to the requirements of most building structures, and flexural and shear reinforcements are computed for the optimum reinforcement quantities appropriate to each size of beam. PC compatible software is generally used, and checks for deflections, crack widths, etc. are made simultaneously with strength calculations.

Extensive testing has, in the main, been carried out over many years to prove the satisfactory performance of precast reinforced and prestressed beams. Examples include a PCI Journal [4.6] and a Magazine of Concrete Research [4.7].

Standardized designs are prepared for beams which vary only in depth, breadth and quantity of reinforcement. Ultimate moments and shear resistances are thus equated with particular project requirements. Families of curves or tabulated data are usually prepared in advance. Standard calculations often allow for service holes up to 50 mm diameter with their axes located near to the neutral axis of the beam.

Design methods are either 'non-composite', in which the insitu concrete infill between the ends of the slab and beam is ignored (Section 4.3.4) or 'composite'. The latter is dealt with in Chapter 6.

### 4.3.4 Non-composite reinforced concrete beams

Non-composite construction utilizes the properties of the basic beam. Ultimate moments and shear resistances are equated with particular project requirements. Load versus span data may be presented as shown in Figs 4.6(a) and (b) for a range of reinforced concrete edge beams and prestressed concrete internal beams shown in Figs 3.24(a) and (b).

L-shape edge beams, otherwise known as 'ledger beams', support non-symmetrical floor loads, usually in the edge of the structure. In this book the part of the beam supporting the floor is called the 'boot' and the main web is the 'upstand'. Edge beams are designed as under reinforced and singly or doubly reinforced because the lacer bars in the top of the boot and in the upstand will increase the moment of resistance. The only complication to design is if the breadth of the upstand varies, but a simple computer program can be written to deal with this. There are two types of edge beam, shown in Figs 4.7 and 4.8:

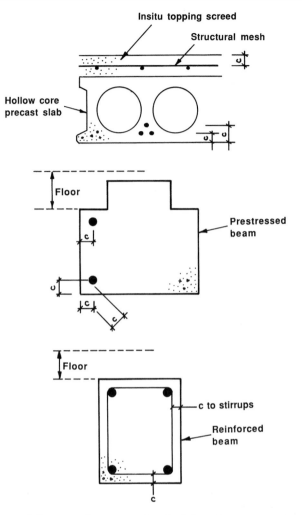

**Fig. 4.5** Cover to reinforcement in prestressed and reinforced components (centroidal cover distance is often used in psc).

- Type I, where a wide upstand is part of the structural section.
- Type II, where a narrow upstand (less than 100 mm) provides a permanent formwork to the floor slab, and is considered monolithic with the insitu concrete infill at the ends of the floor slab. The insitu concrete is adequately confined by the floor slabs.

In Type I beams, the minimum width of the upstand should be approximately $b_w = 160$ mm. This allows for two no. 16 mm diameter bars to be placed within a 12 mm diameter shear stirrup, with adequate cover and bar spacing as shown in Fig. 4.9. The ledge width is worked out by adding together the nominal slab bearing length (75 mm), a fixing tolerance (10 mm) and the clear space for insitu infill (50 mm), giving a total dimension of 135 mm. Thus the minimum breadth of a Type I beam is 295 mm, say 300 mm. The precast concrete upstand width in Type II beams is about 75 to 100 mm, but this is because the insitu concrete is considered as part of the beam. Then $b_w = (b - 75$ mm$)$ and the minimum breadth is about $b = 250$ mm.

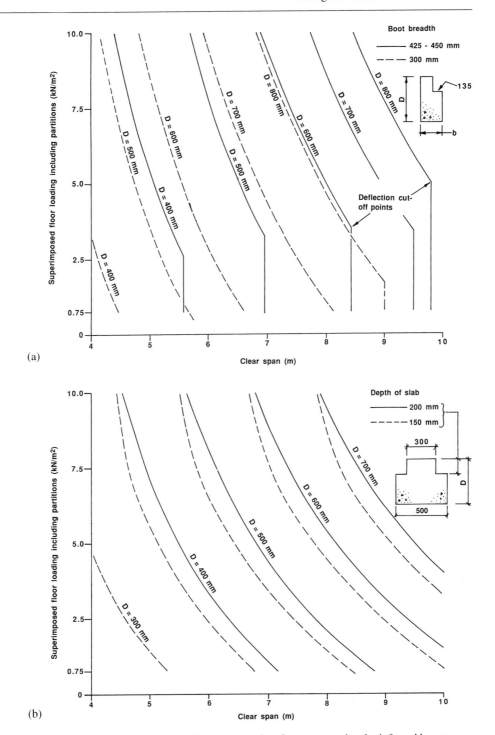

**Fig. 4.6** Typical superimposed load vs span data for prestressed and reinforced beams. (a) Reinforced external L beams. (b) Prestressed inverted-tee beams.

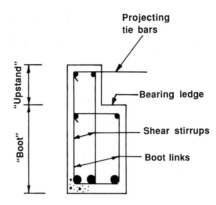

**Fig. 4.7** Type I edge beams.

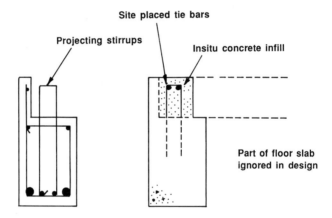

**Fig. 4.8** Type II edge beams.

Minimum depth is often determined by the size of the connector in the end of the beam, but there is no reason why the beam need not be shallower away from the connection. In this case the minimum depth would be equal to the depth of the floor slab ($h_s$) plus the minimum boot depth of 150 mm. In most buildings floor depth is at least 150 mm.

The general approach is to limit the increments to 50 mm for both depth (from 300 to 900 mm) and breadth (from 250 to 450 mm) and compute ultimate moments and shear resistances as follows. Referring to Fig. 4.9, let $\rho = A_s/b\,h$ and assume that in Type I beams the depth to the neutral axis is $X > h_s$. Then:

$$T = 0.87\,f_y\,\rho b h \tag{4.1}$$

$$C_1 = 0.45\,f_{cu}\,(b - 135)\,h_s \tag{4.2}$$

$$C_2 = 0.45\,f_{cu}\,b\,(0.9\,X - h_s) \tag{4.3}$$

$$T - C_1 = C_2$$

hence $X$: check $X < 0.5\,d$.

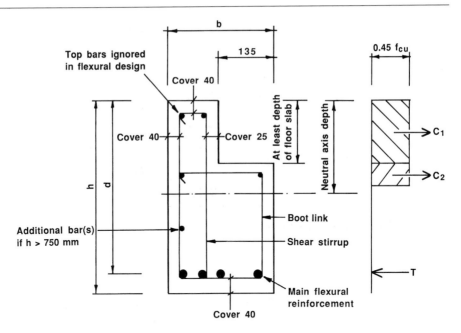

**Fig. 4.9** Definitions of geometry for Type I beams.

$$z_1 = d - \frac{h_s}{2} \text{ and } z_2 = (d - 0.45\,X) - \frac{h_s}{2} \tag{4.4}$$

$$M_r = C_1\,z_1 + C_2\,z_2 \tag{4.5}$$

If $C_1 < T$, then $X < h_s$, and $M_R = T\,(d - 0.45\,X)$ \hfill (4.6)

The design for Type II beams (Fig. 4.8) is similar except that the upstand width is $b - 75$ mm.

Table 4.7 gives examples of the flexural capacities of some typical singly reinforced 300 mm wide L-shape edge beams, using $b_w = 165$ mm, $h_s = 200$ mm, $f_{cu} = 40$ N/mm$^2$, $f_y = 460$ N/mm$^2$, cover to reinforcement = 35 mm, and assumed diameter of links = 12 mm.

Shear links are designed according to BS 8110, Part 1, clause 3.4.5 [4.2] usually using high tensile ribbed steel. Firmer cages may be made using ribbed (rather than plain) bar.

**Table 4.7** Flexural strength of Type I L-shape edge beams.

| Overall depth of beam $h$ (mm) | Bars in bottom of beam | $A_s$ (mm$^2$) | $d$ (mm) | $X/d$ | $M_R$ (kN m) |
|---|---|---|---|---|---|
| 400 | 2 T 25 + 1 T 12 | 1095 | 340 | 0.482 | 117 |
| 500 | 3 T 25 | 1470 | 440 | 0.500 | 201 |
| 600 | 2 T 32 + 1 T 20 | 1924 | 537 | 0.481 | 316 |
| 700 | 3 T 32 | 2410 | 637 | 0.468 | 468 |
| 800 | 2 T 40 + 1 T 25 | 3001 | 733 | 0.474 | 664 |
| 900 | 3 T 40 | 3770 | 833 | 0.492 | 937 |

The links are placed in the upstand of the beam, and not in the boot as shown in Fig. 4.10, to ensure that the reinforcement crosses the plane of potential shear cracking where the shear stress is maximum.

### 4.3.5 Beam boot design

The boot of the beam must be reinforced using links around the full perimeter of the boot. If the depth of the boot is less than 300 mm it should be designed as a short cantilever in bending. Otherwise the behaviour is nearer to the strut and tie action. The bending method gives a slightly greater area of tie back steel (about 5–10 per cent).

As with all projecting nibs it is first necessary to preclude a shear failure at the root of the nib. The enhanced shear stress given in BS 8110, Part 1, clause 3.4.5.9 [4.2] usually takes care of any vertical shear problems. If not, then the depth of the boot should be increased in preference to providing shear reinforcement. Punching shear beneath point loads is dealt with later.

A 'shallow' boot is one where the lever arm '$a$' in Fig. 4.11(a) is greater than $0.6d''$, where $d''$ is the effective depth to the steel in the top of the boot from the bottom of the beam. Otherwise the nib is classed as 'deep'. The design of the boot is very different for the two cases of wide and narrow slab flooring. In the former, e.g. hollow core slab or composite plank, the reactions are uniformly distributed, whereas in the latter, e.g. double-tee or beam and plank, the reactions are point loads. The case of uniformly loaded beams will be discussed first.

If the floor slab is placed in direct contact with a shallow bearing nib a horizontal force resulting from possible contractions or other movement (e.g. thermal effects) of the floor slab relative to beam will develop at the interface. Referring to Fig. 4.11(a), if the floor reaction is $V$ per unit length of beam, the horizontal force is $\mu V$, where $\mu$ is the coefficient of friction between two concrete surfaces (see also Section 4.3.10). Thus the tie force is:

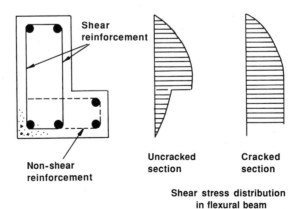

**Fig. 4.10** Positions of shear links and shear stresses in L beams.

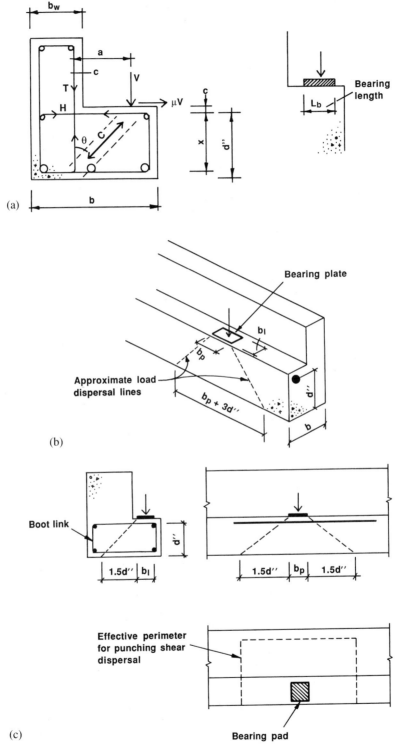

(a)

(b)

(c)

**Fig. 4.11** Design of boot reinforcement in L beams. (a) Strut and tie model; (b) beam geometry for designing for point loads and (c) design parameters for punching shear design in L beams.

$$H = V \tan \theta + \left( \frac{x+c}{x} \right) \mu V \tag{4.7}$$

where $x = (d'' - c)$ is the centre-to-centre distance of the boot link, and $c$ the edge distance to the centroid of the steel bar in the top of the boot. The horizontal bars placed in the top of the boot must satisfy:

$$A_{sh} = \frac{H}{0.87 f_y} \tag{4.8}$$

The bars are formed into links, but do not contribute to vertical shear strength of the beam unless the boot is sufficiently deep that the vertical leg of the boot link extends at least one-third of the depth of the beam above the neutral axis of the section. Because the upstand width is fairly small, typically 150 mm, the bars in the top of the boot must extend a full anchorage length in the rear face of the beam. This means that the bars are stressed beyond a point which is more than four diameters from the corner of the bar, and the bend radius must be checked so that the bursting stresses caused by small bend radii are not a problem. The usual practice is to provide T8 or T10 links at a spacing no greater than 300 mm for mild steel bars and 160 mm for high tensile bars (BS 8110, Part 1, Table 3.30).

The strut force is given by:

$$C = \frac{V}{\cos \theta} \tag{4.9}$$

which must be resisted by a compressive strut in the uncracked part of the nib. The uncracked zone may extend to a point at $0.5d''$ from the bottom of the beam; the limiting compressive strength of the concrete is $0.4 f_{cu}$. Thus the strut capacity is:

$$C = 0.2 f_{cu} d'' \sin \theta \text{ per unit length of beam} \tag{4.10}$$

The vertical force $T$ in the stirrup in the main body of the beam is given by:

$$T = V \left[ 1 + \mu \left( \frac{x+c}{a} \right) \right] \tag{4.11}$$

where $a$ is the centre distance from the stirrup to the line of action of the slab reaction. Then:

$$A_{sv} = \frac{T}{0.87 f_y} \tag{4.12}$$

This steel must be in addition to any shear requirement.

If the floor slab is fully tied to the beam using reinforced insitu strips capable of generating the frictional force $\mu V$, then this force may be ignored in the above design, in which case $T = V$.

In a deep boot, the floor slab reactions would be carried directly into the web of the beam by a diagonal strut action assuming $\theta = 45°$. If the level of the bearing surface is above the neutral axis the only steel required would be the horizontal steel $A_{sh}$.

In fact, the design of all the above reinforcement should be carried in two stages, before and after insitu concrete has been added to the ends of the slab. This is because the insitu concrete increases the bearing length to the full ledge width, and hence reduces the lever arm $a$. Before the insitu concrete is added the lever arm is $a = c + (b - b_w) - l_b/2$, and the slab reaction is due to the self weight of the slab plus the insitu concrete infill. Afterwards $a = c + (b - b_w)/2$, and the slab reaction is due to superimposed dead and live loading.

Considering the case of isolated point loads due to the reactions from double-tee slabs, for example, additional punching shear and bursting checks should be made.

Punching shear is dealt with by considering the first zone shear perimeter $u$ (according to BS 8110, Part 1, clause 3.7.7) spreading into the upstand of the beam in a 45° trajectory on one side, and spreading to the edge of the boot on the other. The first task is to decide what is the effective shear perimeter over which the shear is punching. Using the notation in Figs 4.11(b) and (c), if the beam is sufficiently wide such that the spread of load from the bearing is encompassed on three sides of the beam the shear perimeter $u$ may be taken as:

$$u = [2 \times (b_1 + 1.5d'')] + (b_p + 2 \times 1.5d'') \tag{4.13}$$

If the bearing pad is square (the usual case) $b_p = b_1$ then $u = 3b_p + 6d''$.

However, if the breadth of the beam $b < b_1 + 1.5d''$ the situation is similar to a concentrated load near to a free edge. Then:

$$u = 2b + (b_p + 2 \times 1.5d'') \tag{4.14}$$

Note that a shear perimeter is only allowed if reinforcement crosses each plane.

If the design shear stress $v = V/ud''$ is less than the enhanced concrete shear stress $1.5d''v_c/a_v$ (where $a_v$ is taken as the distance from the edge of the bearing to the shear plane considered), shear reinforcement $A_{shv}$ is not required. Otherwise it is provided according to:

$$A_{shv} = \frac{[v - v_c]ud''}{0.87f_y} \quad \text{for } 1.0\, v_c < v < 1.6\, v_c \tag{4.15a}$$

or

$$A_{shv} = \frac{5[0.7v - v_c]ud''}{0.87f_y} \quad \text{for } 1.6\, v_c < v < 2.0\, v_c \tag{4.15b}$$

but not less than

$$\frac{0.4ud''}{0.87f_y}$$

Note that the link $A_{shv}$ must be in addition to any bending or tie steel $A_{sh}$ provided in Equation (4.8) because the link is being stressed by two separate forces.

The concrete shear strength $v_c$ is a function of the area of steel $A_s$ crossing the shear plane. In beams this is different in the longitudinal and transverse directions. BS 8110 [4.2] does not consider such a situation and so a reasonable approach is to use the average value. It is therefore necessary to have designed the longitudinal steel in the top of the boot and the boot links before a punching shear check is made.

The reaction force $V$ is carried to the beam through a neoprene (or similar) spreader pad (of breadth $b_p$) and distributed into the upstand of the beam on a 45° trajectory from a point near to the edge of the bearing pad. A stiff compression zone will extend from a point about 200 mm either side of the edge of the bearing. Thus the steel $A_{sh}$ in Equation (4.8) must be concentrated into a zone of effective width $b_e$ where:

$$b_e = b_p + 2a + 400 \text{ mm} \tag{4.16}$$

In most cases this distance is about 800 mm. The PCI Manual states that an effective width of $6 \times$ boot depth to either side of the bearing may be used, but this seems to be stretching the limits of load spreading.

The vertical upstand (hanger) stirrups, called $A_{sv}$ above and designed in the same way, must be positioned in the same localized zone according to the forces generated in the compressive strut and tie steel $A_{sh}$, i.e. over the same effective breadth as above. The upstand stirrups are most easily located by tying them to the boot links.

If the boot depth is less than 300 mm, bending steel is required to resist the moment $M = Va$ according to the usual bending theory. Here the effective breadth used in the stress factor equation $K = M/f_{cu} \, b_e \, d''^2$ is according to Equation (4.16).

In all cases (irrespective of the design theory used), at least one boot link should be placed within 50 mm from the edge of the bearing to prevent a shear crack forming under the pad. In between these areas of localized concentrations the boot links should be distributed along the beam as nominal steel at 0.15 per cent $A_c$, typically T 10 bars at 250 mm centres. However, most precasters would continue the links at the closer spacing to simplify cage manufacture. This also has the added advantage that the position of the nib links is independent of the location of the double-tee bearing point. Typical reinforcement cages for this type of beam are shown in Fig. 4.12.

In long span floors where the end slab reaction is greater than about 200 kN a minimum depth of boot of about 250 mm is required to enable a 45° compressive strut to develop according to Equations (4.9) and (4.10), and satisfy local punching shear either side of the bearing point. If the ultimate slab reaction is less than about 150 kN, a minimum boot depth of about 200 mm may be used. To calculate this, Equation (4.10) should be modified as follows:

$$C = 0.2 f_{cu} (b_p + 2a) \, d'' \sin \theta \tag{4.17}$$

No composite action may be used in this case as the insitu concrete infill does not penetrate to the level of the slab bearing. Although some of the end shear due to superimposed loads will be transferred to the beam through the reinforced interface at the level of any structural screed, this is ignored in design and the boot is reinforced for the full end reaction.

Where the floor slab is connected to the supporting beam by mechanical fasteners the frictional force $\mu V$ may be ignored, providing of course that the mechanical fastener is capable of mobilizing this force. This may occur in double-tee slabs where, as explained in Section 5.3.2, fully anchored end plates in the flanges of the slab are fillet welded to plates cast into the beam.

The concentrated point loads cause longitudinal bursting tensile forces in the top and outer surfaces of the boot which must be resisted by a longitudinal bar in the corner of the boot. This bar, of area $A_s''$ is also necessary in forming the cage but its size and type is determined as follows:

$$A_s'' = \frac{0.23V}{0.87 f_y} \tag{4.18}$$

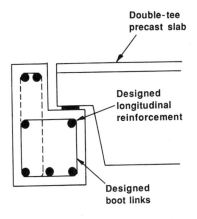

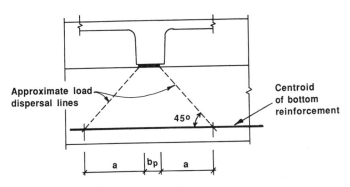

**Fig. 4.12** Reinforcement in the boot of a beam supporting point loads (e.g. double-tee slabs).

where the quotient figure of 0.23 is the maximum value derived from bursting theory as explained in Section 4.3.10. A single longitudinal bar will suffice at not more than 50 mm centroidal distance below the top of the boot. This bar, which is typically T 8 or T 10, does not contribute to the flexural strength of the beam. If the depth of the boot is greater than 500 mm a second longitudinal bar is positioned at 250 mm below the first. (Note these bars are also provided even if wide slab floor is used.)

In reinforcing beams of this kind, some knowledge of the positions of the bearing points must be known in advance. It is therefore extremely important that the floor layout is finalized before the beam detailing is. This is not necessary with wide floor slab.

### 4.3.6   Upstand design

Edge beams with narrow breadth upstands – which are in compression of course – are checked for lateral stability or slenderness in the temporary condition before the insitu concrete connection with the floor is made. It is possible that this is a limiting criteria with respect to span. The maximum span is the lesser of $60\,b_w$ or $250\,b_w^2/d$ [4.2] for type I beams only.

The upstands may be designed to resist vehicle impacts according to BS 6399, Part 1 [4.8]. A horizontal line force is applied at a height of about 400 to 500 mm above the top of the floor slab, and the upstand is designed as a vertical cantilever. The bending moment is transmitted to the slab through the top reinforcement in the structural screed, in the case of double-tee and composite plank units, or through top steel in the broken out cores of hollow core slabs. The structural model is shown in Fig. 4.13. As a precautionary measure, a fixing cleat is provided at each end of the beam at the top connection with the column. This is designed to carry one-half of the impact force, although it will actually be less than this. Horizontal equilibrium is maintained through the beam-to-column connection at the bottom of the beam (see Fig. 3.25).

Edge beams subjected to asymmetrical loading are not always reinforced against torsion on the assumption that the floor plate provides a horizontal prop force and the lateral stiffness of the beam is sufficiently large to prevent excessive horizontal deflections under the action of the propping force. However, some engineers dispute this point of view particularly if the floor loading and/or floor span is large as is often the case where double-tee slabs are specified. It is more difficult to justify the elimination of torsion using the sole propping action of the structural screed because of the problems in establishing a practical shear transfer mechanism in this type of construction.

An example of this concerns the design of edge beams subjected to asymmetrical loading. It would appear that the beam shown in Fig. 4.14(a) should be designed against the torque $T = wLe/2$. The value of $T$ can be in the order of 30 to 40 kNm, resulting in a torsional stress (for a typical $300 \times 600$ mm deep beam) of around 1.5 to 2.0 N/mm$^2$. This is in excess of $v_{tmin}$ and according to BS 8110, Part 2, clause 2.4.6, torsional links are required. However, edge beams such as the one shown in Fig. 4.14(b) are not reinforced against torsion on the assumption that the precast concrete floor plate, which is not monolithic with the beam but is tied into the beam as shown, will provide a horizontal propping force to prevent beam rotations occurring.

Adlparvar [4.9] and Elliott *et al.* [4.10] tested some typical precast edge beams in combined bending, shear and torsion. The beams were 600 mm deep $\times$ 300 mm wide,

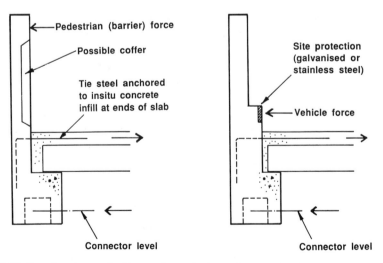

**Fig. 4.13** Reactions in spandrel beams due to horizontal parapet loading.

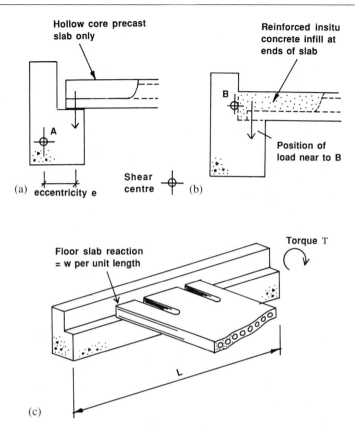

**Fig. 4.14** Response of asymmetrically loaded L beams. (a) Without insitu concrete infill; (b) with insitu concrete infill between the beam and slab and (c) general arrangement.

and were of both Types I and II defined above. Because the emphasis was on torsion the torque was exaggerated to produce torsional failures in the beam. In the first tests the beams were loaded without hollow core floor slabs, i.e. in their temporary condition before the insitu infill concrete had hardened. The torsional stress at cracking, according to the 'sand heap' analogy, was 1.8 N/mm$^2$ as expected, and the beams failed in torsion when the applied torque at either end of the beam was about 60 kNm. However, in the second set of tests when the same beams were tested under asymmetrical loading through 200 mm deep hollow core floor slabs, equivalent floor loads of up to 80 kN/m caused shear failure in the slab. The only cracking in the beam was due to bending at mid-span and shear near to the connector. Figures 4.15(a) and (b) show failure crack patterns for the beam alone, and for the beam when it was tied in to the floor slab, respectively.

In the temporary construction stage the beam is not tied to the slab, and therefore only the friction between the floor slab and bearing ledge of the beam will prevent beam rotations. However, the torque induced from the self weight of the floor slab is only about 0.3 of the total torque in service. This was found not to be great enough to cause torsional cracking. The conclusions from this work were:

• the insitu concrete infill produces an extended bearing, thus reducing the eccentricity of the load, as shown in Fig. 4.14(b)

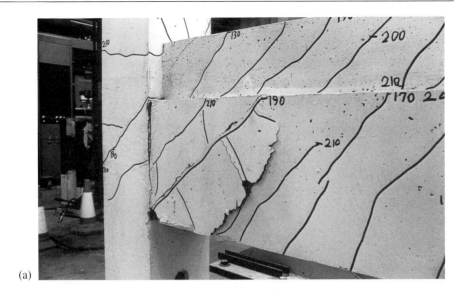

(a)

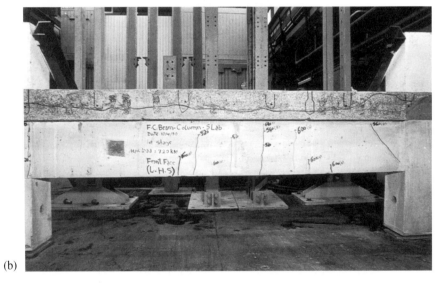

(b)

**Fig. 4.15** Cracking in L beam subjected to asymmetrical load. (a) Without insitu concrete infill and (b) with insitu concrete infill between the beam and slab.

- composite action between the beam and slab alters the position of the shear centre for the beam from point A to B in Fig. 4.14, thus reducing the eccentricity even more
- the beam cannot possibly rotate because of the flexural stiffness of the precast floor slab, causing the equilibrating stresses as shown in Fig. 4.16.

The minimum compressive cube strength of insitu concrete required to produce these results was 25 N/mm$^2$, and the minimum amount of tie steel to the slabs was T 10 bars at 600 mm centres.

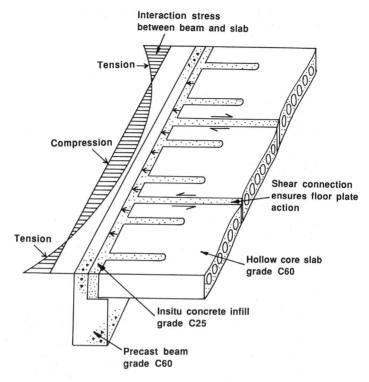

**Fig. 4.16** Horizontal stresses in the interface between floor slabs and L beams subjected to asymmetrical floor load.

## EXERCISE 4.1    Precast L beam design

Design the flexural and shear reinforcement for a precast concrete L shape edge beam over a simply supported effective span of 7.0 m to carry a 200 mm deep hollow core floor slab and an external cavity wall as shown in Fig. 4.17. Check the serviceability limit state for deflection. The characteristic superimposed dead and live beam loading is given in Table 4.8.

The floor slab bearing is 75 mm. Use grade C40 concrete, high tensile flexural reinforcement, mild steel stirrups, moderate exposure and a 1 hour fire resistance.

## Solution

### Section sizes and cover

Exposure – cover to reinforcement (BS 8110, Part 1, Table 3.4) for C40 concrete = 30 mm.
Fire – cover to reinforcement (BS 8110, Part 1, Table 3.5) for 1 hour = 20 mm.
Fire – minimum beam width (BS 8110, Part 1, Fig. 3.2) for 1 hour = 200 mm.
Use $b = 300$ mm wide beam.
Minimum upstand width (assume T 8 stirrups) = $30 + (6 \times 8) + 30 = 108$ mm.
Slenderness check (clause 3.4.1.6); $b_w > L/60 = 7000/60 = 117$ mm
or $\sqrt{L\, d}/250 = \sqrt{(7000 \times 550/250)} = 124$ mm (see Deflection check and Flexural design sections where $d = 550$ mm).
Use maximum possible upstand width $b_w = 165$ mm with 200 mm deep floor recess.

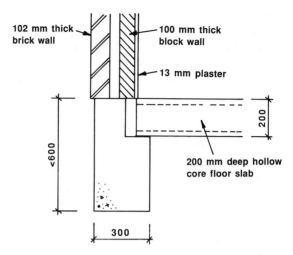

**Fig. 4.17** Details to Exercise 4.1.

**Table 4.8** Characteristic superimposed dead and live beam loading.

|  | Loading (kN/m) |
| --- | --- |
| Floor slab dead, including services and ceiling | 14.0 |
| Floor live, including partitions | 20.0 |
| Brickwork (2.2 kN/m$^2$ × 3.0 m) | 6.6 |
| Blockwork and finishes (1.3 kN/m$^2$ × 2.4 m) | 3.1 |

Effective depth of beam $d$ is controlled by deflection and the design of the flexurally balanced section.

**Deflection check**

Basic factor (BS 8110, Part 1, Table 3.10 [4.2]) for rectangular section = 20.

If beam is to be balanced, then $M/bd^2 = K' f_{cu} = 0.156 \times 40 = 6.24$.

Modification factor (BS 8110, Part 1, Table 3.11) for $f_s = 288$ N/mm$^2$ = 0.77.

$d > 7000/20 \times 0.77 = 454$ mm.

$h > d + (30 + 8 + (\text{say}) 12) = d + 50$ mm.

Try initial depth of beam $h = 600$ mm and $d = 550$ mm.

Self weight = $[(0.4 \times 0.3) + (0.2 \times 0.165)] \times 24 = 3.67$ kN/m.

Maximum ultimate loading on beam = $(1.4 \times 27.37) + (1.6 \times 20.00) = 70.32$ kN/m.

$M_u = 70.32 \times 7.00^2/8 = 430.7$ kN m.

$V_u = 70.32 \times 7.00/2 = 246.0$ kN.

**Flexural design**

Section properties $b = 165$ mm; $d = 550$ mm.

$K = 0.216 > 0.156$ compression reinforcement required to carry $M' = 119.2$ kN m.

Use $h = 600$ mm deep beam.

$d - d' = 550 - 50 = 500$ mm.

Top steel $A_s' = 596$ mm$^2$.
Minimum compression steel (clause 3.12.5) $= 0.2\%\ b_w h = 0.2\% \times 165 \times 600 = 198$ mm$^2$.
Use two T 20 bars (628 mm$^2$) in top of upstand.
For the design of the bottom steel $z/d = 0.775$ because $K = 0.156$.
$A_s = (311.5 \times 10^6/0.87 \times 460 \times 0.775 \times 550) + 596 = 2423$ mm$^2$.
*Use two T 32 plus two T 25 bars (2590 mm$^2$) in bottom of beam, with two T 10 additional bars at top of boot.*

### Curtailment of flexural steel

Reduce top bars to nominal when $M = 311.5$ kN m plus an anchorage length, i.e. at 1.62 m from support.
Anchorage length (clause 3.12.9.1) $= d = 550$ mm.
Minimum steel $= 198$ mm$^2$.
Curtail top steel to two T 12 (226) at 1.07 m from end.
Reduce bottom bars to 50 per cent $A_s$ at $0.08L$ (clause 3.12.10.2) $= 560$ mm.
Curtail bottom steel to two T 32 (1608) at 560 mm from centre line of support.

### Shear design

It is necessary to check the shear at the following positions:
    (1) End of the beam, where $v = 246$ kN
    (2) Curtailment point for the bottom steel, where $V = 246 - (70.32 \times 0.56) = 206$ kN
    (3) Where nominal links are required, where $v = v_c + 0.4$ N/mm$^2$.

*(1) End of the beams where V = 246 kN*
$v = 246 \times 10^3/165 \times 550 = 2.71$ N/mm$^2$.
$100\ A_s/b_w d = 100 \times 1608/165 \times 550 = 1.77$.
Hence $v_c = 0.76 \times [(40/25)^{0.33}] = 0.89$ N/mm$^2$.
$A_{sv}/s_v = (2.71 - 0.89) \times 165/0.87 \times 250 = 1.38$ mm$^2$/mm $= 1380$ mm$^2$/m for two leg stirrup.
Use pairs of R 8 at 125 mm centres (1600) from end of beam to 560 mm from centre of support.

*(2) Curtailment point for the bottom steel, where v = 246 - (70.32 × 0.56) = 206 kN*
$v = 2.27$ N/mm$^2$.
$100\ A_s/b_w d = 100 \times 2590/165 \times 550 = 2.85$.
Hence $v_c = 0.89 \times [(40/25)^{0.33}] = 1.04$ N/mm$^2$.
$A_{sv}/s_v = (2.27 - 1.04) \times 165/0.87 \times 250 = 0.938$ mm$^2$/mm $= 938$ mm$^2$/m for two leg stirrup.
*Use single R 8 at 100 mm centres (1000) from 560 mm from centre of support to point 3.*

*(3) Where nominal links are required, where v = v_c + 0.4 N/mm$^2$*
$v = 0.4 + 1.04 = 1.44$ N/mm$^2$.
Hence $v = 1.44 \times 165 \times 550 \times 10^{-3} = 131$ kN.
This point occurs at 1.635 m from centre of support.
$A_{sv}/s_v = 0.4 \times 165/0.87 \times 250 = 0.304$ mm$^2$/mm $= 304$ mm$^2$/m for two leg stirrup.
But $0.75d = 412$ mm.
*Use R 8 at 300 mm centres (333) from 1.64 m from centre of support.*

Boot reinforcement (ignore frictional force $\mu V$ in this exercise)
Floor slab loading $= (1.4 \times 14.00) + (1.6 \times 20.00) = 51.6$ kN/m.

Boot depth = 400 mm, therefore use strut and tie method.
The triangle of forces (see Fig. 4.11) gives:

$$\theta = \tan^{-1} \frac{(135 - 37 + 34)}{(400 - 34 - 34)} = 22°.$$

Thus, $H = V \tan \theta = 51.6 \times 0.404 = 20.85$ kN/m.
$A_{sh} = 20.85 \times 10^3/0.87 \times 460 = 52$ mm$^2$/m.
But maximum spacing = 300 mm.
*Use T 8 at 300 centres (167) links in boot full length.*

## EXERCISE 4.2    Boot reinforcement under uniform load

Design the boot reinforcement in a 900 mm deep × 400 mm wide L beam which supports 250 mm deep hollow core flooring over a clear span of 10 m. Tie bars T 12 at 600 mm centres are cast in the beam and are to be concreted into the broken out cores of the slabs. The floor slab is to be completed using grade C25 insitu concrete. The slab bearing length is 100 mm. The dimensions of the boot are 250 mm × 300 mm deep. The superimposed dead and live loads are 2.0 kN/m$^2$ and 5.0 kN/m$^2$, respectively.

Use $f_{cu} = 40$ N/mm$^2$ and $f_y = f_{yv} = 460$ N/mm$^2$. Cover to bars in protected faces = 25 mm, and to exposed faces = 40 mm.

### Solution

The boot depth is exactly 300 mm deep; hence use strut and tie design method.

(1) *Before insitu concrete after slabs laid dry*
Length of slab = 10.0 + 0.2 = 10.2 m.
Self weight of slab (see Table 3.1) = 3.5 kN/m$^2$.
Ultimate slab reaction $V_1 = 1.4 \times 3.5 \times 10.2/2 = 25.0$ kN/m.
Beam bearing length $b_1 = 100$ mm, thus $a = 25 + 6$ (say) + 250 − 100/2 = 231 mm.
Boot link $x = 300 - 25 - 40 - 12$ (say) = 223 mm.
Diagonal strut inclination $\theta = \tan^{-1} 231/223 =$ say 45°.

$$H = 25.0 \tan 45° + \left[ \frac{223 + 31}{223} \right] 0.7 \times 25.0 = 42.2 \text{ kN/m.}$$

$$A_{sh1} = \frac{42.2 \times 10^3}{0.87 \times 460} = 106 \text{ mm}^2 \text{ / m.}$$

The diagonal force $C = 25.0/\cos 45° = 35.36$ kN.
The diagonal capacity = 0.2 × 40 (300 − 25 − 6) sin 45° = 1521 kN >> 35.36 kN.

$$T = 25.0 \left[ 1 + \frac{0.7 \times 254}{231} \right] = 44.2 \text{ kN/m.}$$

Hence $A_{sv1} = 110$ mm$^2$/m.

(2) *After insitu concreting*
Effective length of slab = 10.0 + 0.25 + 0.25 = 10.5 m due to extended bearing.
Ultimate slab reaction $V_2 = [(1.4 \times 2.0) + (1.6 \times 5.0)] \times 10.5/2 = 56.7$ kN/m.
Beam bearing length $b_1 = 250$ mm, thus $a = 25 + 6$ (say) + 250/2 = 156 mm.
Boot link $x = 300 - 25 - 40 - 12$ (say) = 223 mm.
Total frictional force = $\mu (V_1 + V_2) = 0.7 (25.0 + 56.7) = 57.2$ kN/m.

Tie steel capacity $= 0.87 \times 460 \times 113 \times (1000/600) \times 10^{-3} = 75.3$ kN $> 57.2$ kN/m.
Hence frictional force is not mobilized in the composite situation.
Thus $H = 39.7$ kN/m and $A_{sh2} = 100$ mm$^2$/m.
Total $A_{sh} = 206$ mm$^2$/m, use T 10 at 300 centres.
Compressive strut not critical by inspection.
Also $T = 56.7$ kN/m and $A_{sv2} = 142$ mm$^2$/m.
Total $A_{sv} = 252$ mm$^2$/m.
*Use T 10 at 300 centres (262).*

## EXERCISE 4.3     Boot reinforcement under concentrated load

Determine the minimum depth of the boot and design the boot reinforcement for a 400 mm wide L beam to support double-tee floor slabs spanning 11 m. The characteristic floor loading is 8.5 kN/m$^2$ dead plus 5.0 kN/m$^2$ live. The floor slabs are seated on $150 \times 150 \times 10$ mm neoprene pads at 1200 mm centres. The fixing tolerance for the slab is 15 mm each end of slab. Immediately after fixing, the end plates in the slabs are welded to side plates in the beam.
  Use $f_{cu} = 40$ N/mm$^2$ and $f_{yv} = 460$ N/mm$^2$. Cover to bars in all faces $= 40$ mm.

### Solution

Ultimate end reaction in floor slab webs $= [(1.4 \times 8.50) + (1.6 \times 5.0)] \times 1.20 \times 11.0/ 2 = 131.34$ kN.
Bearing dimensions $b_1 = b_p = 150$ mm.
Ledge width $=$ bearing pad 150 mm plus $2 \times$ tolerance 15 mm $= 180$ mm.
Upstand width $= 400 - 180 = 220$ mm.
Thus, maximum eccentricity $a =$ cover 40 $+$ (say) 6 $+$ gap 30 $+$ half ledge 75 $= 151$ mm.
Load spread width (Equation (4.16)) $= 150 + (2 \times 151) + 400 = 852$ mm.
To determine required depth of boot, use Equation (4.17).
Thus $d'' = 131.34 \times 10^3/0.2 \times 40 \times 852 \times \sin 45° \times \cos 45° = 39$ mm $< 151$ mm required to develop a 45° strut.
Hence boot depth $= 151 + 40 + $ (say) 6 $= 197$ mm, try 200 mm.
Then $d'' = 154$ mm.
Bursting force beneath bearing pad (refer to Equation (4.18)).
$A_s'' = 0.23 \times 131.34 \times 10^3/0.87 \times 460 = 75$ mm$^2$.
*Use one no. longitudinal T 10 bar (78) in top corner of boot.*

### Bending steel
Bending moment $M = 131.34 \times 0.151 = 19.83$ kN m.

$$K = \frac{19.83 \times 10^6}{40 \times 852 \times 154^2} = 0.024 \ z/d = 0.95.$$

$$A_{sh} = \frac{19.83 \times 10^6}{0.87 \times 460 \times 0.95 \times 154} = 339 \text{ mm}^2.$$

Minimum $= 0.13\% = 260$ mm$^2$/m.
Spacing between bars $<160$ mm.
*Use five no. T 10 at 150 mm centres (392), ensuring that at least one link is within 50 mm of each edge of the bearing pad.*

It is interesting to note that if the 45° strut and tie method is used, the reinforcement in localized zone is

$$A_{sh} = \frac{131.34 \times 10^3}{0.87 \times 460} = 327 \text{ mm}^2. \text{ The answer is about the same.}$$

### Reinforcement between localized zones

Distance between the localized zones = 1200 − 852 = 348 mm.
In this region use nominal steel at 0.15% = 300 mm²/m.
*Use T 10 at 250 mm centres between.*

### Punching shear check

Consider three-sided perimeter because $b_1 + 1.5d'' = 150 + 1.5 \times 154 = 381$ mm < 400 mm beam breadth. Use Equation (4.13).

$u = [2 \times (150 + 1.5 \times 154)] + (150 + 3 \times 154) = 1374$ mm.

$$v = \frac{131.34 \times 10^3}{1374 \times 154} = 0.62 \text{ N/mm}^2.$$

To determine critical value for $v_c$ consider the steel crossing the shear perimeter and use the lesser value.

Longitudinal steel in top of boot is two no. T 10 bars, then

$$100\, A_s/bd'' = \frac{100 \times 2 \times 78}{381 \times 154} = 0.27\%.$$

Boot links T 10 at 150 c/c, then $100\dfrac{A_{sh}}{bd''} = \dfrac{100 \times 78.5}{150 \times 154} = 0.34\%.$

Then $v_c$ (average) = 0.63 N/mm².
But 1.5 $d''/a_v$ = 1 at this point.
Then $v < v_c$, no punching shear reinforcement.

## 4.3.7   Non-composite prestressed beams

### 4.3.7.1   *Flexural design*

The design of prestressed beams is less versatile than in the reinforced beams because the tendons positions are restricted to a predetermined pattern by an array of holes in the jacking heads, which is usually a permanent fixture in a precasting works. The opportunity for architectural freedom is limited to symmetrical sections of modular depth – usually 50 mm increments. A range of rectangular or inverted-tee beams (L beams are not symmetrical) from 200 to 600 mm deep will usually satisfy the flexural and shear requirements in most structures.

Figure 4.18 shows a full array of possible tendon positions in an inverted-tee beam, and an example of a typical tendon layout. Note the symmetry. The tendons are placed at all the corners; a 40 to 50 mm centroidal cover distance being used in most cases. The positions of the tendons are such that they do not interfere with any of the beam end connections, such as shear boxes or shear hanger plates. They can also be arranged to allow service holes to pass through the beam by leaving a clear zone of about 75 mm near to the centroid of the uncracked concrete section.

As shown in Fig 4.18(a), a referencing system of letters (A, B, C, etc.) and numbers (1, 2, 3, 4, etc.) is used to convey information to the manufacturer as to which strands are to be used in a particular beam. Where debonding of strands is required a coded message relating to the requirements is also given, e.g. B 2–800 means the strand at position B 2 is to be sheathed for a distance of 800 mm from the end of the beam. Figure 4.18(b) shows a typical arrangement of strands – note the strand positions which are essential for use.

The minimum breadth of the beam is a function of the type of floor slab to be used. The breadth is equal to twice the recess width, plus the upstand width. The same reasoning as for the L beam is used if floor ties are intended to be placed within the recess and concealed in the depth of hollow core floor slabs. The minimum recess width for this condition is 100–125 mm. If the ties are to be located elsewhere the recess width may be 90–100 mm. The width of the upstand depends largely on accommodating end connections, particularly in shallow beams where the connector may be positioned near to the top of the beam (e.g. see Fig. 4.36). The minimum upstand breadth is therefore 250–300 mm, depending on the type of connector used. Thus the breadth of internal-tee beams varies from between 450 mm and 600 mm.

Beam depths depend on three factors:

    (a)  The flexural/shear capacities
    (b)  The size of the end connector
    (c)  The depth of the boot required to carry the floor loads.

Factor (a) is self-evident from standard designs as mentioned above. Factor (b) will be dealt with in Section 4.3.10.

The process of selecting the prestressing requirements can be optimized by choosing a reinforcement pattern that will simultaneously satisfy transfer at the ends of the beam and working loads at the point of maximum imposed bending moment without having to debond or deflect the tendons. This method is further refined if the permissible stress at transfer $0.5 f_{ci}$ after the initial losses (due to elastic shortening) is made equal (or as close as possible) to the working stress $0.33 f_{cu}$ after all losses. In most beam designs the initial and final losses are about 7 to 8 per cent and 20 to 25 per cent, respectively, meaning that the ratio $f_{ci}/f_{cu}$ should be at least 0.55. In fact the transfer strength for grade C60 concrete is at least 40 N/mm$^2$, and therefore transfer stresses will (nearly) always govern for parallel, unbonded tendons. To overcome this problem it is desirable to debond a small number of strands, say 4 no. in a typical situation.

It is wise to actually restrict the top fibre stress to something less than $0.45\sqrt{f_{ci}}$, say two thirds or even half this value (i.e. $0.225\sqrt{f_{ci}}$), whilst accepting that there will be a small loss in moment capacity. This factor of safety is based on the experience gained from transporting and handling highly stressed prestressed beams, and the need to avoid flexural cracking for the sake of durability. It also allows some freedom in choosing a final tendon arrangement if the eccentricity of the tendons needs to be greater than the calculated design value.

A further factor is that the actual initial pretensioning force achieved in practice may be higher than the design value. In order to avoid under stressing the manufacturer will deliberately over stress by two or three per cent to compensate for slippage in the chucks, losses in the hydraulic stressing system, etc.

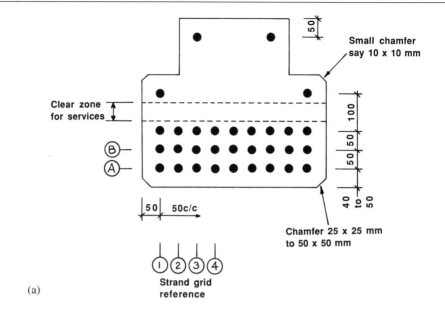

(a)

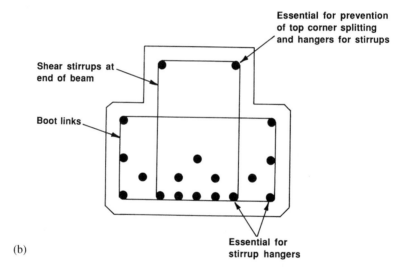

(b)

**Fig. 4.18** Location of tendons in inverted-tee beams. (a) Full array and (b) typical arrangements.

Simple algebra will give a set of relationships to solve for the prestressing force $P$ and the eccentricity $e$, as follows. The geometric centroid of the section is computed for the gross concrete section ignoring the area of reinforcement (which is barely significant). The total initial force (in all the tendons) is therefore governed by the state of stress at transfer and is given by:

$$P_i = \frac{A}{2(1-\xi)}\left[(0.5f_{ci} - 0.225\sqrt{f_{ci}}) + \left((0.5f_{ci} + 0.225\sqrt{f_{ci}})\frac{1-\alpha}{1+\alpha}\right)\right] \qquad (4.19)$$

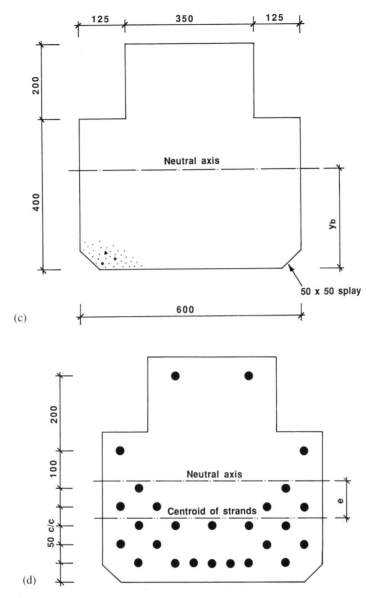

**Fig. 4.18** (*continued*) Location of tendons in inverted-tee beams. (c) Detail of inverted-tee beam used in beam design example and (d) strand pattern to beam design example.

where $\xi$ = initial prestress loss due to elastic shortening (expressed as a decimal fraction), and $\alpha = Z_t/Z_b$. The number of tendons required $N = P_i/\mu\, A_{ps}\, f_{pu}$ (where $\mu$ = degree of prestress usually taken as 0.7 to 0.75 depending on the type of tendon used) should be rounded down to the next integer to prevent overstress, particularly in the tensile fibres at the top of the beam. The eccentricity $e$ is given by:

$$e = \frac{(0.5 f_{ci} + 0.225 \sqrt{f_{ci}})}{(1 - \xi) P_i \beta} \tag{4.20}$$

where $\beta = \dfrac{1}{Z_t} + \dfrac{1}{Z_b}$, and $P_i$ is based on the value obtained from Equation (4.19). The design then follows the procedures given in Section 5.2.8.

The above analysis will give the maximum possible moment capacity for a single arrangement of tendons for the beam, but if the moment capacity is limited by tension in the top fibre then a smaller number of strands will give the same moment of resistance. The designer has to be careful in checking that the above analysis does not lead to an uneconomical solution in using too many strands. This is particularly important in inverted-tee beams where non-symmetry leads to $M_{sr}$ governed by bottom fibre stresses being about 15 per cent greater than $M_{sr}$ governed by the top fibre stress. It is also advantageous to use more than one arrangement of tendons in situations where smaller bending moments are required in beams of the same cross-section. This is achieved by reducing the number of tendons, by say two or three at a time, and finding new values for the eccentricity $e$ that will satisfy maximum fibre stresses as before. The following example shows the basic idea.

Consider a prestressed inverted-tee beam 600 mm deep $\times$ 600 mm wide, having a 200 mm deep $\times$ 350 mm wide upstand as shown in Fig. 4.18(c). In this example, use $f_{cu} = 60$ N/mm$^2$, $f_{ci} = 40$ N/mm$^2$, $E_{ci} = 28$ kN/mm$^2$, $P_i$ per strand $= 0.7 \times 1750 \times 94.2 = 115.5$ kN, $E_s = 195$ kN/mm$^2$ and 2.5 per cent strand relaxation.

Geometric data:

$A = 307.5 \times 10^3$ mm$^2$  $\qquad$  $y_b = 269.8$ mm
$Z_b = 30\ 215 \times 10^3$ mm$^3$  $\qquad$  $Z_t = 24\ 685 \times 10^3$ mm$^3$
$\alpha = 0.817$  $\qquad$  $\beta = 7.36 \times 10^{-8}$ mm$^{-1}$

The limiting bottom and top fibre stresses at transfer are:

$$f_b < +20 \text{ N/mm}^2$$

$$f_t < 0.225 \sqrt{40} = -1.42 \text{ N/mm}^2$$

Assuming that the elastic shortening loss is 0.08 of the initial, from Equations (4.19) and (4.20):

$$P_i = \frac{307.5 \times 10^3}{2 \times (1 - 0.08)} \left[ 18.58 + \left( 21.42 \times \frac{0.183}{1.817} \right) \right] = 3465.6 \text{ kN}$$

$$e = \frac{21.42}{0.92 \times 3465.6 \times 10^3 \times 7.36 \times 10^{-8}} = 91 \text{ mm (to nearest mm)}$$

the number of tendons $N = 3465.6/115.5 = 30.005$, but rounded down to 29 for the reasons explained earlier, the actual initial prestressing force $P_i = 29 \times 115.5 = 3349.5$ kN, and $P_t = 3081.5$ kN after initial losses. The maximum fibre stresses at transfer are:

$$f_{bi} = \frac{3081.5}{307.5} + \frac{3081.5 \times 91}{30\ 215} = 10.02 + 9.28 = 19.30 \text{ N/mm}^2 < +20 \text{ N/mm}^2$$

$$f_{ti} = \frac{3081.5}{307.5} - \frac{3081.5 \times 91}{24\ 685} = 10.02 - 11.36 = -1.34 \text{ N/mm}^2 < -1.42 \text{ N/mm}^2$$

The initial stress at the level of the centroid of the strands $f_{cci} = 14.36$ N/mm$^2$.

The elastic shortening loss $= \dfrac{195 \times 14.36}{28 \times (0.7 \times 1750)} = 8.15\%$ (original assumption of 8%

satisfactory).

The other final losses are as follows (see Section 5.2.8):

- creep $= 1.8 \times 8.15\% = 14.67\%$
- shrinkage $= \dfrac{300 \times 10^{-6} \times 195}{0.7 \times 1750} = 4.77\%$
- relaxation $= 1.2 \times 2.5 = 3.0\%$

Final losses $= 30.6\%$. Thus the final prestress force $= (1 - 0.306) \times 3349.5 = 2324.5$ kN.
The final working fibre stresses are $f_{bc} = +14.56$ N/mm$^2$ and $f_{tc} = -1.01$ N/mm$^2$
Then $M_{sr}$ is the lesser of $M_{sr} = (14.56 + 3.5) \times 30\ 215 \times 10^3 \times 10^{-6} = 545.7$ kN m based
on the bottom fibre stress limit, or

$$M_{sr} = (1.01 + 20) \times 24\ 685 \times 10^3 \times 10^{-6} = 518.6 \text{ kN m}$$
based on the top fibre stress limit.

Clearly there are wasted strands as the section capacity is limited by the tension fibre, as was suggested above.

It is now possible to redesign the prestressing requirements by reducing the number of strands to, say 26, and calculating a new service moment for the beam. Thus if $P_i = 3003$ kN, $\xi = 0.075$ (this may be estimated pro-rata as the number of strands used above), then:

$$e = \frac{21.42}{0.925 \times 3003 \times 10^3 \times 7.36 \times 10^{-8}} = 104.8 \text{ mm}$$

then the maximum fibre stress in the top of the beam at transfer will be exceeded, viz
$f_t = 9.03 - 11.79 = -2.76$ N/mm$^2 > -1.42$ N/mm$^2$. In this case, $e$ must be obtained by
letting $f_t = -1.42$ N/mm$^2$ and solving as follows:

$$f_t = 9.03 + 1.42 = 10.45 \text{ N/mm}^2 = \frac{(1 - \xi)P_i e}{Z_t} = \frac{0.925 \times 3003 \times 10^3 \times e}{24\ 685 \times 10^3}$$

$e = 92.86$ mm

Then the transfer stresses are $f_{bi} = 9.03 + 8.53 = +17.56$ N/mm$^2 < 20$ N/mm$^2$, and $f_{ti} = 9.03 - 10.45 = -1.42$ N/mm$^2$.

The initial stress at the level of the centroid of the strands $f_{cci} = 12.93$ N/mm$^2$.

The elastic shortening loss $= 7.34\%$. The other final losses are creep $= 13.22\%$, shrinkage plus relaxation $= 7.77\%$. Final losses $= 28.33\%$.

Thus the working fibre stresses are $f_{bc} = +13.60$ N/mm$^2$, and $f_{tc} = -1.10$ N/mm$^2$.
Then $M_{sr}$ is the lesser of $M_{sr} = (13.60 + 3.50) \times 30\ 215 \times 10^3 \times 10^{-6} = 516.7$ kN m based
on the bottom fibre stress limit, or

$$M_{sr} = (1.10 + 20.0) \times 24\ 685 \times 10^3 \times 10^{-6} = 520.8 \text{ kN m}$$
based on the top fibre stress limit.

This situation represents a 'balanced' situation in terms of the limiting design stresses being reached simultaneously in the top and bottom fibres.

Paradoxically the service moment is actually less when using 29 strands (518.6 kN m) than 26 strands (519.7 kN m). This is because in the former the limit state is governed by the working stress in the top fibre of the beam, whereas in the latter it is properly governed by the bottom fibre condition.

Figure 4.18(d) shows a possible pattern for the strands to give $e = 92.7$ mm, which is the nearest possible figure to the required value of 92.9 mm. Note that the actual eccentricity derived from the strand pattern should not (but may) exceed the theoretical value. The restricted 'no strand' zones along the vertical centre line of the beam are to enable the end shear connector to be positioned uninhibited. The other restricted zone across the centre of the beam (near to the neutral axis position of 269 mm above the soffit) is to facilitate a 50 mm service pipe.

The arrangement of the strands is very important in order to avoid splitting stresses (although this would be rare in a beam of this shape). The layout must be symmetrical about the vertical centre line, and there should be no large concentration of strands (called bunching) vertically.

The prestressing arrangement may now proceed using a smaller number of strands, say 23, 21 and 19 in this example, until the level of prestress renders the section uneconomic. This will occur when the moment of resistance of a beam with a smaller depth is greater than the beam in question. As mentioned above, the depth increment for prestressed beams is usually 50 mm. Thus considering 550 mm as the next smallest size it is found that the balanced moment of resistance for a 550 mm deep inverted-tee beam is 431 kN m when using 24 strands. The number of strands required in the 600 mm deep beam to give a slightly higher moment capacity than this is found to be 19. This therefore represents the lower bound prestressing arrangement for the 600 mm deep beam, and the overall economics of choosing either the deeper or shallower section must be considered. Ultimate limit state is based on a rectangular section of breadth $b_w$ and effective depth $d$ to centroid of strands which are beneath the neutral axis. Refer to Section 5.2.8.2 for further details.

Table 4.9 shows an example of calculations carried out for a series of 500 mm wide by 300 mm boot depth inverted-tee beams with varying numbers of 10.9 mm diameter low relaxation strands. In Chart 4.1 the moments and shear force capacities are given for a range of beams of varying depth and reinforcement.

### 4.3.7.2   Boot reinforcement

The boot of inverted-tee beams is designed and reinforced in exactly the same way as in the L shaped beam, except that a single link $A_{sh}$ is used to satisfy tension in both sides of the beam. The vertical steel $A_{sv}$ carries the maximum reaction according to Equations (4.11) and (4.12).

In most cases the depth of the recess is equal to, or smaller than the depth of the floor slab, or the depth of the half-joint in double-tee slabs. This means that the depth of the boot on which the floor rests is sufficient for there to be no requirement for horizontal tie steel. Here the concrete is in longitudinal compression and is usually uncracked flexurally. Only in cases where the boot depth is less than about 250 mm, or the beam is heavily loaded using double-tee slabs, are links required in the boot. As shown in Fig. 4.12, links (typically T 8 or T 10) are placed at least one either side of the slab reaction points.

### 4.3.7.3   Shear design

Shear reinforcement is rarely necessary because the ultimate shear resistance of prestressed members is a function of the ultimate flexural requirements. The shear span ($M_u/V_u$) for

**Table 4.9** Specimen inverted-tee beam flexural design.

Notes:  (a)  Beam Geometry

$b = 500$ mm, $b_{upstand} = 300$ mm, $h_{boot} = 300$ mm, $h_{upstand} = 150$ mm.

Bottom corner triangular chamfer $= 50 \times 50$ mm.

$A = 192.5 \times 10^3$ mm$^2$, $y_b = 204.3$ mm, $I = 2874.5 \times 10^6$ mm$^2$.

$$Z_b = 14.069 \times 10^6 \text{ mm}^3, \ Z_t = 11.699 \times 10^6 \text{ mm}^3, \ \alpha = \frac{Z_t}{Z_b} = 0.8316$$

(b)  Material Data

$f_{cu} = 50$ N/mm$^2$ $E_c = 30$ kN/mm$^2$, $f_{ci} = 35$ N/mm$^2$, $E_{ci} = 27$ kN/mm$^2$.

$f_{pu} = 1750$ N/mm$^2$, $A_{ps} = 71.0$ mm$^2$ per strand, $E_s = 195$ kN/mm$^2$.

Degree of initial prestress $\mu = 0.7$. Hence $P_i = 87$ kN per strand.

| N | $P_i$ | Transfer Loss Est*(%) | $P_t$ | $e$ | $\frac{P_t}{A}$ | $\frac{P_t e}{Z_b}$ | $\frac{P_t e}{Z_t}$ | $f_{bi}$ | $f_{ti}$ | $f_{cci}$ |
|----|------|------|--------|------|-------|-------|-------|--------|--------|--------|
| 16 | 1392 | 5.0 | 1322.4 | 72.5 | +6.87 | +6.81 | −8.20 | +13.68 | −1.33 | +9.28 |
| 14 | 1218 | 4.0 | 1169.3 | 74.0 | +6.07 | +6.15 | −7.40 | +12.22 | −1.33 | +8.29 |
| 12 | 1044 | 3.5 | 1007.5 | 76.2 | +5.23 | +5.46 | −6.56 | +10.69 | −1.33 | +7.27 |

\* Est = estimate prior to transfer stress analysis.

| N | Losses (%) Shortening | Creep | Shrink + Relax | Total | $P_t$ | $f_{bc}$ | $f_{tc}$ | $M_{sr}$ (bottom) | $M_{sr}$ (top) |
|----|------|------|------|------|-------|-------|-------|-------|-------|
| 16 | 5.47 | 9.85 | 7.77 | 23.09 | 1070.5 | +11.07 | −1.076 | 200.7 | 205.6 |
| 14 | 4.88 | 8.78 | 7.77 | 21.43 | 957.0 | +10.00 | −1.088 | 185.7 | 205.7 |
| 12 | 4.28 | 7.71 | 7.77 | 19.76 | 837.7 | +8.89 | −1.106 | 170.1 | 206.0 |

| $N_T$ | $A_{ps}$ | $d$ | $\frac{f_{pu} A_{ps}}{f_{cu} bd}$ | $\frac{f_{pb}}{0.87 f_{pu}}$ | $\frac{x}{d}$ | $d_n$ | $M_{ur}$ | $M_{ur}/M_{sr}$ |
|----|------|------|------|------|------|------|------|------|
| 12 | 852 | 380.8 | 0.261 | 0.87 | 0.475 | 81.4 | 337.7 | 1.683 |
| 10 | 710 | 385.0 | 0.215 | 0.90 | 0.420 | 72.8 | 303.7 | 1.635 |
| 8 | 568 | 391.3 | 0.169 | 0.95 | 0.350 | 61.6 | 270.8 | 1.592 |

*Notes:*  $N =$ no. of strands in total, all of which are equally stressed and fully bonded.

$N_T =$ no. of strands in tension zone. In this exercise we assume that 2 no. strands will be placed in the top of the upstand and 2 no. strands in the top of the boot. Hence $N_T = N - 4$.

most beams in precast structures is $L/4$. Thus the shear force at the position of flexural decompression rarely exceeds $0.5 \ V_{cr}$. Deflected tendons may be specified in very special cases but complications to the long-line prestressing outweigh the enhancements in structural performance.

Prestressed beams are highly efficient both structurally and economically. Attempts are now being made to reduce the depth of the downstand below the soffit of the slab. This has inevitably led to composite construction as discussed in Chapter 6.

## EXERCISE 4.4    Inverted-tee beam design

Calculate the flexural serviceability and ultimate moments of resistance, and the uncracked and cracked shear resistances of the class 2 psc inverted-tee beam shown in Fig. 4.19.

Use $f_{cu} = 60$ N/mm$^2$, $f_{ci} = 40$ N/mm$^2$ at transfer, and 12.5 mm diameter low relaxation helical strand stressed to 70 per cent ultimate with $f_{pu} = 1751$ N/mm$^2$ (quoted value).

### Solution

#### Component cross-sectional properties

| | |
|---|---|
| Overall width = 500 mm | Upstand width = 300 mm |
| Overall depth = 500 mm | Upstand depth = 200 mm |
| Area = 210 000 mm$^2$ | $I = 4003.57 \times 10^6$ mm$^4$ |

Height to neutral axis (NA) from bottom $y_b = 221.4$ mm

$Z_b = 18.080 \times 10^6$ mm$^3$ $\qquad\qquad\qquad Z_t = 14.37 \times 10^6$ mm$^3$

Self weight = 5.15 kN/m.

#### Concrete data

Characteristic cube strength at 28 days = 60 N/mm$^2$, $E_c$ (28 days) = 32 kN/mm$^2$.
Release (or transfer) strength = 40 N/mm$^2$, $E_{ci} = 28$ kN/mm$^2$.

#### Reinforcement data

No. of tendons = 17 at 12.5 mm diameter helical strand.                  Area per strand = 94.2 mm$^2$.

Relaxation type 2 = 2.5 per cent loss.
Centroid of strands from bottom = 143 mm.
Eccentricity from neutral axis $e = 78.4$ mm.
$A_{ps} = 1601.4$ mm$^2$                                                                $f_{pu} A_{ps} = 2805.0$ kN.
$E_s = 195$ kN/mm$^2$.

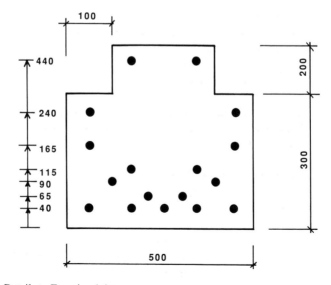

**Fig. 4.19** Details to Exercise 4.4.

### Serviceability limit state of bending

Initial prestressing stress $= 0.7 \times 1751 = 1225$ N/mm$^2$.
Initial prestressing force $P_i = 0.7 \times 2805 = 1963.5$ kN.
Prestress at transfer assuming 6% initial loss:

$$f_{bc} = 16.80 \text{ N/mm}^2 \text{ (bottom)}, f_{tc} = -1.29 \text{ N/mm}^2 \text{ (top)}$$
$$f_{cc} = 11.63 \text{ N/mm}^2 \text{ (centroid of strand)}.$$

Losses

| | | |
|---|---|---|
| Elastic shortening $= 11.63 \times 195/28 = 81.0$ N/mm$^2$ | = | 6.61% |
| Creep loss $= 1.8 \times$ elastic shortening loss | = | 11.90% |
| Steel relaxation quoted at 2.50 per cent at 100 h | = | 3.00% |
| Concrete shrinkage at 300 $\mu\varepsilon = 58.5$ N/mm$^2$ | = | 4.77% |
| Total | = | 26.28% |

Effective prestressing force after losses at transfer $= 1833.7$ kN.
Prestress at transfer:
$$f_b = +16.68 \text{ N/mm}^2 < 20 \text{ N/mm}^2$$
$$f_t = -1.28 \text{ N/mm}^2 < -2.85 \text{ N/mm}^2 \text{ (using } 0.45 \sqrt{f_{ci}}).$$
Effective prestressing force after final losses $= 1447.5$ kN.
Prestress after losses:
$$f_{bc} = +13.17 \text{ N/mm}^2$$
$$f_{tc} = -1.01 \text{ N/mm}^2$$
$$f_{cc} = +9.12 \text{ N/mm}^2 \text{ (strand centroid)}$$
$$f_{cp} = +6.88 \text{ N/mm}^2 \text{ (section centroid)}.$$
For class 2 elements, allowable tension (for $f_{cu} = 60$ N/mm$^2$) $= 3.50$ N/mm$^2$.
Service resistance moment, $M_s = (13.17 + 3.50) \times 18.08 = 301.4$ kN m.
Allowable compression (for $f_{cu} = 60$ N/mm$^2$) $= 20$ N/mm$^2$.
Service resistance moment, $M_s = (1.01 + 20.00) \times 14.37 = 301.9$ kN m.
**Design value $= 301.9$ kN m.**

### Ultimate limit state of bending

Number of strands below neutral axis in tension zone $= 13$.
$A_{ps} = 1224.6$ mm, $f_{pb}/f_{pu} = 0.735$.
$F_s = f_{pb} A_{ps} = 1287 \times 1224.6 = 1576$ kN.
$d = 417.7$ mm, $b = 300$ mm, $f_{cu} = 60$ N/mm$^2$.
$F_c = 0.45 f_{cu} b \, 0.9 \, X$.
Hence $X = 216$ mm, $>$ upstand but ignore additional breadth in boot $d_n = 97$ mm.
Ultimate resistance moment, $M_u = 1576 \times (417.7 - 97) = 505.4$ kN m.
Thus, ratio $M_u/M_s = 1.67$, i.e. serviceability limit state will be critical.

### Ultimate limit state of shear

(a) *Uncracked*, at the level of the intersect between the upstand and the boot:

| | |
|---|---|
| $b_v = 300$ mm | $A = 300 \times 200 = 60\,000$ mm$^2$. |
| $y_t = 500 - 221.4 = 278.6$ mm | $y = 278.6 - 100 = 178.6$ mm. |

Then $I \, b_v/A \, y = 112\,080$ mm$^2$.
$f_{cp}$ (at intersect plane) $= +4.66$ N/mm$^2$     $f_t = 1.859$ N/mm$^2$.
$f_{cpx}$ (at 300 mm from end) $= +3.91$ N/mm$^2$
Ultimate shear resistance, $V_{co} = 112.08 \times 3.05 = 341.8$ kN.

(b) *Cracked* (see also Section 5.2.8)

$\sqrt{60} = 7.74$ N/mm$^2$.

Decompression moment $M_0 = [(0.37 \times 7.74) + (0.8 \times 13.17)] \times 18.08 = 242.8$ kN m.

$b_v = 500$ mm, because the cracked part is below the neutral axis.

$d = 417.7$ mm.

$V_{cr} = (0.037 \times 7.74 \times 500 \times 417.7) \times 10^{-3} + (242.8 \times V_{cr}/1.5 \times 301.9)$

$\qquad = 59.81 + 0.536 \, V_{cr} = 128.9$ kN.

But not less than $0.1 \times 500 \times 417.7 \times 7.74 = 161.6$ kN.

### 4.3.8   Beam end shear design

Special attention is given to shear reinforcement near to the end connections. Stirrups and bent-up reinforcing bars are provided to ensure the transfer of shear in the critical region. Bent-up bars may be used to resist approximately one-half of the shear force. In some instances a prefabricated shear box, comprising welded angles, channels, composite plates, or RHS sections, partially or wholly replaces the stirrup cage. These are used to:

- transfer the shear forces to a point in the beam where the stirrups are considered to be fully effective
- prevent bursting of the concrete at ends of the beam
- provide a steel bearing plate with which to make a steel-to-steel connection to the supporting member.

Extensive testing has, in the main, been carried out to prove the shear capacity of beams with steel inserts (e.g. [4.11 and 4.12]).

Recessed or pocketed beam ends, shown in Figs 4.20 and 4.21, are used where the beam is to be supported on a member projecting from the face of the column. The structural connection is made entirely within the depth of the beam. The generalized structural model, presented here in Fig. 4.22, is similar in principle to the design of reinforced concrete corbels. The dotted lines represent principle stress trajectories as determined using a finite element solution [4.13]. The stresses fan out in a 45° manner and principle compressive and tensile forces are in equilibrium. The concrete provides the compressive force, providing it is restrained against later bursting by the binding steel, and tie bars provide the tensile force resistance.

One of the most comprehensive dissertations on this subject is by Martin and Korkosz [4.14], which cites 147 references (up to 1982) and studies the analytical and experimental aspects of many of the connections given in the various PCI design manuals on precast structures [4.15] and connections [4.16].

Beam to beam connections may also be made as shown in Fig. 3.29 where it is not possible to position a column at the end of a beam. The design of the end shear reinforcement is no different to the above. Additional reinforcement is required in the primary beam to cope with the combined effects of bending, shear, torsion and bearing stresses. The design method involves the resolution of actions into shear or axial forces, and the superposition of resulting stresses in the connection. Steel inserts are most widely used, particularly if the ultimate beam reaction is greater than about 200 kN. The design of this connection is discussed in Section 7.12.

**Fig. 4.20** Recessed beam end.

**Fig. 4.21** Pocketed beam end.

### 4.3.9   Recessed beam ends

Recessed beam ends are used where the beam is to be supported on a member projecting from the face of the column, and the structural connection is made entirely within the depth of the beam. The design methods presented here are similar in principle to the

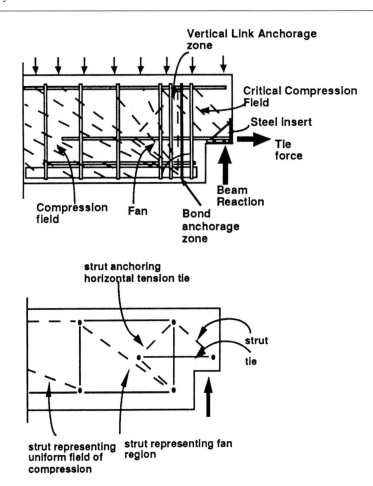

**Fig. 4.22** Generalized structural model of halving joint in beams [4.13].

halving joint or corbels given in BS 8110, Part 1, Clause 5.2.7 [4.2]. The general dimensions for a recessed beam are given in Fig. 4.23.

A special type of beam end, shown in Figs 4.21 and 4.24, is the pocketed end type. Two concrete wings, Y in the diagram, are reinforced to allow the main flexural steel in the bottom of the beam to develop full anchorage without the use of anchor plates or bent-up bars.

In all cases the ends of the beams are reinforced at the ultimate limit state using an arrangement of vertical links and diagonal or horizontal bars to cater for tie forces and bursting pressures in the compression zones, respectively. The horizontal bars in direct contact with the bearing surface are used in all cases to avoid the risk of local spalling and crushing and to ensure uniformity of bearing pressures in this highly stressed zone.

The serviceability limit state is equally as important in this region, and the design should check that service strains, based on a short-term modulus of rupture of $0.556 \sqrt{f_{cu}}$ [4.17] are not exceeded. A method of calculating the service load stresses, strains and crack widths is proposed by Clark and Thorogood [4.18] (see Fig. 4.25). By limiting

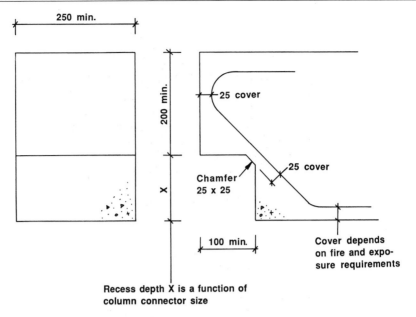

**Fig. 4.23** Geometry of recessed beam end.

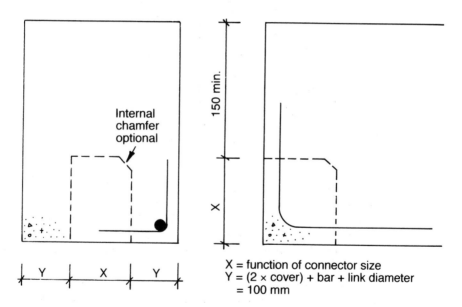

**Fig. 4.24** Geometry of pocketed beam end.

either the strain in the concrete $\varepsilon_c$ (to say 0.0018) or steel $\varepsilon_s$ (to say 0.0020) considered along the cracked section OA, it is possible to determine the forces $F_c$ and $F_s$ and hence the serviceability force $V_s$. The most significant factor in controlling service load stresses is the inclined reinforcement across the inside corner of the recess. Mattock and Chan

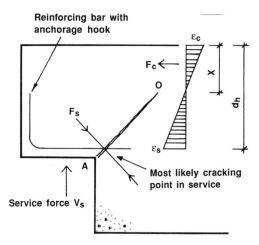

**Fig. 4.25** Strains in half joint.

also measured crack widths in 'dapped-end' beams subjected to vertical end shear forces, and vertical plus horizontal end forces [4.19].

The inclination of the shear crack ($\alpha$ to the horizontal) propagating from the root depends on the relative magnitude of the shear and normal stress at that point. Away from the local stress concentrations this ratio is a function of the reduced depth $h'$ to the full depth $h$ of the beam. Windisch [4.20] states that extensive linear elastic finite element analysis has shown:

$$\alpha = \frac{90h'}{h}(°) \tag{4.21}$$

and that the strength of the diagonal bar placed at the root should be down rated if the crack is not normal to the bar. In most cases the angular discrepancy will not be more than about 20° such that the down rating is only 10 per cent [4.20].

These types of joints are also subject to large stress concentrations in the vicinity of the bearing surface. In the beam end test shown in Fig. 4.26, 4 no. T 12 diagonal bars were provided to resist a vertical end shear force of 80 kN (service) and 125 kN (ultimate), together with 1 no. T 12 U-bar to resist lateral bursting. The reinforcement cage is shown in Fig. 4.26(a) and the grade of concrete was C40. The first sign of cracking was found at 90 kN, but it did not occur at the half joint as shown in Fig. 4.26(b). Instead the cracks propagated from a point directly over the edge of the steel bearing plate, and occurred when the bearing stress was approximately 12 N/mm$^2$ or 0.33 $f_{cu}$. Figure 4.26(c) shows the ultimate test condition.

Plates cast into the beam are sized to ensure that the average bearing stress in the plate does not exceed 0.60 $f_{cu}$. In the more highly stressed situations reinforcement is sometimes welded to the plates to ensure that concrete in contact with the plate is confined. It is clear that the failure mode of these beams is affected by the width of the bearing plate as the shear crack in the half joint began above the edge of the bearing plate. Narrow plates will produce the type of failure shown in Fig. 4.27. Cracks above the end plate will propagate rapidly into shear cracks in the sides of the beam.

In the test the ultimate limit state was reached following yielding of the links directly above the recess. The average failure shear stress in the beam was 1.20 N/mm$^2$ in the main body of the beam and 1.75 N/mm$^2$ in the half joint.

The width of the bearing plate $b_p$ in the beam should not be greater than $0.4b$ [4.5], to avoid punching through the concrete above the plate, and be not greater than $(b - 100)$ mm to give adequate cover, with tolerances to the sides of the beam. In most instances $b_p = b/3$. This thickness of the plate is determined from its punching shear and lateral bursting capacity so that the plate remains perfectly flat and does not deform over the bearing point. Plate thickness is at least 8 mm to 10 mm, and up to 16 mm in the more heavily loaded connections (>250 kN). In most cases mild steel grade 43 or 50 is used for the plate and grade E43 or E51 electrodes for welding reinforcement to it.

The end cover to the edge of the plate should not be greater than 25 mm, with some designers preferring to take the plates to the end of the beam to prevent spalling. The stress distribution over the surface of the plate is assumed to be triangular (trapezoidal would be nearer the reality) and therefore the resultant of the force $V$ acts at a distance $b_l/3 + 25$ mm (end cover distance) from the end of the beam. Accumulative deviations in manufacture and erection will give the most onerous eccentricity, and therefore the gap between beam and column should be increased by the possible site fixing tolerances of up to 15 mm, or 7.5 mm each end. Thus the total eccentricity from the face of the column may be calculated. In practice, dimensional accuracy for precast beams (particularly those cast in steel moulds) is much less than the values quoted in BS 8110, Part 1, clause 5.2.4 [4.2] of 3 mm per metre distance between supports.

### 4.3.10   Design methods for end shear

Two methods are used depending on whether the end of the beam is recessed or pocketed, and on the depth of the recess as a fraction of the overall depth of beam. In all cases truss action develops in the beam and adequate shear friction and shear bond anchorage is provided either mechanically or physically. Reinforcement normal to the potential cracking plane provides the normal force required to maintain the ultimate shear resistance by shear friction.

Although two different shear transfer mechanisms exist and may be analysed separately, it is common practice to add the shear resistances in the following:

- vertical links and compressive struts
- diagonal bars,

to obtain the total ultimate shear capacity. This is permissible only if the ductility factors for high tensile bars are not exceeded, i.e. the shear capacity obtained by the two different methods does not differ by more than about 40 per cent in favour of the diagonal bars. This is the most economical solution because it is necessary to include the vertical links over the bearing surface in order to mobilize the forces in the anti-bursting bars in this region.

The best way of dealing with the kind of problem given above is to calculate the steel requirements for lateral bursting, which has to be provided in all cases, and then to calculate the vertical shear resistance of the vertical links placed above the bearing surface.

(a)

(b)

**Fig. 4.26** Beam end shear test. (a) Reinforcement cage; (b) first crack at 90 kN end reaction and (c) ultimate failure at 180 kN end reaction (design value = 125 kN).

The remaining reaction force, which should not be greater than 0.5 $V$ (because bent-up bars should not be used to carry more than 50 per cent of shear forces), may then be carried by the diagonal bars.

(c)

**Fig. 4.26** (*continued*)

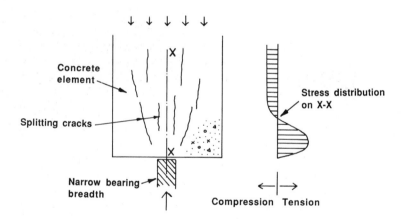

**Fig. 4.27**  Narrow plate beam failure.

Figure 4.28(a) [4.5] shows the first of the two structural models and typical details for this connection, where the recess is about one-half of the total depth. The breadth of the beam is $b$, the effective depth of the beam is $d$, and the recessed depth is $d_h$.

The first task is to check that the bearing stress between the beam and the bearing surface does not exceed $0.6 f_{cu}$. Bearing stresses $f_b = V/b_p b_l$ are calculated in accordance with the recommendations given in Section 7.6 and BS 8110, Part 1, Clause 5.2.3.4 [4.2]. The assumption is that the distribution of bearing stresses is trapezoidal such that the maximum ultimate stress is approximately equal to $f_{cu}$. If $f_b > 0.6 f_{cu}$ a steel bearing plate is provided and the concrete in contact with the plate is confined. An ultimate stress of $0.8 f_{cu}$ may be used in contact with the plate. The plate is fully bonded to the beam by

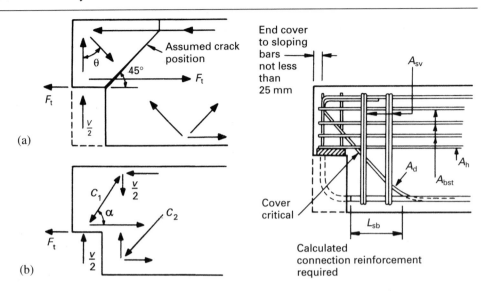

**Fig. 4.28** Structural action and typical reinforcement arrangements. (a) Truss action in half joint with deep pocket [4.5] and (b) truss action in half joint with shallow pocket [4.5].

virtue of tie bars welded to the inside surface of the plate. The thickness of the plate is determined from two factors:

- tensile force $T$ in the tie bars
- punching shear capacity of the plate, but only if the width of the plate is greater than the width of the bearing, otherwise the force is transferred directly through the plate.

If $f_b < 0.6 f_{cu}$ the concrete beam may be bedded directly onto a felt, or neoprene (or similar) pad, of 10 mm minimum thickness, and covering the whole of the bearing surface. This is to allow small rotations, in the order of 0.001 radians, to take place without causing damage to the edges of the bearings.

The second task is to check that the shear stress in the recessed section of the beam does not exceed $0.8\sqrt{f_{cu}}$. This is to ensure that the section will not fail by shear even though reinforcement is providing the entire shear resistance.

When a precast concrete beam is positioned firmly onto a bearing surface, axial deformations in the beam will induced axial tension or compression at the point of contact. This is due to the fact that the beam cannot slide on the bearing until the frictional force is overcome. The axial strain required to induce the frictional force is quite small in the order of 50 $\mu\varepsilon$ for most types of precast components. Thus a horizontal frictional force $T = \mu V$ is generated over the bearing at the interface, where $\mu$ = shear friction coefficient as follows:

- 0.7 for plain concrete to concrete surfaces
- 0.4 for steel to concrete surfaces
- 0.4 for steel to steel surfaces.

The reinforcement $A_h$ required to provide this force is given by:

$$A_h = \frac{\mu V}{0.87 f_y} \qquad (4.22)$$

In the (rather unlikely) event that the peripheral stability tie force $F_t$ (see Chapter 9) is carried by the connector itself, then $A_h$ should be increased to:

$$A_h = \frac{\mu V + F_t}{0.87 f_y} \qquad (4.23)$$

At least two mild steel bars should be used for $A_h$ and they should preferably be welded to the bearing plate as it is unlikely that full anchorage would be achieved by other means. If no plate is required a single bar should be formed into a U shape and placed directly over the bearing with 15 mm to 20 mm cover. The size of this bar should not be greater than 25 mm diameter in order to achieve the correct position. It is important that the internal radius of the bend should be checked for bursting forces. If more than one bar is required the vertical separation distance should not exceed 50 mm.

Although codes of practice do not require the steel $A_h$ to carry flexural forces, they will undoubtedly exist if the nib is long, say greater than one effective depth. It is therefore worth checking that the bars provided will resist any bending moment at the root of the nib.

According to BS 8110 (Part 1, equation 50) [4.2] the internal radius $r$ of a bar of diameter $\phi$ which extends for more than four diameters beyond the point at which it is required to resist the full force $F$ is given by:

$$r = \frac{F(1 + 2(\phi / a_b))}{2 f_{cu} \phi} \qquad (4.24)$$

where $a_b$ = the distance from the inside face of the bar to the nearest free surface, or the clear distance between bars.

Cover to the end of the beam and to the internal root of the pocket should be not less than 25 mm. The area of diagonal bars $A_d$ is given by:

$$A_d = \frac{0.5 V \sec \theta}{0.87 f_y} \qquad (4.25)$$

where $\theta$ is usually 40° to 45°.

If truss action is to develop, these bars must be fully anchored in both the top and bottom of the beam. The bend at the top is usually 135° and so the bending radius must not be contravened because the concrete is already in a 'high risk' zone and localized effects must not be allowed to exacerbate the situation. For this reason T 25 bars are considered to be the largest practical bar size for $A_d$. The diagonal bar must continue a full anchorage length from the root of the recess. The cover to this bar is critical; 25 to 30 mm being the correct cover to use. A small triangular chamfer, say $25 \times 25$ mm, is often used to reduce the possibility of cracks developing at the internal corner.

The remaining $0.5 V$ is carried by the first diagonal compressive strut, assumed to be inclined at $\alpha$ to the horizontal. The compression in the strut is:

$$C_1 = 0.5\,V\cosec\alpha \tag{4.26}$$

The limiting compressive force in this strut is given by:

$$C_1 = 0.4\,f_{cu}\,b\,0.5\,d_h\cos\alpha \tag{4.27}$$

as shown in Fig. 4.28(b). This ensures that not more than half the effective depth is in compression. If the compression is exceeded, two courses of action are possible:

- either the depth of the recess is reduced or $d_h$ is increased, or
- the 40 per cent ductility allowance is called upon and the proportion of the force carried by the compressive strut is reduced to 0.42 $V$ (with a corresponding increase in the force carried by the diagonal bars = 0.58 $V$).

The vertical force $V$ gives rise to a lateral horizontal bursting force $H_{bst}$ across the end of the beam, which may be obtained from BS 8110, Part 1, Table 4.7 'end block' theory – the most appropriate method [4.2]. Thus $H_{bst} = \zeta\,V$, where $\zeta$ is the bursting force coefficient, typically 0.20 to 0.23 for $b_p/b = 0.2$ to 0.4. The BS table is reproduced here in Table 4.10. Linear interpolation is used.

**Table 4.10** Bursting force coefficients to BS 8110 (code uses $y_{po}/y_o$ instead of $b_p/b$).

| $b_p/b$ | 0.3 | 0.4 | 0.5 | 0.6 | 0.7 |
|---------|-----|-----|-----|-----|-----|
| $\zeta$ | 0.23 | 0.20 | 0.17 | 0.14 | 0.11 |

The derivation of the above data may be seen from Fig. 4.29 where the dispersal of forces away from a bearing plate may be given by the resolution of the triangulation of forces. The angle subtended to the direction of load is given by:

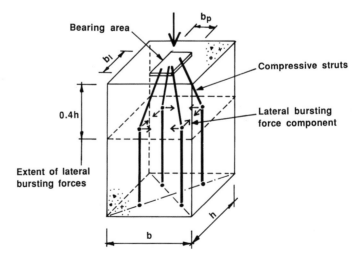

**Fig. 4.29** Dispersal of forces above bearing plates.

$$\theta = \frac{\sqrt{2}}{4} \frac{b - b_p}{0.4b}$$

(4.28)

The tensile bursting force coefficient $\zeta$ between two such struts is given by:

$$\zeta = \frac{\theta}{2\sqrt{2}} = 0.312\left[1 - \frac{b_p}{b}\right]$$

(4.29)

The results are given in Table 4.11 which are in good agreement with those in Table 4.10. In the beam end, horizontal hairpins are provided such that:

$$A_{bst} = \frac{H_b}{0.87 f_y}$$

(4.30)

across the end face of the beam.

PCI recommendations [4.15] give $A_{bst} = A_h/2$, and here the result is about the same as carrying out the above analysis. Part of this steel may be consumed as part of the U-bars provided for $A_h$ above. However, where $A_h$ need only be provided directly above the bearing, bursting steel must be positioned over the full height of the end of the beam. Thus at least two, and preferably three, hairpin bars should be provided. T 8 or T 10 rebars at 50 mm centres are the norm. The bars should extend across the full face of the beam and enclose any vertical bars. The anchorage length, which is standard, should start at 25 mm beyond the face of the recess. Two or three vertical stirrups should be used to support the hairpins over the top of the bearing plate. The area of the vertical bar in each face should be not less than $A_{bst}$.

The horizontal component of the force taken by bars $A_d$ is given by $H = 0.5\,V \tan \theta$. This is a compressive force in the top of the beam. The horizontal component of the force $C_1$ is $H = C_1 \cos \alpha$. The steel required for the total $H$ is given by:

$$A_s' = \frac{H}{0.87 f_y}$$

(4.31)

The bars $A_s'$ must not be less than the minimum percentage steel for concrete in compression, and they must be surrounded by links up to the end of the beam in order to be fully effective. The bars should extend to within a cover distance from the end of the beam and be lapped to the diagonal bars a distance equal to 1.4 times the basic lap length. Occasionally the top bars are welded to the first vertical link to provide the necessary anchorage at the end of the cage.

To complete the shear cage the first full depth vertical links $A_{sv1}$ designed using the truss action, shown in Fig. 4.28(a), should be placed at one cover distance from the inside face of the recess. The area of the links is:

**Table 4.11** Bursting force coefficients using strut and tie methods.

| $b_p/b$ | 0.3 | 0.4 | 0.5 | 0.6 | 0.7 |
|---|---|---|---|---|---|
| $\zeta$ | | 0.22 | 0.19 | 0.16 | 0.13 | 0.09 |

$$A_{sv1} = \frac{0.5\ V}{0.87 f_y} \tag{4.32}$$

Note that the concrete shear stress $v_c$ is ignored.

It is likely that pairs of links are required. The links should **enclose** the diagonal bars $A_d$, the hairpin bars $A_{bst}$, and the tie bars $A_h$. Although in theory the full vertical shear resistance of the section does not depend entirely on these first links, they are nevertheless provided to ensure the gradual transfer of shear from the truss to stirrup actions.

The vertical force in these links is equilibrated by a second compressive strut $C_2$, inclined at 45° to the horizontal, where:

$$C_2 = 0.5\ V \cosec 45° \tag{4.33}$$

The limiting value of $C_2$ is given by:

$$C_2 = 0.4 f_{cu}\ b\ 0.5\ d \cos 45° \tag{4.34}$$

As before, the horizontal component of the force at the bottom end of $A_d$ produces a tension $H = 0.5\ V \tan \theta$. In addition to this the horizontal component of the force $C_2$ is $H = C_2 \cos 45°$. The steel required for the total $H$ is given by:

$$A_s = \frac{H}{0.87 f_y} \tag{4.35}$$

The bars $A_s$ are, of course, the same bars used in the curtailment of flexural steel.

Finally, where the diagonal bar meets the bottom steel a second set of links $A_{sv2}$ should be provided to carry the full shear force $V$ within the nodal distance $L_{sb}$ so that:

$$A_{sv2} = \frac{V}{0.87 f_y} \tag{4.36}$$

Note that the nodal distance $L_{sb}$ is much less than that which would be calculated by applying anchorage bond stresses. This is because it is assumed that the force is transferred not only by bond, but also by friction between the links which are tied tightly to the inclined bar. Beyond $L_{sb}$, the development of shear resistance is subsequent catered for by the links and concrete according to BS 8110, Part 1, clause 3.4.5 [4.2].

Where the recess is smaller, about one-third of the depth, a diagonal compressive strut develops alone instead of the tension tie above. Figure 4.28(b) shows the structural model for this second case. Note the differences compared with Fig. 4.28(a). The similarities are that horizontal hairpin bars are placed directly over the top of the pocket as a restraint against lateral bursting forces $F_{bst}$. The bars are held in position by an equal amount of fully anchored vertical steel. They form the basis of the shear reinforcement cage in which the spacing of the links are gradually increased as in normal reinforced concrete design.

The main differences with the previous case are that the net depth of concrete above the bearing surface is sufficiently large to enable a compressive diagonal strut $C_3 = V \cosec \alpha$ to form. Here $\alpha$ depends on the geometry of the recess, and is usually about 60°–70°. The simple resolution of forces enables the reinforcements of area $A_s'$, $A_h$, $A_{bst}$, $A_{sv}$ and $A_s$ to be determined as before. The main difference here is that the longitudinal bars $A_h$ carry an additional force to the previous case, the horizontal component of the diagonal force $V \sec \alpha$. Thus if the peripheral tie force is included, then:

$$A_h = \frac{\mu V + F_t + V \cot \alpha}{0.87 f_y} \qquad (4.37)$$

Proper anchorage is required to carry the forces in the diagonal truss member by ensuring that the forces in the horizontal steel $A_h$ is fully active. This means that the end of the horizontal bar $A_h$ may have to be welded to the inside of the bearing plate or fully anchored as discussed above. The remote end of the bar enters into an area of compression as shown in the figure.

Cook [4.13, 4.21] developed a non-linear finite element program called 'FIELDS' which uses compression field theory to study the strut and tie action in these, and similar types of joints, i.e. corbels (see Fig. 4.22). The dashed lines represent principal stress trajectories. Full scale testing on precast beams with deep recessed ends, i.e. 350 mm in a total beam depth of 600 mm, has verified the computational work and shows that reinforcing beams using the strut and tie analogy gives conservative estimates of the ultimate shear resistance. Factors of safety against ACI code equations are about 1.2, but the failures occur in a very ductile manner.

A detailing problem occurs at point X in Fig. 4.28(b) where proper anchorage is required for the longitudinal steel and vertical links. The first link will be displaced beyond its effective position because of the radius of the hook required to anchor the reinforcement at the end of the recess. There are two possible solutions to this:

- To weld an anchor plate or angle to the end of the main steel as shown in Fig. 4.28(b). This would require fire protection as is drawn here.
- To provide an additional straight bar in the corner of the cage for the sole purpose of providing a link hanger. This bar would be additional to the curtailment requirements for the main longitudinal reinforcement.

### 4.3.11   Hanging shear cages for wide beams

The analysis considered in Section 4.3.10 is used in narrow beam design where it is not possible to carry the end shear forces in the plane of the connection. In wider beams, as shown in Fig. 4.30, where the net cross-section of concrete at point B is sufficient to carry the shear force, hanger steel may be used to transfer the end reaction from A to B. In this case the diagonal bars $A_d$ in Fig. 4.28(a) are replaced with hanger bars $A_{dd}$. The most important section occurs in inverted-tee beams where the breadth of concrete at section z-z in the figure is small, i.e. less than 150 mm. Referring to Fig. 4.30, the area of hanger steel is given by:

$$A_{dd} = \frac{V}{0.87 f_y \cos \theta_x \cos \theta_y} \qquad (4.38)$$

where $\theta_x$ and $\theta_y$ are the inclination of the bar to the vertical in the x and y planes. Bars inclined to both axes are *extremely* difficult to bend, particularly where accuracy is very important, as is the case here. Ideally the hanger bar(s) should be vertical in the y plane for ease of cage manufacture.

In practice this reinforcing method may only be used in pocketed beams where the breadth of the beam either side of the pocket is at least 200 to 300 mm depending on the shear force. Thus the overall breadth of the beam is likely to be at least 600 mm.

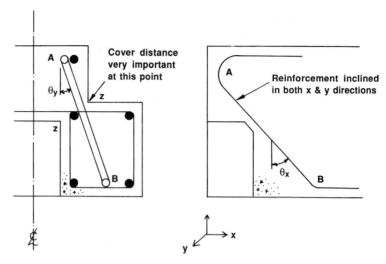

**Fig. 4.30** Hanger reinforcement in wide pocketed end beam.

## EXERCISE 4.5   Beam end shear design

Beam end shear reinforcement is required in a 600 mm deep by 300 mm wide beam to carry an ultimate shear force of 200 kN. The shape of the end of the beam is to be recessed across the full width of the beam to a depth of 200 mm and length 130 mm. A steel bearing plate is to be used in the horizontal bearing surface. The width of the supporting steel billet is 80 mm.

Given the following data, design the end shear reinforcement and bearing plate. The peripheral tie force is taken care of elsewhere. Use $f_{cu} = 40$ N/mm$^2$, $f_y = 460$ N/mm$^2$ (250 N/mm$^2$ for welded bars), $p_y$ plate = 275 N/mm$^2$ for grade 43 steel; $f_{weld} = 215$ N/mm$^2$ × 6 mm CFW. Cover to links = 25 mm.

### *Solution*

#### Bearing plate

$b/3 < bp < (b - 100)$ mm or 0.4 $b$.

Try $b_p = 100$ mm.

If a steel plate is to be used the design ultimate average bearing stress = $0.8 f_{cu} = 32$ N/mm$^2$:

$$b_1 \geq \frac{200 \times 10^3}{100 \times 32} = 62.5 \text{ mm}$$

Allow 20 mm end tolerance plus a further 10 mm fixing tolerance:

use $b_1 = 100$ mm.

Check the maximum possible bearing stress assuming triangular stress distribution (a conservative assumption because the minimum stress will not be zero):

$$= \frac{200 \times 10^3}{100 \times 100} \times 2 = 40 \text{ N/mm}^2 \approx f_{cu} \text{ for satisfactory confined concrete}$$

**Plate thickness**

The plate is subjected to double shear (design stress $0.6\,p_y$) from steel bearing corbel. Use effective bearing length of 80 mm (allowing for 10 mm tolerances and site fixing deviations).

$$t > \frac{200 \times 10^3}{2 \times 0.6 \times 275 \times 70} = 8.65 \text{ mm}$$

Also the plate is subjected to an axial tension force $= \mu V = 0.4 \times 200 = 80$ kN.

Cross-sectional area of plate $= \dfrac{80.0 \times 10^3}{275} = 291 \text{ mm}^2 \qquad t > \dfrac{291}{100} = 2.9 \text{ mm.}$

*Use $100 \times 100 \times 12$ plate.*

**Reinforcement**

Using the compressive strut and tensile tie analogy, shown in Figs 4.28(b) and 4.31(a), it is first necessary to determine the angle $\theta$ of the primary strut $C_1$, which from Fig. 4.31(a) is:

$$\theta = \tan^{-1}\left(\frac{110}{340}\right) = 18°$$

Thus

$$\text{Force } C_1 = \frac{V}{\cos\theta} = \frac{200}{0.95} = 210.3 \text{ kN.}$$

The width of the compressive strut $w_1$ must be checked to ensure that it does not extend beyond the end of the beam. The depth of the compressive block is also limited to $d_h/2$, where $d_h$ is the effective depth to the horizontal tie bars attached to the plate. Referring once more to Fig. 4.31(a), $d = 380$ mm, from which $w_1$ max $= (380/2)\sin\theta = 59$ mm. The actual value of $w_1$ is given by:

$$w_1 > \frac{210.3 \times 10^3}{300 \times 0.4 \times 40} = 44 \text{ mm} < 59 \text{ mm}$$

The effect of the reaction creates two types of tensile forces. The first is a lateral bursting force across the end of the beam and is a function of the width of the plate divided by the width of the beam, i.e. end block theory mentioned in the text. Using $y_{po}/y_o = 0.33$ (see Table 4.10), the bursting force is given by $F_{bst} = 0.22 \times 200 = 44$ kN, and the area of the horizontal bars to resist this force is:

$$A_{bst} = \frac{44 \times 10^3}{0.87 \times 460} = 110 \text{ mm}^2$$

*Use three no. T 8 bars at 100 mm centres,* i.e. within the lower half of the end face of the beam. Also provide 110 mm$^2$ of steel vertically in the form of closed links, use 2 T 8 links.

The second force is a longitudinal tie force as given by $F_h = \mu V + V \tan\theta$. Thus:

$$A_h = \frac{(0.4 \times 200 + 200 \times 0.325) \times 10^3}{0.87 \times 250} = 667 \text{ mm}^2$$

*Use two no. R 25 bars welded to the plate using electrodes grade E43.*
The length of the bars should be a bond length of 31 diameters = 775 mm. The length of a 6 mm double-sided fillet weld required to resist the force of 145 kN distributed equally between two bars is:

$$l_w = \frac{145 \times 10^3}{4 \times 6 \times 215} = 28 \text{ mm} + 12 \text{ mm run outs. Use 50 mm weld length.}$$

The vertical force *V* in the stirrups gives:

$$A_{sv} = \frac{200 \times 10^3}{0.87 \times 460} = 500 \text{ mm}^2$$

*Use three no. T 12 stirrups at 50 mm centres*, the first of which should be placed as close to the end of the beam as possible, i.e. one cover distance. This means that two additional bars (say two T 12 bars) must be placed in the bottom corners of the beam to carry the first stirrup.

The horizontal compressive force *H* in the top of the beam is $H = V \tan \theta = 65$ kN, and $A_s' = 162$ mm$^2$.
*Use two no. T 12 bars.*
Length of bar from nodal point = 32 × 12 = 385 mm, or length of bar from end of beam = 385 + 160 = 545 mm.

The maximum diagonal compressive force $C_2$ is given by $V/\sin 45° = 282$ kN. Using the method as before, the maximum effective width of concrete in compression is (550/2) sin 45° = 194 mm, and the actual width is 58 mm and satisfactory.

Finally, the tension in the bottom of the beam is $T = V$ (because the compressive strut $C_2$ is assumed to act at 45°). Hence $A_s = 500$ mm$^2$.
*Use two no. T 20 bars with a full hook end of internal radius 60 mm.*
Length of bar from nodal point = 32 × 20 = 640 mm. These bars should be properly anchored to a corner angle using a fillet weld designed as before (for direct tension), or the bars should be fully anchored using a standard end hook for simply supported beams. Note that the area of this steel may be increased due to the curtailment requirements of the main flexural reinforcement. Figure 4.31(b) shows the reinforcement cage.

**EXERCISE 4.6**

Repeat Exercise 4.5 where the depth of the halving joint recess is increased from 200 mm to 350 mm.

**Solution**

The size of the bearing plate will be as before. The reduction in the half joint depth will create diagonal tension across the face of the joint as shown in Fig. 4.28. The shear stress at this point is $v = 2.90$ N/mm$^2$ and, when combined with the horizontal tensile stresses and the stress concentration at the sharp 90° corner, is surely going to cause a diagonal failure. One way of reducing the stress concentration is to chamfer the corner with a 45° 20 mm × 20 mm fillet.

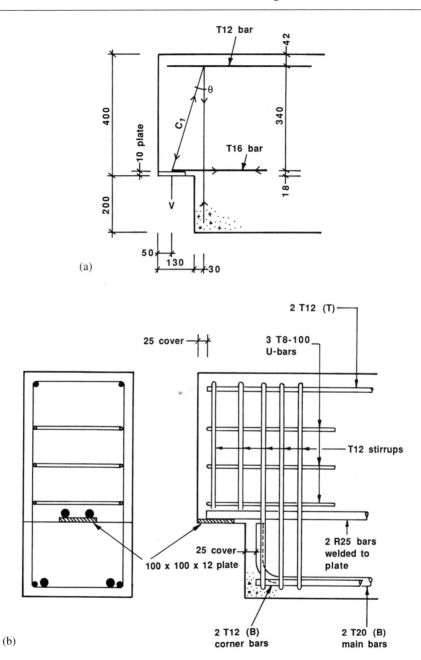

**Fig. 4.31** Details to Exercise 4.5. (a) Strut and tie forces used and (b) reinforcement cage.

**Reinforcement**

Referring to Fig. 4.32, it is possible to place the diagonal bar at 45° to the direction of the force *V* to resist the diagonal force resulting from 0.5 *V*. Hence:

$$F_{\rm d} = \frac{100}{\sin 45°} = 142 \text{ kN, from which } A_{\rm d} = 354 \text{ mm}^2$$

The preference is to use bars which may be bent easily and will not violate the bearing stresses inside bends (BS 8110, Part 1, clause 3.12.8.25 [4.2]). Therefore use four no. T 12 bars at 70 mm apart, in preference to two no. T 16 bars. To satisfy bearing stresses, where $a_{\rm b} = 60$ mm and $F_{\rm bt} = 142/4 = 35.5$ kN per bar, the internal bar radius should be greater than:

$$r = \frac{35.5 \times 10^3 (1 + 2(12/60))}{2 \times 40 \times 12} = 52 \text{ mm}$$

The bars should be fully anchored a full bond length of $32 \times$ diameter = 384 mm from the corner of the halving joint or the nodal point as shown in Fig. 4.32.

The design for lateral bursting and vertical shear is as before. The design for longitudinal tie steel anchored to the plate is given by $F_{\rm h} = \mu V$. Thus:

$$A_{\rm h} = \frac{0.4 \times 200 \times 10^3}{0.87 \times 250} = 367 \text{ mm}^2$$

*Use two no. R16 bars* welded to the plate in the same manner as before.

The weld leg length may be kept at 50 mm. Anchorage length = 496 mm. The bars are to be positioned on the inside of the inner diagonal bars.

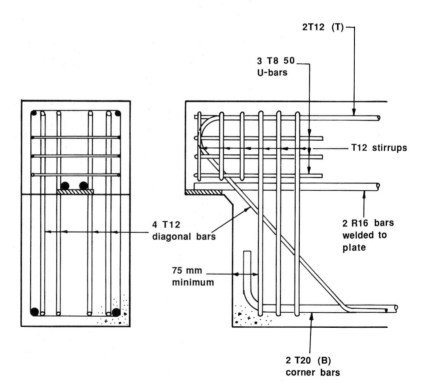

Fig. 4.32 Diagonal reinforcement in half joint shear cage in Exercise 4.6.

### 4.3.12    Prefabricated shear boxes

An alternative to the reinforced shear cage is the so-called 'shear-box' approach in which a solid plate, rectangular hollow section (RHS) or other, structural steel section projects from the end of the beam (see Figs 4.33 to 4.35). Shear boxes are necessary in situations where the depth of the beam becomes prohibitively large because of the large amount of reinforcement required in the end cage. Depending upon the circumstances a steel box section or two channel sections back-to-back can reduce the depth of the beam by 100 mm. This may be greater in L shaped beams where non-symmetrical boxes form part of the upstand. A typical RHS section size of $150 \times 100 \times 6$ (grade 43 steel) will carry shear forces in the region of 250 kN. About 75 mm cover distance is required to the top of the insert, and links must pass over the top of the insert in order to carry tensile forces to the bottom of the beam. The minimum bearing length is 60 mm.

Design recommendations are based on adequate bearing stresses both in the plate and concrete beam, the prevention of spalling, bursting and splitting, and an adequate tie back in the concrete beam. The ultimate shear capacity of the section is based on the shear capacity of the shear box itself, and is gradually transferred into the rc beam. Tie

**Fig. 4.33** Photograph of a beam shear box.

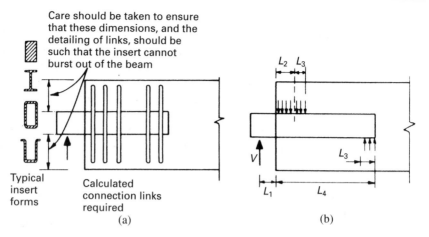

**Fig. 4.34** Prefabricated beam shear box design (taken from [4.5]).

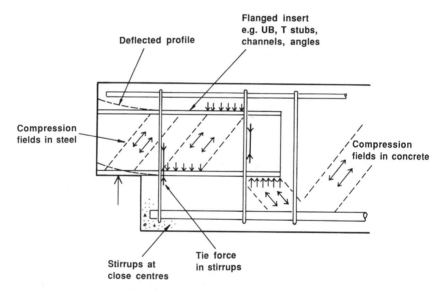

**Fig. 4.35** Beam shear box design principles.

back forces are distributed into the concrete beam either by an appropriate concentration of vertical stirrups, bent bars, or by welding a wide plate (or similar) to the bottom of the shear box.

### 4.3.12.1 Wide box design

The wide box design is based on three-point bending, and is therefore statistically determinate. The section is completely filled by concrete and surrounded by rc to prevent local wall buckling, etc. The concrete is otherwise ignored up to a point near to the end of the box where truss action takes over as in a normal beam.

End reactions $V$ are transferred to the box either directly, providing the local bearing stresses allow and the connection is stable in the temporary fixing condition, or through a flat plate welded to the bottom of the box. If the position of the box is near to the top of the beam, i.e. less than 200 mm, the box is restrained vertically and prevented from bursting out of the top of the beam by a tension hanger in the form of reinforcement of plate straps, as shown in Fig. 4.36. The strap, which generates a force $T$, may consist of a bent plate, 6 mm thick $\times$ 50 mm to 100 mm wide depending on the required capacity, welded to a bottom bearing plate. A compressive strut force must be allowed to develop above the bottom plate acting at 45° at least to the horizontal. Thus the 'clearance' between the underside of the wide shear box and the bottom plate must be at least equal to the length of the strap, plus a little extra, say 25 mm.

Alternatively, rebars may be welded to the sides (or bottom) of the box and provided with a full anchorage length and correct bend radius (Equation (4.24)). In no circumstances should roughened surfaces of the steel box be used to generate shear friction – a ductile, mechanical connection is required.

The equilibrium compression force $C = T - V$ is provided at the remote end of the box, either by direct bearing or through an additional bearing plate. The line pressure may be taken as $0.8 f_{cu} b_p$ providing that the concrete beneath the bearing is confined laterally. Links are provided in the beam at a spacing of not more than 150 mm in readiness to carry the shear force as explained above, and these are usually more than adequate for this purpose. Taking moments about the centre of the tension strap, then referring to Fig. 4.37.

$$V = \frac{0.8 f_{cu} b_p L_3 (L_4 - 0.5 L_3)}{L_1} \tag{4.39}$$

The length of the box $(L_4 + L_1)$ is in the order of 500 mm for connector capacities of about 250 kN, reaching 700 mm for 400 kN capacity connectors. Resolving vertically:

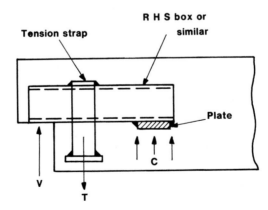

C : Compression under plate
T : Tension carried by strap
V : T-C

**Fig. 4.36** Beam shear box with additional steel plate strap hangers.

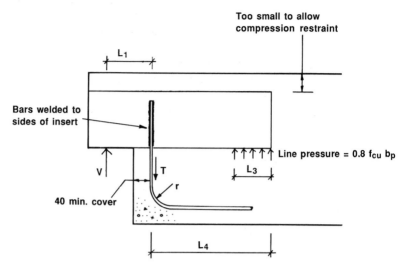

**Fig. 4.37** Beam shear box with additional rebar hangers.

$$V = T - 0.8 f_{cu} b_p L_3 \tag{4.40}$$

Hence $T$ may be computed. The area of reinforcement is

$$A_s = \frac{T}{0.87 f_y} \quad \text{for rebar}$$

or

$$A_s = \frac{T}{p_y} \quad \text{for steel plate} \tag{4.41}$$

The thickness of the plate at the bottom of the strap is based on the shear capacity of the plate, and a sufficient thickness to be welded to the strap, typically 6 mm. Figures 4.36 and 4.37 show examples of the completed shear box using steel straps and rebars, respectively.

**EXERCISE 4.7　Beam end shear box design**

Design a RHS or square hollow section (SHS) shear box to carry an ultimate end reaction of 250 kN in a 300 mm wide reinforced concrete beam. Determine the minimum depth of beam required to support the shear box, and design the shear reinforcement. Assume that the beam bearing is on to a mild steel plate. Use two methods for the tie force: (a) rebars, (b) straps.

Use $f_{cu} = 40$ N/mm$^2$, $f_{yv} = 460$ N/mm$^2$, $f_y = 250$ N/mm$^2$, $p_y = 275$ N/mm$^2$, $p_{bearing} = 190$ N/mm$^2$, $p_{weld} = 215$ N/mm$^2$. Cover to reinforcement $= 40$ mm.

*Solution*

Breadth of box $= b/3 = 100$ mm.

Length of bearing $= 250 \times 10^3/190 \times 100 = 13.2$ mm, plus 10 mm fixing tolerance $= 25$ mm.

Use minimum recommended length = 60 mm.
Beam rebate length = 60 = 20 mm grouting gap = 80 mm.

(a) Using reinforcement hanger (see Fig. 4.37)

$L_1 = 80 - 60/2 + 40$ cover + say 12 = 102 mm.

Try $L_4 = 300$ mm.

Line pressure under box = $0.8 \times 40 \times 100 = 3200$ N/mm.

Moments about tie $250 \times 10^3 = 3200 L_3 (300 - 0.5 L_3)/102$.

Hence $L_3 = 27.8$ mm, say 28 mm.

$C = 3200 \times 28 \times 10^{-3} = 89.6$ kN.

$T = 250 + 89.6 = 339.6$ kN.

$A_s = (339.6 \times 10^3)/(0.87 \times 250) = 1560$ mm$^2$.

*Use two no. R 32 bars (1608) welded to side of box.*

(1) Bend radius to R 32

Cover distance + diameter = 100 mm.

$$r = \frac{169.8 \times 10^3 (1 + 2\,(32/100))}{2 \times 40 \times 32} = 108 \text{ mm, say } 110 \text{ mm.}$$

Minimum required depth beneath box = 110 + 32 + 10 link + 40 cover = 192 mm, use 200 mm.

(2) Confinement of concrete beneath box

$F_{bst} = 0.22 \times 89.6 = 19.7$ kN.

$A_{bst} = 49$ mm$^2$.

*Use two no. T 8 bars (100) at 50 mm centres at end of box.*

(3) RHS design

$M_{max} = 250 \times 102 = 25\,500$ kN mm.

Plastic modulus $S_{xx} > 25\,500/275 = 92.7$ cm$^3$.

$V_{max} = 250$ kN.

Web area $d_t > 250 \times 10^3/165 = 1515$ mm$^2$.

*Use 100 × 100 8 RHS grade 43 steel ($S_{xx} = 99.9$, $d_t = 1600$)*

(4) Weld design to side of box

Maximum length available = 100 mm. Assume 8 mm leg length, then:

Weld length = $(339.6 \times 10^3)/(2 \times 2 \times 5.6 \times 215) = 70.5$ mm.

*Use 10 mm × 100 mm long CFW to tie bars.*

Shear links in concrete between tie and end of box, and beyond end of box for 1 × effective depth.

Total depth of beam = 50 top cover to box + 100 box + 200 bottom cover = 350 mm.

$b = 300$ mm, $d = 350 - 40 - 10 - 12 = 288$ mm.

$100 A_s/bd = 100 \times 1608/300 \times 288 = 1.86\%$.

$v = 250 \times 10^3/300 \times 288 = 2.89$ N/mm$^2$.

$v_c = 0.83$ N/mm$^2$.

$A_{sv} = ((2.89 - 0.99) \times 300)/(0.87 \times 460) = 1.42$ mm$^2$/mm = 1420 mm$^2$/m.

*Use T 10 links at 100 mm centres.*

(b)  Using plate strap hanger (see Fig. 4.36)

Because the plate strap occupies a greater length than a single rebar above, the lever arm $L_1$ is increased by a distance equal to half the length of the plate minus the radius of the bar. Using the tie force above as a guide, adding about 10 per cent extra for the additional lever arm, try 6 mm thick × 110 mm long strap.

$L_1 = 80 - 60/2 + 40$ cover $+ 110/2 = 145$ mm.

Try $L_4 = 300$ mm as before and increase if necessary.

Moments about tie $250 \times 10^3 = 3200\ L_3\ (300 - 0.5\ L_3)/145$.

Hence $L_3 = 40.5$ mm, say 41 mm.

$C = 3200 \times 41 \times 10^{-3} = 131.2$ kN.

$T = 250 + 131.2 = 381.2$ kN.

$A_{plate} = 381.2 \times 10^3/275 = 1386$ mm$^2$.

Length of strap $= 1386/2 \times 6 = 115.5$ mm.

*Use two no. 6 mm thick × 120 mm long straps.*

(1)  RHS design

$M_{max} = 250 \times 145 = 36\ 250$ kN mm.

Plastic modulus $S_{xx} > 36\ 250/275 = 131.8$ cm$^3$.

$V_{max} = 250$ kN.

Web area $2dt > 250 \times 10^3/165 = 1515$ mm$^2$.

*Use 150 × 100 × 6.3 RHS grade 43 steel ($S_{xx} = 148$, $2dt = 1890$).*

(2)  Weld design to side of box

Maximum length available $= 150$ mm. Assume 8 mm leg length, then:

Weld length $= (381.2 \times 10^3)/(2 \times 2 \times 5.6 \times 215) = 79$ mm.

*Use 8 mm × 100 mm long CFW to side strap.*

(3)  Width of bottom plate $= 100 + (2 \times 6) + 2 \times$ weld leg, say 8 mm $= 128$ mm, use 130 mm.

Bearing capacity above bottom plate $= 0.8 \times 40 \times 120 \times 130 \times 10^{-3} = 499.2$ kN $> 381.2$ kN required.

Shear thickness $t = (381.2 \times 10^3)/(2 \times 120 \times 165) = 9.6$ mm.

Use 130 × 120 × 10 mm thick bottom plate.

Weld leg length to bottom plate (assuming 8 mm run outs):

$= (381.2 \times 10^3)/(2 \times 2 \times 5.6 \times 215) = 79$ mm.

*Use 10 mm weld to bottom plate.*

(4)  Depth of concrete or clearance between RHS box and bottom plate $= 120 + 25 = 145$ mm.

(5)  Shear links in concrete between tie and end of box, and beyond end of box for 1 × effective depth

Total depth of beam $= 50$ top cover to box $+ 150$ box $+ 145$ clearance $+ 10$ plate $+ 50$ bottom cover $= 405$ mm, say 400 mm.

Assume two no. T 25 bars in end of flexural cage.

$b = 300$ mm, $d = 400 - 40 - 10 - 12 = 348$ mm.

$100\ A_s/bd = 100 \times 982/300 \times 348 = 0.94\%$.

$v = 250 \times 10^3/300 \times 348 = 2.40$ N/mm$^2$.

$v_c = 0.75 \text{ N/mm}^2.$

$A_{sv} = ((2.40 - 0.75) \times 300)/(0.87 \times 460) = 1.23 \text{ mm}^2/\text{mm} = 1230 \text{ mm}^2/\text{m}.$

*Use T 10 links at 125 mm centres.*

### 4.3.12.2   Narrow plate design

A narrow shear plate is defined as one where the width (or thickness) of the plate $t$ is in the range $0.1\, b < t < 75$ mm. Figure 4.38 shows three details of design methods using a narrow plate insert instead of the wider shear box. The narrow plate is used mostly with the welded plate column connector (see Fig. 7.42(c)) where the plate is fillet welded to a steel billet cast in the column. If a wide bearing plate is added to the end of the plate the beam may be connected to many of the other types of column connectors, e.g. the cleat type. In order to transfer the end reaction safely to the beam the narrow plate is supplemented with any of the following:

(a) additional bearing plates welded to the underside of the vertical plate at a distance of 75 mm from the end of the beam, and if necessary a plate welded to the top of the remote end of the plate (Fig. 4.38(a))

(b) additional (2 no. minimum) rebars welded to the sides of the plate at a centre distance of 75 mm from the end of the beam, and a bottom plate welded to the underside of the vertical plate (Fig. 4.38(b))

(c) rebars as (b), but with a vertical end plate and horizontal tie back steel bars welded to the bottom of the plate (Fig. 4.38(c)). This option is used where the length of the plate is restricted, for example by a service hole or check-out in the beam.

As with the wide shear box the narrow shear plate is designed in three-point bending, and is therefore statistically determinate. The plate is completely surrounded by concrete to prevent twisting and buckling, etc. The concrete is otherwise ignored up to a point just beyond the tie bars where truss action takes over as in a normal beam. Unlike the wide box connector which may be embedded in a deep concrete beam without additional ties, the plate *must* be restrained vertically and prevented from splitting through the concrete, no matter how deeply embedded it is, either by the additional bearing plates or by closely spaced links near to the reaction points.

Option (a) is used if the distance from the bottom of the beam to the underside of the plate is small, say less than about 75 mm, such that insufficient compression can develop over the top of the steel insert. In this context the concrete at the top of the beam should be at least capable of resisting punching shear and of generating a diagonal compressive strut to the steel in the top of the beam. The cover distance depends on the magnitude of $V$, but if the distance $L_4$ is large the end reaction is obviously small and a cover distance of about 150 mm is adequate.

Referring to Fig. 4.38(a), the depth of the plate $d$ is calculated from the greater of:

$$d = \frac{V}{0.6\, p_y t} \tag{4.42}$$

or

$$d = \sqrt{\frac{4V[L_1 + 0.5\, L_2]}{p_y t}} \tag{4.43}$$

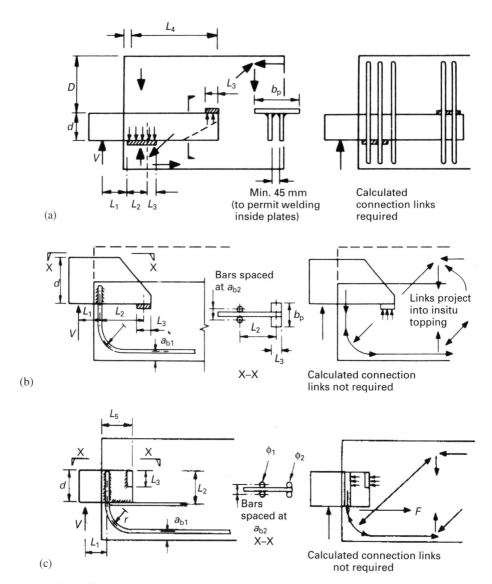

**Fig. 4.38** Narrow plate beam end design. (a) With additional bearing plate [4.5]; (b) with additional rebar and remote end plate [4.5] and (c) with additional remote end plate and rebar [4.5].

The equilibrium compression force $C = T - V$ is provided at the remote end of the plate. The bearing pressure may be taken as $0.8 f_{cu}$ providing that the concrete around the plate is confined laterally, although some designers prefer to restrict the stress to $0.6 f_{cu}$ to guard against cracking at the corner of the plate. Small diameter links are provided in the beam at a spacing of not more than 150 mm as explained above. The overall difference in the design compared with the wide box connector is directly related to the position of the plate in the beam, or more specifically the amount of concrete cover beneath the plate.

The first step is to choose a length $L_4$ for the narrow shear plate. As an initial guess, try $L_4 = 300$ mm for $V < 150$ kN, increasing to $L_4 = 500$ mm for $V < 300$ kN (assume $b > 300$ mm, and $f_{cu} > 40$ N/mm$^2$). The width of the bearing plate should be approximately $b_p = b/3$. Then:

$$V = 0.8\, f_{cu}\, b_p\, L_2 \tag{4.44}$$

Assuming that both plates have the same width, $L_3$ is calculated from equilibrium of moments:

$$L_3\,(L_4 - L_2 - L_3) = L_2\,(L_1 + 0.5\, L_2) \tag{4.45}$$

Check that the bearing stresses do not overlap such that $L_2 + 2\,L_3 < 0.9\,L_4$.

The force in the end plate must be upheld by steel stirrups $A_{sv}$ as follows:

$$A_{sv} = \frac{0.8 f_{cu} b_p (L_2 + L_3)}{0.87 f_{yv}} \tag{4.46}$$

equally spaced over a distance $L_2 + L_3 + 125$ mm from the end of the concrete. Notice that $0.8\, f_{cu}\, b_p\,(L_2 + L_3) > V$, and therefore the stirrups are required to carry a force greater than the end reaction.

The force at the end of the shear plate $0.8\, f_{cu}\, b_p\, L_3$ is resisted by a diagonal compressive strut $C$ acting at an assumed angle of $45°$:

$$C = \frac{0.8 f_{cu} b_p L_3}{\sin 45°} \tag{4.47}$$

and the horizontal component of that force $C \cos 45°$ is carried into the bottom reinforcement, whilst the vertical component $C \sin 45°$ is absorbed into the shear stirrups at the end of the plate which carry the full reaction $V$.

Where no remote end bearing plate is used, taking moments about the centre of the tension tie steel, and referring to Fig. 4.38(b):

$$V = \frac{0.6 f_{cu} t L_3 (L_4 - 0.5 L_3)}{L_1} \tag{4.48}$$

Typical sizes for the plate are in the order of 20 mm thick × 500 mm long for connector capacities of about 250 kN, reaching 25 mm × 700 mm for 400 kN capacity connectors. Resolving vertically:

$$V = T - 0.6 f_{cu}\, t\, L_3 \tag{4.49}$$

Hence $T$ may be computed. The area of reinforcement is:

$$A_s = \frac{T}{0.87 f_y} \tag{4.50}$$

Option (b) is also used where the narrow plate is projecting above the top of the beam, or is very close to the top, i.e. less than 50 mm cover. This design follows the methods used for the wide shear box, and the design principle is based on the so-called 'Cazaly hanger', developed by Lawrence Cazaly of Ontario, Canada, in which the top plate supports the bars which carry the shear to the beam. Tests on this device were carried out by Ife *et al.* [4.22]. In all cases the anchor rebars must be formed over large radii mandrels (typically 150 to 200 mm diameter for 25 mm diameter bars) to avoid excessive local bearing stresses causing diagonal tensile splitting directly beneath the steel section where the shear stress is large. Variations of the Cazalay hanger have been designed in which the anchor steel is connected to the underside of a channel or back-to-back steel angles. The bars are either fillet or butt welded to the steel sections (depending upon the force required) or are threaded and pass through holes in the section. The same design equations (Equations (4.44) through to (4.47)) are used here.

In using Option (c), there must be sufficient top and bottom cover to the plate to enable the reaction forces to generate compressive struts and bond strengths, respectively. If a bearing plate of length $X$ and width $b_p$ is used, Equation (4.48) is altered to:

$$V = \frac{0.8 f_{cu} b_p X (L_4 - 0.5X)}{L_1} \tag{4.51}$$

Hence $X$ is found. Then:

$$V = T - 0.8 f_{cu} b_p X \tag{4.52}$$

and the analysis proceeds as before.

Finally, factors of safety for the shear capacities of precast beams using shear cages and bearing plates, wide shear boxes, or narrow plates are usually obtained by individual manufacturers by full scale laboratory testing. Most of this work remains unpublished but it is known that the results show that ultimate failure is a function of the shear reinforcement as in ordinary rc despite excessive local cracking near to the bearing plate. Shear cages and boxes are usually rated in increments of 50 kN up to 300 kN, and at 100 kN or 150 kN thereafter. Shear cages and boxes are capable of transmitting end shear forces up to about 700 kN. At greater load magnitudes the practical construction of such a connection is difficult to achieve in the normal course of events.

## 4.4  COLUMNS SUBJECTED TO GRAVITY LOADS

The design of columns for overturning is given in Chapter 8.

### 4.4.1  General design

A column is specified where the ratio of the greater to the lesser cross-sectional dimension is less than four, otherwise the component is designed as a wall. Precast column design is no different to the design of ordinary rc columns and walls once all the aspects of manufacture, different types of structural connections, and temporary stability have been resolved. In fact, with a few isolated cases where columns have been prestressed axially to enable very long units to be pitched without (flexural) cracking, designs are based on the recommendations given in BS 8110 [4.2] for reinforced columns subjected to combined axial compression and bending (uniaxial or biaxial).

Precast columns and walls are manufactured horizontally, often in very accurate steel moulds. The standards of control are therefore greater than in vertically cast insitu work and congested arrangements or reinforcement, particularly at column splices and foundations, can be specified with confidence in the knowledge that full compaction of concrete and correct spacing of bars will always be achieved. It is also possible to precast a concrete column having up to ten per cent reinforcement (the recommended maximum value) at the level of the splices, although this quantity of reinforcement is rarely used in preference for a larger gross section. Figure 2.8 shows some reinforcement details in column sections. The characteristic cube strength of the concrete is usually 50 N/mm$^2$ but because of the early strength required for lifting in the factory, actual characteristic strengths are in the range 60 to 70 N/mm$^2$.

The design commences with an assessment of structural stability, and of the axial loadings and bending moments at each floor level. As shown in Figs 1.23 and 4.39(a) and (b), column bending moments are the result of eccentric loading in the connection. The eccentricity varies with the type of connection, corbel or haunch. Typical values are $e = (h/2 + 50)$ mm to $(h/2 + 150)$ mm.

In edge columns supporting a single beam the ultimate reaction $V$ produces a bending moment at the connector node $M = Ve$. At internal columns supporting a (near) symmetrical arrangement of beams the moment is obtained by the summation of moments at each side of the column. Patch loading is used where the maximum and minimum beam reactions on each side of the column are:

- $V_{max} = 1.4 \times$ dead $+ 1.6 \times$ live load
- $V_{min} = 1.0 \times$ dead.

The net overturning moment is obtained from the worst possible scenario when the construction tolerance $\Delta$ is added to the eccentricity of the greater load and deducted from the eccentricity of the smaller. Thus if the distance from the centroid of the column to the centre of the beam reaction is $e$, the net column moment is:

$$M_{net} = M_{max} - M_{min} = V_{max}(e + \Delta) - V_{min}(e - \Delta) \tag{4.53}$$

The net eccentricity is given by:

$$e_{net} = M_{net}/(V_{max} + V_{min}) \tag{4.54}$$

The analysis may now proceed in the same manner as for the single-sided beam.

A similar approach is adopted for three-way and four-way beam connections where bi-axial bending moments are present in the column.

The resulting bending moments are distributed in the column in proportion to the stiffness $EI/L$ of the column between adjacent floor levels, Fig. 4.39(b) (where the stiffness factors $4EI/L$ and $3EI/L$ assume equal storey heights), and the column is designed accordingly to BS 8110 [4.2] using the clauses reproduced below for column effective length factors and second order moments.

#### 4.4.1.1   *Design rules in BS 8110 for columns in precast structures*

Effective column height according to Part 2, Clause 2.5.

(a) Braced columns: the effective height for columns in framed structures may be taken as the lesser of:

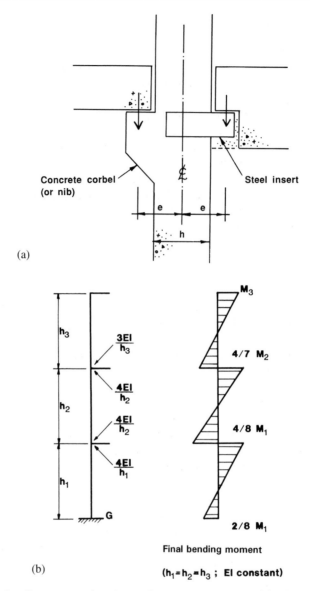

**Fig. 4.39** Bending moments in columns due to connector eccentricity. (a) Definition of eccentricity and (b) distribution of bending moments in columns.

$$l_e = l_o (0.7 + 0.05 (\alpha_{c1} + \alpha_{c2})) < l_o \qquad (4.55)$$

$$l_e = l_o (0.85 + 0.05 \alpha_{cmin}) < l_o \qquad (4.56)$$

(b)  Unbraced columns: the effective height for columns in framed structures may be taken as the lesser of:

$$l_e = l_o (1.0 + 0.15 (\alpha_{c1} + \alpha_{c2})) \qquad (4.57)$$

$$l_e = l_o (2.0 + 0.3\ \alpha_{cmin}) \tag{4.58}$$

where:

$\alpha_{c1}$ = ratio of the sum of the column stiffnesses to the sum of the beam stiffnesses at the lower end of a column

$\alpha_{c2}$ = ratio of the sum of the column stiffnesses to the sum of the beam stiffnesses at the upper end of a column

$\alpha_{cmin}$ = lesser of $\alpha_{c1}$ and $\alpha_{c2}$

$l_e$ = effective height of a column in the plane of bending considered

$l_o$ = clear height between end restraints.

In the calculation of $\alpha_{cmin}$, $\alpha_{c1}$ and $\alpha_{c2}$ only members properly framed into the end of the column in the appropriate plane of bending should be considered. The stiffness of each member equals $I/l_o$, where $I$ is the uncracked second moment of area for the section considered.

In specific cases of relative stiffness the following simplifying assumptions may be used:

(b) simply-supported beams framing into a column: $\alpha_c$ to be taken as 10

(c) connection between column and base designed to resist only nominal moment: $\alpha_c$ to be taken as 10

(d) connection between column and base designed to resist column moment: $\alpha_c$ to be taken as 1.0.

For columns in precast concrete structures effective length factors are based on $\alpha_c = 10$ for pinned connections and actual frame geometry. The value of $\alpha_c = 10$ leads to $\beta = 1.35$ in a braced structure, and so $\beta = 1.0$ is used, and $\beta = 2.3$ in an unbraced structure. Although the designated value of $\alpha_c = 10$ for the relative stiffness of pin-jointed connections is somewhat arbitrary, it is adopted to justify the design of unbraced columns.

(c) Deflection induced moments in solid slender columns, Part 1, clause 3.8.3.

... account has to be taken of the additional moment induced in the column by its deflection. The deflection of a rectangular or circular column under ultimate conditions may be taken to be:

$$a_u = \beta_a\, K\, h \tag{4.59}$$

where

$$\beta_a = \frac{\left( (l_e/b)^2 \right)}{2000} \tag{4.60}$$

where $K$ = a reduction factor that corrects the deflection to allow for the influence of axial load, $h$ = overall depth of a column in the plane considered, and $b$ = smaller dimension of a column.

In some precast structures there are only beams (except at gable ends or around lifts or stairwells) in one of the two orthogonal directions. In this case the clear height is taken to the centre depth of the first floor slab based on the ground to first floor height. Subsequent upper storey columns are designed using the same effective length factor as before because the column is structurally continuous in these types of structures. Bending moments due to eccentric loading, horizontal forces and second-order deflections ($a_u$) effects are combined to give the most onerous design condition. Deflection-induced moments $M_{add}$ are distributed throughout the structure in proportion to the stiffness of all the columns.

This assumes that the floor plate – a rigid diaphragm in any multi-storey structure – is capable of transmitting small prop forces which are the result of reactions from differential additional moments in various parts of the structure. An example of the application of this analysis is where large additional moments are created in only a few columns surrounding a roof level lift motor or plant room. The natural structural response of the columns below this level ensures the full distribution of additional moments throughout the structure by generating a prop force in the floor and thereby slightly altering the notional magnitude of $a_u$ (i.e. initial value) at each floor level.

The resulting bending moments are usually large and this precludes the use of column splices and pinned bases. The magnitudes of the forces and moments also restrict the capacities of most columns to three storeys – particularly if the ground to first floor height exceeds about 3.5 m. Here the design value of the deflection-induced eccentricity (according to BS 8110 [4.2]) is in the order of 80 mm to 110 mm for 300 mm to 400 mm minimum column dimensions, respectively. It is difficult to visualize the reality of this without the column experiencing cracking far in excess of that which is normally associated with ultimate failure. Coupled with the bending moments generated by horizontal forces the dimensions of the columns easily exceed architectural expectations for the sizes of columns in three-storey buildings, i.e. $300 \times 300$ mm. This leads to the use of braced or partially braced structures.

### 4.4.2   Column in braced structures

Braced columns in so-called 'no-sway frames' (small-sway frames would be more precise) are checked for slenderness; an effective length factor of 1.0 being used in all upper floors and 0.9 between a fixed base and first floor. There is still inadequate analytical or experimental evidence to show the beneficial effects of moment rotation stiffnesses in the beam-to-column connection in reducing column effective lengths. Base plates and pocket foundations may be considered as fully moment resisting, but the stiffness coefficient $\alpha_c = 1.0$ means that they are not fully rigid in the interpretation of the code. Sometimes the base is often considered as a pinned connection (despite the moment fixity) in order to simplify column manufacture and foundation design.

Braced slender columns are analysed in the usual manner taking into consideration the second order deflection (or additional) bending moments as appropriate. The design axial load $N$ is considered in combination with eccentric connector moments $Ve$.

### 4.4.3   Columns in unbraced structures

In unbraced structures stability is shared between the columns in proportion to the flexural stiffness of each column and degree of moment fixity provided at the foundation.

Most unbraced columns are slender, although deep columns acting as 'wind posts' are deliberately proportioned so as to be designed as short. The effective length factor for cantilever columns in pin-jointed structures founded on moment resisting bases is 2.3.

Bending moments due to eccentric loading, horizontal forces and second-order deflections are combined to give the most onerous design condition.

### 4.4.4   Columns in partially braced structures

According to BS 8110 [4.2] columns in the unbraced part of the structure are designed as cantilevers with an effective length ratio of 2.3. This is a conservative value because some of the columns immediately adjacent to the stabilizing walls may have an effective length factor of 2.0. This is because their connection to the braced structure can be considered as infinitely rigid (although BS 8110 does not recognize such an end condition). A special case, shown in Fig. 4.40, is where columns are supporting raking steelwork, which is connected at its other end to a rigid (no-sway) part of the structure. These columns are designed as propped cantilevers with an effective length factor of 1.6.

An issue of particular concern arises in partially braced structures because the lower end conditions for the columns, which are not in direct contact with shear walls, is not

**Fig. 4.40** Example of propped cantilever column.

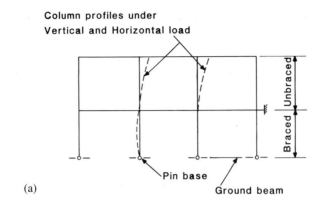

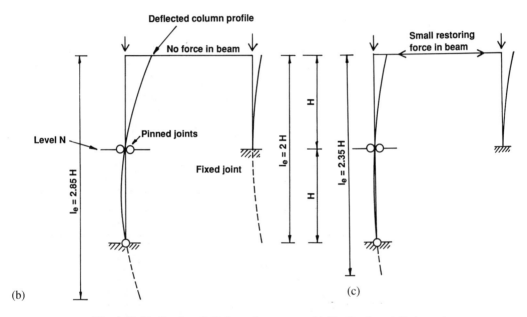

**Fig. 4.41** Idealized partially braced structures. (a) Idealized partially braced structure. (b) Column buckling considering each column separately and (c) column buckling in framed structure.

defined in BS 8110. The structure can be idealized as shown in Fig. 4.41. In Fig. 4.41(b) the deflected profile of a column is held in position but not in direction at level $N$, and a free cantilever up to level $(N + 1)$ is shown. The effective length of the column at level $(N + 1)$ is 2.85. Thus the true manner of slenderness-induced deflections would be as shown in Fig. 4.41(c) where the effective length of *all* columns is 2.35. This has been computed from critical buckling loads obtained using stability functions given by Elliott, Davies and Mahdi [4.23, 4.24] and following the work carried out by Cranston [4.25]. The shear forces at the base of the unbraced columns are carried in the floor plate to the stiffening elements in accordance with their stiffnesses and position. However, bending moments resulting from sway in the unbraced part are carried over into the braced part of the structure, diminishing to zero with distance to the level of the floor plate below. The effective length factor for the columns in this lower region is 1.0.

The situation where the ground floor area is free of walls for architectural reasons is not a common option for the following reasons:

(a) The columns will be founded in moment resisting bases. In the one-storey unbraced version the column effective length factor is 1.15 (because according to BS 8110, Part 2, clause 2.5 [4.2] this gives $\alpha_{c1} = 1$ and $\alpha_{c2} = 0$). In the two-storey unbraced version it is 2.3 in the ground-first floor column and 1.15 in the ground-second floor column. The major problem is that the columns surrounding the upper floor walls are heavily loaded axially because the overturning moments are concentrated there.

(b) This gives rise to very large second-order bending moments in the columns with the inherent problems that these bring to foundation design.

## EXERCISE 4.8    Column design

The four-storey braced column shown in Fig. 4.42 carries beams on either side in a symmetrical arrangement. The distance from the face of the column to the point load $V$ is nominally 60 mm. The construction allowance is 10 mm. The beam end reactions $V$ (kN) are given in Table 4.12.

Given that $f_{cu} = 50$ N/mm$^2$, steel $f_y = 460$ N/mm$^2$ and cover to 10 mm diameter links = 35 mm:

**Table 4.12** Beam end reactions $V$ (kN).

|        | Dead | Live |
|--------|------|------|
| Roof   | 80   | 27   |
| Floors | 108  | 90   |

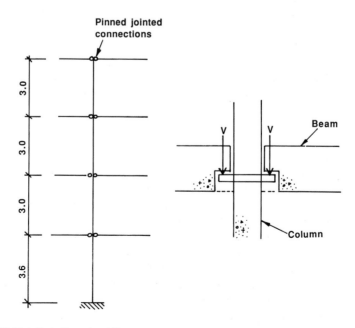

**Fig. 4.42** Details to Exercise 4.8.

- calculate maximum and minimum axial loads and moments
- draw bending moment and axial force diagrams
- design column reinforcement and column size
- check reinforcement for factory lifting if $f_{cu} = 15$ N/mm$^2$
- check site pitching if $f_{cu} = 30$ N/mm$^2$.

### Solution

Try $300 \times 300$ mm column, this being the smallest recommended size for a four-storey building.

Self weight $= 2.16 \times 12.6 = 27.2$ kN.

Maximum axial load = 1.4 dead load + 1.6 live load.

$$N_{max} = \begin{array}{l} 1.4\ (27.2 + 160 + (3 \times 216)) = 1169.3 \\ 1.6\ (54 + (3 \times 180)) \qquad\quad = \underline{\ \ 950.4\ } \\ \qquad\qquad\qquad\qquad\qquad\qquad 2120\ kN \end{array}$$

$N_{min} = 1.0 \times$ dead $\qquad\qquad = 835$ kN

$M$ due to min. eccentricity $= 0.05\ h$, $N = 0.05 \times 0.300 \times 2120 = 31.8$ kN m.

Column effective length factor for ground to first floor = 0.9.

$$l_e = \frac{0.9 \times 3600}{300} = 10.8 < 15, \text{ hence short column.}$$

Maximum bending moment occurs due to patch loading where $V = 1.0$ dead on one beam and $V = 1.4$ dead + 1.6 live on the other beam. Eccentricity of $V$ about centre line of column $e$:

$$= 0.5\ h + 60 \pm \text{construction allowance of 10 mm}$$
$$= 200 \text{ mm min and 220 mm max.}$$

Maximum moment $= 0.22\ (1.4$ dead $+ 1.6$ live$) - 0.200 \times 1.0$ dead.

At roof: $M = 34.15 - 16.00 = 18.15$ kN m
At floors: $M = 64.95 - 21.60 = 43.35$ kN m.

Moment distribution factors:

- at roof $\qquad\qquad\qquad\qquad = 1.0$ and 0
- at 3rd floor $= \dfrac{3}{3+4} \qquad = 0.428$ and 0.571
- at 2nd floor $= \dfrac{4}{4+4} \qquad = 0.5$ and 0.5
- at 1st floor $= \dfrac{3.6}{3.0+3.6} = 0.545$ and 0.455

with 50 per cent carry over to foundation.

The bending moment and axial force diagrams are given in Fig. 4.43. Maximum moment = 22.63 kN m at the 1st floor, where $N = 1069$ kN.

### Reinforcement design

$b = 300$ mm; $h = 300$ mm; $d = 300 - 35 - 10 -$ say $12 = 243$ mm.

$d/h = 0.81$. Use $d/h = 0.8$ given in BS 8110, Part 3, Chart 47 [4.2].

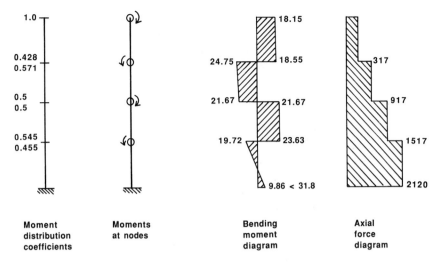

| Moment<br>distribution<br>coefficients | Moments<br>at nodes | Bending<br>moment<br>diagram | Axial<br>force<br>diagram |

**Fig. 4.43** Moments and axial forces in Exercise 4.8.

(a) Case 1 at foundation

$$\frac{N}{bh} = \frac{2120 \times 10^3}{300 \times 300} = 23.55 \text{ and } \frac{M}{bh^2} = \frac{31.8 \times 10^6}{300 \times 300^2} = 1.18$$

$A_{sc} = 1.15\% = 1035 \text{ mm}^2$

*Use four no. T 20 main bars and R 6 links at 240 centres.*

(b) Case 2 at first floor

$$\frac{N}{bh} = \frac{1069 \times 10^3}{300 \times 300} = 11.9 \text{ and } \frac{M}{bh^2} = \frac{23.62 \times 10^6}{300 \times 300^2} = 0.87$$

$A_{sc} = 0.4\% < \text{above.}$

**Factory lifting (see Section 10.2.2 before proceeding)**
Self weight of column plus 50 per cent suction and impact allowance = 3.24 kN/m.
Lifting points at $0.2L = 0.2 \times 12.6 = 2.52$ m from ends.

$$M = 1.4 \times 3.24 \times \frac{2.52^2}{2} = 14.4 \text{ kNm.}$$

$$K = \frac{14.4 \times 10^6}{15 \times 300 \times 240^2} = 0.06 \text{ giving } Z = 224 \text{ mm.}$$

$$A_s = \frac{14.4 \times 10^6}{0.87 \times 460 \times 224} = 160 \text{ mm}^2 \text{ for two bars} < \text{above.}$$

**On-site pitching**
Add 25 per cent to self weight for impact = 2.70 kN/m.
Pitching point at $0.3L$ from end = 3.78 m.

$$M_{max} = 1.4 \times 2.70 \times \frac{3.78^2}{2} = 27.0 \text{ kN m.}$$

$$K = \frac{27.0 \times 10^6}{30 \times 300 \times 240^2} = 0.052 \text{ giving } Z = 225 \text{ mm.}$$

$$A_s = \frac{27.0 \times 10^6}{0.87 \times 460 \times 225} = 300 \text{ mm}^2 \text{ for two bars} < \text{above.}$$

## 4.5   STAIRCASES

### 4.5.1   Reinforced concrete staircases

Staircase design is by rc or psc solid slab or T-beam analysis. The main features which distinguish it from traditional insitu design are the manufacture (i.e. early age tilting and lifting), layout (i.e. allocation of flights, landings and half-landings per single item as described in Section 3.2.3) and the design of the joints.

Precast staircases are manufactured with the tread facing either up, down or to. In the former, so-called 'ski-slope' moulds made of high quality steel are used so that the unit is manufactured in its final orientation, i.e. treads horizontal. Three of the four faces and the risers are ex-mould – only the treads are trowelled. In the latter a variable

**Fig. 4.44**  Staircases cast using an adjustable tread and riser mould.

**Fig. 4.45**  Prestressed concrete staircase production.

tread mould is used so that the inclination of the stair is fully adjustable (see Figs 4.44 and 4.45). Flexural reinforcement is provided for lifting and ultimate service conditions. Tread-down casting is more ideally suited to the manufacture of flight units only. The unit is likely to be doubly reinforced because the soffit is cast face up. Individual moulds are used only for flights of similar inclination and therefore tend to be made of timber.

The flights are designed as single-way spanning solid slabs. Typical reinforcement consists of a square cage using T 12 to T 16 bars in the longitudinal direction of the span, and T 8 or T 10 bars transversely. An alternative economical design for long-span units (exceeding 6 m) uses stringer beams designed as rc tee beams (Fig. 4.46). The steps cantilever transverse to the flight span and are reinforced with transverse steel in the top of the flight. The stringer beam is checked for torsion for the case of non-symmetrical superimposed loading.

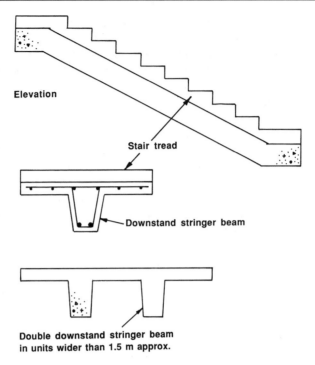

**Fig. 4.46** Stringer beam design for staircases.

Landings, whether simply supported or cantilevered, are designed as rc solid slabs. Spans rarely exceed 4 m and it is therefore possible to design units not exceeding 250 mm in depth. It is usual for the landing to be shallower than the stairflight, and this also allows for a levelling screed to be used over the landing. Composite action with a structural screed is used where appropriate (but rarely). Precast prestressed flat planks or hollow core slabs are used if correct details at the connections can be found.

Other specialist precast rc components such as lift motor slabs, partition walls, roof gangways with a bund wall, steps, ramps or ad-hoc bases for machinery, etc. are usually classified as 'non-structural' and dowelled or bolted to the appropriate part of the structure.

### 4.5.2 Prestressed concrete staircases

The main benefit in prestressing stair flight units compared with rc is in the ease of production and the reduced depth of unit (waist) and amount of reinforcement required. Depths of 150 mm are commonplace for spans of up to 3.5 m. The reinforcement is simply in the form of several 5 mm or 7 mm diameter wires (the pretensioning force is usually too small for the use of strands), typically 6 no. per 1 m wide unit as shown in Fig. 4.45. The grade of concrete is C50 or C60. Because the maximum stair flight length is governed by the maximum number of steps in a single flight, the flexural moments are small in comparison with other types of prestressed work.

### 4.5.3   Staircase and landing end reinforcement

The ends of staircase units may be shaped into scarf joints, or halving joints. These may be visualized as two identical continuous bearing nibs in which the forces acting on each are equal and opposite in direction and magnitude. The design is based on a short cantilever nib principle, rather like the beam bearing nib design, but where accuracy in the positioning of the flexural reinforcement is of utmost importance. In order to increase the depth at the root of the nib the bearing ledge is sloping at a gradient $\alpha$, about 1 in 7 to 1 in 10. This naturally affects the resultant force, as the frictional force $\mu V \cos \alpha$ in Fig. 4.47 is acting down the slope, where $V$ is the uniformly distributed ultimate stair light end shear force. In concrete–concrete joints $\mu = 0.7$ and so the resultant force is inclined at $\theta \simeq \tan^{-1} \mu \cos \alpha + \alpha$ relative to the surface of the bearing ledge. Typically $\theta = 40°$–$45°$. Figure 4.47 shows the notation and triangle of forces at the reaction point which gives the following forces and areas of reinforcement:

$$H = \frac{V}{[\sin \alpha + (\cos \alpha \tan \beta)]} + \mu V \cos \alpha \left( \frac{x+c}{x} \right) \tag{4.61}$$

and by the resolution of forces at the load point:

$$C = \frac{V}{(\cos \beta \tan \alpha) + \sin \beta} \tag{4.62}$$

and

$$T = V + \mu V \cos \alpha \sin \alpha \tag{4.63}$$

The corresponding areas of reinforcement are given by:

$$A_{sh} = \frac{H}{0.87 f_y} \tag{4.64}$$

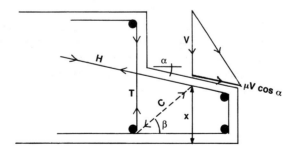

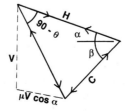

**Triangle of forces**

**Fig. 4.47**  Scarf joint geometry in staircases.

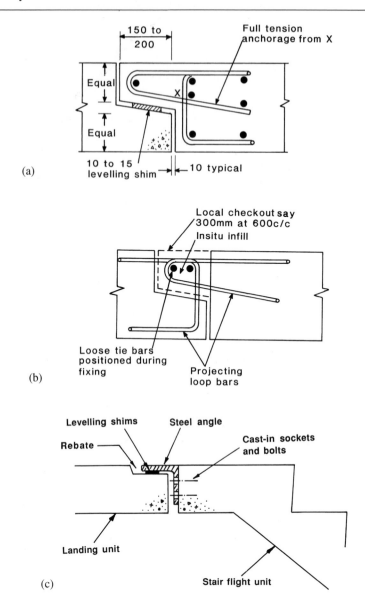

(a)

(b)

(c)

**Fig. 4.48** Staircase to landing joint details. (a) Reinforcement in stair scarf joint; (b) intermittent scarf joint and (c) steel shelf angle method.

$$A_{sv} = \frac{T}{0.87 f_y}$$
(4.65)

Note that $T$ should not be less than $V$.

Small diameter bars at frequent intervals, say T 8 at 100 mm centres, are preferred to larger bars where bending restrictions may violate cover requirements. Figure 4.48(a) shows typical dimensions and ideal reinforcement details. Neoprene pads, or steel packing shims with a soft mortar bed are used to ensure the correct bearing. Although ultimate design pressures rarely exceed about 1.0 N/mm$^2$ (for typical spans of 3 m and 5 kN/m$^2$ superimposed loading) the line of action of loading is very important. This should pass through a point which is well behind the line of the longitudinal bar in the front of the nib, and not less than about 30 mm from the ends of either the upper or lower half of the joint.

Continuous end details of this type do not provide continuity between precast units. This may be achieved using intermittent scarf joints, as shown in Fig. 4.48(b) in which pockets of reinforced insitu concrete tie the precast units together. These units are obviously more expensive to manufacture. Welded connections made between fully anchored plates (similar to double-tee floor slabs) may also be specified if structurally required. Concrete cover to the welding is required for fire protection and cosmetic purposes.

Shelf angles are a popular alternative to concrete scarf joints in that production is simplified by the simple provision of cast-in sockets for the bolted connection. The detail is 'inverted' (relative to the normal use of a shelf angle) as shown in Fig. 4.48(c). A small check-out in the supporting (landing) member is prepared to receive the flange of the angle attached to the flight unit. At least two angles are used in the connection. The assembly is designed on the basis of an eccentric vertical load acting at the most onerous position. Flexural and shear stresses are considered at the root of the angle, and pull-out forces and shear stresses are used in designing the bolt and socket. Friction grip bolts in slotted holes are required for fixing tolerances. Concrete cover is required to give fire protection to the angle. Typical details of angle and bolt are given in Fig. 4.48(c).

Hollow core floor slabs may be used to replace solid rc landing units. The depth of the hollow core slab is about 50 mm less than that of the precast stair flight to enable a structural finishing screed to cover the steel angle (hollow core units cannot be recessed). A check should be made to ensure that the shear stress in the region of the hollow core is not excessive, as indicated in Fig. 4.49(a). If the hollow core slab unit is cut to the required landing width, the stair flight should not bear solely on the top flange of the slab. Figure 4.49(b) shows the correct practice.

**EXERCISE 4.9    Staircase half joint design**

Design the reinforcement for the halving joint shown in Fig. 4.50 for an ultimate landing reaction of 30 kN/m run. Allow a 10 mm vertical gap between the units.
Use $f_{cu} = 40$ N/mm$^2$, $f_y = 250$ N/mm$^2$, and cover to reinforcement = 25 mm.

*Solution*

Using Fig. 4.47 notation:
$\mu = 0.7$, $\alpha = \tan^{-1} 20/150 = 7.6°$
Bearing length = $150 - 10$ gap = 140 mm
Centre of bearing = $10 + 140/2 = 80$ mm from face of nib

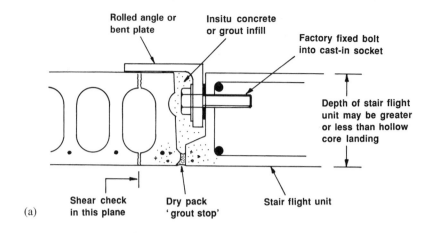

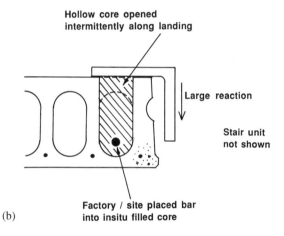

**Fig. 4.49** Hollow core floor slabs used as landings. (a) With unfilled edge core and (b) with filled edge core.

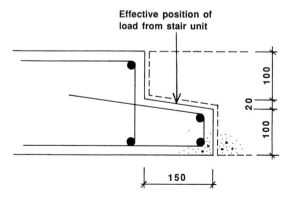

**Fig. 4.50** Details to Exercise 4.9.

Lever arm to tie steel = 80 + 25 cover + 5 say = 110 mm

$x = 110 - 25 = 85$ mm, $c = 25$ mm.

Then $\beta = \tan^{-1} 85/110 = 37.7°$. Consider 1.0 m wide unit, then:

$$V = 30 \text{ kN/m}$$

$$H = \frac{30}{[\sin 7.6° + (\cos 7.6° \tan 37.7°)]} + 0.7 \times 30 \times \cos 7.6° \left( \frac{85 + 25}{85} \right) = 60.3 \text{ kN/m run}$$

$$C = \frac{30}{(\cos 37.7° \tan 7.6°) + \sin 37.7°} = 41.83 \text{ kN/m}$$

and

$$T = 30 + (0.7 \times 30 \times \cos 7.6° \times \sin 7.6°) = 32.75 \text{ kN/m}$$

Thus:

$$A_{sh} = \frac{60.3 \times 10^3}{0.87 \times 250} = 277 \text{ mm}^2/\text{m}$$

*Use R 8 at 150 mm centres (333) U shape bars in nib.*

$$A_{sv} = \frac{32.75 \times 10^3}{0.87 \times 250} = 151 \text{ mm}^2/\text{m}$$

*Use R 8 at 300 mm centres (166) U shape bars in end of landing.* Provide R 8 transverse steel as shown in Fig. 4.48(a).

**Chart 4.1** Example moments and shear force capacities for inverted-tee beams.

Geometric data:  Overall width = 500 mm             Upstand width = 300 mm
Bearing ledge width = 100 mm         Upstand depth = 200 mm
Boot depth = variable 300 to 600 mm

Material data:  Concrete      $f_{cu} = 60 \text{ N/mm}^2$      $f_{ci} = 40 \text{ N/mm}^2$
Strand        $A_{ps} = 94.2 \text{ mm}^2$      $f_{pu} = 1750 \text{ N/mm}^2$
Initial prestress $\eta = 0.70$           $P_i = 115.5$ kN per strand
Transmission length = 475 mm

Fixed losses:  Shrinkage = 58.5 $\text{N/mm}^2$ = 4.77%     Strand relaxation = 1.2 × 2.5% = 3.00%

| Beam ref. | Boot depth (mm) | No. strands | Final $f_{bc}$ (N/mm$^2$) | Final $f_{tc}$ (N/mm$^2$) | $M_{sr}$ (kN m) | $M_{ur}$ (kN m) | $V_{co}$ (kN) | Min. $V_{cr}$ (kN) |
|---|---|---|---|---|---|---|---|---|
| TB300 | 300 | 13 | +11.1 | −1.0 | 263 | 459 | 309 | 92 |
|  |  | 15 | +12.3 | −1.1 | 286 | 485 | 321 | 99 |
|  |  | 17 | +13.5 | −1.1 | 303 | 509 | 333 | 106 |
| TB400 | 400 | 15 | +10.4 | −1.0 | 363 | 638 | 380 | 106 |
|  |  | 18 | +12.2 | −1.1 | 409 | 692 | 399 | 119 |
|  |  | 21 | +13.7 | −1.1 | 448 | 734 | 417 | 129 |
| TB500 | 500 | 17 | +10.2 | −1.0 | 488 | 851 | 449 | 123 |
|  |  | 21 | +12.0 | −1.1 | 555 | 913 | 477 | 138 |
|  |  | 25 | +13.8 | −1.1 | 617 | 1001 | 502 | 152 |
| TB600 | 600 | 20 | +10.4 | −1.0 | 655 | 1122 | 526 | 144 |
|  |  | 25 | +12.4 | −1.1 | 747 | 1211 | 562 | 162 |
|  |  | 30 | +14.2 | −1.1 | 832 | 1300 | 593 | 181 |

# Chapter 5

# Design of Precast Floors used in Precast Frames

*The options for using precast floors in frames is given together with the fundamental design methods for individual units and complete floors.*

## 5.1 FLOORING OPTIONS

Precast concrete flooring offers an economic and versatile solution to ground and suspended floors in any type of building construction. It has been widely used in commercial, industrial and domestic building, offering both design and cost advantages over traditional methods such as insitu concrete, composite and timber floors. There is a wide range of flooring types available to give the most economic solution for all loadings and spans. The floors give maximum structural performance with minimum weight and may be used with or without structural topping screeds, non-structural finishes (such as granolithic screed), or with raised timber floors. The main types of flooring are shown in Figs 5.1 (a) to (e):

- hollow core slab
- double-tee slab
- beam and block
- solid composite plank
- beam and composite plank.

There is a market for each type of slab, which is dictated by load capacities, spans, preferred construction methods, integrity of the floor plate, and cost. The main building functions of these slabs was discussed in Chapter 3 in which Table 3.1 summarizes the performance criteria.

The two main criteria in the design of floors are strength and stiffness, both in the vertical direction under gravity loading and in the horizontal direction under wind (or earthquake) loading. In each case, in the event of overload or accidental damage, the floor system must be ductile. Other design criteria include fire resistance, thermal and acoustic properties, vibrations, durability, handling, and construction methods.

Hollow core floors may be used without a topping screed because the individual floor units are keyed together over the full surface area of their edges. Thus, vertical and horizontal load transfer is effective over the entire floor area, and the units are classified according to Section 4.3.1 as non-isolated. This is not the case with all the other types of precast floor where a structural (i.e. containing adequate reinforcement) topping must be used either for horizontal load transfer, flexural and shear strength, or simply to complete the construction.

More than 90 per cent of all precast concrete used in flooring is prestressed. The slabs are designed in accordance with BS 8110 [5.1] and other selected literature produced by

206

IStructE [5.2], BCA [5.3], FIP [5.4], CEN [5.5], PCI [5.6] and individual research (e.g. [5.7–5.10]).

All slabs are designed as single spanning. Lateral load distribution perpendicular to the span is only permitted if the longitudinal joint is capable of transmitting a shear stress of 1 N/mm$^2$. This is not possible with unscreeded beam and block floors, but in the case of hollow core slabs  loads may be distributed across several slabs . Heavy point loads or line loads are distributed over at least three individual slabs. See Section 5.2.7 for further details.

Figure 5.2 shows, diagrammatically, the five different criteria for the capacity of flooring units. These are, from short to long spans, respectively:

- bearing capacity
- shear capacity
- flexural capacity
- deflection limits
- handling restriction.

Shear is usually only critical in very short (e.g. < 3 m) or heavily loaded spans, particularly where large line loads (e.g. > 100 kN/m) are present.

Thus, for each type of floor slab, a family of standardized cross-sections and reinforcement patterns to cover the widest possible combination of floor loading and spans are produced. A computerized output is submitted as the standard calculations are easy to commit to software. Section sizes are selected at incremental depths, usually 50 mm, and a set of reinforcement patterns are chosen. From these data a predetermined set of bending moment and shear force capacities are compared with project design requirements. Usually the engineer has at least two, and sometimes three, different depths of unit to choose from. The economical choice is usually the shallowest and most heavily reinforced unit, although unacceptable deflections may rule this one out. The additional advantage is that the structural floor zone is kept to a minimum.

Although cambers and deflections are easy to calculate, actual measurements are smaller than calculated values by up to 25 per cent. This is because of unreliable data for Young's modulus and variations in material properties due to temperature and humidity fluctuations throughout the casting, curing and manufacturing cycle. The precamber should not exceed span/300. Deflections are usually critical where the span to depth ratio is more than 40:1.

## 5.2  HOLLOW CORE SLABS

### 5.2.1   General

A general introduction to hollow core slab units (hcu) is given in Section 3.2.2.1. They are available as either rc or psc units. The span to weight ratio is better in a prestressed slab, but the unit is less versatile when it comes to forming cantilevers, dealing with point loads, or reinforcing around large holes, etc. Although the cross-sections and versatility of these units varies widely, the differences are largely due to the manufacturing process.

There are two main proprietary types:

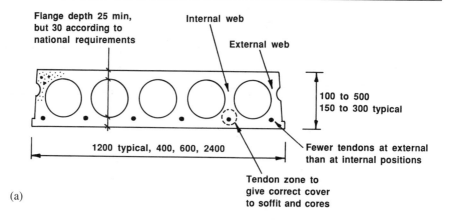

(a)

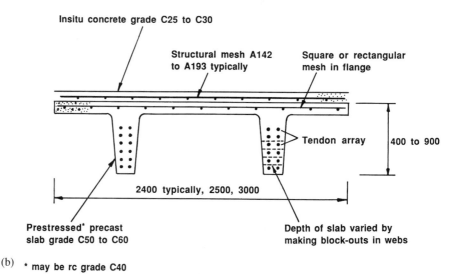

(b) * may be rc grade C40

**Fig. 5.1** Types of precast concrete flooring. (a) Hollow core slab, (b) double-tee slab.

- Type A, with oval or non-circular voids, shown in Fig. 5.3, are produced by the slip forming technique
- Type B, with mainly circular voids, shown in Fig. 5.4, are produced by the extrusion technique.

The manufacturing process is shown in Figs 5.5 and 5.6. The concrete is a dry mix with zero slump [5.11]. The water/cement ratio is about 0.36. 10 mm coarse aggregate gravel or crushed limestone is used together with medium grade sand. Lightweight aggregates are not used because of the excess wear on the machinery, and the density of aggregate required to achieve the necessary strength is quite high. Recycled concrete, crushed down from waste production to 10 mm aggregate is used at not more than nine

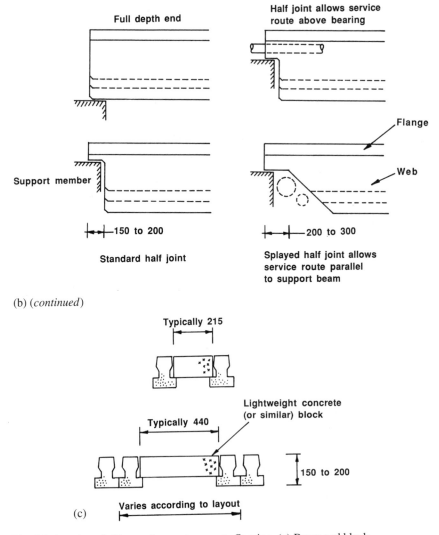

**Full depth end**

**Half joint allows service route above bearing**

Flange

Web

Support member

—150 to 200

**Standard half joint**

—200 to 300

**Splayed half joint allows service route parallel to support beam**

(b) (*continued*)

Typically 215

Lightweight concrete (or similar) block

Typically 440

150 to 200

Varies according to layout

(c)

**Fig. 5.1** (*continued*) Types of precast concrete flooring. (c) Beam and block.

per cent of the total aggregate content. New research is considering adding short steel fibres to the mix to improve concrete toughness and bonding performance.

Type A units are slip-formed by feeding concrete around steel formers. A profiled form is moved during placement and the concrete is vibrated around these forms without too much pressure. Figure 5.5 shows the operation by using the *Roth* technique. The machine lays the concrete in three layers – the bottom flange, the webs, and the top flange. The shape of the former is designed to give the optimum flexural and shear performance as well as providing adequate cover to the reinforcing tendons.

In the extrusion process the concrete is pressed out by screws during compaction into the required cross-section. Figure 5.7 shows the result of the extrusion technique – handling takes place within 12 hours of casting. The voids are circular in profile, with the result

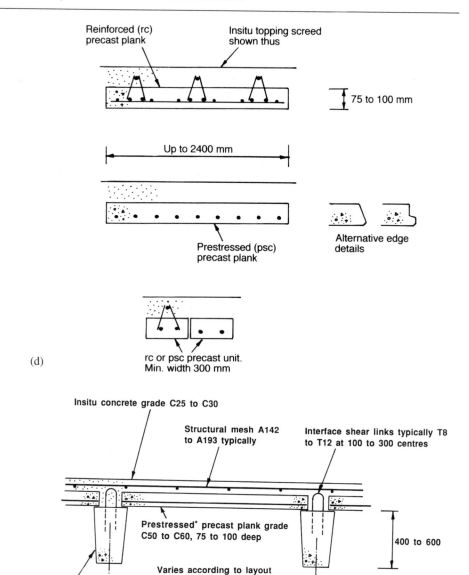

(d)

(e)   * may be rc grade C40

**Fig. 5.1** (*continued*) Types of precast concrete flooring. (d) Solid composite plank;
(e) beam and composite plank.

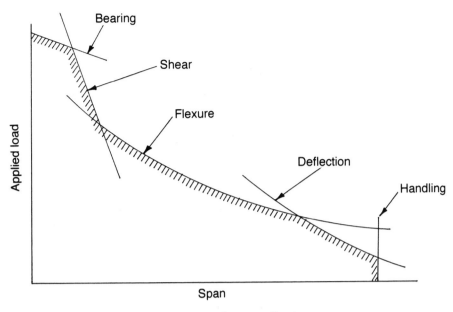

**Fig. 5.2** Load vs span criteria for prestressed concrete flooring.

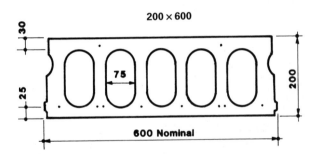

**Fig. 5.3** Types of hollow core units – Type A: slip-formed with non-circular voids.

that in order to optimize the cross-sectional properties of the slab, the thickness of the web at the mid-height of the voids is usually only about 30 mm.

In comparing the two techniques, slip-forming machines tend to be noisy but have lower maintenance costs. One problem, however, is that the slip-forming machine may move forward in jerks, leaving a rippled surface to the sides (and sometimes the tops) of the units, with the result that the width of the unit is 2 to 3 mm too large. Ironically this profile may actually improve the horizontal diaphragm capacity of the floor slab. Any new investors in hollow core machinery seem to prefer the slip-forming process.

Wet cast units (being less popular, and somewhat easier to analyse) are dealt with in Section 5.2.13.

Cross-section, concrete strength, fire resistance and surface finish are standard to each system of manufacture. Small variations, which may be included at an extra cost, should be discussed early in the design appraisal. These include increased fire resistance, provisions for vertical service holes, opening of cores for special fixings, cut-outs at columns, etc.

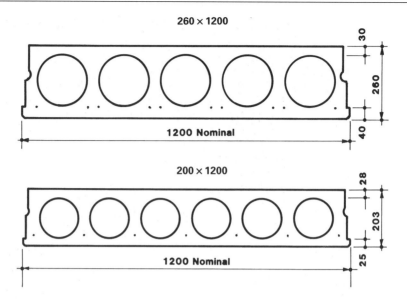

**Fig. 5.4** Types of hollow core units – Type B: extruded with circular voids.

**Fig. 5.5** Production of hollow core slab – the slip-forming machine.

The edges of the slab are profiled to ensure that an adequate shear key of insitu concrete (10 mm size aggregate), rather than grout, is formed between adjacent units. The extrusion process is not sympathetic to providing projecting reinforcement across the joint. The capacity of the shear key between the units is sufficient to prevent the adjacent slabs from differential movement. Despite a slight roughening of the surfaces during the

**Fig. 5.6** Production of hollow core slab in Belgium.

**Fig. 5.7** Production of hollow core slab – extrusion through helical screws.

manufacturing process where indentations of up to 2 mm are present, the surface is classified in BS 8110 Part 1, clause 5.3.7 [5.1] as 'smooth', as opposed to being 'roughened'.

### 5.2.2   Design

Hollow core slabs are designed as a series of I-sections with semi-circular or parallel sided webs. The concrete is grade C50 or C60. The section is structurally very efficient and the design is well documented [5.4]. The maximum span-to-depth ratio is 55, and the holes guarantee the maximum possible radius of gyration in addition to reducing the self-weight of the slab by 40 to 50 per cent.

The shapes of the voids are usually circular, oval, or parallel sided with triangular and/or semi-circular ends, illustrated in Fig. 5.8 from various manufacturers product catalogues. The danger here is that the webs become slender and have been known to collapse by sideways buckling in 300 mm deep units. The problem is rare owing to the dryness of the mix. There is a tendency for manufacturers to make the voids as large as possible for obvious reasons; less concrete and improved performance-to-weight ratio. A measure of this latter parameter is expressed in terms of the radius of gyration ($r = \sqrt{I/A}$) for the section. Table 5.1 gives approximate data for a range of extruded (e.g. 'Spiroll') and slip-formed (e.g. 'Roth' and 'Spancrete') units.

Owing to the absence of stirrups, the minimum web thickness is governed by limitation of the tensile stresses due to spalling upon release of prestress from the tendons to the concrete, and shear. The maximum spalling stress is the criterion, with a resulting web thickness of 30 mm, giving a total web to unit breadth ratio of between 0.25 and 0.35.

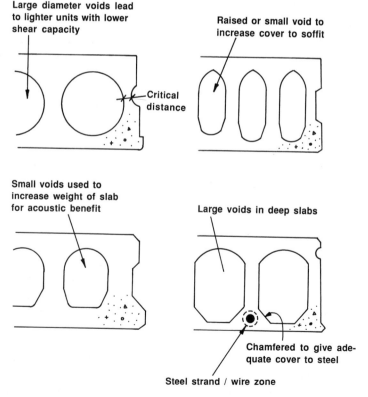

**Fig. 5.8** Shapes of voids and edge profiles in hollow core units.

**Table 5.1** Some section properties of hollow core slabs per 1.2 m width.

| Type of unit | Depth (mm) | Second moment of area $I \times 10^6$ (mm$^4$) | Cross-sectional area $A \times 10^3$ (mm$^2$) | Radius of gyration $r$ (mm) |
|---|---|---|---|---|
| Extruded | 260 | 1500 | 188 | 90 |
| | 203 | 670 | 135 | 70 |
| | 152 | 305 | 117 | 51 |
| Slip-formed | 250 | 1148 | 169 | 82 |
| 'Roth' | 200 | 628 | 142 | 66 |
| | 150 | 298 | 120 | 49 |
| Slip-formed | 305 | 2400 | 230 | 102 |
| 'Spancrete' | 254 | 1450 | 201 | 85 |
| | 203 | 750 | 166 | 67 |
| | 152 | 350 | 154 | 47 |

The rules for the size of the longitudinal gap are shown in Fig. 5.9. They are based on the premise that tie steel might be placed in the joint. FIP Recommendations [5.4] for edge profiles are also included in Fig. 5.9.

It is surprising how much attention is devoted to the shear capacity of hollow core slabs, when in 95 per cent of cases a mid-span flexural failure would occur. The reasons for this include:

- shear failures are catastrophic due to an absence of stirrups, and
- shear behaviour, particularly in the transmission zone, is not fully understood.

Design is carried out to the normal prestressed concrete clauses in BS 8110 [5.1], but level of prestress in the transmission zone is obtained from the FIP Recommendations [5.4].

Despite the large number of propriety dry-cast units available, the two major types are distinguishable only in the shape of the voids - circular or oval. In the former group are 'Spiroll', 'Flexicore' and 'Tembo'. In the latter are 'Roth', 'Dynaspan' and 'Spancrete'. Typical load vs span data for these units are given in Fig. 5.10 (a).

## 5.2.3  Design of cross-section

The cross-sectional dimensions of hcus should be optimized to enable the unit to function in service without local deformations, cracking, spalling, etc. The profile must also allow for the concrete to be correctly placed, compacted and cured uniformly. The geometry of the unit includes size, shape and spacing of voids, web thickness, depth of flanges, concrete cover to tendons and edge profile.

Although the size and shape of the voids depend on the manufacturing process, some general rules as regards size and spacing of voids may be given. The height of voids should not exceed $h - 50$ mm, where $h$ is the overall depth of the unit. The diameter of circular voids is usually $h - 60$ to 75 mm. The minimum flange thickness depends on the overall depth of the unit $h$, given by $1.6\sqrt{h}$. This gives 25 mm for a 200 mm deep unit,

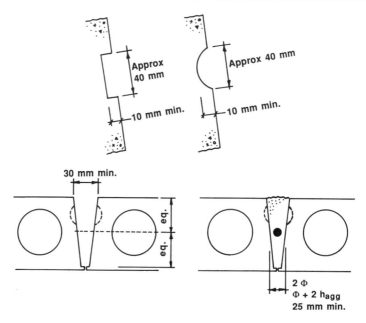

**Fig. 5.9** Rules for determining the size of the longitudinal joint between units.

etc. However, because of cover requirements it is usually necessary for the bottom flange to be at least 30 mm thick. The width of the void (i.e. clear distance between webs ) is governed by the maximum stresses which may develop in the transverse direction. To prevent localized collapse of the flange, the ratio of the width of the (non-circular) void to flange depth is about 3:1 or 4:1.

### 5.2.4   Web thickness

The critical region is near to the ends of the units where web thickness is governed by the limitations in the tensile stress capacity of the concrete at the onset of the transfer of prestress to the concrete. These stresses manifest themselves in spalling, splitting and bursting of the concrete in what is known as the 'transmission zone'. This is shown in Fig. 5.11 as distance $l_t$. Although the most important of these is the splitting, due to bond failure around the prestressing tendon, spalling in the webs must be avoided by following some fairly comprehensive design rules by Den Uijl [5.12] as adopted in the FIP literature [5.4].

Figures 5.12(a) and (b) show the results of a finite element study leading to a lower bound solution for the spalling stress $\sigma_{sp}$ in terms of the prestressing force $P_o$, transmission length $l_t$, and cross-sectional geometry eccentricity $e_o$, depth $h$, web breadth $b_w$, and $k$ (the so-called 'core distance') $= l/0.5\ hA$, defined in Fig. 5.12(a). Designers of hollow core sections show that the resulting value of $\sigma_{sp} < 0.24\sqrt{f_{cu}}$.

As for the breadth of the web, the usual result for a grade C60 concrete in fully stressed units of up to 300 mm in depth is a minimum $b_w = 30$ mm.

Excessive prestressing (the result of trying to increase the strength capacity along the length of the unit) in these relatively thin regions has resulted in horizontal cracking

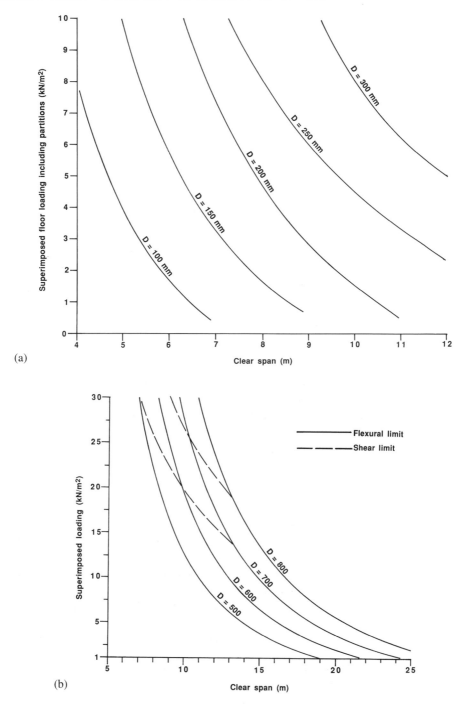

**Fig. 5.10** Applied load vs clear span charts for prestressed concrete slabs. (a) Hollow core slabs, (b) double tee slabs.

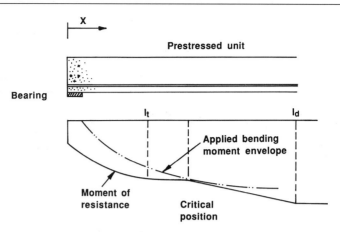

**Fig. 5.11** Development and anchorage lengths of helical strand in hollow core units.

along the length of the unit due to shear-compression. The cracks propagate from the ends of the units with the appearance of a crocodile's mouth. Hence the term 'crocodiling' was coined. Thicker webs, greater than 30 mm, prevent this from happening.

### 5.2.5  Edge profiles

Standard edge profiles have evolved to ensure an adequate transfer of horizontal and vertical shear between adjacent units. The main function of the joint is to prevent relative displacements between units. In hcus these objectives are achieved using structural grade insitu concrete (C25 minimum) compacted by a small diameter poker in dampened joints. Typical profile geometries are shown in Fig. 5.9.

The sides of the units have natural random roughness with indentations up to 2 mm in depth. The design ultimate horizontal shear stress, given in BS 8110 [5.1] for the type of joint conforming to Fig. 5.9, is 0.23 N/mm². For a typical 6 m long 200 deep unit, $V_{uh} = 275$ kN; a value which is rarely critical. Vertical shear capacity is based on single castellated joint design with minimum root indentation 40 mm × 10 mm deep. An ultimate shear stress of 1.3 N/mm² yields an ultimate shear resistance equivalent to 45 kN/m² floor loading for 1.2 m wide units.

If reinforcing bars are placed in the joint, the joint should be wide enough at the level of the bar to enable insitu concrete to be compacted around the bar. This should not be less than 25 mm if 10 mm maximum size of aggregate is used. The diameter of the bar should not be greater than half the gap width.

### 5.2.6  Reinforcement

Hollow core units have, in general, no reinforcement other than longitudinal prestressing tendons (helical strand or wire) anchored by bond. It is therefore desirable to take advantage of the tensile strength of concrete in the determination of $M_{sr}$ and $V_{co}$.

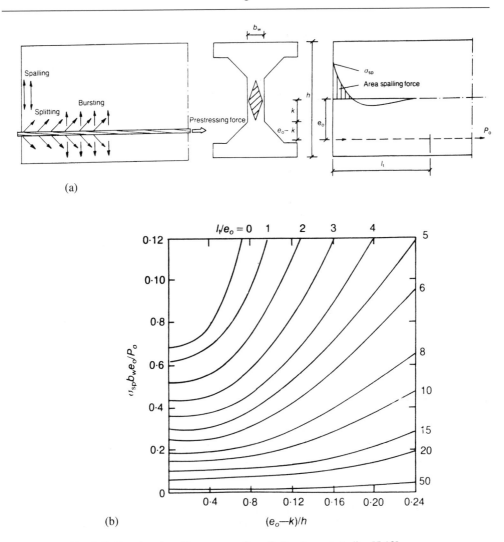

**Fig. 5.12** Results of spalling stresses from finite element studies [5.12].

The low (class 2) relaxation strand is of the 7-wire helical form, and the wire is plain or crimped according to the specification given in Section 4.2.4. Test certificates for the ultimate tensile strength are available from all manufacturers. Stressing is restricted to between 70 and 75 per cent of the characteristic strength. Tendons are exposed at the ends of the member as shown in Fig. 5.13. The pull-in of the tendon should be between 1.5 mm and 2 mm. This is measured using a depth gauge.

The percentage area of steel is relatively small particularly in comparison with an rc section of comparable flexural capacity. Steel ratio $\rho = A_{ps}/bd$ varies from 0.10 to 0.25 per cent in most of the units manufactured in the UK. The failure mode for these slabs is excessive deflections, flexural cracking in the soffits and finally, rupture in the tendons.

**Fig. 5.13** Close-up of helical strand in hollow core units.

### 5.2.7  Lateral load distribution

The question often arises whether hcus can safely distribute line and point loads to adjacent units, and to what extent a slab 'field' exists.

Patch loads and line loads parallel with the direction of span give rise to bending moments in a direction at right angles to the direction of span, and to vertical shear forces in the longitudinal joints between adjacent units. Similar situations occur where steel trimmer angles are used to frame out large floor voids. The problem is that hollow core floors are not provided with transverse reinforcement in the precast units or in the joints between the units. Load transfer depends on adhesion, bond and friction in the shear key between the units, and the possible shear-torsion failure of the hcu is likely to be sudden and without warning.

Although the behaviour of thin plates is well understood, the behaviour of hollow core slabs is complicated by the partial moment interaction between adjacent units. The joints between individual hcus behave as hinges capable of finite moment and full shear transfer. However, the former is not well understood and relies extensively on the quality of the infill grout. In the interest of safety the moments are ignored. The point or line load in Fig. 5.14 produces a shear reaction in the longitudinal edge of the adjacent units, and this induces torsion in the slab. The capacity of the hollow core slab to carry the torsion is limited by the tensile capacity of the concrete. The magnitude of the shear reaction depends on the torsional stiffness and the longitudinal and transverse stiffnesses of all the adjacent slabs. Low stiffnesses result in low load sharing. The precast units themselves may be assumed to be cracked longitudinally in their bottom flanges, but shear friction generated by transverse restraints in the floor grillage ensures integrity at the ultimate state. The deflected profile of the total floor slab is computed using finite strips and differential analysis. The cross-section of each floor element is considered as a solid rectangular

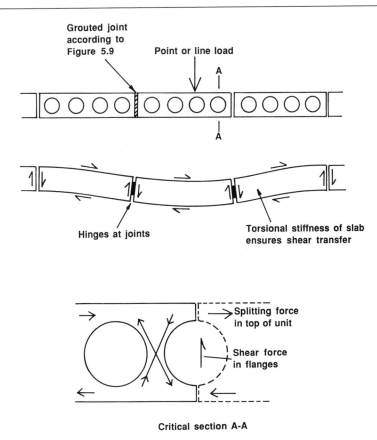

**Fig. 5.14** The torsional hinge concept to lateral load distribution in one-way spanning elements.

element and the circular (or oval) voids are ignored. The result is unconservative and therefore reduction factors are applied to the shear reactions.

The transverse moments and shear forces may be distributed (in accordance with BS 8110, Part 1, clause 5.2.2.2, [5.1]) over an effective width equal to the total width of three precast units, or one quarter of the span either side of the loaded area. The equivalent uniformly distributed loading on each slab unit may thus be computed. This is a conservative approach as data given in FIP Recommendations [5.4] show that for spans exceeding 4 m up to five units are effective, Fig. 5.15. The data in Fig. 5.16 also show that for edge elements, e.g. adjacent to a large void, only two slabs contribute significantly in carrying the load.

A similar analytical or empirical approach is made in dealing with point loads. BS 8110 calls for test results to justify exceeding the quarter span criteria exacted above. This information is not readily available, and therefore bending moment distribution factors, given in Fig. 5.17, may be used to show that the effective width is equal to the width of at least three slabs, irrespective of the position of the load.

Design rules for determining the lateral distribution of concentrated vertical loads in decks made from hollow core type slabs were derived using the finite strip method [5.13]. The analysis considered the torsional, shear and flexural stiffness of units up to 1.2 m

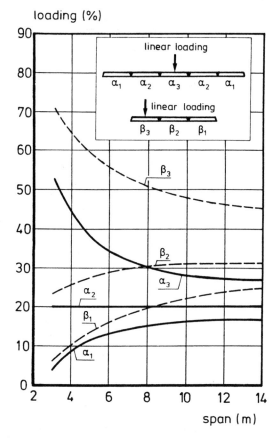

**Fig. 5.15**  Lateral load distribution coefficients for a central load [5.4].

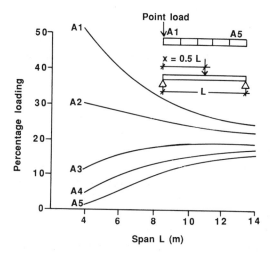

**Fig. 5.16**  Lateral load distribution coefficients for an edge load [5.4].

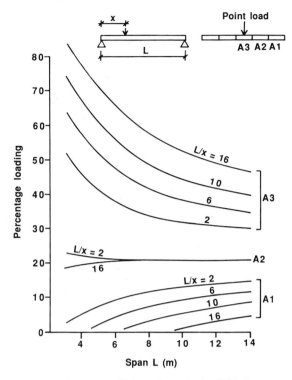

**Fig. 5.17** Lateral load distribution coefficients for point loads [5.4].

wide containing circular voids. The longitudinal joint between the units was considered as a rotational hinge with a shear capacity equal to the minimum shear capacity of the precast unit. The results showed that the lateral bending moment, vertical shear and vertical deflection responses were all different; the shear response being the most concentrated around the loaded unit. Because the curves shown in Fig. 5.18 for bending moment factors $\alpha$ are closed-form solutions (i.e. require iterations to obtain the final result) and seeming difficult to handle in design situations, adopting a value of $\mu = 1$ gives a good approximation and enables a realistic value to be assigned to the effective breadth of the slab field $= 0.125\ b/\alpha$.

Stanton [5.14] found distribution widths for bending, shear and deflection resulting from point loads not only near to the centre of the floor area but also close to a free longitudinal edge. This is the effective width over which a point load may be distributed. In summary Stanton's equations suggest that if a point load is acting at mid-span over a length $L$, then the distribution width is $0.35\ L$ either side of the centre of the loaded unit. If the point load is at $\frac{1}{4}$ span, the distribution is $0.28\ L$ either side. Applying these results to spans of 6 to 8 m, the loads are being distributed over five units of 1.2 m width. These results are in good agreement with the FIP values presented in the figures.

Shear is the most complex response because the effects of torsion must be separated from vertical shears. The distribution of shear is largely dependent on the separation distance between the load and response positions. Shear distribution widths increase rapidly with this increasing distance, but are seldom greater than $0.125\ L$, and the distribution curve is triangular.

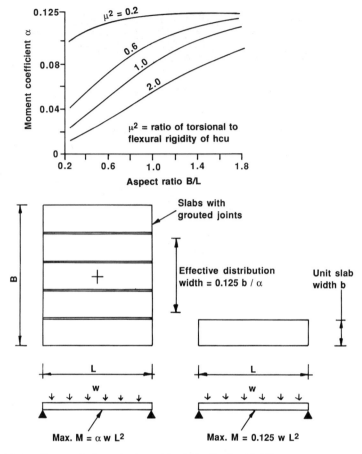

**Fig. 5.18** Bending coefficients for lateral load distribution [5.13].

Aswad and Jacques [5.15] tested some 203 mm deep *Dycore* units to determine allowable magnitudes for a single point edge load. Two modes of failure, in which the slabs failed without warning, were found:

- by edge punching shear under the load at mid-span
- shear-torsion failure at the corners of the units when the load point was nearer to the support (see Fig. 5.19).

With the load point at mid-span the lowest failure load was 51.2 kN, whilst the lowest failure load for a load point at quarter-span was 44.5 kN. The former failed by punching shear, whilst the latter by shear-torsion. Factors of safety of 2.80 {1.60/0.85 × 1.5} were applied to these results, which gave good agreement (within 5 per cent) with the FIP equation in the Technical Report [5.16].

### 5.2.8 Flexural capacity

The flexural capacity of a prestressed unit is checked at both the serviceability and ultimate

**Fig. 5.19** Edge failures in tests by Aswad and Jacques [5.15] (courtesy of PCI Journal).

limit states. It is usual for the ratio of the ultimate moment of resistance $M_{ur}$ to the serviceability moment of resistance $M_{sr}$ to be in the order of 1.7 to 1.8. Thus, with the use of the present load factors (1.4 for dead and 1.6 for superimposed), the serviceability condition will always be critical. The problem of cracking in the unreinforced zones of hcus is particularly important with regard to the uncracked shear resistance. It is therefore necessary to ensure that tensile stresses are not exceeded.

Prestressed concrete is well documented and the reader should refer to standard texts such as Kong and Evans [5.17]. Figure 5.20 shows the basic ideas. Units are classified *class 2 members* to BS 8110 [5.1] with respect to the serviceability limit state of cracking. This means that a certain amount of tension is allowed. For spans exceeding about 4 m it is usual for the governing design criteria to be the flexural tensile stresses at the serviceability limit state.

### 5.2.8.1   *Serviceability limit state of flexure*

The serviceability flexural resistance $M_{sr}$ is limited by the following:

(a)  flexural tensile and compressive stresses in the concrete
(b)  precamber
(c)  deflections.

Items (b) and (c) are particularly relevant to the more highly stressed units exceeding about 7 m in length. Item (a) is applicable to all units of all ages and under all possible loading conditions (including handling and temporary conditions). The amount of flexural tensile stress $f_{ct}$ allowed classifies the units as follows:

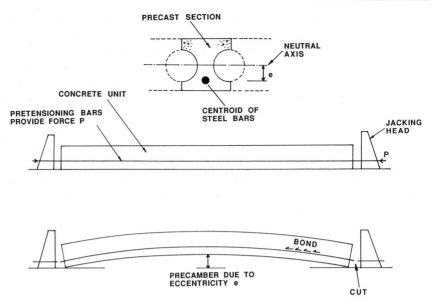

**Fig. 5.20** Basic prestressing principles.

- Class 1:   No flexural tensile stress.
- Class 2:   Flexural tensile stresses allowed not exceeding $0.45\sqrt{f_{cu}}$ or 3.5 N/mm$^2$, whichever is the smaller, but no visible cracking.
- Class 3:   Flexural tensile stresses allowed but for crack widths less than 0.1 mm in units in a severe environment the stress is $(1.34\sqrt{f_{cu}} - 2.2$ N/mm$^2)$ or 6.3 N/mm$^2$, whichever is the smaller, and with crack widths less than 0.2 mm in units in all other environmental condition the stress is $(1.34\sqrt{f_{cu}} - 3.2$ N/mm$^2)$ or 7.3 N/mm$^2$, whichever is the smaller. Class 3 tensile stresses are considered 'hypothetical' in that, because the concrete is cracked and the actual stresses close to the cracks are virtually zero, an equivalent average stress is used. This would be the tensile stress immediately prior to cracking. Depth factors and certain enhancements to the basic tensile stress, given in codes of practice, may be used.

Most designers specify Class 2, but occasionally Class 1 if deflection is excessive. This means a higher prestressing level is required to compensate for the zero tension in the soffit – which is nearly always the limiting case. The flexural compressive stress in the top surface of simply supported units (the only condition used in precast work) is limited to $0.33 f_{cu}$. It is rarely critical in slabs other than the temporary condition in the prestressed solid plank units.

It is necessary to calculate stresses due to the loading arrangement for the following cases:

(a)  immediately after the transfer of prestress
(b)  after all the losses in the prestressing tendons have occurred
(c)  as (b) but in service using $\gamma_f = 1.0$.

The transfer stress, expressed in the usual manner as the characteristic cube strength $f_{ci}$, is a function of the final concrete strength $f_{cu}$. For $f_{cu} = 60$ N/mm$^2$ (the typical strength), $f_{ci}$ should be 38 to 40 N/mm$^2$. For $f_{cu} = 50$ N/mm$^2$, $f_{ci}$ 35 N/mm$^2$.

Use of rapid hardening cements, semi-dry mixes and humid indoor curing conditions are conducive to early strength gain. The transfer strength is achieved in less than 24 hours. The design compressive and tensile strengths for Class 2 members are, respectively, $0.5 f_{ci}$ and $0.45\sqrt{f_{ci}}$. Assuming that the initial pretensioning force is 70 per cent of ultimate (in other countries between 60 and 75 per cent is used), the magnitude of the prestressing force per wire or tendon is taken as $0.7 (1 - \eta) A_{ps} f_{pu}/\gamma_m$, where $\gamma_m = 1.0$ in this situation, and $\eta$ is the losses in the prestressing tendons. If low relaxation wires are used $\eta = 0.04$ at transfer, i.e. half the final value of the 8 per cent relaxation loss given in the specification. If low relaxation helical wire strands are used $\eta = 0.013$ at transfer, i.e. half of $2\frac{1}{2}$ per cent final relaxation loss. In the majority of cases $f_{pu} = 1750$ N/mm$^2$, although super strand $f_{pu} = 1860$ N/mm$^2$ with 3.5 per cent relaxation at 75 per cent stress level may be obtained in certain situations.

The state of stress after all losses have occurred is similar to the above except that $\eta$ now refers to the total losses. These are:

(a) Prestressing steel relaxation losses, as given above
   The long-term loss for Class 2 wire and strand is multiplied by a factor of 1.2 times the 1000 hour relaxation test value provided by manufacturers (or as given in BS 5896 [5.18])

(b) Concrete shrinkage losses
   This is the product of the shrinkage per unit length (taken as $300 \times 10^{-6}$ for indoor manufacture) and modulus of elasticity of the tendons (taken as 200 kN/mm$^2$, although 195 kN/mm$^2$ is more applicable to helical strand which has a slight tendency to unwind when stretched). This gives a shrinkage loss of 60 N/mm$^2$, or 4.89 per cent at 70 per cent prestress using $f_{pu} = 1750$ N/mm$^2$, i.e. $100 \times 60/0.7 \times 1750$.

(c) Elastic shortening losses
   This is based on the initial short-term modulus of elasticity for the concrete at transfer taken from the usual database. An average value of 28 kN/mm$^2$ may be taken for this with jeopardizing accuracy. Elastic shortening is calculated using the prestress in the concrete $f_{cc}$ at the level of the tendon and therefore the extreme fibre stresses are first required. Hence the elastic loss in the steel is $f_{cc} E_s/E_{ci}$.

(d) Creep of concrete losses
   This is proportional to the stress in the concrete at the level of the tendon. A specific creep strain (i.e. creep per unit length per unit of stress) of 1.8, for indoor curing and loading at 90 days, is used in the UK. Hence, the creep loss is 1.8 times $f_{cc} E_s/E_{ci}$.

Exact solutions for each of these are easy to compute (and will be given in a design exercise). However, typical values for strands or wires stressed to 70 per cent of their ultimate are:

- 2 to 8 per cent for relaxation
- 5 per cent shrinkage
- 5 to 7 per cent for elastic shortening
- 8 to 12 per cent for creep loss.

Total losses range from 19 to 26 per cent for minimum to maximum levels of prestress. The design effective prestress in the tendons after all losses is given by $f_{pe}$.

The section is considered uncracked and the net cross-sectional area $A$ and moment of inertia $I$ are used to compute maximum fibre stresses $f_{bc}$ and $f_{tc}$ in the bottom and top of the section. The section is subjected to a final prestressing force $P_f = 0.7 f_{pe} A_{ps}$ acting at an eccentricity $e$ from the geometrical neutral axis. The design method is standard prestressed concrete procedure. Using the usual notation, the service moment of resistance $M_{sr}$ is given by the lesser of:

$$M_{sr} = (f_{bc} + 0.45\sqrt{f_{cu}})\, Z_b$$

or

$$M_{sr} = (f_{tc} + 0.33 f_{cu})\, Z_t \tag{5.1}$$

where

$$f_{bc} = P_f \left( \frac{1}{A} + \frac{e}{Z_b} \right) \quad \text{and} \quad f_{tc} = P_f \left( \frac{1}{A} - \frac{e}{Z_t} \right)$$

### 5.2.8.2   Ultimate limit state of flexure

The ultimate flexural resistance $M_{ur}$ when using bonded tendons is limited by the following:

- ultimate compressive strength of concrete, $0.45 f_{cu}$
- the design tensile stress in the tendons, $f_{pb}$.

The depth of the (strain responsive) neutral axis $X$ is obtained by considering the equilibrium of the section. The tensile strength of the steel depends on the net prestress $f_{pe}$ in the tendons after all losses and initial prestress levels have been considered. In most hollow core production the ratio $f_{pe}/f_{pu} = 0.50$ to 0.55. Values for $X/d$ and $f_{pb}$ may be obtained from BS 8110, Part 1, Table 4.4 [5.1]. The ultimate moment of resistance of a (rectangular) section containing bonded tendons, all of which are located in the tension zone at an effective depth $d$, is given by:

$$M_{ur} = f_{pb} A_{ps} (d - 0.45\, X) \tag{5.2}$$

If the compressive stress block is not rectangular, as in the case of hollow core slabs where $X >$ cover to cores, the depth to the neutral axis must be found by geometrical or arithmetic means (see Section 5.6). An allowance for other non-prestressed reinforcing bars in the tension zone is made by replacing the area of reinforcement $A_s$ by an equivalent area $A_s f_y/f_{pu}$, about $0.25 A_s$ for high tensile bars. This additional area should be used in the calculation of $X$, $f_{pb}$ and hence $M_{ur}$.

### 5.2.8.3   Strength sensitivity exercise

A sensitivity analysis was carried out on 200 mm deep *Roth* type hollow core floor slabs to determine the effects that 5 mm deviations in both the position of the centroid of the tendons and the overall depth of the slab have on strength. The results are given in Table 5.2 for flexural serviceability and ultimate strengths.

The results shows some interesting statistics at the serviceability limit in that lowering the tendons by 5 mm increase the moment by only one percentage point, and yet raising the tendons by the same amount reduced the moment by three percentage points. The

**Table 5.2** Effects of dimensional deviation on flexural serviceability and ultimate strength of hollow core floor slab.

| Limiting state | Dimension depth deviation (mm) | Relative flexural strength for deviation of positions of tendons* | | |
|---|---|---|---|---|
| | | −5 mm | 0 | +5 mm |
| Serviceability | −5 | 0.92 | 0.95 | 0.96 |
| | 0 | 0.97 | 1.00 | 1.01 |
| | +5 | 1.02 | 1.04 | 1.05 |
| Ultimate | −5 | 0.93 | 0.96 | 1.00 |
| | 0 | 0.97 | 1.00 | 1.03 |
| | +5 | 1.00 | 1.03 | 1.06 |

*Note*: *−ve value means nearer to the neutral axis of the section. A mean value of 40 mm was taken for the zero deviation. All prestressing parameters constant for each size of beam. Concrete strength $f_{cu} = 60$ N/mm².

results for the ultimate condition showed that the effects of negative slab depth and positive tendon deviations, and vice versa, cancel each other. Where the 5 mm deviations are cumulative the reductions or increases are in the order of eight percentage points for both limiting states.

### 5.2.9   Precamber and deflections

The calculated values for precamber and deflections are based on a stiffness $E_c I$ using an average value for the long-term modulus obtained from the design cube crushing strength. Inevitable variations in $E_c$ lead to variations in the deflection stiffness. However, the precamber δ is calculated using the following approximation:

$$\delta = \frac{P_f e L^2}{8E_c I} \qquad\qquad (5.3)$$

where $P_f$ includes *all* losses. The camber in hollow core slabs is about $L/300$.

In-service deflections are calculated in the usual manner taking into consideration the support conditions and loading arrangement. For normal office loading, deflections are in the order of 10 mm to 20 mm for 6 m to 9 m simply supported spans. Deflections are limited to span/500 or 20 mm where brittle finishes are to be applied or span/350 or 20 mm for non-brittle finishes. The net deflection, i.e. imposed deflection minus precamber, is less than span/1500. The usual requirements of limiting the span to effective depth ratio to control deflection is not applicable to prestressed concrete because of the major influence of the prestressing.

### 5.2.10   Shear capacity

Shear is not usually a problem in hcus under uniformly distributed loading. The only time shear is critical is in simply supported units of less than about 3 m in span. Large point or line loads must be considered on an individual basis. An extensive discussion on this is given by Girhammer [5.19].

As in any prestressed concrete elements the design shear capacity is calculated for two conditions; the uncracked section and the cracked section in flexure. The uncracked ultimate shear resistance $V_{co}$ is greater than the ultimate cracked value $V_{cr}$ because the full section properties are considered and a small amount of diagonal tension in the concrete is permitted. The ratio of $V_{co}/V_{cr}$ is typically 2.0 to 2.4 for 200 mm deep units. This is advantageous because the maximum design shear, which usually occurs where the flexural moments are small, may be equated with $V_{co}$ – the greater shear capacity. The position of the cracked section depends very much on the loading conditions, but for uniformly distributed loadings it is usually far enough into the span of the slab so as to render a check against $V_{cr}$ as unwarranted. Strictly speaking, $V_{cr}$ should always be checked.

It is not possible to specify shear reinforcement in hcus, and therefore the prestressed concrete must do the work. A note of caution must be added here; it is not economical to increase the number of prestressing wires merely to satisfy shear. For example, increasing the area of prestressing reinforcement by a factor of 4 increases the flexural capacity by about 2.5 and the shear capacity by only 1.2. The most economical method of increasing the shear capacity is to make solid the ends of the units by opening the tops of the cores and filling them with structural grade concrete (from the same mix as in the units) [5.20]. A less popular alternative is to increase the depth of the slab.

Some hcus have a greater shear capacity, with all things being equal, than others. The slip-formed type floor with the large number of webs has a shear capacity about 30 per cent greater than its extruded counterpart.

Two modes of shear failure are possible: flexural shear and shear tension. The occurrence of each is not merely a function of the shear span ratio, although the latter do usually occur when $a_v/d < 2$ to 3.

Flexural shear occurs if a flexural crack develops into a shear crack. Because of the relatively large amount of prestress in hcus, it is usually the first (or second) flexural crack which causes the shear failure. Failure occurs where the shear force exceeds the shear compression capacity and a single flexural crack initiates shear failure.

Shear tension failures are also possible and embrittled diagonal cracks are found close to the ends of the units propagating through the unprestressed and unreinforced regions of the slab (see Fig. 5.21). Figure 5.22 (from [5.7]) shows the development of prestress in a simple rectangular section. Stresses in hollow core slabs are further complicated by the cross-sectional shape of the unit. It is further assumed that the extremity of the support bearing pressure spreads at an angle of about 45° from the inner bearing point. Walraven [5.7] suggests that an increment equal to half the depth of the slab is required for a shear-tension crack to develop. The direct transmission of forces, by strut and tie action, does not occur in prestressed members which rely on the development of bond to satisfy truss analogy. BS 8110 [5.1] does not recognize this fact; the equation for the uncracked shear resistance given in Section 5.2.10.1 is considered adequate in all respects.

### 5.2.10.1   *Shear capacity in the uncracked region,* $V_{co}$

The term 'uncracked' refers to flexural cracking, and where this exists $V_{co}$ must not be used. The ultimate shear capacity is given by:

**Fig. 5.21** Shear cracking in hollow core slabs.

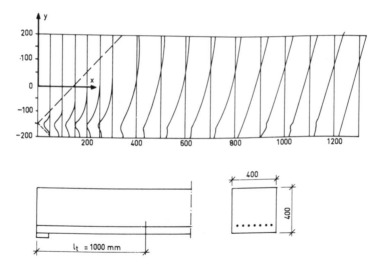

**Fig. 5.22** Contours of prestress development in ends of hollow core units [5.7].

$$V_{co} = 0.67 b_v h \sqrt{f_t^2 + 0.8 f_{cp} f_t}$$ (5.4)

where $f_t = 0.24 \sqrt{f_{cu}}$ .

Although this expression is derived using a rectangular section and not the actual flanged section appropriate to hcus, the difference is accepted as being small. The term

0.67 $b_v h$ may be replaced by the more common $I\, b_v/S$ if the neutral axis falls below the top line of the cores by more than approximately 5 mm. A safety factor is built into the expression because only 0.8 of the compressive stress on the centroidal axis may be used, and $f_{cp}$ is the prestress after losses. The value for $f_t$ is also conservative.

Pisanty [5.10] has found that a reduction of 10 per cent in the calculated value of $V_{co}$ given in BS 8110 [5.1] brings the design value to good agreement with full scale experimental test results. In this case the reduction is necessary because 0.67 $b_v h$ is less than $I\, b_v/S$ for the shape of the hollow core web. See also Table 5.3 where the mean value of the ratio of the test failure load divided by 0.9 $V_{co}$ is 1.05.

Design rules recognize the fact that the critical shear plane may occur in the prestress development zone where $f_{cp}$ is not fully developed. It is shown [5.7] that prestressing forces develop parabolically and therefore a reduced value $f_{cpx}$ is used. It may be shown that $f_{cpx}$ is given by:

$$f_{cpx} = \frac{x}{l_p}\left(2 - \frac{x}{l_p}\right)f_{cp} \tag{5.5}$$

where $l_p$ is the development length (see Section 5.2.11) and $x$ is from the end of the unit measured at 45° to the inner bearing edge. All these factors have been taken into account in the design data given at the end of this chapter.

Lin Yang [5.21] has proposed a design method to calculate the shear web capacity. The equation allows for changes in cross-section, material properties and prestress levels. Web shear is critical at a point at the narrowest part of the web between the voids on a diagonal line inclined at 35° to the horizontal from the edge of the support. This is typically 0.72 $D$ from the support, where $D$ is the depth of the slab. Comparisons with 118 tests at VTT in Finland found that a 95 per cent confidence limit could be used in this equation if the value so obtained is decreased by (approximately) 50 kN.

### 5.2.10.2   Shear capacity in the region cracked in flexure, $V_{cr}$

The value of $V_{cr}$ is calculated using the following (semi-empirical) equation:

**Table 5.3** Summary of shear tests by Pisanty [5.10]

| Test failure load (kN) | Predicted $V_{cr}$ (kN) | Predicted $V_{co}$ (kN) | Test/0.9 $V_{co}$ |
|---|---|---|---|
| 178.50 | 133.63 | 188.60 | 1.050 |
| 172.50 | 133.84 | 189.75 | 1.008 |
| 174.00 | 133.73 | 189.17 | 1.021 |
| 173.50 | 133.54 | 187.45 | 1.027 |
| 47.50 | 41.92 | 69.44 | 0.760 |
| 67.00 | 41.98 | 69.71 | 1.068 |
| 78.00 | 41.92 | 69.44 | 1.248 |
| 53.75 | 41.77 | 68.88 | 0.867 |
| 47.00 | 36.57 | 47.36 | 1.103 |
| 57.25 | 36.54 | 47.04 | 1.352 |

$$V_{cr} = \left(1 - 0.55\frac{f_{pe}}{f_{pu}}\right) - v_c b_v d + M_o \frac{V}{M} \tag{5.6}$$

The derivation of this equation is long, tedious and well documented. One of the problems in using the above equation is that the decompression moment $M_o$, the shear force and bending moment values, $V$ and $M$, must be known before $V_{cr}$ may be computed. This is an obvious drawback in the preparation of predetermined shear capacity data tables much favoured in the precast industry. A short cut is possible here by using the expression given in the Handbook to BS 8110 as follows:

$$V_{cr} = 0.037\sqrt{f_{cu}}\,b_v d + \left[0.37\sqrt{f_{cu}} + 0.8 f_{pt}\right] - Z_b \frac{V}{M} \tag{5.7}$$

All values except $V$ and $M$ are known for specified section geometry and prestressing levels. It is assumed that at the critical section the design shear $V$ cannot exceed $V_{cr}$, and so therefore $V = V_{cr}$. Similarly the ultimate design moment $M$ cannot exceed $\gamma_f M_s$. In most cases floor dead and live loads are roughly equal and so therefore $M = 1.5 M_s$ approximately. Thus, a unique value for $V_{cr}$ exists.

It appears from tests carried out on 1.93 m long × 300 mm deep extruded slabs in three-point bending [5.10] that Equation (5.7) is conservative. Table 5.3 summarizes the test data and comparisons with BS 8110 [5.1].

### 5.2.11  Anchorage and bond development lengths

The notion of anchorage and bond transfer lengths in prestressed concrete is well established. However, in cases where helically wound strand is used as the prestressing medium, the unique properties give rise to increased stress transfer lengths, known as 'development lengths'. These distances are indicated in Fig. 5.11. Development lengths may be of the order of 1.5 to 2.0 m for 12.5 or 12.9 mm diameter strand and therefore must form part of this study so that the designer can assess the implications of flexural cracking in this region. The situation is exacerbated by the fact that shear, which is most critical at about 0.5 $D$ from the support at the narrowest point in the web, relies entirely on the development of prestress.

When the tendon is released at transfer, frictional bond stresses develop around the tendon. Equilibrium is reached when the total bond force equals the prestressing force in the tendon, and the tendon stops slipping. The length of tendon necessary to reach the state of equilibrium is called the 'transfer length', $l_t$. Typical values for $l_t$ are 50 to 80 diameters, and so 70 diameters is used in design.

If a flexural crack occurs within the transfer length it is not possible to develop the full prestressing force because the equilibrium has been destroyed. Thus, in flexurally cracked concrete an additional length is necessary to ensure that the ultimate stress in the tendon is developed. This additional distance is called the 'flexural–bond length' and the total length is the 'development length', $l_d$.

The stress at the end of the transfer length is equal to the prestress after losses. The stress at the end of the development length is equal to the rupture stress in the tendon. Thus if a flexural crack occurs at a distance $x$ from the end of the slab (Fig. 5.11), then if:

- $x > l_d$, full flexural capacity is achieved
- $l_t < x < l_d$, the capacity is larger than the flexural cracking moment but smaller than the ultimate moment

- $x < l_t$, the anchorage capacity is smaller than the cracking moment, so that failure occurs upon flexural cracking. Also the cracking moment is reduced due to incomplete development of prestress in that zone.

Walraven and Mercx [5.7] have reported that the ACI Code [5.22] formulae for:

$$l_t = f_{pe} \frac{\phi}{21} \tag{5.8}$$

$$l_d = l_t + \left[ (f_{pu} - f_{pe}) \frac{\phi}{7} \right] \tag{5.9}$$

are adequate, where $\phi$ = bar diameter. High concrete strengths do not guarantee good anchorage capacity, it depends mainly on good compaction around the tendon. There is more laitance around the strands in units produced by the extrusion process (compared with slip forming), which may have a slight effect on the bond characteristics.

BS 8110, Part 1, clauses 4.3.8.4 and 4.10.3 [5.1] gives $l_d$ = greater of $D$ or $K_t \phi / \sqrt{f_{ci}}$ where $K_t = 240$ for strand, 400 for crimped wire and 600 for plain wire.

Recent finite element work by Akesson [5.23] has simulated the bond around 7-wire 12.9 mm diameter helical strands in 215, 265 and 380 mm deep hollow core slab sections shown in Fig. 5.23. The concrete strength was 47.5 N/mm². The main objectives of the study were to determine the extent of cracking in the interface, measure transmission lengths, and compare strand pull-in with actual measured values by Tassi [5.24] and Gylltoft [5.25]. The strands were stressed to $f_{pe} = 0.54 f_{pu}$ to $0.60 f_{pu}$ and the results show transfer lengths were the order of 340 mm, i.e. 26 $\phi$. In other words the denominator in Equation (5.8) would read 38, not 21. The strand pull-in was about 0.8 mm. A failure analysis was also carried out using $f_{pe} = f_{pu}$ whereupon $l_t$ increases to 719 mm and the pull-in increases to 2.19 mm in the worst case.

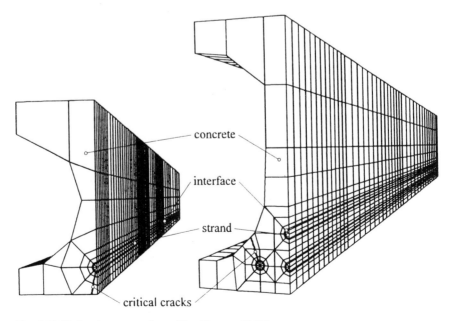

**Fig. 5.23**  Finite element mesh used by Akesson [5.23].

Figure 5.24 shows contours of the splitting stresses through the centre of the web. The maximum stress is 2.18 N/mm², i.e. 0.32 $\sqrt{f_{cu}}$ which under normal circumstances would probably not lead to cracking. However, in this case cracking was found in the concrete closest to the interface over a distance of between 172 and 191 mm.

### 5.2.12   Cantilever design using hollow core slabs

Cantilever spans of up to about 1.5 m long for units 200 to 250 mm deep may be designed using one of two methods. If no structural topping screed is used, one solution is to provide top steel in the cantilevered part and in the main span for a considerable distance to provide a tie-back with an adequate factor of safety of 2. Figure 5.25 shows an example of 750 mm long cantilever units with additional top steel. The disruption to production is considerable, as seen in Fig. 5.26. The most onerous partial load factors are used.

The second solution is to use debonded strands so that no upward prestressing force is applied in the cantilevered portion, but this may be wasteful if the cantilever is short compared with the main span. Where debonded tendons are used, the transmission lengths for top steel should be increased by 50 per cent due to the varying nature of concrete close to the tops of units.

In either case there is considerable disruption to the manufacturing procedures and a cost increase in the order of 10 to 15 per cent.

### 5.2.13   Wet cast reinforced hollow core flooring

Not all hollow core flooring units are manufactured using the long line extrusion or slip forming methods described above. These types of units are described as *dry cast* because of the very low water content in the mixes. However, wet cast units are manufactured to the typical profiles shown in Fig. 5.27. Up to 40 per cent of the North European market for domestic hollow core flooring is wet cast into moulds of fixed dimensions and stock piled for building contractors to purchase as the need arises. This is not the case with long-line manufacture which in the main is made to order.

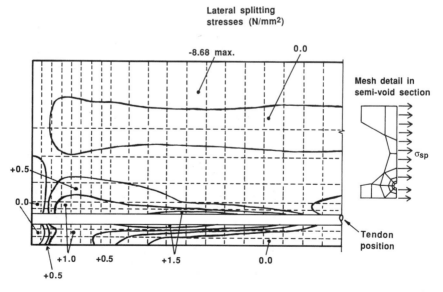

**Fig. 5.24** Principle tensile stress contours around tendons by Akesson [5.23].

**Fig. 5.25** Cantilevered hollow core slabs.

**Fig. 5.26** Making cut-outs in to the tops of hollow core units to form reinforced concrete slots to resist hogging bending moments.

Section properties and moment and shear capacities are comparable to the dry cast units and self weights are very low in the deeper units, e.g. 300 mm. The big advantage with the wet cast method is that additional reinforcement may be added to standard cages for any special requirement, e.g. point loads, large voids, high shear, cantilevers, etc. The cost penalty is small.

The American trade name for this type of unit is 'Span-Deck', and they have been

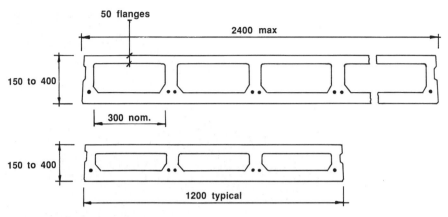

**Fig. 5.27**  Wet cast hollow core profiles.

used for floors, walls and self-finished cladding panels, as illustrated in Fig. 5.28. Manufactured in a two-part process, a 50 mm thick (and sometimes lightweight concrete of low strength) soffit is first cast, followed by the webs (which contain the prestressing tendons) and the top flange. Voids are formed using fibreglass or timber formers which are drawn out 24 hours after casting.

*Note*: The data in Charts 5.1, 5.2 and 5.3 are derived for particular cross-sections of slabs, and as such should not be used to design floor slabs of the same name, but which may have different sectional and material properties.

## EXERCISE 5.1

Design a 200 mm deep non-composite dry-cast hollow core slab to carry a superimposed loading of 5.0 kN/m$^2$ over a simply supported clear span of 6.6 m. End bearing lengths are 75 mm nominally, and the units are supported by grade 40 precast concrete beams. Allow 1.0, 1.0 and 0.6 kN/m$^2$, respectively, for partitions, finishes, services and suspended ceiling, respectively. A two hour fire rating is required. Check the unit for flexure, shear and bearing.

### Solution

Use Spiroll Unit 20SS. See also Table 5.4.
Effective span = 6600 + 75/2 + 75/2 = 6675 mm.

**Table 5.4**  Data for Exercise 5.1.

| Loading | | Service | Ultimate |
|---|---|---|---|
| Dead loads: | Self weight of slab* | 2.67 | 3.74 |
| | Partitions, finishes, services and ceiling | 2.60 | 3.64 |
| | | 5.27 | 7.38 |
| Live loads: | Superimposed | 5.00 | 8.00 |
| | Total | 10.27 | 15.38 |

* Ignore the insitu concrete in the longitudinal gaps between the units.

**Fig. 5.28** Wet cast hollow core units used for floors as well as walls (courtesy of Span-Deck, US).

### Design in flexure
$M_s = 10.27 \times 6.675^2/8 = 57.2$ kNm/m $\times$ 1.2 (for unit width) $= 68.7$ kNm per unit.
$M_u = 15.38 \times 6.675^2/8 = 85.7$ kNm/m $\times$ 1.2 (for unit width) $= 102.8$ kNm per unit.

### Design in shear
$V_u = 15.38 \times 6.675/2 = 51.3$ kN/m $\times$ 1.2 (for unit width) $= 61.6$ kN per unit.
Referring to design data Chart 5.1, use unit ref. 20SS77, requiring 7 No. 10.9 mm diameter strands of total $A_{ps} = 497$ mm$^2$ (0.204% $A_c'$).

### Design in bearing (see also Section 6.7)
Net bearing length (clause 5.2.3.3) = 600 mm.

**Chart 5.1** Example data for 150 to 260 mm deep 'Spiroll' prestressed concrete hollow core slabs, non-composite section properties and capacities.

*Basic section properties*

$b = 1200$ mm

$A_c(\text{net}) = 135\,030$ mm$^2$

$I_{xx} = 678.45 \times 10^6$ mm$^4$

$f_{cu} = 60$ N/mm$^2$

$h = 200$ mm

$y_t = 97.96$ mm

$Z_t = 6.72 \times 10^6$ mm$^3$

No. of circular voids = 6

$y_b = 102.04$ mm

$Z_b = 6.65 \times 10^6$ mm$^3$

| Unit type | Width (mm) | $M_s$ (kN m) | $M_u$ (kN m) | $V_{co}$ (kN) | $100\,A_{ps}/A_c'$ |
|---|---|---|---|---|---|
| 15S3322 | 600 | 17.40 | 22.83 | 43.8 | 0.172 |
| 15S56 | 600 | 24.09 | 37.74 | 53.6 | 0.316 |
| 15S57 | 600 | 28.65 | 45.15 | 55.2 | 0.394 |
| 15SS7322 | 1200 | 34.87 | 45.57 | 93.1 | 0.171 |
| 15SS7622 | 1200 | 41.64 | 61.51 | 106.9 | 0.242 |
| 15SS96 | 1200 | 45.44 | 69.10 | 112.2 | 0.285 |
| 15SS97 | 1200 | 54.08 | 87.01 | 116.1 | 0.355 |
| 20S43 | 600 | 28.28 | 33.91 | 42.4 | 0.128 |
| 20S46 | 600 | 34.45 | 47.36 | 49.7 | 0.187 |
| 20S47 | 600 | 40.98 | 62.10 | 51.5 | 0.233 |
| 20SS3643 | 1200 | 57.37 | 70.07 | 95.7 | 0.129 |
| 20SS76 | 1200 | 64.65 | 83.74 | 105.6 | 0.164 |
| 20SS77 | 1200 | 75.58 | 110.71 | 109.4 | 0.204 |
| 20SS78 | 1200 | 88.29 | 140.29 | 113.0 | 0.270 |
| 26SS68 | 1200 | 121.99 | 171.37 | 144.0 | 0.181 |
| 26SS88 | 1200 | 144.05 | 223.33 | 152.52 | 0.241 |
| 26SS108 | 1200 | 163.51 | 271.72 | 159.7 | 0.302 |

$A_c' = $ gross cross-sectional area including voids, e.g. $1200 \times 200$ mm$^2$. Serviceability moments are for Class 2 elements to BS 8110, clause 4.3.4.3(b) [5.1] using $f_{ct} = 3.50$ N/mm$^2$. Ultimate moments are calculated in accordance with BS 8110, clause 4.3.7.3 using $f_{pb} = 0.87 f_{pu}$. Shear forces are calculated in accordance with BS 8110, clause 4.3.8.4 using $f_{cpx}$ measured at approximately 175 mm from end of unit and using a transmission length = 515 mm.

The code is: 15, 20, 26 = depth in cm; S = 600 mm wide, SS = 1200 mm wide.

Following digits give number and type of bar:

- type 2 = 5 mm dia. wire (area = 19.6 mm$^2$)
- type 3 = 7 mm dia. wire (area = 38.5 mm$^2$)
- type 6 = 9.3 mm dia. strand (area = 52.3 mm$^2$)
- type 7 = 10.9 mm dia. strand (area = 71.0 mm$^2$)
- type 8 = 12.5 mm dia. strand (area = 94.2 mm$^2$).

Net bearing width (clause 5.2.3.7) = 75 – 15 = 60 mm.

Bearing stress = $61.6 \times 1000/600 \times 60 = 1.71$ N/mm$^2$ < $0.4 f_{cu} = 16$ N/mm$^2$.

## EXERCISE 5.2

Repeat Exercise 5.1 using Roth type 200 mm deep × 1200 mm wide slabs.

## Solution

Additional loading due to increased self weight of Roth units = 2.94 – 2.67 = 0.27 kN/m$^2$.

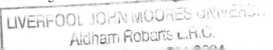

**Design in flexure**

$M_s = 10.54 \times 6.675^2 / 8 \times 1.2$ wide $= 70.4$ kN m per unit.

$M_u = 15.76 \times 6.675^2 / 8 \times 1.2$ wide $= 105.4$ kN m per unit.

**Design in shear**

$V_u = 15.76 \times 6.675/2 \times 1.2$ wide $= 63.6$ kN per unit.

Referring to design data Chart 5.2, use unit ref. 2R (0 + 29) unit, requiring 29 No. 5 mm diameter wires of total $A_{ps} = 569$ mm$^2$ (0.237% $A_c'$). Chart 5.3 gives the tendon locations and prestressing values for these slabs .

## 5.3  DOUBLE-TEE SLABS

### 5.3.1  General

A general introduction to double-tee slabs is given in Section 3.2.2.2. Double-tees are cast in long steel moulds and, in more than 90 per cent of cases, are prestressed for

**Chart 5.2** Example data for 200 mm deep 'Roth' prestressed concrete hollow core slabs, non-composite section properties and capacities.

*Basic section properties*

| | | |
|---|---|---|
| $b = 1200$ mm | $h = 200$ mm | No. of voids = 11 |
| $A_{c(net)} = 142\,311$ mm$^2$ | $y_t = 102.7$ mm | $y_b = 97.3$ mm |
| $I_{xx} = 628.1 \times 10^6$ mm$^4$ | $Z_t = 6.1154 \times 10^6$ mm$^3$ | $Z_b = 6.4558 \times 10^6$ mm$^3$ |
| $f_{cu} = 50$ N/mm$^2$ | | |

| Unit type | Width (mm) | $M_s$ (kN m) | $M_u$ (kN m) | $V_{co}$ (kN) | 100 $A_{ps}/A_c'$ |
|---|---|---|---|---|---|
| 2R (0 + 8) | 1200 | 36.54 | 35.94 | 129.8 | 0.065 |
| 2R (0 + 10) | 1200 | 40.35 | 44.58 | 133.0 | 0.082 |
| 2R (0 + 12) | 1200 | 44.09 | 53.10 | 136.1 | 0.098 |
| 2R (0 + 14) | 1200 | 47.64 | 61.33 | 139.0 | 0.114 |
| 2R (0 + 16) | 1200 | 51.13 | 69.43 | 141.8 | 0.131 |
| 2R (0 + 19) | 1200 | 56.55 | 81.88 | 145.8 | 0.155 |
| 2R (0 + 21) | 1200 | 59.84 | 89.60 | 148.4 | 0.172 |
| 2R (0 + 23) | 1200 | 63.14 | 97.24 | 150.9 | 0.188 |
| 2R (0 + 25) | 1200 | 66.30 | 104.6 | 153.2 | 0.204 |
| 2R (0 + 27) | 1200 | 69.46 | 111.8 | 155.6 | 0.221 |
| 2R (0 + 29) | 1200 | 72.56 | 117.6 | 157.8 | 0.237 |
| 2R (0 + 31) | 1200 | 75.53 | 122.6 | 160.0 | 0.253 |
| 2R (0 + 33) | 1200 | 78.05 | 125.6 | 162.1 | 0.270 |

*Note:* $f_{bc} = ((M_s/6.4558) - 3.2)$ N/mm$^2$ where $M_s$ is in kN m units.

$A_c' = $ gross cross-sectional area including voids, i.e. $1200 \times 200$ mm$^2$.

Serviceability moments are for Class 2 elements to BS 8110, clause 4.3.4.3(b) using $f_{pt} = 3.2$ N/mm$^2$. Ultimate moments are calculated in accordance with BS 8110, clause 4.3.7.3 using $f_{pb} = 0.87 f_{pu}$. Shear forces are calculated in accordance with BS 8110, clause 4.3.8.4 using $f_{cpx}$ measured at approximately 175 mm from end of unit and using a transmission length = 285 mm.

**Chart 5.3** Example prestressing data for 200 mm deep 'Roth' prestressed concrete hollow core slabs. Although Chart 5.2 gives the most widely used design data, further information is often required to enable a designer to check the behaviour of the floor slab if more then just the service and ultimate moments are required. For example, in a composite design the designer needs to be able to check the stresses at the different loading stages, and to calculate the depth to the centroid of the tendons. Therefore to complete the design data file, prestressing magnitudes and tendon locations are given here.

| Unit type | No. of and position of tendons | | | | | | Pre-stressing force (kN) | Ecc. (mm) | $f_{tc}$ (N/mm$^2$) | $f_{bc}$ (N/mm$^2$) |
| | Row 1 | | Row 2 | | Row 3 | | | | | |
| | No. | Ht. | No. | Ht. | No. | Ht. | | | | |
|---|---|---|---|---|---|---|---|---|---|---|
| 2R (0 + 8) | | | 8 | 42.5 | | | 158.7 | 54.8 | −0.31 | +2.46 |
| 2R (0 + 10) | | | 10 | 42.5 | | | 196.6 | 54.8 | −0.38 | +3.05 |
| 2R (0 + 12) | | | 12 | 42.5 | | | 233.7 | 54.8 | −0.45 | +3.63 |
| 2R (0 + 14) | 1 | 27.5 | 12 | 42.5 | 1 | 62.5 | 270.2 | 54.4 | −0.51 | +4.18 |
| 2R (0 + 16) | 2 | 27.5 | 12 | 42.5 | 2 | 62.5 | 306.1 | 54.2 | −0.56 | +4.72 |
| 2R (0 + 19) | 4 | 27.5 | 12 | 42.5 | 3 | 62.5 | 358.2 | 54.8 | −0.69 | +5.56 |
| 2R (0 + 21) | 5 | 27.5 | 12 | 42.5 | 4 | 62.5 | 392.3 | 54.6 | −0.74 | +6.07 |
| 2R (0 + 23) | 6 | 27.5 | 12 | 42.5 | 5 | 62.5 | 425.8 | 54.4 | −0.79 | +6.58 |
| 2R (0 + 25) | 7 | 27.5 | 12 | 42.5 | 6 | 62.5 | 458.7 | 54.2 | −0.84 | +7.07 |
| 2R (0 + 27) | 8 | 27.5 | 12 | 42.5 | 7 | 62.5 | 490.9 | 54.1 | −0.89 | +7.56 |
| 2R (0 + 29) | 9 | 27.5 | 12 | 42.5 | 8 | 62.5 | 522.5 | 53.9 | −0.94 | +8.04 |
| 2R (0 + 31) | 10 | 27.5 | 12 | 42.5 | 9 | 62.5 | 553.4 | 53.8 | −0.98 | +8.50 |
| 2R (0 + 33) | 10 | 27.5 | 12 | 42.5 | 11 | 62.5 | 585.1 | 52.7 | −0.93 | +8.89 |

Note the sign convention; compression is positive. Ht. = height, Ecc. = eccentricity.

Thus, slab unit reference 2R(0 + 19) contains 19 No. 5.0 mm diameter tendons of which 4 No. are located at a centroidal distance of 27.5 mm from the soffit of the slab, 12 No. at 42.5 mm, and 3 No. at 62.5 mm from the soffit. The resulting centroid of the entire group of tendons, all of which are *assumed* to be equally stressed, is at 42.5 mm from the soffit, or at an effective depth of $d = 157.5$ mm from the top of the slab. If the depth to the centroidal axis of the concrete is $y_t = 102.7$ mm, the eccentricity of the tendon group is $d - y_t = 54.8$ mm. The prestressing force given in Chart 5.3 includes for the losses described in the text. The resulting prestress is calculated from the equations also given in the text. The tensile stress in the top of the unit should not exceed +3.2 N/mm$^2$, and the compressive stress in the bottom should not exceed −16.5 N/mm$^2$.

Note that the value chosen for the centroidal distance to the majority number of tendons, i.e. 42.5 mm, ensures that the average cover to the tendons is equal to or exceeds 40 mm. This is the cover distance necessary to achieve a two hour fire resistance.

strength. Reinforced double-tees are used, but long-term deflections have precluded their use for spans in excess of about 8 m. Concreting 'trains' driven by two operatives are used in some high production factories, Fig. 5.29 (a) and (b). The concrete is of medium workability (75 to 100 mm slump) and is pumped or skipped to the mould. The top surface is hand-tamped to give up to 6 mm furrows in the surface (Figs 5.30 and 5.31). This provides an excellent shear key to the insitu concrete topping, and no interface shear reinforcement is necessary. The design strength is at least 50 N/mm$^2$. Detensioning

(a)

(b)

**Fig. 5.29** Manufacture of double-tees. (a) Using concrete pouring arms and (b) using concreting train.

takes place between 12 and 24 hours after casting and at a strength as required by the design prestress, which is usually 35 to 40 N/mm² after steam curing.

The ends of the prestressing tendons require protection because they will be exposed in the building, particularly in car parks. A cold bitumen paint is used, as shown in Fig. 5.32. Flame cut tendons do not require additional protection.

**Fig. 5.30** Two part manufacturing methods - the web is poured and vibrated first, followed by the flanges.

**Fig. 5.31** Surface finishes to the tops of the units are tamped rough.

### 5.3.2   Design

The slabs are designed as T sections. The flexural and shear strength derives from the deep webs, with the thin flanges providing lateral stability to the units. Deflected or

**Fig. 5.32** Half joint in the ends of double-tee units.

debonded tendons are used in some cases to overcome transfer stress problems in long span units.

The depth of the section is reduced at the ends of the units to facilitate a recessed bearing on to the supporting beam or wall. The concrete is heavily reinforced at this point because it is subjected to shear, bending and bearing. In addition, stress concentrations occur at the internal corner which is why triangular chamfers are used. A truss analogy, or strut and tie method of analysis (similar to Section 4.3.9) is used to calculate the forces. Figure 5.33 shows a typical end shear cage in which the vertical bars are provided for the ultimate shear resistance, and the horizontal U-bars are for the lateral bursting resistance.

Welded connections between adjacent double-tee units, or between the units and supporting member, shown in Fig. 5.34, provide an ultimate shear resistance of about 25 kN/m, but which is ignored in the final design. Electrodes grade E43 are used to form short continuous fillet welds between fully anchored mild steel plates (stainless steel plates and electrodes may be specified in special circumstances). A small saw cut is made at the ends of the cast-in plate to act as a stress reliever to the heated plate during welding.

Double-tee units are either designed compositely with a structural screed, in which case the flange thickness is 50 to 75 mm, or are self-topped with thicker flanges around 120 mm. In the former, vertical and horizontal shear are transferred entirely in the insitu structural screed using a design value for shear stress of 0.45 N/mm$^2$. In the latter, projecting reinforcement loops, typically T8 at 200 mm c/c and lapped to the mesh inside the flange, are provided in recesses. These are insitu grouted on site, and as such are a source of major concern with regard to thermal and shrinkage effects, and waterproofing in parking structures.

Double-tee units often contain web openings at positions that will not coincide with the longitudinal strands and are preferably near to the geometric centroid of the unit.

**Fig. 5.33** Shear cage in the ends of double-tee units.

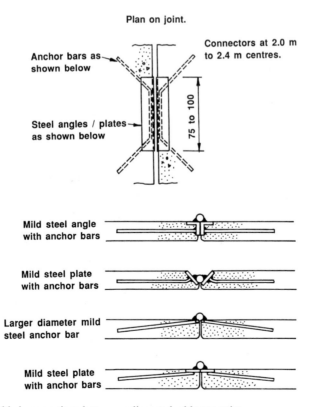

**Fig. 5.34** Welded connections between adjacent double-tee units.

Web openings do not have a deleterious effect on strength or stiffness providing that a few simple rules are observed [5.26, 5.27]. It is found that prestressed units with openings behave as Vierendeel trusses and the shear at the opening is distributed between the top and bottom chords in proportion to the chords' stiffness.

Tests carried out by Savage *et al.* [5.28] on 24 in. deep units have shown that where voids are placed directly beneath the flange and are not greater than half the depth of the unit, as shown in Fig. 5.35, the openings did not affect the behaviour of the voided slab until the applied bending moment was equal to 1.4 times the service moment. The general rules are as follows:

- web openings should not be placed in high shear regions, typically in the first quarter span
- web openings should not be placed nearer than 50 mm from the nearest prestressing strand
- the centre of the void should coincide with the centroid of the unit
- reinforcement, either in the form of stirrups or inclined links, should be placed on each side of the opening to aid crack control
- the corners of the openings should be chamfered or rounded.

### 5.3.3   Flexural and shear capacity, precamber and deflections

The analysis given in Sections 5.2.8 to 5.2.11 is also adopted for double-tee units with

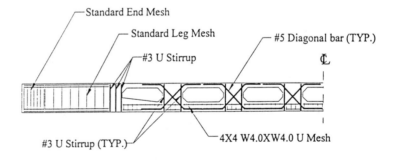

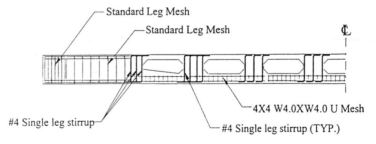

**Fig. 5.35** Reinforcement around large openings in webs in double-tee slabs.

due consideration for the cross-sectional geometry and the concentration of the main flexural reinforcement in the webs. Additional shear capacity is provided where tendons are deflected. Typical load vs span data for these units are given in Fig. 5.10(b).

### 5.3.4   Special design situations

Double-tees may be used as cantilever balconies by adding additional reinforcement in the top flange. It is not possible to use halving joints over cantilever support beams and therefore the structural zone is often a prohibitive factor in this design. The main problems are in the detailing and the avoidance of cold bridging if the units remain physically exposed. Precast rib end closure pieces may be used to complete the precast assembly.

## 5.4   COMPOSITE PLANK FLOOR

### 5.4.1   General

A general introduction to composite plank floor is given in Section 3.2.2.3. The precast concrete is grade C35 if reinforced or C50 if prestressed. The insitu topping need only be grade C25 to C30, and a lightweight aggregate concrete grade C25 is often used. The precast planks are manufactured from normal and lightweight aggregate concrete. There are several proprietary types of plank floor, including 'Omnia' and 'Katzenburger'. The precast part varies in depth from 65 to 100 mm, although some European producers manufacture 40 mm deep planks. Land [5.29, 5.30] gives a good account of many aspects of the use of these units which are beyond the scope of this book. Plank floor lends itself to automated manufacture where robots are used to measure out, select and assemble the reinforcement, and the concreting and lifting operations are carried out using minimal labour (Fig. 5.36).

Figure 3.17 showed the construction of the composite plank floor, and in terms of robustness this floor is considered equal to a cast insitu floor. This floor system has advantages in using a prefabricated soffit unit (smooth finish, no formwork, rapid fixing up to 100 m$^2$ per hour), but carried performance penalties on span and self-weight; see Table 3.1.

Solid plank floor is prestressed or reinforced as shown in Fig. 5.37. The triangulated lattice girders are manufactured using drawn wire spot welded in semi-automatic production. The lattices provide four important functions:

- ensure mechanical bond between the precast and insitu concrete
- provide the precast plank with vertical stiffness in the temporary condition
- the lower bars provide the flexural reinforcement
- the lattice makes a convenient lifting point.

In prestressed units the lattices are positioned next to the prestressing wires. Although the longitudinal bars in the lattices are ignored at the serviceability condition, they *may* be included in the flexural design at ultimate.

**Fig. 5.36** Manufacture of plank floor.

### 5.4.2  Design

The lattice girder is manufactured from 5 or 7 mm diameter hard drawn wire or 8 to 20 mm diameter ribbed high tensile bar with a characteristic 0.2 per cent proof stress of 460 N/mm$^2$. It consists of:

- small diameter bars in pairs, typically 5 or 6 mm diameter mild steel, bent into a 45° zig-zag girder to provide the shear strength in the temporary condition, and the shear resistance at ultimate
- a single top bar, typically T 8 to T 20, depending on the span of the floor and the propping methods used
- at least two bottom bars of similar size, depending on the tensile strength in the reinforcement at ultimate.

The lattices are cut so that a complete stirrup is left intact at both ends of the girder. Any additional high yield reinforcement is attached to the lattice prior to placement in the mould. The cover to the bottom reinforcement is 20 mm for internal exposure and 1 hour fire resistance for a simply supported slab. If the slab is designed continuously this may be increased to 1.5 hours. The clear distance between the top of the precast surface and the bar in the top of the girder should be 20 mm.

The floor slab is designed as single-way spanning, despite the introduction of transverse reinforcement in the topping. The two-way spanning capabilities have not been fully exploited to date.

In the temporary stage, the precast plank is simply supported. The unit may be designed so that unpropped spans of up to 4 m are possible, usually by increasing the number of lattices to increase shear stiffness, but not necessarily increasing the number of bottom bars. The top bar is in compression, but is firmly restrained both vertically and horizontally by the inclined bars making the lattice. By adjusting the size of the top bar and the

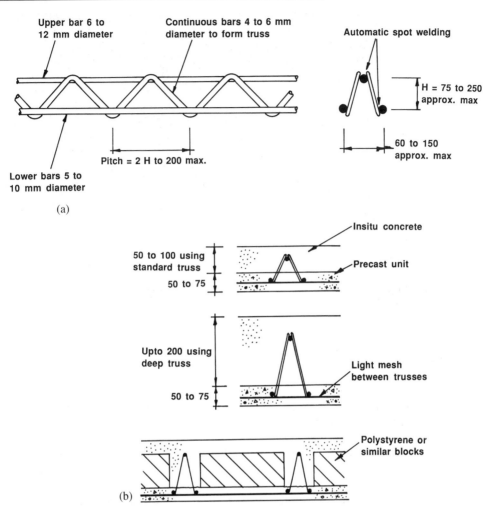

**Fig. 5.37** Details of composite plank floor. (a) Lattices made using automatic production; (b) construction details.

spacing between the lattices it is possible to vary the strength of the plank alone. The unit is most critical when the self-weight of the wet insitu concrete is added to the self-weight of the precast plank. An allowance for construction traffic of up to 1.5 kN/m² is added to the temporary loading when calculating the sizes of bars required.

In the permanent situation the hardened insitu concrete provides the compressive resistance. The flexural sagging resistance at mid-span is governed by the strength of the bottom reinforcing bars, as in an ordinary under-reinforced situation. The flexural hogging moment over the supports is provided by a combination of precast and insitu concrete, as shown in Fig. 5.38, together with some additional site placed reinforcement in the top of the slab.

If the maximum bending moment in the temporary condition is $M_1$ (inclusive of the construction traffic), then the area of top steel in the lattice is given by:

$$A_{s1} = \frac{M_1}{z_1 0.87 f_y} \tag{5.10}$$

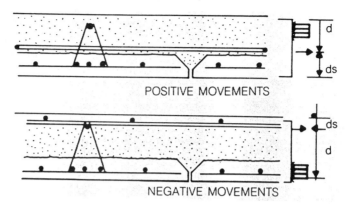

**Fig. 5.38** Flexural resistances in sagging and hogging modes.

where $z_1$ is the centre distance between bars in the lattice. The area of the bottom steel is specified after full service load is considered, but without the effects of the construction load as this load will have been removed when in service. Hence, if the net temporary moment is $M_1'$ and the ultimate moment due to superimposed loading is $M_2$, the area of bottom steel in the lattice is given by:

$$A_{s1} = \frac{M_1'}{z_1 0.87 fy} + \frac{M_2}{z_2 0.87 fy} \tag{5.11}$$

where $z_2$ is the lever arm obtained from the flexural design.

The shear reinforcement in the lattices is designed by taking the vertical component of the axial force in the inclined bars as the only shear resistance against the temporary shear force $V_1$. The area of the lattice's 'shear links' is given by:

$$A_{s1} = \frac{\sqrt{2} V_1}{0.87 f_{yv}} \tag{5.12}$$

Two lattice bars are used to provide the shear reinforcement.

The diagonal bars are also used to transfer shear forces due to superimposed loads $V_2$ in the precast insitu interface. Thus, if the applied shear stress $\tau$ given by:

$$\tau = \frac{V_2}{bd} \tag{5.13}$$

is greater than the limiting value in BS 8110, Table 5.5 [5.1], then the area of shear steel should be increased to carry the whole shear force $F = \tau b$.

The precast unit may be propped at one or two positions. The design of propped slabs is dealt with in Section 6.5.3. In this case spans of up to 6.5 m are possible.

Vertical shear is rarely critical because of the large cover width at the support. It may become critical in the presence of large voids where the effective width of the slab at the support is less than about one quarter of the full width. Additional shear reinforcement is used to take care of the ultimate shear and the problem of stress concentrations near to the corners of voids. The effects of point loads or line loads (e.g. walls around voids) are dealt with in the same manner as for any insitu concrete floor. The only point to note is that the additional bottom steel has to be cast into the precast plank at the factory.

Interface shear is checked according to the details given in Section 6.3. Unless the unit width is greatly reduced for the same reasons as given above, the reinforcement provided by the lattices is sufficient for all purposes. Indeed it is designed to be that way. The top surface of the plank is tamped to give an undulating surface with furrows up to 2 to 3 mm deep.

The resulting load vs span performance of these units is given in Table 5.5.

### 5.4.3   Voided composite slab

Lightweight void formers – typically, but not always made of expanded polystyrene – are pressed into the top surface of the precast plank during manufacture, Fig. 5.39. The

**Table 5.5** Typical clear span data for composite slab.
(a) Units unpropped

Construction

| Unit depth (mm) | Total depth (mm) | Clear span (m) for imposed loading (kN/m²) | | | | | | | | | |
|---|---|---|---|---|---|---|---|---|---|---|---|
| | | 0.75 | 1.5 | 2.0 | 2.5 | 3.0 | 4.0 | 5.0 | 7.5 | 10.0 | 15.0 |
| 75 | 115 | 3.75 | 3.75 | 3.75 | 3.75 | 3.75 | 3.75 | 3.75 | 3.75 | 3.50 | 3.09 |
| | 150 | 3.75 | 3.75 | 3.75 | 3.75 | 3.75 | 3.75 | 3.75 | 3.75 | 3.75 | 3.49 |
| | 175 | 3.75 | 3.75 | 3.75 | 3.75 | 3.75 | 3.75 | 3.75 | 3.75 | 3.75 | 3.61 |
| | 200 | 3.75 | 3.75 | 3.75 | 3.75 | 3.75 | 3.75 | 3.75 | 3.75 | 3.75 | 3.65 |
| 100 | 150 | 5.00 | 5.00 | 5.00 | 5.00 | 5.00 | 5.00 | 5.00 | 5.00 | 4.78 | 4.42 |
| | 200 | 5.00 | 5.00 | 5.00 | 5.00 | 5.00 | 5.00 | 5.00 | 5.00 | 4.78 | 5.42 |
| | 250 | 5.00 | 5.00 | 5.00 | 5.00 | 5.00 | 5.00 | 5.00 | 4.88 | 4.75 | 4.51 |
| | 300 | 4.90 | 4.87 | 4.85 | 4.83 | 4.82 | 4.78 | 4.74 | 4.66 | 4.58 | 4.43 |

(b) Units propped at mid-span

Construction

| Unit depth (mm) | Total depth (mm) | Clear span (m) for imposed loading (kN/m²) | | | | | | | | | |
|---|---|---|---|---|---|---|---|---|---|---|---|
| | | 0.75 | 1.5 | 2.0 | 2.5 | 3.0 | 4.0 | 5.0 | 7.5 | 10.0 | 15.0 |
| 75 | 115 | 6.58 | 6.14 | 5.89 | 5.67 | 5.47 | 5.13 | 4.84 | 4.28 | 3.87 | 3.32 |
| | 150 | 7.50 | 7.36 | 7.08 | 6.83 | 6.60 | 6.21 | 5.88 | 5.24 | 4.77 | 4.12 |
| | 175 | 7.50 | 7.50 | 7.50 | 7.43 | 7.20 | 6.79 | 6.45 | 5.78 | 5.28 | 4.58 |
| | 200 | 7.50 | 7.50 | 7.50 | 7.50 | 7.50 | 7.24 | 6.90 | 6.21 | 5.69 | 4.96 |
| 100 | 150 | 8.08 | 7.56 | 7.27 | 7.01 | 6.78 | 6.37 | 6.03 | 5.37 | 4.89 | 4.22 |
| | 200 | 9.07 | 8.57 | 8.29 | 8.03 | 7.79 | 7.38 | 7.02 | 6.32 | 5.79 | 5.04 |
| | 250 | 9.40 | 9.23 | 8.96 | 8.72 | 8.49 | 8.08 | 7.72 | 7.01 | 6.46 | 5.67 |
| | 300 | 9.40 | 9.40 | 9.39 | 9.16 | 8.94 | 8.55 | 8.20 | 7.50 | 6.95 | 6.13 |

**Fig. 5.39**  Voided composite plank floor (courtesy of MWE Precast Concrete Bhd.).

adhesion and friction between the polystyrene and wet precast concrete is sufficient to hold the block in position throughout the erection and insitu concreting operations. (This is quite important because the blocks are buoyant in the wet insitu concrete.) The reduction in self-weight of a voided floor compared to a solid floor is 30 to 60 per cent for depths of slab from 200 to 350 mm, respectively.

**EXERCISE 5.3**

Design a composite plank floor to carry a characteristic superimposed live load of 5 kN/m$^2$ over a simply supported effective span of 3.5 m. No propping is allowed.

Use concrete $f_{cu} = 40$ N/mm$^2$ for the precast, $f_{cu} = 25$ N/mm$^2$ for the insitu. Use drawn wire for the bars in the precast plank, and a square wire mesh in the insitu topping, using $f_y = 460$ N/mm$^2$. Cover to all reinforcement = 25 mm.

Check the design at both the temporary and permanent stages for flexural and vertical shear only. (Interface shear will be dealt with in Chapter 6.)

*Solution*

Deflection control. Basic span/$d$ = 20. Modification factor for an initial estimate $M/bd^2 = 1.0$ is 1.38.

Thus $d = 3500/20 \times 1.38 = 127$ mm.

$h = 127 + 25 +$ say 8 = 160 mm.

Consider 1 m width of floor, using 50 mm deep precast with 110 mm deep insitu topping.

**Loading on lattice in temporary condition**

|  | kN/m$^2$ |
|---|---|
| Self-weight of 50 mm deep precast unit = $0.050 \times 24$ | = 1.20 |

Self-weight of 110 mm deep wet concrete = $0.110 \times 24$    = 2.64
Construction traffic allowance    = 1.00

Total = 4.84

$M_{u1} = 1.4 \times 4.84 \times 3.5^2/8 = 10.37$ kN m/m.
$V_{u1} = 1.4 \times 4.84 \times 3.5/2 = 11.86$ kN/m.
Assume lattice top and bottom bar size = 16 mm.
Lever arm = $160 - 33 - 33 = 98$ mm.
Force in top and bottom bars = $10.37 \times 10^3/98 = 106$ kN/m run.

$$A_s = \frac{106 \times 10^3}{0.87 \times 460} = 264 \text{ mm}^2/\text{m} \times 0.6 \text{ m centres} = 159 \text{ mm}^2 \text{ per lattice.}$$

*Use one no. T 16 top bar (201) in lattices at 600 mm c/c.*
Bottom bars will be specified after full service loads considered. Subtract the effects of the construction load when calculating the force in bottom steel (as this load will have been removed when in service). Hence $A_s = (3.84/4.84) \times 264 = 210$ mm²/m.
Shear per lattice = $0.6 \times 11.86 = 7.12$ kN.
Lattice bars at 45° inclination, thus force in diagonal bar = $7.12/\sin 45° = 10.0$ kN.

$$A_s = \frac{10 \times 10^3}{0.87 \times 250} = 46 \text{ mm}^2/2 \text{ no. bars} = 23 \text{ mm}^2.$$

*Use double R 6 diagonal lattice bars (inclined at 45°).*

### Service loading
When the insitu concrete has hardened it is effectively stress free because the deflections have all occurred whilst the concrete was wet. Therefore the only stresses in the insitu topping derives from superimposed loads = 5.00 kN/m².
$M_{u2} = 1.6 \times 5.00 \times 3.5^2/8 = 12.25$ kN m/m.
$V_{u1} = 1.6 \times 5.00 \times 3.5/2 = 14.0$ kN/m.

### (1) Flexural design:

$$K = \frac{12.25 \times 10^6}{25 \times 1000 \times 127^2} = 0.03 < 0.156.$$

Then $z/d = 0.95$, and the area of the bottom steel is:

$$A_s = \frac{12.25 \times 10^6}{0.95 \times 127 \times 0.87 \times 460} = 254 \text{ mm}^2/\text{m}$$

plus 210 mm²/m from the construction stage = 464 mm²/m $\times$ 0.6 m centres = 278 mm² per lattice.
*Use two no. T 16 bottom bars (402) in lattices at 600 mm c/c.*

### (2) Shear design:

$v = 14.0 \times 10^3/1000 \times 127 = 0.11$ N/mm² < minimum value in code, no additional reinforcement to the lattice required.

## 5.5   PRECAST BEAM AND PLANK FLOORING

### 5.5.1   General

The beam and plank floor is commonly known as the Precast Beam Composite (PBC) system, and is shown generally in Fig. 5.1 and in detail in Fig. 5.40 (a)–(c). It is a three-part construction as follows:

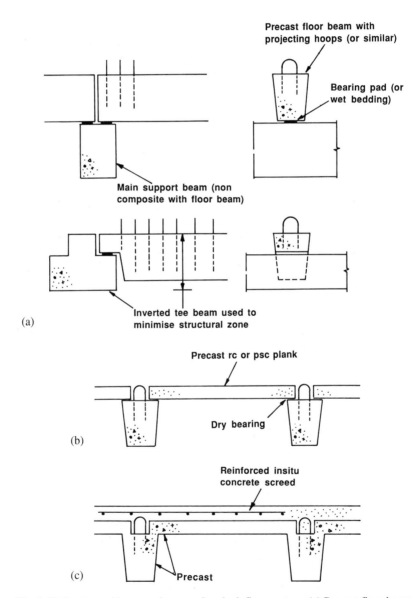

**Fig. 5.40** Prestressed beam and composite plank floor system. (a) Precast floor beam positioned; (b) precast plank fixed and (c) insitu topping added.

(1) Prestressed beams are the major structural components which span between the primary beams in the structure. They are available in depths of up to 550 mm, and weigh up to 4.3 kN/m run. The spacing between the beams varies between 900 and 2400 mm. The maximum span for roofs is 20.9 m, and for office floors is 18.2 m.

(2) Precast planks, which are between 50 and 75 mm deep and of the same specification as above, span transversely to the beams. The width of the plank may be up to 2.4 m, but smaller units, say 600 mm wide and weighing about 50 kg, may be man-handled into their final position which is clearly beneficial from a construction point of view. The design of the plank is as given in Section 5.4.2.

(3) A 40 to 50 mm thick grade C40 insitu concrete topping is added to the top of the planks, making the total construction span-depth, inclusive of all three items, in the order of 25:1 for offices, and 30:1 for domestic buildings. Voids may easily be formed between the prestressed beams, up to about 2.0 m wide. All the details given in Section 5.4 above may be incorporated into this system.

The PBC method is more versatile geometrically than the hollow core or double-tee methods and may be adapted to suit a wide range of plan configurations. The floor system may be used in insitu concrete, precast concrete, or steelwork structures.

### 5.5.2   Design of prestressed beams in the beam and plank flooring system

The beam's cross-section is trapezoidal with a 2.5° side splay. A single tapered mould may be used for any depth of unit, although the manufacturers prefer to work to a small number of depths, usually 400, 450, 500 and 550 mm. Similarly the width of the beam is restricted in size, usually to 320 mm at the bottom.

The beams are designed according to the basic prestressing principles. The depth of the ends of the beams may be reduced to form a half joint in the same manner as for double-tee units. Cantilever balconies may be formed either by reinforcing the beam for the negative moment, or by utilizing the top reinforcement in the structural topping. The hogging moment induced from the prestressing must be nullified (or reversed) in the cantilever part.

## 5.6   DESIGN CALCULATIONS

### 5.6.1   Hollow core flooring

The following calculations appear courtesy of *Richard Lees Ltd.*

(1) Non-composite design for unit reference 20SS78

This is a nominally 1200 mm wide by 200 mm deep unit as shown in Fig. 5.41. The unit contains 6 no. cores each of 150 mm in diameter.

(2) Unit properties per 1200 mm width

Width = 1200 mm (bottom), 1168 mm (top), actual depth = 203 mm.

Area = 135 030 $mm^2$, $I = 678.45 \times 10^6$ $mm^4$.

Height to neutral axis = 102.04 mm, depth of concrete to core = 28.0 mm.

$Z_b = 6.65 \times 10^6$ $mm^3$, $Z_t = 6.72 \times 10^6$ $mm^3$.

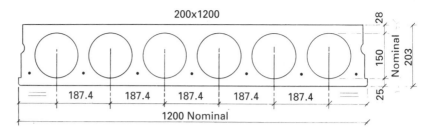

**Fig. 5.41** Cross-section of hollow core slab used in design calculations.

(3) Concrete data

Characteristic cube strength at 28 days = 60 N/mm$^2$, $E_c$ (28 days) = 32 kN/mm$^2$.
Release (or transfer) strength = 35 N/mm$^2$, $E_{ci}$ (release) = 27 kN/mm$^2$.

(4) Reinforcement data

No. of tendons = 7 @ 12.5 mm diameter helical strand.
Relaxation type 2 as quoted by strand supplier = 1.6%.
Prestress to 70 per cent ultimate.
Bottom cover to tendons = 40 mm.
Eccentricity from neutral axis = 55.79 mm.
$A_{ps}$ = 651 mm$^2$, $f_{pu}$ = 1763.4 N/mm$^2$, $f_{pu}, A_{ps}$ = 1148.0 kN.
$E_s$ = 200 kN/mm$^2$ (quoted by supplier).

(5) Serviceability limit state of bending

Initial prestressing force = 0.7 × 1148 = 803.6 kN.
Prestress before losses $f_{bc}$ = 12.69 N/mm$^2$ (bottom), $f_{tc}$ = –0.72 N/mm$^2$ (top).
$f_{cc}$ = 9.64 N/mm$^2$ (at tendon centroid).
Elastic loss = 9.64 × 200/27 = 71.4 N/mm$^2$ = 5.78%.
Creep loss = 1.8 × elastic loss = 10.40%.
Shrinkage loss = 300 × 10$^{-6}$ × 200 × 10$^3$ = 60.0 N/mm$^2$ = 4.86%.
Relaxation = 1.60 (quoted) × 1.2 = 1.92%.
Total losses = 22.97%.
Effective prestressing force after losses = 618.9 kN.
Prestress after losses $f_{bc}$ = 9.78 N/mm$^2$ (bottom), $f_{tc}$ = –0.55 N/mm$^2$ (top).
For class 2 elements, allowable tension ($f_{cu}$ = 60 N/mm$^2$) = 3.50 N/mm$^2$.
Service resistance moment $M_s$ based on stress condition in bottom of unit (Equation (5.1)),

$$M_s = (9.78 + 3.50) \times 6.65 = 88.29 \ kNm.$$

$M_s$ based on top stress = (0.55 + 0.33 × 60) × 6.72 = 136.75 kNm > above, not critical.

(6) Ultimate limit state of bending

$A_{ps}$ = 651 mm, $f_{pb}/f_{pu}$ = 0.87, $F_s = f_{pb} \times A_{ps}$ = 998.76 kN.
All tendons in tension zone, thus $d$ = 203.0 – 46.25 = 156.75 mm.
$F_c$ = 0.45 $f_{cu}$ $b$ 0.9 $X$, hence $X$ = 35.2 mm > 28 mm.
Hence neutral axis falls beneath top of circular core.
By geometry, $X$ = 36.67 mm and $d_n$ = 16.28 mm.
Ultimate resistance moment, $M_u$ = 998.76 × (156.75 – 16.28) = 140.29 kNm.

(7)  Ultimate limit state of shear

*Uncracked*

Transmission length = $240 \times 12.5/\sqrt{35} = 507.1$ mm.

$f_{cp} = 4.58$ N/mm$^2$.

$f_{cpx}$ at a distance 75 (bearing) + 102 height to neutral axis = 177 mm from end = 2.64 N/mm$^2$.

$b_v = 1168 - (6 \times 150) = 268$ mm, $f_t = 0.24\sqrt{60} = 1.86$ N/mm$^2$.

$Ay'$ 4 372 323 mm$^3$.

Ultimate shear resistance, $V_{co} = I b_v (\sqrt{f_t^2 + 0.8 f_{cp} f_t})/Ay' = 41.59 \times 2.71 = 113.01$ *kN.*

*Cracked*

$100 A_s/ b_v d = 1.55$, $v_c = 1.08$ N/mm$^2$.

$M_o = 0.8 \times 9.78 \times 6.65 = 52.01$ kN m.

$f_{pe}/f_{pu} = (1 - 0.2297) \times 0.7 = 0.539$.

Minimum ultimate shear resistance,

$V_{cr \, (min)} = 31.96 + (52.01 \times V_{cr \, (min)}/140.29) = 50.94$ *kN.*

(8)  General data

Fire resistance = 2 hours, thermal resistance = 0.163 m$^2$ °C/W.

Self-weight = 2.67 kN/m$^2$.

# Chapter 6

# Composite Construction

*The design and construction aspects of using precast concrete elements compositely with insitu concrete are described.*

## 6.1  INTRODUCTION

Composite construction offers many advantages in precast concrete design, particularly in enhancing the flexural and shear strength of psc beams where greater axial stresses may be generated in the precast units than in ordinary non-composite designs. To most minds composite construction means adding insitu concrete on top of precast components to form a single unit acting as though it were one. However, in the context of precast design there are many ways in which insitu concrete is used compositely in the structure. For example, composite action is used mainly to:

- increase flexural and shear strength of floor slabs
- tie floor slabs to beams, thereby ensuring a secure bearing, and increasing the flexural and shear strength of beams
- provide the compressive and/or shear transfer between adjacent precast units, e.g. between walls, shear walls and columns, and at column foundations
- ensure floor diaphragm action, with or without structural screeds (see Section 8.2),
- anchor stability tie steel in to precast components (see Section 9.2)

In all cases insitu concrete surrounds the precast components to form a monolithic structure. Shear and compressive forces are carried through the insitu concrete by shear friction, wedging and/or bearing. Tension is effected by fully anchored rebars, or other mechanical means, so that the concrete is confined to prevent lateral splitting. Design values at the interface vary over a wide range depending on the surface characteristics of the joining faces, the loading, and the mode of failure, where non-ductile situations attract higher partial safety factors.

The strength of the two concretes may be different; usually the precast concrete is grade C40 to C60 and the insitu concrete is grade C25 or C30 (see Table 6.1), but this is taken into account in the analysis for both the service and ultimate limit states. Reference is also made to the design recommendations in BS 8110, Part 1, clause 5.4 [6.1].

Non-structural finishing screeds may be applied directly on to precast concrete slabs, but allowances for precamber should be made in calculating overall floor depth. It is only in the presence of very large line or point loads, or in cases where the dynamic or acoustic characteristics of the precast slab are judged to be inadequate, that a structural insitu rc screed might be required. Structural screeds are nearly always necessary where double-tee units are used and are an obvious prerequisite for flat plank construction. For a screed to act compositely with the precast slab, in a structural sense, the concrete must be reinforced and unbroken by service chases, etc.

**Table 6.1** Strengths and short-term elastic modulus for typical concrete used in composite construction [6.1].

| Type of concrete | $f_{cu}$ (N/mm$^2$) | $f_{ct}$ (N/mm$^2$) | $E_c$ (kN/mm$^2$) |
|---|---|---|---|
| Insitu | 25.0 | – | 25 |
| Insitu | 30.0 | – | 26 |
| Precast reinforced | 40.0 | – | 28 |
| Prestressed | 50.0 | 3.2 | 30 |
| Prestressed | 60.0 | 3.5 | 32 |

The two main areas where composite construction is carried out is in floor slabs and beams (Fig. 6.1(a)–(e)). The structural function of some precast elements, e.g. precast planks (Fig. 6.1(c)) such as 'Omnia' floor (Fig. 5.1(d)) rely implicitly on composite action. However, composite action in other elements, e.g. hollow core slabs and beams (Figs 6.1(a) and (e)) is optional and may be used at the discretion of the designer wishing to increase flexural and shear capacities, stiffness, fire resistance and vibration characteristics. Composite construction may also be used to create extended bearings at the ends of units as shown in Fig. 4.3.

The design is carried out in two stages, before and after the insitu concrete has reached its design strength. The main design criteria are:

- flexural and shear strength, serviceability and ultimate states
- confinement or reinforcing of insitu concrete to avoid separation, called 'delamination', from the precast concrete
- interface shear transfer
- constraint of insitu concrete shrinkage.

These items will be considered for the two cases of composite floors and composite beams discussed in Sections 6.5 and 6.7, respectively.

## 6.2  TEXTURE OF PRECAST CONCRETE SURFACES

### 6.2.1  Classification of surface textures

The types of surface which a precast unit may have, prior to receiving the insitu concrete to form a composite section, may be classified according to the following types given by the FIP Recommendations [6.2]. These alternatives are presented in order of increasing roughness:

(1)  smooth, as obtained by casting the unit against a steel or timber shutter
(2)  trowelled or floated to a degree where it is effectively as smooth as (1)
(3)  tamped so that the fines have been brought to the top, but where some small ridges, indentations or undulations have been left
(4)  achieved by slip-forming and vibro-beam screeding
(5)  produced by extrusion technique

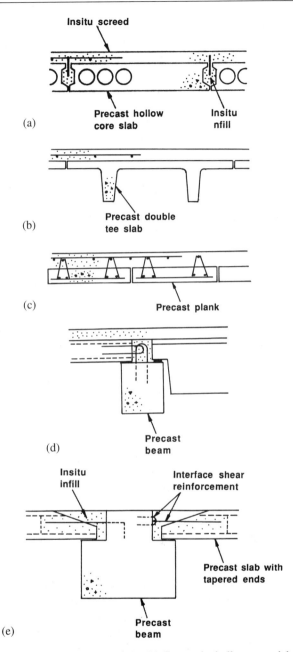

**Fig. 6.1** Composite slab and beam sections. (a) Composite hollow core slab; (b) composite double-tee slab; (c) composite plank floor; (d) composite beam action with a structural screed and (e) requirements for composite beam action without a structural screed.

(6) textured by brushing the concrete when wet ideally to a specified depth of ridge
(7) brushing by a transverse screeder, by combing with a steel rake or tamping with a former faced with a suitable expanded metal
(8) the concrete has been thoroughly compacted but no attempt has been made to

smooth, tamp or texture the surface in any way leaving a rough surface with coarse aggregate protruding (but firmly fixed in the matrix)

(9) the concrete has been sprayed wet, to expose the coarse aggregate without disturbing it

(10) provided with mechanical shear keys.

Type (1) rarely occurs in the types of composite section illustrated in Fig. 6.1 (a)–(d). Type (2) is almost impossible to achieve and would rarely be attempted in practice. Type (3) is possibly the most common type of smooth surface considered in design, although in practice it is usually rough to some extent. The remaining types all have some measure of *roughness* whether naturally occurring or deliberately induced.

With some extrusion processes, Type (5), high vibration can cause bleeding and produce a very smooth surface, but generally this type of surface would have the same level of roughness as Type (6). Types (6) and (7) are possibly the most common textured surfaces, Type (7) probably being at least as effective as Type (9) (which is specifically mentioned in many codes of practice).

Type (8) can produce a very rough surface but there can be problems with it since the concrete near the top may be poorly compacted.

Some indication is given in the literature [6.2] of the degree of roughness covered by these descriptions but there are no clearly defined boundaries between these different types, and what is actually achieved will depend very much on workmanship in individual cases. For example, the effectiveness of the smoother surfaces can be influenced by the presence and type of mould oil. All surfaces not roughened deliberately can be affected by casting procedures and whether or not use is made of super-plasticizers. Whatever approach is adopted, it is essential that the surface is not weakened by destroying the aggregate bond.

Surfaces (5) to (9) inclusive could be generally considered as being at least equivalent to the definition given in the first paragraph of this section of a surface for a construction joint in insitu construction, at least in terms of roughness. It is also impossible in practice to obtain this equivalence with Types (3) and (4) surfaces.

## 6.2.2  Surface treatment and roughness

In designing the interface to ensure composite action, most previous codes have considered surface roughness to be the principal factor giving limiting stresses for different defined categories of roughness – which have not always related to the types of surface which can realistically be achieved in practice. Test data, for example the work by Vesa [6.3] and Gustavsson [6.4], indicate that the treatment of the precast surface is at least as important as the degree of roughness. Factors such as cleanliness, compaction, curing and wetting of the surface have a major influence on the shear strength of the interface. Indeed, with the optimum combination of these, and with good mix design and workmanship for the insitu concrete, it is possible to develop a strength with a smooth surface (Types 3 and 4, and possibly 2 for example) which is equal to, or even greater than, that obtained with a rough surface (Types 7 and 8), where less attention has been paid to surface treatment.

For these reasons, the FIP Recommendations [6.2] places great emphasis on surface treatment and on workmanship requirements for the insitu concrete topping. The major principle in following these Recommendations is that any treatment of the surface – to clean it, or to produce the required level of roughness – should not impair the interface strength. This means causing cracks in the matrix or aggregate.

Roughness can influence the potential bond which may be achieved between the two concretes in the following ways:

- the increase in the interface surface area caused by roughness,
- the fact that natural debonding factors such as laitance skin, dust, water, and so on are concentrated at the bottom zones of the surface, whereas the tops tend to be less affected; the rougher the surface, the less susceptible it is to the quality of the workmanship in cleaning and preparation,
- the shape and inclination of the ridges should ideally lead to a definite mechanical grip (interlocking) between the two concretes. The amplitude and pitch of the ridges or high spots (and their shear strength) should be such as to permit this interlocking to take place. It should also be noted that during curing, ridges will often be exposed to less favourable conditions than the rest of the concrete.

These characteristics of the adhesion between the two concretes are important in considering the deformations involved when failure occurs along an interface. Chemical bond will be destroyed first, but the interlock effect can only be destroyed if all the ridges and high spots fail in shear. Even if this does happen, then if the two concretes are physically prevented from moving apart (by reinforcement, for example) it will be possible to mobilize shear resistance due to friction, where the rougher surface will give a greater frictional resistance. The opportunities to obtain a strength equivalent to a monolithic section (the maximum possible) are therefore considerable.

The qualities which affect bond and shear transfer at the interface surface are surface roughness, surface strength, and cleanliness. Where interfaces are to be bonded, the designer should prescribe the relevant category of roughness and, if bond is particularly important structurally, should consider whether or not to specify methods of achieving the required standard. As an aid to the specification, it is of considerable help if models, photographs or specimens are supplied to the factory.

Various kinds of micro-roughness may be achieved for surfaces which are cast against a mould and which, in general, are characterized as plane or smooth. The mould material itself, as well as the demoulding agent used, has a considerable influence. Thus wood, steel and resin moulds result in different kinds of surface which may be differentiated further by the demoulding agents used. An extreme example is the use of retarders, resulting in exposed aggregate interfaces after surfaces have been washed and wire brushed.

Mechanical means to achieve suitable roughness vary from fibre and steel brushes to rake-like devices, all dependent on the concrete consistency and roughness pattern intended. It is advisable to pre-test the mechanical device chosen to ensure a surface pattern which conforms to the surface category required. The treatment should preferably be done in such a way that resulting ridges are directed normally to the main shearing forces to be transferred at the interface.

## 6.3  CALCULATION OF STRESSES AT THE INTERFACE

Shear at the interface need only be checked for the ultimate limit state. The design method used, based on experimental evidence, will ensure that serviceability conditions are satisfied.

The average ultimate shear stresses at the interface may be calculated using Equation (6.1). The design values in Table 6.2 are based on the use of this formula, and allowance has been made for the small errors in defining the shear stresses that occur in the equation:

$$v_{ave} = \frac{F_v}{bL_z} \qquad (6.1)$$

where $v_{ave}$ = the average shear stress at the cross-section of the interface considered at the ultimate limit state

$F_v$ = the design force in the concrete to one side of the interface

$b$ = the transverse width of the interface

$L_z$ = distance between the points of minimum and maximum bending moment.

If the interface is in a compression zone, then $F_v$ is equal to the compression force in the insitu concrete only, i.e. above the interface. If the interface is in a tension zone, then $F_v$ is equal to the total compression or tension calculated from ultimate loads. The force is distributed evenly over the contact interface breadth and over the length of the beam between points of maximum and zero moment, thus giving the average interface shear stress $v_{ave}$. The average stress is then distributed in accordance with the magnitude of the vertical shear at any section, to give the design shear stress $v_h$. Thus, for uniformly distributed superimposed loading (self-weight does not create interface stress), the maximum stress $v_h = 2 v_{ave}$. For a point load at mid-span $v_h = v_{ave}$ and so on.

Horizontal interface shear stresses $v_h$ are checked for the uncracked section (BS 8110, Part 1, Clause 5.4.7.2 [6.1]) against values in Table 6.2 (reproduced from BS 8110, Part 1, Table 5.5). If $v_h$ is greater than the ultimate stress in Table 6.2 then reinforcement (per 1 m run) is provided (according to Equation (62) in BS 8110), as follows:

$$A_f = \frac{1000 \, b \, v_h}{0.87 \, f_y} \qquad (6.2)$$

**Table 6.2** Design ultimate horizontal shear stress at interface (N/mm²) [6.1].

| Precast unit | Surface type | Grade of insitu concrete | | |
| --- | --- | --- | --- | --- |
| | | 25 | 30 | 40+ |
| Without links | As cast or as extruded | 0.40 | 0.55 | 0.65 |
| | Brushed, screeded or rough tamped | 0.60 | 0.65 | 0.75 |
| | Washed to remove laitance, or treated with retarding agent and cleaned | 0.70 | 0.75 | 0.80 |
| Nominal links projecting into insitu concrete | As cast or as extruded | 1.20 | 1.80 | 2.00 |
| | Brushed, screeded or rough tamped | 1.80 | 2.00 | 2.2 |
| | Washed to remove laitance, or treated with retarding agent and cleaned | 2.1 | 2.2 | 2.5 |

The reinforcement should be adequately anchored on both sides of the interface. If loops are used, as shown in Fig. 6.2, the clear space beneath the bend should be at least 5 mm + size of aggregate. It is found that the bend radius need not comply with bursting requirements, only the minimum of $3\phi$ is required.

The bars are uniformly distributed along the length of the interface, although the spacing could in fact be reduced towards the point of zero shear. Nominal links should be at least equal to 0.15 per cent of the contact area. The spacing of links should not be too large, with 1.2 m being typical for hollow core slabs. Where links are provided in ribs of T-beams the spacing should not exceed four times the minimum thickness of the insitu concrete, nor 600 mm.

The permissible interface shear stress for hollow core and double-tee units is therefore 0.4 N/mm$^2$ and 0.6 N/mm$^2$, respectively, for normally produced units. In short spans where the shear is large (compared with flexural requirements) interface links can be left projecting in the longitudinal joints between hcus using loops (T 10 at 1.2 m centres for example) as shown in Fig. 6.2.

The definition of $b$ in Equations (6.1) and (6.2) is clear in most cases but, for example, in the situatons given in Fig. 6.3, $b$ must be defined on its merits because it is not practical to define specific rules which cover all cases. It can, however, be recommended that those parts of the interface which are covered by insitu concrete of thickness less than 30 mm should normally not be considered as effective interface area, except in sections of the type shown at G–H in Fig. 6.4. Because of the compacting pressure during casting, 20 mm or the maximum aggregate size can be considered to be the minimum thickness on the side of beams.

## 6.4  LOSSES AND DIFFERENTIAL SHRINKAGE EFFECTS

### 6.4.1  Losses in prestressed composite sections

It is difficult to make an assessment of the losses of prestress that occur in a composite section, as it obviously depends on when the insitu concrete is added to the psc, normally between one and four months. Although it may be assumed that most of the losses occur

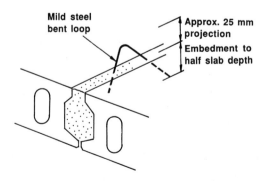

**Fig. 6.2**  Projecting loops placed in the longitudinal joints between hollow core units.

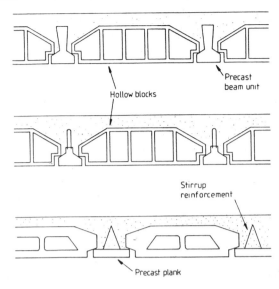

**Fig. 6.3** Definition of interface contact breadth [6.2].

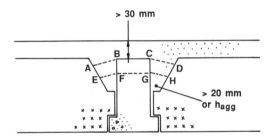

**Fig. 6.4** Examples of special cases for consideration in determining contact breadth.

before this addition, in many cases this will not be strictly correct and the effects of differential shrinkage and creep should be at least taken into account.

If little or no losses have taken place in the psc before the screed is placed (i.e. in less than seven days), then the shortening of the precast unit will be restrained to some extent by the added concrete, though the amount of the actual restraint will depend on the shape of the section and the quantity of the insitu concrete used.

Any bending moment that is induced in the composite section as a result of differential shrinkage affects only the elastic stress conditions, and does not affect the ultimate behaviour (in the same way that the level of prestress has a small influence on the ultimate strength). Because it is difficult (and probably unnecessary) to make a 100 per cent correct assessment of the differential deformations due to shrinkage and creep, it seems sensible to first compute the shrinkage and creep movement taking place in the precast beam, and then to add the effects of the relative movement in the topping concrete. To proceed in this manner will usually lead to slightly conservative results as the relative shrinkage strain in the interface will be greater than if all the shrinkages are assumed to occur simultaneously.

### 6.4.2  Design method for differential shrinkage

The forces due to differential shrinkage may be calculated by the following method. In this analysis, the term 'differential shrinkage' is used to describe the difference in the free strains due to the shrinkage of the insitu concrete and the combined shrinkage and creep of the precast concrete. At the end of the span, the stress in the precast concrete due to dead load and prestress is small and the differential strain should normally be taken as the difference in the free shrinkage values of the two concretes.

If the beam or slab reinforcement is placed non-symmetrically, a bending moment is induced by shrinkage due to non-uniform restraint by the reinforcement and, as a result, shrinkage increases the curvature and consequently the flexural deflection of the components.

In the following equivalent tensile force method for estimating shrinkage-induced stresses, applied loads and shrinkage forces are resisted by an uncracked, cracked or partially cracked member. In the case of the cracked or partially cracked member, the assumption is made that the shrinkage occurring prior to cracking is insignificant. It is therefore possible to treat shrinkage-induced stress in a similar way to load-induced stress at the serviceability limit state.

Consider a unit length of the composite precast beam of precast depth $h$, and topping screed depth $h_s$, shown in Fig. 6.5(a) in which, after any interval of time following the casting of the insitu concrete flange, the free shrinkage of the flange is $\varepsilon_f$, and the combined free shrinkage and creep of the beam is $\varepsilon_b$ at the centroid, with values of $\varepsilon_{bt}$ and $\varepsilon_{bb}$ at the top and bottom fibres, respectively. Refer also to the design guidance given in BS 8110, Part 2, Clause 7.4 [6.1]. The analysis considers that the concrete member is free to shrink, and when this happens the compression and tension steels are compressed by fictitious force $\varepsilon_b A_s' E_s$ and $\varepsilon_b A_s E_s$ respectively, where $A_s'$ and $A_s$ are the areas of the compression steel and tension steel, respectively, and $E_s$ is the Young's modulus of the steel bars or tendons. When these loads are released, it is equivalent to eccentric tension loads $\varepsilon_b A_s' E_s$ and $\varepsilon_b A_s E_s$ applied at the steel level to the entire transformed area of the member, as shown in Fig. 6.5(b). These forces produce a bending moment and consequently a curvature and flexural deflection of the concrete member. If $A_s > A_s'$ then the deflection is downwards. These forces also produce an interface shear stress as follows.

The bending moment $M_b$ produced by the effects of beam shrinkage is given by:

$$M_b = \varepsilon_b E_s [A_s (d - x) - A_s' (x - d')] \tag{6.3}$$

where $d$ and $d'$ are the depths of the tension and compression reinforcement respectively, and $x$ is the distance from the top of the screed to the neutral axis of the transformed composite section (see Section 6.5 for an explanation of transformed sections). In multi-layered reinforcement use the sum of each layer in the term $A_s (d - x)$.

The strains in the top and bottom surfaces of the beam are given by Equations (6.4) and (6.5) (assuming $\varepsilon_{bt} > \varepsilon_{bb}$):

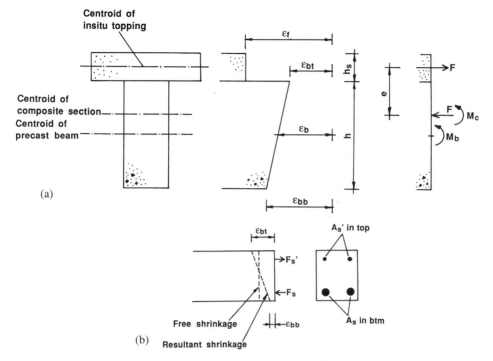

**Fig. 6.5** Theoretical approach to shrinkage-induced deflections in composite construction. (a) Definitions; (b) shrinkage effects in reinforced sections.

$$\varepsilon_{bt} - \varepsilon_{bb} = \frac{M_b h}{E_c I_c} \tag{6.4}$$

where $E_c$ is the modulus of elasticity of beam concrete, and $I_c$ is the second moment of area of the transformed composite section. Also, the strain at the interface between the precast and screed:

$$\varepsilon_{bt} = \frac{M_b(x - h_s)}{E_c I_c} + \varepsilon_b \tag{6.5}$$

The relative shrinkage strain between the flanged screed ($\varepsilon_f$) and the top surface of the precast concrete beam ($\varepsilon_{bt}$) is given by:

$$\varepsilon_s = \varepsilon_f - \varepsilon_{bt} \tag{6.6}$$

The modification factor for the restraining effect of the mesh in the screed $(1 + K\rho)$ (where $\rho = 0.13$ to $0.24$ per cent, and $K = 15$ to $25$, typically) is not significant in screeds.

A tensile force $F$ is applied at the centroid of the flange to overcome the strain differential between the flange and the beam (Fig. 6.5(c)) and is given by:

$$F = \varepsilon_s E_c m A_f \tag{6.7}$$

where $m = E_c'/E_c$ is the modulus ratio, and $A_f$ is the area of flange section. In composite beams the effective width of the flange is the same as for any T section according to BS 8110, Part 1, clause 3.4.1.5 [6.1].

The two concretes may now be joined together (theoretically) and the external actions released if a compressive force $F$ and a balancing moment $M_c$ are applied to the composite section, such that:

$$M_c = F\,e + M_b \tag{6.8}$$

where $e$ is the distance between the centroid of the flange and the centroid of the composite section. The contribution of $M_b$ is usually not more than about five per cent of the total moment $M_c$. The resultant longitudinal force $F_v$ (of these two actions) to one side of the interface (tension in the flange and compression in the beam) is given by:

$$F_v = F + M_c\,\frac{S_c}{I_c} \tag{6.9}$$

where $S_c$ is the first moment of area to one side of interface about the centroid of the transformed composite section and $I_c$ is the second moment of area of the transformed composite section. This is the value for $F_v$ that must be used in Equation (6.1).

The resulting induced bending moment $M_c$ causes a sagging curvature deflection $\delta_s$ (Fig. 6.5(d)) over a simply supported span $L$ of:

$$\delta_s = \frac{M_c L^2}{8\,E_c I_c} \tag{6.10}$$

and so on depending on the beam geometry and loading conditions. Note that in the

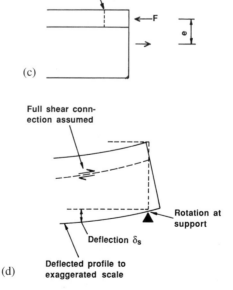

**Fig. 6.5** (*continued*) Theoretical approach to shrinkage-induced deflections in composite construction. (c) effective eccentric force due to differential shrinkage and (d) shrinkage-induced deflection.

unlikely event of $\varepsilon_f < \varepsilon_{bt}$, then $F$ will be negative. In this case use $F = 0$, and so $M_c = M_b$ and $F_v = M_b\, S_c/I_c$.

Shear lag effects in the insitu concrete will reduce the relative shrinkage strains. This will be noticeable in depths of screed greater than about 100 mm. However, in most applications in precast work where thicknesses of insitu concrete are usually less than 75 mm, the simplified version given above is adequate for design purposes.

A further complication in the analysis is the changing thickness of insitu screeds on precambered prestressed beams and slabs. In a typical situation the depth of screed used on a hollow core or double-tee floor slab is 50 mm at the crown (mid-span) of the floor slab and up to 80 mm at the supports. Thus the composite section properties calculated at the crown of the floor are not appropriate to the situation at the supports where the effects of differential shrinkage are the greatest. For the same reason neither are the section properties at the support correct. The usual practice is to calculate the interface stresses based on the composite section properties at the position of the mean depth of the screed.

Where non-rectangular beams, e.g. L or inverted-tee beams, are designed compositely with floors and screeds, the situation is more complicated because the contact area between the precast and insitu concretes includes the sides of upstand. This may be ignored if the width of the insitu infill is less than 30 mm. In Type II edge beams (Section 4.3.4) where the insitu is much wider, the critical interface is at the ledge rather than at the top of the beam.

## EXERCISE 6.1    Shrinkage-induced forces in rc sections

The rc inverted-tee beam shown in cross-section in Fig. 6.6 is simply supported over an effective span of 10 m. The beam supports 200 mm deep hollow core floor slabs, and the mean depth of the topping screed over the line of the beam is 75 mm. Given that after the screed has been cast, the free shrinkages in the beam and screed are 80 µε and 150 µε, respectively, calculate the average interface shear stress between the top of the beam and the screed.

Use $f_{cu} = 40$ N/mm$^2$ for the beam, $f_{cu}' = 25$ N/mm$^2$ for the screed, $E_s = 200$ kN/mm$^2$, cover to the 10 mm diameter links = 35 mm.

## Solution

Considering free shrinkage of beam acting over transformed composite section.

Effective breadth of screed = 300 + 10 000/5 = 2300 mm.

$E_c = 28$ kN/mm$^2$, $E_c' = 25$ kN/mm$^2$ (from Table 6.1).

Modular ratio = 25/28 = 0.893.

Depth to neutral axis of transformed section (ignore steel area) = 220 mm.

$e = 220 - 37.5 = 182.5$ mm.

$I_c = 1.3 \times 10^{10}$ mm$^4$.

Bottom reinforcement (4 T 32 bars) $A_s = 3216$ mm$^2$.

$d = 575 - 35 - 10 - 16 = 514$ mm.

Mid reinforcement (2 T 10 bars) $A_s' = 157$ mm$^2$.

$d = 275 + 35 + 10 + 5 = 325$ mm.

Top reinforcement (2 T 10 bars) $A_s' = 157$ mm$^2$.

$d = 75 + 35 + 10 + 5 = 125$ mm.

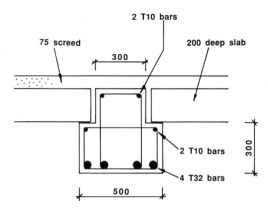

**Fig. 6.6** Details to Exercise 6.1.

$M_b = 80 \times 10^{-6} \times 200 \times 10^3 \times [3216\ (514 - 220) + 157\ (325 - 220)$
$-157\ (220 - 125)] \times 10^{-6} = 15.15$ kN m.

$$\varepsilon_{bt} = \frac{15.15 \times 10^6 \times (220 - 75)}{28\ 000 \times 1.3 \times 10^{10}} + 80 \times 10^{-6} = 86 \times 10^{-6}$$

Considering free shrinkage of screed relative to top of beam:
$\varepsilon_s = 150 \times 10^{-6} - 86 \times 10^{-6} = 64 \times 10^{-6}$
$F = 64 \times 10^{-6} \times (0.893 \times 2300) \times 75 \times 28\ 000 \times 10^{-3} = 276$ kN
$M_c = 276 \times 0.1825 + 15.15 = 65.52$ kN m
Now $S_c = (0.893 \times 2300) \times 75 \times 182.5 = 28.1 \times 10^6$ mm$^3$

Then $F_v = 276 + \dfrac{15.15 \times 10^3 \times 28.1 \times 10^6}{1.3 \times 10^{10}} = 308.8$ kN

Average interface shear stress over half the length of the beam $= \dfrac{308.8 \times 10^3}{300 \times 5000}$

$= 0.2$ N/mm$^2$.

### 6.4.3   Cracking in the precast and insitu concrete

The question often arises as to whether the insitu concrete at the interface cracks when the uniaxial cracking strain is exceeded, or whether the precast concrete unit restrains the insitu concrete. Some restraint is given to weaker concrete in these situations. If the bond at the interface is sufficient for the two concretes to act monolithically then a linear strain distribution may be assumed throughout the entire section. Also flexural cracking at the interface propagates simultaneously in the precast and insitu concrete [6.6].

The precast unit, which covers the entire bottom surface of the insitu concrete, behaves in a similar manner to reinforcing bars, in that it eliminates any concentration of strain at any section where the concrete quality is below average for the specimen; thus the average strain before cracking occurs is greater than that for an unrestrained plain concrete having the same overall properties.

It has been seen that the assumption of linear strain distribution depends on the fact that the connection between the two concretes that make up the composite section is strong enough to ensure that the longitudinal shearing forces cause no relative movement at the interface. A rough tamped top surface of the precast unit will be sufficient to ensure that a horizontal shear failure does not occur, though it may take place as a secondary effect after some other factor has caused the primary failure. As far as cracking is concerned, a rough surface by itself will also prove to be a better interlocking medium than the corresponding smooth surface that also has either castellations or projecting steel stirrups.

## 6.5  COMPOSITE FLOORS

### 6.5.1  General considerations

This consists of a precast rc or psc floor unit with insitu concrete topping, or screed (Fig. 6.7). The screed may be floated (power or hand) smooth to give a finished floor, but this prevents the placement of services in the screed. First, horizontal diaphragm action resulting from horizontal wind or earthquake loads may be transmitted through the screed acting as a deep beam or truss. The resulting diaphragm stresses are only in the order of $0.5 \text{ N/mm}^2$ and do not affect the flexural behaviour of the screed. Second, the reinforcement in the screed may be used to satisfy the requirements for stability ties, in which case the reinforcement should be continuously lapped across the whole floor.

The types of floor slabs considered here are hollow core floors, double-tee units and plank. In designing a composite floor, reference must be made to the basic section and

**Fig. 6.7** Laying insitu concrete screeds onto hollow core floors.

material properties of the precast unit itself, i.e. $b$, $h$, $d$, $b_w$, $A$, $I$, $Z$, $f_{cu}$, $f_t$, $v_c$, etc. The precast unit serves two functions, by providing permanent shuttering to the screed as well as contributing to the flexural and shear strength to the completed floor.

An enhancement in the ultimate moment of resistance is only usually possible in spans up to about 10 m because of the deleterious effects of additional dead-weight. The optimum thickness of screed from a strength point of view is typically about 50 mm at mid-span. Figure 6.8 shows the enhanced load vs span characteristics of a range of 400 and 600 mm deep double-tee units using 75 mm thickness structural screeds. The increased performance of composite hollow core slabs gives rise to increases in superimposed loading of between 20 and 40 per cent when using a 40 mm thick screed, and between 60 and 110 per cent when using a 100 mm thick screed. However, when these increased loadings are compared with the basic non-composite data the efficiency of the flooring, expressed in terms of the volume of concrete used, is reduced by about 8 per cent. This is an obvious result as a prestressed unit is structurally more efficient than an equivalent reinforced section. Further comparisons are made in Section 6.7.

Structural screeds can be made of ordinary Portland cement and a coarse aggregate with a size that should not exceed half of the minimum thickness. The thickness of the screed $h_s$ should not be less than 40 mm, and the minimum grade of concrete is C25. It is important that the workability should be specified to suit the delivery (pumped or skipped) and compaction methods (mechanical vibrating bar or poker) used. A slump of 75 mm is usually suitable. When calculating the average depth of the screed, allowances for precamber should be made as shown in Fig. 6.9. The normal levelling tolerances given in BS 5606 [6.5] are specified; this usually means a maximum deviation in thickness of 5 mm.

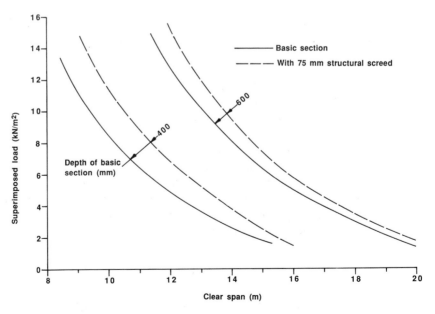

**Fig. 6.8** Load span characteristics of double-tee units using structural screeds.

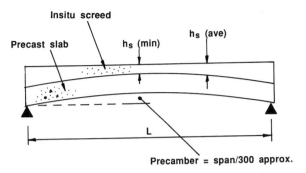

**Fig. 6.9** Effect of precamber on depth of screed.

The screed is reinforced using a mesh of minimum area 0.13 per cent $A_c$, or greater for the requirements for stability ties. The screed is reinforced using a single sheet *square* mesh with a bar spacing of 200 mm (typically A142 or A193) placed at mid-depth to resist shrinkage stresses. The mesh is lapped to peripheral projecting reinforcement in the edge beams and continuous over internal beams. Care has to be taken when lapping mesh and no more than three sheets of the mesh should meet at any point. A minimum lap length of 300 mm is required.

### 6.5.2 Flexural analysis for prestressed concrete elements

#### 6.5.2.1 At serviceability state

Permissible service stresses are checked at two stages of loading – Stages 1 and 2 – before and after hardening of the insitu screed, respectively. Stresses and deflections are calculated by the superposition method taking into account the following loads and section properties:

- Stage 1 for the self-weight of the precast slab plus the self-weight of insitu concrete screed, plus an allowance for construction traffic of up to 1.5 kN/m$^2$
- Stage 2 for superimposed loading.

Bending moments $M_1$ and $M_2$, and section properties $Z_{b1}$ and $Z_{t1}$, and $Z_{b2}$ and $Z_{t2}$ are calculated for each stage. The support conditions are simple, or cantilevered. A modular ratio based on elastic response $E_c$ (insitu)/$E_c$ (precast) is used to calculate the effective section properties at Stage 2 for the service state (Fig. 6.10 and [6.7]). Stresses are compared with permissible values according to BS 8110, Part 1, clauses 4.3.4 and 5.4.6 [6.1] as follows, see Fig. 6.11(a):

$$\frac{M_1}{Z_{b1}} + \frac{M_2}{Z_{b2}} \leq f_{ct} + f_{bc} \text{ in the soffit of the precast slab} \tag{6.11}$$

$$\frac{M_1}{Z_{t1}} + \frac{M_2}{Z_{t2}} \leq 0.33 f_{cu} + f_{tc} \text{ in the top of the precast slab}$$

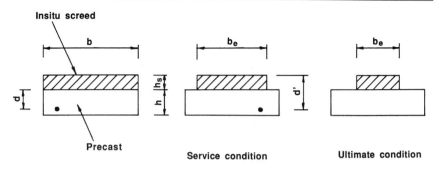

**Fig. 6.10** Transformed sections.

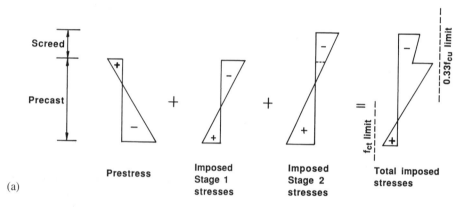

(a)

**Fig. 6.11** Composite section design. (a) State of stress in a prestressed composite section and (b) composite action at ultimate limit state.

$$\frac{M_2}{Z_{t2}'} \leq 0.33 f_{cu} \text{ in the top of the insitu screed}$$

where $Z_{t2}'$ is the section modulus at the top of the insitu concrete, $f_{bc}$ and $f_{tc}$ are the prestress values after losses in the bottom and top of the beam, respectively, and $f_{ct}$ is the permissible flexural tensile stress allowed in Class 2 structures given in BS 8110, Part 1, Table 4.1.

The critical situation occurs in the bottom of the slab because the new position of the neutral axis is close to the top of the precast unit. Equation (6.1) can be written:

$$M_2 = (f_{ct} + f_{bc}) Z_{b2} - [M_1 \times (Z_{b2}/Z_{b1})] \tag{6.12}$$

Thus, for specified geometry and prestress, values of $f_{ct}$, $f_{bc}$, $Z_{b1}$ and $Z_{b2}$ are known. Similarly $M_1$ may be computed for a given span $L$ from which the imposed load moment $M_2$ may be computed.

The construction traffic loading allowance must be considered as part of $M_1$ when checking Stage 1 service stresses. However, it does *not* have to be included in the final calculations in Equation (6.12) because it is not a permanent load.

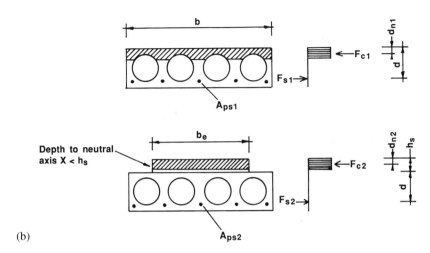

(b)

**Fig. 6.11** (*continued*)

### 6.5.2.2    At ultimate limit state

The design at ultimate is also a two-stage process, with the flexural stresses resulting from the self-weight of the precast element plus any wet insitu concrete being carried by the precast unit alone. The lever arm in this calculation is obviously the same as in a non-composite design, i.e. $d$. The method is to calculate the area of steel, $A_{ps1}$, required in Stage 1, and to add the area, $A_{ps2}$, required in Stage 2 using an increased lever arm.

Some designers choose not to separate the stresses and moments as given in Equation (6.13a), preferring to use Equation (6.13b) for the total ultimate moment. The differences in using either method are not great if the superimposed moment $M_{u2}$ is more than twice the self weight moment $M_{u1}$.

The effect of the structural screed is to increase the lever arm to the steel reinforcement by an amount equal to the thickness of the screed $h_s$, proving the depth to the neutral axis is less than $h_s$ (Fig. 6.11(b)). In Stage 2 the effective breadth of the insitu screed is taken as $b_e = b f_{cu}'$ (insitu)/$f_{cu}$ (precast), where $b$ is the breadth of the compression flange of the precast unit. Then:

$$M_{u1} = f_{pb} A_{ps1} (d - d_{n1}) \tag{6.13a}$$

$$M_{u2} = f_{pb} A_{ps2} (d + h_s - d_{n2}) \tag{6.13b}$$

where each of the two values for $f_{pb}$ and $d_n$ are obtained from BS 8110, Part 1, Table 4.4 [6.1] for specific levels of prestress and strength ratios $f_{pu} A_{ps}/f_{cu} b_e (d + h_s)$. Thus, before carrying out a composite design it is necessary to know stresses in the precast unit.

The difficulty in calculating a single moment of resistance is that the Stage 1 stresses must first be established. These are span-dependent and so the superimposed moment capacity $M_{u2}$ is a function of span and Stage 1 loads, i.e.

$$M_{u2} = f_{pb}\left[A_{ps} - \frac{M_{u1}}{f_{pb}(d - d_{n1})}\right](d + h_s - d_{n2}) \tag{6.14}$$

### 6.5.3  Propping

During the normal course of construction it is very likely that the section will be unpropped, and that the slab will be positioned as it is meant to be used in the future – simply-supported with no intermediate props of any kind. However, additional benefits may accrue if the precast unit is propped whilst the insitu concrete is hardening, as shown in Fig. 6.12. The benefit derives from the fact that the Stage I moments due to the weight of the wet concrete screed are determined over a continuous double span, each of $L/2$. When the props are removed the prop reaction $R$ creates a new moment (for a point load at mid-span or $\frac{1}{3}$ points depending on the number of props used) which is carried by the composite section. Finally, the superimposed loads are added as previously as shown in Fig. 6.12.

Under the action of the prop the hogging moment is $-0.03125\,wL^2$ (where $w$ is self weight of insitu screed and $L$ = effective floor span). The prop reaction $R = 0.625\,wL$,

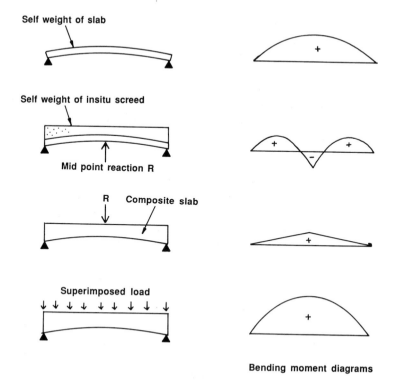

Bending moment diagrams

**Fig. 6.12**  Bending moments in propped composite construction.

such that when the prop is removed the sagging moment induced is $+0.15625\ wL^2$. (See Exercise 6.3.)

The benefits of the additional structural performance must be balanced against the cost and disruption in propping. Precast construction teams prefer not to prop slabs as each unit (1.2 m wide) must be propped in an identical manner, which is not always easy to achieve in practice. Precise instructions must be relayed to site if propping has been assumed in design. Clear instructions should be visible on drawings.

## 6.5.4   Design calculations

*Composite design for unit reference 20SS78 using grade C25 insitu concrete screed*
The section is identical to the non-composite hcu design in Section 5.6. General equations are derived in which the depth $T$ of the screed is variable. In the following, $M_1$ = maximum bending moment due to self-weight of precast unit plus insitu concrete, and $M_2$ = maximum bending moment due to superimposed loading (including partitions, finishes, ceilings, services and live loads).
*Unit properties per 1200 mm width*
   Basic section as above. $Z_{b1}$ = section modulus of precast unit.
   Composite values obtained using modular ratio method.
   $Z_{b2}$ = section modulus of composite section.
*Insitu concrete data*
   $f_{cu}$ at 28 days = 25 N/mm$^2$ minimum.
   $E_c$ insitu = 25 kN/mm$^2$.
   Modular ratio at service $m_s$ = 0.781.
   Modular ratio at ultimate $m_u$ = 0.417.
*Serviceability limit state of bending*
   Effective breadth of insitu concrete $b_e = 1200 \times 0.781 = 937.5$ mm.
   Considering bottom fibre stresses: $f_{bc} - (M_1/Z_{b1}) - (M_2/Z_{b2}) = -3.5$ N/mm$^2$.
   Service moment, $M_{s2} = (13.28 \times Z_{b2}) - (M_1 \times Z_{b2}/Z_{b1})$.
*Ultimate limit state of bending*
   $F_s = 0.87 \times 1148 = 998.76$ kN, $d = 203 - 40 - 6.25 = 156.75$ mm.
   $F_c$ (insitu) = $0.45\ f_{cu}$ (insitu) $b$ $0.9X$, hence $X = 82.2$ mm.
   If $T > 0.9 \times 82.2 = 74$ mm, then lever arm = $156.75 + T - 0.45X = 119.76 + T$ mm.
   Ultimate moment, $M_{u2} = 998.76 \times (119.76 + T) \times 10^{-3}$ kN m.
   If $T < 74$ mm, then the above expression for the lever arm is not valid.
   Hence $F_c$ (psc) = $0.45\ f_{cu}$ (psc) $b$ $(0.9\ X - T) = F_s - F_c$ (insitu).
   Hence $X$ is found.
   Then $M_{u2} = [F_c$ (insitu) $(d + 0.5T) + [F_c$ (psc) $(d - 0.45\ X + 0.5T)]$.
   Refer to Table 6.3.

The precast concrete units designed in the previous example will each have an upward camber of approximately span/300, or 20 mm. This must be taken into account when determining the minimum and maximum thickness of screed.

### Ultimate limit state of shear
Although ultimate shear capacities in the composite section are increased by as much as

**Table 6.3** Properties and moment capacities of composite section per 1200 mm width.

| Depth of insitu screed (mm) | Dist. to neutral axis (mm) | $I \times 10^6$ (mm$^4$) | $Z_{b2} \times 10^6$ (mm$^3$) | $M_{s2}$ (kN m) | $M_{u2}$ (kN m) |
|---|---|---|---|---|---|
| 40 | 128.3 | 1113 | 8.672 | $115.2 + 1.304\,M_1$ | 164.1 |
| 50 | 134.5 | 1240 | 9.222 | $122.5 + 1.387\,M_1$ | 171.9 |
| 75 | 140.5 | 1598 | 10.691 | $142.0 + 1.608\,M_1$ | 194.5 |
| 100 | 163.9 | 2018 | 12.310 | $163.5 + 1.851\,M_1$ | 219.5 |

Increasing the compressive strength of the insitu concrete screed has little effect on the performance of the composite floor. For example, using a 40 N/mm$^2$ strength and 75 mm depth of concrete increases the serviceability and ultimate moments of resistance by approximately 2 and 7 per cent, respectively. In the above example, increasing the depth of the screed from 40 mm to 100 mm increased the total self-weight of the floor by 40 per cent, yet the increased moment capacity is only 33 per cent.

15 per cent for a depth of screed of 100 mm, non-composite values are conservatively used mainly because the ultimate limit state of shear is not usually critical. However, individual values may be computed if an additional shear resistance is required.

In studying composite floors it is necessary to deal first with the non-composite case (Section 5.2.6).

**EXERCISE 6.2    Composite floor design**

Repeat Exercise 5.2 using grade C30 × 50 mm minimum thickness insitu concrete structural screed. Check the design for serviceability and ultimate moments, ultimate shear both cracked and uncracked in flexure, and interface shear stresses between the precast and insitu concrete.

*Solution*

|  | *Service* | *Ultimate* |
|---|---|---|
| (1) Stage 1 loading: self-weight precast slab | 2.94 | 4.12 |
| structural screed* | 1.44 | 2.01 |
|  | 4.38 | 6.13 |

* average $h_s = 60$ mm due to 20 mm camber in precast floor unit.
Stage 1 maximum bending moment $M_{s1} = 29.3$ kN m per 1.2 m wide unit.

| (2) Stage 2 loading: partitions, services and ceiling | 1.60 | 2.24 |
|---|---|---|
| superimposed live | 5.00 | 8.00 |
|  | 6.60 | 10.24 |

Stage 2 maximum bending moment $M_{s2} = 44.1$ kN m per 1.2 m wide unit.
Total ultimate bending moment    $M_u = 109.4$ kN m per 1.2 m wide unit.
Total ultimate shear force        $V_u = 65.6$ kN per 1.2 m wide unit.

**Section properties of composite slab at serviceability limit state (see Chart 5.2 and Table 6.4)**

Effective breadth of screed $b_e = E_c$ (insitu) $\times$ 1200 mm/$E_c$ (precast) = $26 \times 1200/30$ = 1040 mm.

$y_b$ (net) = 25 546 860/194 311 = 131.5 mm.

$Z_{b2} = 9.58 \times 10^6$ mm$^3$.

$Z_{t2}$ (top of precast) = $18.39 \times 10^6$ mm$^3$.

$Z_{t2}$ (top of insitu) = $10.63 \times 10^6$ mm$^3$.

**Stresses at service**

$$f_{bc} \geq \frac{M_{sl}}{Z_{bl}} + \frac{M_{s2}}{Z_{b2}} - 3.2 - 3.2 \ (\text{N/mm}^2)$$

$$\geq \frac{29.3 \times 10^6}{6.456 \times 10^6} + \frac{44.1 \times 10^6}{9.58 \times 10^6} - 3.20 = 5.94 \ \text{N/mm}^2$$

$$f_{tc} \geq 0.33 f_{cu} - \frac{M_{sl}}{Z_{tl}} - \frac{M_{s2}}{Z_{t2}} \ (\text{N/mm}^2)$$

$$\geq 16.5 - 4.78 - 2.41 = 9.31 \ \text{N/mm}^2 \ (\text{compressive satisfactory})$$

$$f_c \ (\text{insitu}) \geq \frac{M_{s2}}{Z_{t2}'(\text{insitu})} = 4.15 \ \text{N/mm}^2 < 0.33 \times 30 = 9.9 \ \text{N/mm}^2$$

Referring to Chapter 5, Chart 5.3, use unit ref. 2R (0 + 21) ($A_{ps} = 412$ mm$^2$).

**Ultimate limit state**

Try unit ref. 2R (0 + 21) where $A_{ps} \times f_{pu} = 21$ No. at 30.8 kN per wire = 646.8 kN.

Effective breadth $b_e = 30 \times 1200/50 = 720$ mm.

New effective depth = total depth minus distance to centroid of prestress wires (from Chart 5.3):

$$d = 200 + 50 - 42.7 = 207.3 \ \text{mm}$$

To determine depth of neutral axis $d_n$ refer to BS 8110, Part 1, Table 4.4 [6.1]:

$$\frac{A_{ps} f_{pu}}{f_{cu} b_e d} = \frac{646\,800}{50 \times 720 \times 207.3} = 0.0867$$

i.e. the section is under-reinforced. Also for:

**Table 6.4** Calculation of composite section properties.

| Component | $A$ (mm$^2$) | $y_b$ (mm) | $A\,y_b$ (mm$^3$) | $I_{oo}$ $\times 10^6$ (mm$^4$) | $A\,y'^2$ $\times 10^6$ (mm$^4$) |
|---|---|---|---|---|---|
| Precast unit | 142 311 | 97.3 | 13 846 860 | 628.10 | 166.5 |
| Insitu screed | 52 000 | 225.0 | 11 700 000 | 10.83 | 454.6 |
| Total | 194 311 | | 25 546 860 | $I_{x-x} = 1260 \times 10^6$ | |

$$\frac{f_{pe}}{f_{pu}} = \frac{392.34}{646.8} = 0.606$$

$$\frac{X}{d} = 0.191$$

hence $X = 39$ mm $<$ depth of screed

$$d_n = 0.45 \times 0.191 \times 207.3 = 17.8 \text{ mm}$$
$$M_u = 0.87 \times 646.8 \times (207.3 - 17.8) \times 10^{-3} = 106.6 \text{ kN m} < 109.4 \text{ kN m required.}$$

Repeat above using 2R (0 + 23) where $A_{ps} \times f_{pu} = 708.4$ kN and $d = 207.1$ mm.

$$\frac{A_{ps}f_{pu}}{f_{cu}b_e d} = \frac{708\,400}{50 \times 720 \times 207.1} = 0.095$$

$$d_n = 19.5 \text{ mm}$$
$$M_u = 0.87 \times 708.4 \times (207.1 - 19.5) \times 10^{-3} = 115.6 \text{ kN m} > 109.4 \text{ kN m}$$
required.

It is unusual for the ultimate limit state to be critical in prestressed concrete design. However, in the transformed composite section the breadth of the screed is reduced by a much larger factor in ultimate (where $f_{cu}$ is used) than in service (where $E_c$ is used) (Fig. 6.9).

### Ultimate shear

(1) $V_{co}$ – for shear uncracked in flexure, use non-composite value for unit ref. 2R (0 + 23) = 151 kN $\gg$ 65.6 kN required.

(2) $V_{cr}$ – for shear resistance of section cracked in flexure, use BS 8110, Equation 55 [6.1]:

$$\frac{f_{pe}}{f_{pu}} = 0.601$$

$$\text{hence} \left[ 1 - 0.55\frac{f_{pe}}{f_{pu}} \right] = 0.67$$

$$b_v = 1200 - 55 - (11 \times 58) = 507 \text{ mm}$$
$$d = 207.3 \text{ mm}$$

$$100\frac{A_{ps}}{b_v d} = \frac{100 \times 23 \times 19.62}{507 \times 207} = 0.43$$

hence $v_c = 0.60$ N/mm$^2$.

The cracking moment $M_o$ may be determined from the service calculations and the stresses in Chart 5.3. The moment required to produce a stress in the bottom fibres of $0.80 f_{bc}$ is given by:

$$M_o = 0.8 \times 6.58 \times 9.58$$
$$= 50.4 \text{ kN m}$$

$M_o = 50.4$ kN m occurs at 0.885 m from the support.

The critical section $= 0.885 + 0.207/2 = 0.99$ m from support

Thus, at the critical section $V_u = 1.2 \times 16.37 \times \left[ \left( \dfrac{6.675}{2} \right) - 0.99 \right] = 46.1$ kN.

Also at the critical section $M_u = 64.9$ kN m.

Hence $V_{cr} = (0.67 \times 0.60 \times 507 \times 207 \times 10^{-3}) + \left( \dfrac{50.4 \times 46.1}{64.9} \right)$

$\qquad = 78.0$ kN $> 46.1$ kN required.

### Interface shear
This is checked at the ultimate limit state:

$$F_v = 0.45 \, f_{cu} \, b_e \, 0.9 \, X = 0.45 \times 50 \times 720 \times 0.9 \times 43.3 = 631.3 \text{ kN per unit.}$$

The effective interface width is 1200 mm (not 720 mm) × half span of slab:

$$v_h(\text{ave}) = \frac{631.3 \times 10^3}{1200 \times 3337} = 0.16 \text{ N/mm}^2$$

$v_h(\text{max}) = 2 \, v_h(\text{ave}) = 0.32 \text{ N/mm}^2 < 0.55 \text{ N/mm}^2$ from BS 8110, Part 1, Table 5.5 [6.1] for extruded floor slabs.

**EXERCISE 6.3    Propped composite slab**
Repeat Exercise 6.2 with the slab propped at mid-span. Check for flexure only.

*Solution*

The negative moment over a central prop is $-0.03125 \, wL^2$, where $w =$ self-weight of the screed and $L$ is the effective span of the slab.
Stage 1 loading for the precast slab self-weight minus the propped screed gives:

$M_{s1} = 16.37 - 2.00 = 14.37 \times 1.2 = 17.24$ kN m/unit.
Prop reaction $= 0.625 \times 1.44 \times 1.2 \times 6.675 = 7.2$ kN/unit.

Stage 2 loading for superimposed load and prop reaction at mid-span gives:

$M_{s2} = 44.1$ (as before) $+ 12.0 = 56.1$ kN m/unit.

Stresses at service:

$f_{bc} > 2.66 + 5.86 - 3.20 = 5.32$ N/mm$^2$.

Referring to Chart 5.3, use 2R(0 + 19).

## 6.6   ECONOMIC COMPARISON OF COMPOSITE AND NON-COMPOSITE HOLLOW CORE FLOORS

A comparison was made between the *approximate* costs, applicable only in the UK, of providing a fully grouted unscreeded precast floor, and a composite insitu precast floor. Cost data were collected from hollow core producers for 150 mm, 200 mm and 250 mm (or 260 mm) deep × 1200 mm wide units. The data, which are presented in relative terms in Tables 6.5 and 6.6, were based on the following specification:

floor clear span = 6.0 m
loading: superimposed = 5.0 kN/m$^2$
partitions = 1.0 kN/m$^2$
services, etc. = 0.5 kN/m$^2$
fire resistance = 1 hour
28 day strength $f_{cu}$ = 60 N/mm$^2$.

The slabs are of the proprietary type, being either extruded (*Spiroll, Tembo, Dycore*) or slip-formed (*Roth*) and prestressed in the usual manner. Stability ties to BS 8110 [6.1] are provided in the unscreeded floor in the form of peripheral tie bars. The insitu concrete is grade C25.

In Table 6.5 the data are expressed in terms of the cost of an unscreeded floor of between 1000 m$^2$ and 5000 m$^2$ floor area. Although the actual costs supplied by the producers varied by as much as 20 per cent of the mean value, it was clear that the additional percentage cost of a structural screed was about 45 per cent. The increased

**Table 6.5** Relative cost of delivered and screeded floor compared with unscreeded floor.

| | Depth of precast unit (mm) | | |
|---|---|---|---|
| | 150 | 200 | 250 |
| Slab delivered to site | 0.75–0.83 | 0.76–0.85 | 0.80–0.86 |
| Slab fixed and grouted with tie steel* | 1 | 1 | 1 |
| Slab fixed and 50 mm screed laid with A142 mesh[†] | 1.42–1.48 | 1.39–1.46 | 1.36–1.40 |

\* for floor area < 500 m$^2$ add 8 to 10 percentage points.
[†] for floor area < 500 m$^2$ add 12 to 15 percentage points.

**Table 6.6** Relative cost of 150 mm and 250 mm deep units compared with 200 mm deep units.

| | Depth of precast slab (mm) | | |
|---|---|---|---|
| | 150 | 200 | 250 |
| Slab delivered to site | 0.92–0.95 | 1 | 1.15–1.20 |
| Slab fixed and grouted with tie steel | 0.93–0.95 | 1 | 1.10–1.18 |
| Slab fixed and 50 mm screed laid with A142 mesh | 0.95–1.00 | 1 | 1.05–1.08 |

cost due to materials was 30 per cent, with labour and plant adding a further 15 per cent. For small projects of less than 500 m² the labour charges per square metre increase rapidly such that the average increased cost of the screed is in the order of 55 per cent.

Table 6.6 shows that the cost of unscreeded floors increases rapidly for depths up to and including 250 mm, whereas the cost of a screeded floor does not vary by more than −5 to +8 per cent over the range considered.

**Table 6.7** Comparison of self-weight of unscreeded and screeded floor (kN/m²).

|  | Depth of precast slab (mm) | | |
| --- | --- | --- | --- |
|  | 150 | 200 | 250 |
| Precast slab only | 2.35 | 2.93 | 3.66 |
| Precast slab with 50 mm screed | 3.79 | 4.37 | 5.10 |
| Ratio | 1.61 | 1.49 | 1.39 |

The conclusion is that in a typical four storey commercial building of about 2500 m² floor area, the structural screed will cost £22k (1994 prices), it will add a further 3000 kN to foundation loads and increase the height of the building by about 300 mm.

## 6.7   COMPOSITE BEAMS

### 6.7.1   Flexural design

There is more potential for enhancement in flexural capacity in composite beams than with composite slabs because the breadth of the compression flange can, if correctly specified, be increased to the maximum permitted value as in monolithic construction, i.e. $b_w + l_z/5$ (BS 8110, Part 1, clause 3.4.1.5 [6.1]). The minimum strength of insitu concrete is again taken as 25 N/mm² and so the appropriate modular ratios for strength and stiffness are used in ultimate and serviceability checks. Early research work by Kajfasz *et al.* [6.8] showed the potential for economic design. There is also a greater potential for the use of propped composite beam design, and this will be given in Section 6.7.2.

As with composite floors, service and ultimate stresses are checked at two stages of loading (or three if a structural screed is added to the floor) and superimposed in the final analysis. If structural screeds are used the loading conditions are as given in Table 6.8.

**Table 6.8** Loading conditions for structural screeds.

| Stage | Loading | Section properties based on |
| --- | --- | --- |
| 1 | Self-weight beam and slab | Precast beam |
| 2 | As Stage 1 plus self-weight screed | As Stage 1 plus insitu near beam only |
| 3 | Superimposed | As Stage 2 plus screed |

The essential features of the design procedure are given in Fig. 6.13. Stage 1 stresses exist in the precast beam only (Fig. 6.13(a)). These are the result of prestress and relaxation (if the beam is prestressed), self-weight of the beam, slab, and wet concrete. Stage 2 stresses exist in the precast beam and insitu infill (Fig. 6.13(b)), and are due to the self-weight of the wet screed plus a notional allowance (of about 1.5 kN/m$^2$ if the amount of insitu concrete to be placed is extensive, as in a structural screed) for construction traffic. Final Stage 3 stresses in the composite beam (Fig. 6.13(c)), which are in addition to the above, are the result of superimposed, services and partition loading, differential shrinkage and the total creep relaxation after hardening of the insitu concrete. Ultimate moment and shear resistances are computed in the usual manner, noting that differential shrinkage and creep strains are ignored at the ultimate limit state due to the effects of plasticification.

An important point requiring clarification is the *effective breadth* of the compression flange, particularly if the cores in certain types of hollow core floor slabs are filled only intermittently, as is the common practice using slip-formed units (Fig. 6.14). Here the minimum section is through the unfilled hollow core. This comprises only the top and bottom flanges of the slab – each usually 20 to 30 mm thick. Although interface shear transfer between beam and slab may be fully effective at the positions of the reinforced

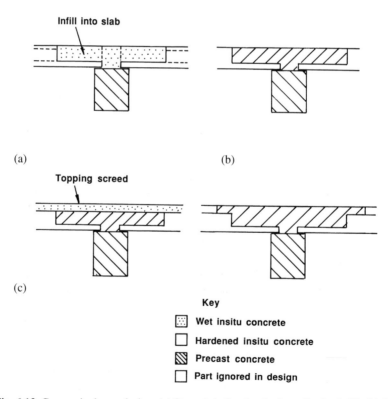

**Fig. 6.13** Composite beam design. (a) Stage 1, before hardening of insitu infill; (b) Stage 2, after hardening of insitu infill and (c) Stage 3, before (left) and after (right) hardening of structural screed.

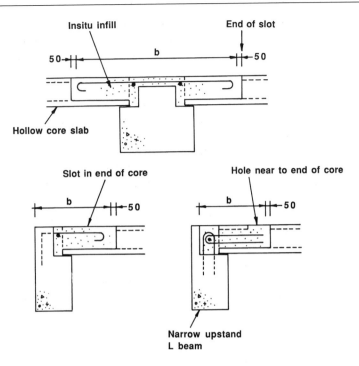

**Fig. 6.14** Effective breadth using hollow core slabs.

insitu filled cores (representing possibly only $\frac{1}{12}$ of the breadth of the slab), there is no analytical or experimental evidence to suggest continuous horizontal shear stresses exist between these points. In such cases the effective breadth of the compressive flange is equal to the breadth of the supporting precast beam.

In instances where several or every core is opened, the vertical interface shear capacity must be justified so that the effective breadth of the flange (based on the full depth of the insitu concrete) may be taken as equal to the actual distance between the extremities of the opened core (even though the insitu concrete may penetrate up to 100 mm beyond these points). Openings in cores are usually about 600 mm in length, resulting in effective breadths typically 1350 or 700 mm wide in symmetrical or non-symmetrical arrangements, respectively.

Figure 6.15 shows a photograph of a test where composite action between the prestressed beam and insitu concrete placed in the broken out cores of some hollow core slabs was studied. The main objectives were to determine the effective widths of the insitu–precast top flange, and to reduce the overall construction depth to give a span:depth ratio in the region of 18 to 1. The tests have shown that full interaction between the precast components and the insitu concrete may be taken in design, providing that minimum percentage interface shear reinforcement is present, and that the effective width of the insitu flange is equal to the actual width provided, i.e. 1.0 m in this case.

Effective breadths for the insitu concrete screed used with solid plank flooring are taken as for monolithic rc construction, i.e. span/10 either side of the supporting beam.

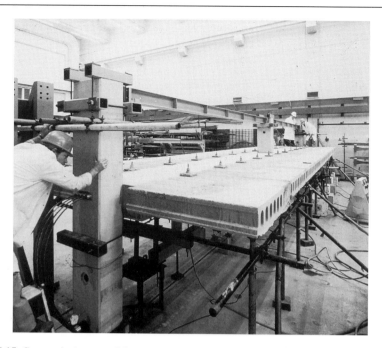

**Fig. 6.15** Composite beam – slab tests at London University (courtesy of Costain Building Products).

Composite beam design is not usually carried out where double-tee flooring units and a relatively thin screed are used. The enhancement in strength is marginal and designers prefer to hold a 'little in hand' in these situations.

### EXERCISE 6.4    Composite reinforced edge beam in flexure

Determine the allowable imposed loading on the edge beam shown in Fig. 6.16. The beam is supporting hollow core floor slabs which weigh 3 kN/m$^2$ and span 8 m. The beam is simply supported over 6 m effective span.

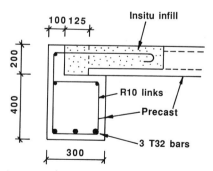

**Fig. 6.16** Details to Exercise 6.4.

Use $f_{cu\ (pc)} = 40$ N/mm$^2$, $f_{cu\ (insitu)} = 25$ N/mm$^2$, $f_y = 460$ N/mm$^2$, cover to 10 mm diameter links = 35 mm.

### Solution

**Stage 1 Loading**

|  |  | kN/m |
|---|---|---|
| Self-weight of beam | = | 3.36 |
| Slab 3.00 × 8/2 | = | 12.00 |
|  |  | 15.36 |

$$M_1 = 1.4 \times 15.36 \times \frac{6.0^2}{8} = 96.8 \text{kN m}.$$

(1) **Stage 1 properties**: $b = 100$ mm,

$d = 600 - 35 - 10 - 16 = 539$ mm, say 540 mm,

$f_{cu} = 40$ N/mm$^2$.

$$K_1 = \frac{96.8 \times 10^6}{40 \times 100 \times 540^2} = 0.083 \qquad z_1 = 484 \text{ mm} \qquad X_1 = 124 \text{ mm}$$

$$A_{s1} = \frac{96.8 \times 10^6}{0.87 \times 460 \times 484} = 500 \text{ mm}^2$$

As the total steel area is 2413 mm$^2$, the remainder 1913 mm$^2$ may be used to carry imposed loads using the composite section Stage 2.

(2) **Stage 2 properties**: $b = 125 + 500$ slot in slab $= 625$ mm,

$d = 540$ mm,

$f_{cu}' = 25$ N/mm$^2$.

$F_{s2} = 0.87 \times 460 \times 1913 \times 10^{-3} = 765$ kN

$F_{c2} = 0.45 f_{cu}' b\, 0.9 X_2$

$$X_2 = \frac{765 \times 10^3}{0.405 \times 25 \times 625} = 121 \text{ mm}$$

As $X_2 = X_1$, then strain compatibility is satisfactory, i.e. neutral axis is at same point.

$z_2 = 540 - (0.45 \times 121) = 485$ mm

$M_2 = 765 \times 485 \times 10^{-3} = 371.5$ kN m

$$\text{Hence } w_2 = \frac{8M}{1.6L^2} = \frac{8 \times 371.5}{1.6 \times 36} = 51.5 \text{ kN/m}$$

### 6.7.2  Propping

The basic concept for propping composite sections before the insitu concrete is added was described in Section 6.5.3. The potential for structural enhancement is greater for beams because the Stage 2 stresses due to the self-weight of the floor slab, insitu infill and screed can be virtually eliminated by the use of two or three props.

The ratio of the Stage 1 moment in the mid-span propped and unpropped situation is given by the ratio $\beta$:

$$\beta = \frac{M_1 \text{ propped}}{M_1 \text{ unpropped}} = \frac{(w_0 - 0.25\, w_1)}{(w_0 + w_1)}$$

where $w_0$ = self weight of beam
       $w_1$ = self-weight of precast slab and infill insitu concrete.
Typical values for $\beta$ are between 0.3 and 0.1.
Then $M_1 = 0.125\, \beta\, (w_0 + w_1)\, L^2$
     Prop force $R = 0.625\, w_1\, L$
and   $M_2 = 0.125\, (w_2 + 1.25\, w_1)\, L^2$
The resulting stresses are

$$f_{bc} \geq \frac{M_{s1}}{Z_{b1}} + \frac{M_{s2}}{Z_{b2}} - f_{ct}\,(\text{N/mm}^2), \text{ and so on as before.}$$

It is worthwhile propping beams of over 6 to 10 m span with two or three props, respectively, particularly if a structural screed is being used and the ratio of the composite properties to the basic section properties exceeds 1.50. There are considerable advantages in keeping the possibility of propping in reserve in case of the unexpected request to drill a hole in the beam, or an error has occurred in the calculation. Propping reduces beam erection rates by about 15 to 25 per cent, and so the delicate balance between material saving and time lost on site has to be considered. Only the precasting companies can give exact information on this.

### 6.7.3   Horizontal interface shear

The integrity of composite construction relies on the long-term continuity between the precast and insitu concrete. Although floor loading in buildings is predominantly static, there may be instances where it is variable and cyclic in nature resulting in a fluctuation in shear stress levels. For this reason primary composite beams are designed with interface shear links or other similar mechanical fasteners. Shear friction is not recommended for beams.

As with slabs, the ultimate interface shear stress $v_h$ is checked for the uncracked section (BS 8110, Part 1, clause 5.4.7.2 [6.1]) against values in Table 5.5 of BS 8110. Nominal interface shear links should always be provided. The minimum area of links is 0.15 per cent of the contact area. If $v_h$ is greater than the values given in the design table, typically 1.2 N/mm$^2$, shear steel is provided according to BS 8110, Part 1, equation 62 to carry the *entire* shear force.

### 6.7.4   Shear check

This is usually not a problem in beams of span greater than about 4 m. It is therefore sufficient to check the applied ultimate shear against the basic shear capacity of the beam.

### 6.7.5   Deflections

Stage 1 deflections due to beam and slab self weight, which are determined from the *uncracked* moment of inertia of the beam alone, are added to the upward camber produced by prestressing. At this stage the net deflection is about $L/1000$ upwards. Stage 2 deflections are based on the composite section and are usually controlled by permissible ceiling movement of no more than 20 mm.

### EXERCISE 6.5   Composite beam calculation

The following exercise shows the calculation for an inverted-tee beam designed for both the basic section and the composite section. The construction details are shown in Fig. 6.17. The precast beam is grade C60 and the insitu infill is C30.

| Loading: superimposed live | $= 5.00 \text{ kN/m}^2$ |
|---|---|
| 200 mm deep hollow core (Roth units) | $= 2.93 \text{ kN/m}^2$ |
| partitions | $= 1.00 \text{ kN/m}^2$ |

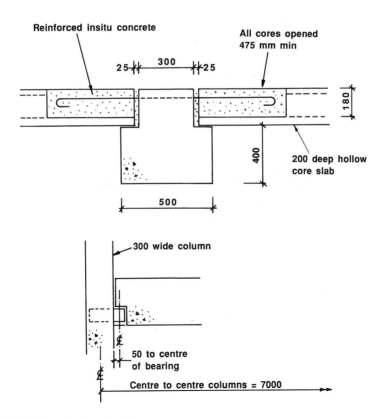

**Fig. 6.17** Details to Exercise 6.5.

| ceiling and services | $= 0.60 \text{ kN/m}^2$ |
|---|---|
| Overall span of beam | $= 7000 \text{ mm}$ |
| Effective span of beam $= 7000 - [2(150 + 50)]$ | $= 6600 \text{ mm}$ |
| Overall floor span supported by beam | $= 7200 \text{ mm}$ |
| Centre to centre distance between beams | $= 7500 \text{ mm}$ |

### *Non-composite design*

Try 500 mm wide × 400 mm deep inverted-tee beam (300 wide × 200 mm upstand).
Self-weight = 6.24 kN/m.

| *Beam loading* | *Service* | *Ultimate* |
|---|---|---|
| Dead $= (2.93 \times 7.2) + (1.6 \times 7.5) + 6.24$ | $= 39.34$ | $55.07$ |
| Live $= 5.00 \times 7.5$ | $= 37.50$ | $60.00$ |
| | Total $= 76.84$ | $115.07$ |

$$M_s = \frac{76.84 \times 6.600^2}{8} = 418.4 \text{ kN m}$$

$$M_u = \frac{115.07 \times 6.600^2}{8} = 626.5 \text{ kN m}$$

$$V_u = \frac{115.07 \times 6.600}{2} = 379.7 \text{ kN m}$$

Referring to Chart 4.1 for capacities, use TB400/21, i.e. 21 no. 12.5 mm diameter strands ($A_{ps} = 21 \times 94.2 = 1978 \text{ mm}^2$) with a service moment capacity of $M_s = 448 \text{ kN m}$, ultimate moment capacity $M_u = 734 \text{ kN m}$, and ultimate shear capacity $V_{co} = 417 \text{ kN}$.

### *Composite design*

Try same section inverted-tee beam as before.

| *Temporary beam loading for Stage 1* | *Service* | *Ultimate* |
|---|---|---|
| Dead $= (2.93 \times 7.2) + 6.24$ | $= 27.34$ | $38.27$ |
| Construction allowance $= 1.50 \times 7.5$ | $= 11.25$ | $15.75$ |
| | Total $= 38.59$ | $54.02$ |

$$M_{s1} = \frac{38.59 \times 6.600^2}{8} = 210.1 \text{ kN m}$$

$$M_{u1} = \frac{54.02 \times 6.600^2}{8} = 294.1 \text{ kN m}$$

$$V_{u1} = \frac{54.02 \times 6.600}{2} = 178.3 \text{ kN m}$$

#### Section properties of basic section

$$Z_b = 26.08 \times 10^6 \text{ mm}^3, \quad f_{b1} = \frac{210.1}{26.08} = 8.05 \text{ N/mm}^2 < 10.4 \text{ N/mm}^2 \text{ for beam TB400/}$$
15 (Chart 4.1)

$$Z_t = 21.22 \times 10^6 \text{ mm}^3, \quad f_{t1} = \frac{210.1}{21.22} = 9.90 \text{ N/mm}^2 < 20.0 \text{ N/mm}^2 \text{ for C60 concrete}$$

*Permanent beam loading for Stage 1*

$$M_{s1} = \frac{27.34 \times 6.00^2}{8} = 148.9 \text{ kN m}$$

$f_{b1} = 5.71 \text{ N/mm}^2$ and $f_{t1} = 7.02 \text{ N/mm}^2$

| *Beam loading for Stage 2* | *Service* | *Ultimate* |
|---|---|---|
| Dead = (1.6 × 7.5) | = 12.00 | 16.80 |
| Live = 5.00 × 7.5 | = 37.50 | 60.00 |
| | Total = 49.50 | 76.80 |

$$M_{s2} = \frac{49.5 \times 6.6^2}{8} = 269.5 \text{ kN m}$$

$$M_{u2} = \frac{76.8 \times 6.6^2}{8} = 418.2 \text{ kN m}$$

$$V_{u2} = \frac{76.8 \times 6.6}{2} = 253.4 \text{ kN m}$$

## Section properties of composite beam at serviceability limit state (see Table 6.9).

Actual breadth of infill concrete = 1000 mm.

Depth of insitu infill = 200 − 20 bottom flange = 180 mm.

Effective breadth of infill $b_e = \dfrac{bE_c{}'(\text{insitu})}{E_c(\text{precast})} = 26 \times 1000/32 = 813 \text{ mm}$

## Section properties of composite section

$$y_b = \frac{144.58 \times 10^6}{406\ 340} = 356 \text{ mm}$$

$Z_{b2} = 36.07 \times 10^6 \text{ mm}^3, \quad f_{b2} = \dfrac{269.5}{36.07} = 7.47 \text{ N/mm}^2.$

$Z_{t2} = 52.62 \times 10^6 \text{ mm}^3, \quad f_{b2} = \dfrac{269.5}{52.62} = 5.12 \text{ N/mm}^2.$

**Table 6.9** Calculation of composite section properties.

| Component | $A$ (mm²) | $y_b$ (mm) | $A\,y_b$ ×10⁶ (mm³) | $I_{oo}$ ×10⁶ (mm⁴) | $A\,y'^2$ ×10⁶ (mm⁴) |
|---|---|---|---|---|---|
| Precast unit | 260 000 | 269.23 | 70.00 | 7020 | 1957 |
| Insitu infill | 146 340 | 510.00 | 74.58 | 394 | 3470 |
| Total | 406 340 | | 144.58 | $I_{x-x} = 12\ 841 \times 10^6$ | |

Combining permanent Stage 1 plus Stage 2 stresses gives $f_b = 13.18$ N/mm$^2$ and $f_t = 12.14$ N/mm$^2$. These values are equated with the basic prestress limits, for Class 2 members with a permissible tensile stress of 3.5 N/mm$^2$. The critical section is at the bottom of the beam.

Hence $f_{bc} > 13.18 - 3.50 = 9.68$ N/mm$^2$ and referring to the beam properties in Chart 4.1, use TB400/15 ($f_{bc} = 10.4$ N/mm$^2$). The ultimate moment of resistance of this basic beam is 638 kN which is in excess of the design moment $M_u = 626.5$ kN m, and the shear resistance is 380 kN which is (just) greater than the design shear force $V_u = 379.7$ kN, and so no further checks are necessary. The saving in strands is 6 no.

The exercise may be repeated for the effects of propping the beam prior to fixing the floor slab. The likely outcome of this action would be a reduction in the size of the beam from 400 to 350 mm depth.

# Chapter 7

# Design of Connections and Joints

*Connections form the most vital part of precast concrete structures. There are many varied types and methods of analysis. This chapter gives a full description of connections and joints, with numerous worked examples and test results.*

## 7.1 DEVELOPMENT OF CONNECTIONS

Connections form *the* vital part of precast concrete construction. Insitu concrete joints were used extensively in the early development of precast structures due mainly to the simplicity of manufacture and the more relaxed attitudes to safe and rapid site progress. A temporary bearing, of about 40 mm to 50 mm contact width, was used to support the beam and carry only its self-weight, as shown in Fig. 7.1, and an extended permanent bearing was formed using insitu concrete. The obvious disadvantage was in shrinkage and flexural cracking in the top of the precast–insitu interface, which led to the use of non-shrink grouts and latex bonding admixtures. Projecting reinforcement from the lower column was lapped onto similarly offset projecting reinforcement in the upper column, but there were reported difficulties in stabilizing the upper column due to the flexibility of the open reinforcement cage. The use of latticed column cages to stabilize upper storey columns is still quite frequent.

Methods were sought to obtain continuity at the connections. The Ministry of Public Buildings and Works in the UK, as was in the early 1950s, copied an American solution called the H-frame, whereby pin-jointed site connections were made at the positions of frame contraflexure; i.e. mid-column height and 0.2 × beam span (see Fig. 7.2). The solution removed one problem by shifting the connections to points of zero (or small) bending moment, but created other difficulties elsewhere in manufacturing and transportation by making large two-dimensional units.

The most robust solution was found in the mechanical connection, brought about by relying on the precast concrete to anchor rolled or fabricated steel sections firmly in position so that a direct steel-to-steel joint was made. The number of different solutions was as great as the number of precast concrete frame manufacturers in the day – about 20. Figure 7.3(a) and (b) shows two examples of types of connectors that have proved to be costly and difficult to position on site.

The reasons for the large number of variations stem partly from the fact that the greater their load carrying capacity, the more complex the connector became in being able to handle the resulting stresses. The main cost elements are fabrication and material costs, and these rise steeply with increases in connector capacity. Precast designers preferred the complicated aspects of the connector to be under the control of the factory, whilst the site operations were reduced to simple dowelling, bolting or welding.

The influence of patented connections in the 1960s and 1970s stifled the widespread use of the most efficient type of connections, and led designers towards ingenious designs that occasionally 'failed', not structurally but on practical grounds. Connector design is an iterative process in which all aspects of strength, compatibility, materials and tolerances

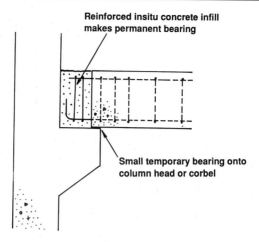

**Fig. 7.1** Early type of connections using cast insitu extended bearings.

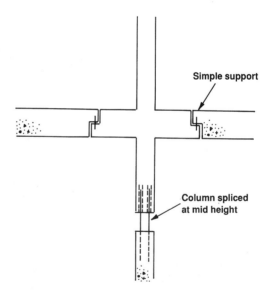

**Fig. 7.2** Connections to H-frames made at points of contraflexure.

cannot be considered in isolation. The example shown in Fig. 7.4 illustrates how easy it is to lose sight of the functional principles.

Stanton *et al.* produced a special study for the PCI [7.1] on moment resistant and simple connections, and stated:

> 'The economic and functional success of a precast concrete structure depends to a great extent on the configuration and properties of its inter element connections. Flexibility of the connections affects the distribution of creep, thermal, and shrinkage strains and determines joint performance over time. Strength of the connections

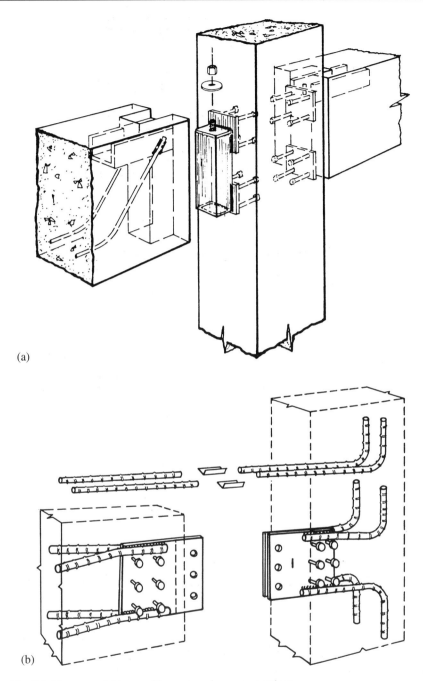

(a)

(b)

**Fig. 7.3** Unsuccessful types of beam-to-column connections.

remaining after the demands of volume change, gravity loads and dimensional or alignment corrections have been satisfied determines whether the structure will deform permanently under action of extreme wind or seismic forces. Ductility of the

**Fig. 7.4** Beam-to-column connections designed with little regard for some of the practicalities associated with precasting.

connections determines whether the permanent deformations will take place by a safe redistribution of load and dissipation of energy or by brittle connection failure.'

This statement captures the essence of precast frame design in knowing *which* effects are to be considered in design, and not simply carrying out the mechanics of stress equilibrium and strain compatibility.

## 7.2   DESIGN BRIEF

The structural requirement is the transfer of vertical shear, transverse horizontal shear, axial tension and compression, and occasionally bending moments and torsion between one precast component and another in a safe and structurally sound manner.

All connections will resist compression but it is necessary to check for the possible addition of stresses resulting from axial forces due to temperature effects, shrinkage and creep. These are becoming increasingly important as beam and floor spans are increased, and the lower age at which some precast components are tied into the structure. Connector forces due to the effects of fire, impact and explosions, etc. have been virtually ignored in favour of alternative load path design methods or the use of peripheral and internal ties (see Section 9.4). Torsion has also been ignored and this has led to fears of rotational instability in the temporary construction stage.

Whilst acknowledging that these forces do actually exist, the designer should not aim to design a connection which will resist every conceivable stress and strain, but allow other parts of the structure to accommodate some of the effects.

One other functional requirement which is frequently ignored is the capacity to rotate and accept beam deflections. The problem is based around the concept of a pinned-joint. The difficulty is that if a joint were designed to behave as a 'true' pin the connection may be deemed unserviceable due to cracking and deformation at the end of the beam. On the other hand, if the connection is capable of developing moment restraint, local stresses in the vicinity of the connection may give rise to other problems. An example of this is shown in Fig. 7.5 where shear forces resulting from moment transfer across a beam–column connection caused a shear failure in the column.

The structural behaviour of connections is often complex and not conducive to normal rc design methods. The simplest form of connection is a simple abutment type bearing where, as shown in Fig. 7.6, there is no restriction on the construction zone, and the transfer of forces from the beam to column is quite obvious from a single photograph. The need for shallower construction depths lead to the concealed steel insert connection, where a combination of steel plates and rolled sections are used to carry extraordinarily high forces. Connections of the type shown in Fig. 7.7 are capable of carrying a beam reaction in the order of 500 kN within a 500 mm depth of connection. This equates to a shear stress of about 3 N/mm$^2$. If the mode of failure is concrete crushing around the connector, then a lower safety factor may be used than if the concrete failure is due to tension or diagonal shear. This is because compressive failure may be controlled by the introduction of ductile reinforcement across the failure plain.

Local stress levels around the steel inserts become very important – they occur in small zones in areas where full concrete compaction may not be easy to achieve, and their distributions are often unknown. The anchoring method is the greatest problem, and full scale testing should not be overlooked in research and development. It is not

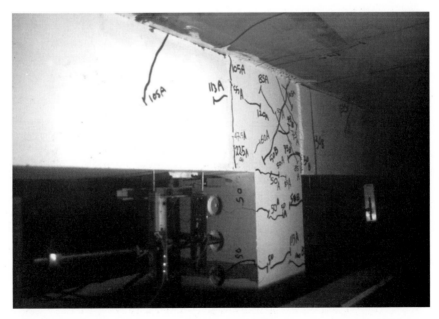

**Fig. 7.5** Shear cracking in inadequately reinforced column due to moment transfer from the beam.

**Fig. 7.6** RAI Centre in Rome, 1990 (courtesy of BF+T, Wiesbaden).

**Fig. 7.7** Solid mild steel billet cast into columns.

uncommon for steel plates to be welded to reinforcement cages, as shown in Fig. 7.8, in order to realize the full capacity of the connection.

The role of connector testing has been criticized in some circles, e.g. Coleman [7.2], due to the unavoidable restrictions in being unable to test the connections in a real frame environment, complete with sway and large deflections, and being able to study the effects of moment distribution after initial cracking. An example of this is the phenomenon of reverse end rotation resulting from high shear force in the traditional shear end tests, as shown in Fig. 7.9. With this arrangement the deformation in the joint is greater than in the beam, resulting in a set of stresses in the test specimens that are not representative of real behaviour. Under circumstances such as these, there is a danger of

**Fig. 7.8** Steel plates being added to reinforcement cages at the factory.

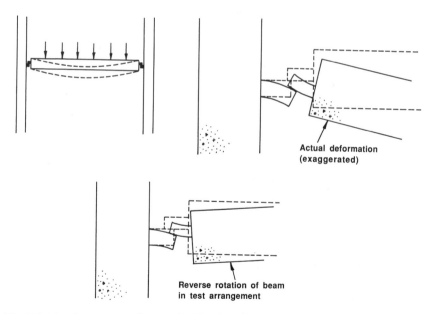

**Fig. 7.9** The phenomenon of reverse bending in column connector tests.

an unsafe extrapolation of test results to other component and connector geometries and materials.

Designers should not be insensitive to the problems of tolerances. A lack of flatness, alignment and squareness will inevitably lead to the accumulation of physical dimensional

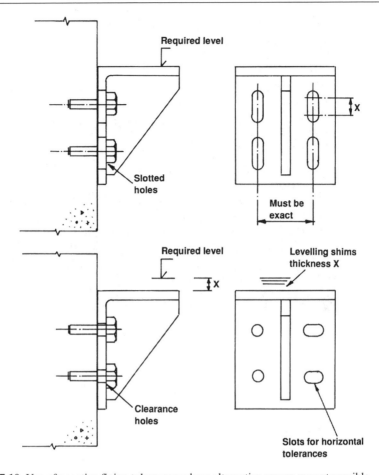

**Fig. 7.10**  Use of negative fixing tolerances where alternative means are not possible.

deviations. Their effect on the strength and behaviour must be positively allowed for in component schedules. Large tolerances may not be compatible with the proposed connector design; for example, if vertical slots are not provided in the connector in Fig. 7.10 then packing must be used and the connector set to a negative height tolerance.

## 7.3   JOINTS AND CONNECTIONS

Much confusion surrounds the difference between a 'connection' and a 'joint'. The correct definition is that a connection is the total construction including the ends of the precast concrete components which meet at it, and a joint comprises the individual parts which form the connection. For example, in a beam-to-column situation, a bearing joint is made between the precast beam and precast column, but when the assembly is completed by the use of insitu grout, etc. the entire construction is called a connection. These terms are often transposed in casual conversation, but for the sake of clarity we shall use these definitions in this book.

It is not always necessary for loads to be transferred through a joint, for example in the case of a sealing joint for water proofing purposes, or an expansion joint for thermal movements, relative movement is allowed across the joint, but one would hardly call this a connection. However, a connection will always have a structural function.

Joints are made in several ways, for instance by:

- dowelling and securing with insitu concrete or grout
- welding and protecting with same
- bolting and protecting with same
- using resin anchors in combination with the above.

From a structural point of view there may be no preference for each. However, construction rates can be dramatically affected in some buildings (tall and with small plan area) if time is needed for the development of strength. The first step is to describe the main criteria for the design and construction of joints, and then to discuss the design of the following main types of connections in skeletal structures:

- beam–slab connections
- beam–column connections
- wall–frame connections
- column splices, including foundations.

## 7.4   CRITERIA FOR JOINTS AND CONNECTIONS

### 7.4.1   Design criteria

The four rules to satisfactory joint design are:

- that the components may resist ultimate design loads in a ductile manner
- the precast members may be manufactured economically and be erected safely and speedily
- the manufacturing and site erection tolerances do not adversely affect intended structural behaviour, or are catered for in a 'worst case' situation
- the final appearance of the joint must satisfy visual, fire and environmental requirements.

Design engineers have satisfied these main criteria by ensuring that the intended behaviour of the joint is guaranteed by the careful attention to detail. In some instances, slight cracking is accepted as an indication that spurious load paths are not being created, and in others reinforcement is deliberately avoided or under-designed in certain areas to prevent restraining moments from forming. Vambersky [7.3] has summarized the main criteria for the serviceability performance of joints in terms of:

- strength
- influence of volume changes
- ductility
- durability, including corrosion and fire protection

- simplicity in fabrication and erection
- temporary loading conditions
- economy and appearance.

Some of these are summarized below with references to other sections where detailed design information may be found.

### 7.4.1.1   Strength

A connection must have the strength to resist the forces to which it will be subjected during its lifetime. Some of these forces are apparent, caused by dead and live gravity loads, wind, earthquake, and soil or water pressure. Others are not so obvious and are frequently overlooked. These are the forces caused by restraint of volume changes in the precast components and those required to maintain stability.

Instability can arise due to eccentric gravity loading and horizontal loads. In addition, further capacity may be provided to resist unanticipated, accidental or abnormal loads. These are due to foundation settlement, explosion, collision, change of building usage or others. These 'abnormal' loads may be accommodated in connections by a capacity for overload and ductility within the connection, or by building in redundancies in the total structure, such as stability ties.

Joint strength can be categorized by the type of induced stress, such as compressive, tensile, flexural, torsional and shear. Connections may have a high degree of resistance to one type of stress, but little or no resistance to another. In many cases it may be unnecessary, or even undesirable, to provide a high capability to resist a combination of stresses. A better solution would be to utilize more than one joint type to achieve the same overall result. See Sections 7.7 to 7.13 for design methods.

### 7.4.1.2   Influence of volume changes

The combined shortening effects of creep, shrinkage and temperature reductions can cause tensile stresses in precast concrete components and their connections. If the connections restrain movement the stresses must be considered in the design, unless some movement is allowed to take place, thus relieving the stresses. See Section 8.5 for design methods.

### 7.4.1.3   Ductility

This is the ability of a connection to undergo large deformations without failure, which is measured by the amount of deformation that occurs between first yield and ultimate failure. Ductility in building frames is usually associated with moment resistance, but in a precast structure, where the connections are designed using pinned joints between the components, semi-rigid behaviour is considered as being adequate. See Section 9.2 for further details. The design should ensure that flexural yielding takes place at the chosen plastic hinge position so that the post-elastic deformation capacity of the entire structure may be controlled.

### 7.4.1.4   Durability

Connections exposed to an aggressive environment should be periodically inspected and maintained. Evidence of poor durability is usually exhibited by corrosion of exposed steel elements, or by cracking and spalling of concrete. Connections which will be exposed to weather should have steel elements adequately covered with concrete, or should be painted or galvanized. If not, non-corrosive materials should be used. The preferred method of protecting exposed steel connection elements is to cover with concrete or grout. Mix proportions are as shown in Table 4.3. No chlorides should be used.

Most precast concrete components are of high quality, and flexural cracking is seldom a serious problem, provided tensile stresses are kept within code limits. However, local cracking or spalling can occur when improper details result in restraint of movement or stress concentrations.

## 7.5   TYPES OF JOINTS

### 7.5.1   Compression joints

Compression is transmitted between precast concrete components either by direct bearing, or through an intermediate medium such as insitu mortar or concrete. The distinction is made depending on tolerances and the importance of the accuracy of the load transfer. For example, vertical load transfer between two columns, one above the other, requires concentricity between member axes. This can only be achieved by using an intermediate medium of reasonable size.

On the other hand, direct contact between precast concrete components with no intermediate padding material may only be used where great accuracy in manufacture and installation is obtained and where bearing stresses are less than about $0.2 f_{cu}$ of the weaker concrete. Only those parts of the components that are solid should be considered in any compression joint analysis; this rules out hollow core slabs unless the ends are concreted solid. Particular attention should be given to detailing reinforcement in the precast components, especially if large cover distances, e.g. 50 mm, are required for severe fire or exposure categories, and a large diameter high tensile bar (with large bend radii) is used. The minimum dimension for a direct contact area should be 50 mm (or 75 mm when using 20 mm size aggregates in both components). The minimum contact area depends on the magnitude of the stress, but experience suggests that it should not be less than about 8000–12 000 mm$^2$.

In the transfer of forces through joint materials such as insitu mortar, neoprene bearing pads and other types of bedding, the aim is to ensure that irregular surfaces can transmit forces without damage to the contact surfaces. Unintentional eccentricities, and spurious shear forces and moments, etc. lead to problems such as spalling and crushing, and splitting of the joined members. The thickness of the jointing material should be as small as possible, but not impede the normal tolerances. The recommended thickness is 10 to 15 mm. The elastic response of the bearing medium should be similar (i.e. Young's modulii differing by not more than about 20 per cent) to that of the precast concrete to avoid localized contraction and splitting forces shown in Fig. 7.11(a). This is very important where the thickness of the joint is greater than 50 mm because the effects of shear lag and lateral tensile stress will reduce concrete compressive stress capacity to less than $0.8 f_{cu}$, Fig. 7.11(b). The significance of this will be apparent later in the text, e.g. Table 7.1.

Bedded bearings are usually unreinforced. The mode of failure is by crushing of the mortar, or splitting of the precast components in contact with it. In spite of the fact that the mortar is highly confined and would, in a perfectly plain stress condition, achieve compressive strengths greater than $f_{cu}$, a low design strength is used because the edges of

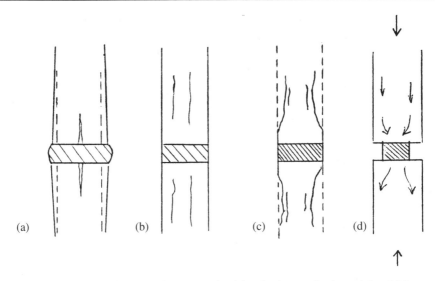

**Fig. 7.11** Force transfer through compression joints having an elastic modulus. (a) Less than the precast; (b) equal to precast; (c) greater than precast and (d) as (c) but with reduced breadth.

the mortar bed tend to spall off and this leads to a non-uniform distribution of stress as shown in Fig. 7.11(c). Poor workmanship may exacerbate the problem.

Stress concentrations are created at narrow bearings, as shown in Fig. 7.11(d), where localized shear failures may occur beneath the edge of the bearing, especially in the unreinforced cover concrete.

Vambersky [7.4] shows relationships between the bearing capacity of an unreinforced mortar joint in terms of the geometry of the joint, i.e. width to thickness ratio $b/v$, and compressive strengths of the precast concrete component $f_{cu}$ and insitu mortar $f_{cu}'$. Two reduction factors are implemented – $\eta_o$ reflecting the trapped air void content, and $\eta_m$ recognizing that site mixed mortar will be different from the conditions under which the laboratory tests were conducted. For the case of precast components placed *on* to a mortar bed $\eta_o = 0.3$, and for fluid or dry packed mortars placed after the component is in place $\eta_o = 0.7$. The values for $\eta_m$ are 0.75 if site cubes are tested, or 1.0 if cores are cut and tested. The former is more likely.

Interpreting Vambersky's results it is clear that a design stress of $0.4 f_{cu}'$ (of the mortar) may be adopted as the bearing capacity for the joint providing that the $b/v$ ratio exceeds 8 to 10 and that the difference in strength between the mortar and precast materials does not differ by more than 25 per cent. In common practice both of these criterion are satisfied.

Bljuger [7.5] presents the information in terms of an effective compressive strength $\eta_1 f_c$ where $f_c$ is the uniaxial compressive strength of the precast members, and $\eta_1$ is a factor depending on the thickness of the joint $v$ and the ratio of the joint strength $f_c'$ to $f_c$. The data are given in Table 7.1.

A further strength reduction factor is due to eccentricity, defined as $e$ in Fig. 7.12(a). The effect of $e$ is to reduce the compressive strength of a joint by the factor $\eta_2$, where:

$$\eta_2 = 1 - 2\, e/b \tag{7.1}$$

**Table 7.1** Compressive joint strength ratios $\eta_1$ for mortar infill.

Ratio of mortar strength to concrete strength ($f_c'/f_c$)

| $v$ (mm) | 0.1 | 0.3 | 0.5 | 0.7 | 0.95 | >1.0 |
|---|---|---|---|---|---|---|
| 10 | 0.87 | 0.93 | 0.94 | 0.95 | 0.95 | 0.95 |
| 30 | 0.60 | 0.78 | 0.85 | 0.90 | 0.93 | 0.95 |
| 50 | 0.37 | 0.68 | 0.81 | 0.88 | 0.93 | 0.95 |

Similarly the effect of lack of verticality, given by $e'$ in Fig. 7.12(b), reduces the strength by a factor $\eta_3$ as follows:

$$\eta_3 = 1\,(1 + e'/b) \tag{7.2}$$

In summarizing the kind of information found above, BS 8110 suggests (in Part 1, clause 5.3.6) [7.6] that in calculating compressive strengths the area of concrete in the joint should be the greater of:

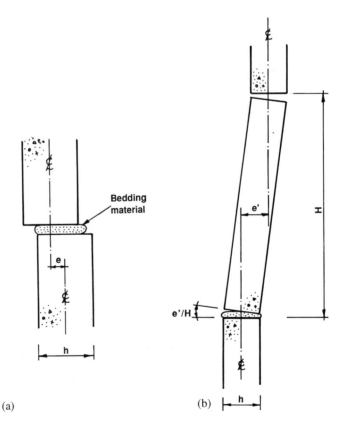

(a)

(b)

**Fig. 7.12** Load reduction effects in compression joints. (a) Eccentricity in a compression joint and (b) lack of verticality in a compression joint.

- the area of the insitu concrete ignoring the area of any intruding component, but not greater than 90 per cent of the contact area
- 75 per cent of the contact area.

The deformability of a compression joint also depends on the nature and number of contact surfaces between different media. In the joints shown in Fig. 7.13, type (a) has two insitu–precast contact surfaces, whilst (b) has one precast–precast and one insitu–precast contact. The deformability of these joints is different due to various lateral splitting and frictional characteristics. The effect is also different in columns and walls where splitting is unconfined in two directions in a column, but only in one direction in a wall.

Generally, Table 7.2 gives the values for the basic deformability $\lambda_o$ of different types of joints [7.7] expressed in linear elastic deformation per unit stress. These data may be used to determine the total deformability of joints and components by the method of addition of strains. For the purpose of structural analysis the members are considered homogenous of equivalent modulus $E'$, and we say that the strain in the equivalent material is the same as in the composite one. The deformability of a joint(s) reduces the stiffness of the connection such that the net value for Young's modulus is given by:

$$E' = \frac{1}{1/E_c + \lambda_m/H} \tag{7.3}$$

where $E_c$ = Young's modulus for the infill concrete

$H$ = storey height

$\lambda_m$ = total joint deformability over the height of the storey obtained by summing the individual joint deformabilities $\lambda_{mi}$ as follows.

Where an interface consists of different parts, e.g. Fig. 7.13(a) and (b), the deformity of the entire interface is given by:

$$\lambda_{mi} = \frac{1}{n_1/\lambda_{o1} + n_2/\lambda_{o2} + \text{etc.}} \tag{7.4}$$

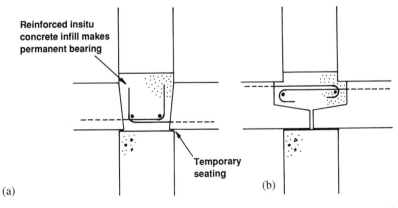

(a)

Reinforced insitu concrete infill makes permanent bearing

Temporary seating

(b)

**Fig. 7.13** Compression joints with discontinuities. (a) Two insitu–precast contact surfaces and (b) one precast–precast and one insitu–precast surface.

**Table 7.2** Deformability $\lambda_o$ of different joints in compression (mm/N/mm$^2$).

| Type of joint | Diagram | Grade of mortar | | |
| --- | --- | --- | --- | --- |
| | | <1 N/mm$^2$ | 5 N/mm$^2$ | >10 N/mm$^2$ |
| Wall to wall | Precast panels / Mortar | 0.10 | 0.06 | 0.04 |
| Wall to wall | Cast in situ concrete / Precast panel | — | — | 0.02 |
| Column splice | Column / Mortar | — | — | 0.02 |

where $n_{1,2}$, etc. is the number of interfaces having a basic deformability $\lambda_{o1,2}$, etc. as given in Table 7.2. Each value of $\lambda_{oi}$ is related to the width of the bearing divided by the width of the interface $b_1/h$, $b_2/h$, etc.

The author has carried out tests on precast–insitu concrete compression joints using unreinforced 100 mm cube sections representing scaled-down versions of 300 mm size components. The specimens comprised a sandwich of grade C40 precast concrete with grade C20 infill of 100, 50, 25 and 0 mm thickness. The maximum size of aggregates used was 10 mm, and the sand was finely graded to allow for the scale factor. Figure 7.14 shows the failure of the 50 mm thick infill specimen, where the predominant cracking is seen in the lower strength infill. Figure 7.15(a) shows stress–strain data for the various tests, and Fig. 7.15(b) the relationship between $\lambda_m/H$ and $b/v$. The data show that the elastic compressive deformability of 25 to 50 mm deep joints of grade C20 concrete between grade C40 precast components equates to a value between 0.03 and 0.06 mm per 1.0 N/mm$^2$ compressive stress for a storey height of 3.0 m. The results from this work are in good agreement with that of Bljuger.

Similar tests were carried out by the author but the compression was induced over the precast–insitu interface by bending a prismatic section. These results are independent of the thickness of the infill and give a single value of 0.02 mm per unit compressive stress for a storey height of 3.0 m, i.e. the flexural joint deformability is less than in the axial compression mode.

**Fig. 7.14** Compression test where precast is twice insitu strength.

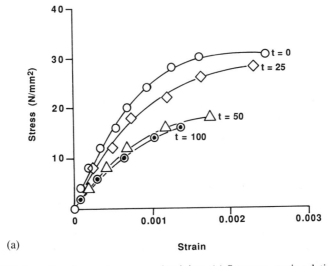

(a)

**Fig. 7.15** Deformation factors in compression joints. (a) Stress vs. strain relationships at precast–insitu concrete interfaces and (b) values for $\lambda_m$ vs. $b/v$ in compression tests.

**EXERCISE 7.1**

Determine the effective Young's modulus for the compression joint shown in Fig. 7.16. The infill concrete and the mortar is grade C40 and C20, respectively.

*Solution*

The deformability of the upper mortar bedding $\lambda_o = 0.04$ mm/N/mm$^2$.

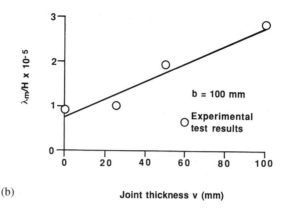

(b)

**Fig. 7.15** (*continued*)

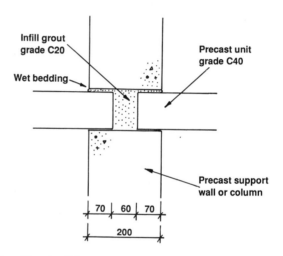

**Fig. 7.16** Details to Exercise 7.1.

The deformability of the insitu infill $\lambda_o = 0.02$ mm/N/mm$^2$.
$E_c$ for the infill concrete $= 24$ kN/mm$^2$ (BS 8110, Part 2, Table 7.2 [7.6]).
For the slab bearing: $n_1 = 2$ for the mortar bedding, with $b_1/h = 70/200 = 0.35$
$\qquad\qquad\qquad n_2 = 1$ for the insitu infill concrete with $b_2/h = 60/200 = 0.30$.

Then $\lambda_{mi} = \dfrac{1}{(0.35 \times 2/0.04) + (0.30 \times 1/0.02)} = 0.031$

The total deformation is the summation $\lambda_m = 0.04 + 0.031 = 0.071$.
From which:

$$E' = \frac{1}{1/24\,000 + 0.071/3000} = 15\,306 \text{ N/mm}^2 = 15.3 \text{ kN/mm}^2$$

i.e. a reduction of 36 per cent compared with the modulus of the insitu infill. Such differences in strain response between members connected in different locations and with different details lead to shear lag between adjacent components. These must be considered in the design.

### 7.5.2  Tensile joints

Lapping of reinforcement bars is often used to connect precast members as shown in Fig. 7.17. The precast units have projecting bars which are to be embedded insitu after erection. A full anchorage length is provided for the embedded bar, and this is calculated according to the same rules as insitu concrete. The projecting bars are usually hooked a full 180°, otherwise the lap becomes ridiculously large. Despite the full anchorage provided for the bars embedded in the precast and insitu concrete, bond stresses quickly break down close to the interface and the two halves of the joint may be considered separately. A tensile crack resulting from elastic deformation in the bar and slippage is formed in the interface and the joint's tension deformability may be calculated in the same manner as for the compression joint. Values given in Table 7.3 are based on the work of Bljuger [7.8]. These data may be used to determine crack widths in flexurally loaded precast–insitu interfaces.

The main problem with vertical lapping is to ensure that the insitu concrete forms a full and positive bond with the steel bars. Pressurized grout is inserted through a hole beneath the level of the lap, and the appearance of the grout at a vent hole above the top of the lap is used as an indication of complete filling as illustrated in Fig. 7.18. The annulus should be at least 6 mm clear on all sides of the bars.

Use clean water to flush out any debris inside the hole before grouting. The water should wet the sides of the hole without ponding. The grout should be non-shrinkable and be sufficiently flowable to allow pressure grouting through a 20 mm diameter nozzle using a manually powered hand pump. A 2:1 sand cement mix containing a proprietary expanding agent is used to give 24 hour strengths of 20 N/mm$^2$ and 28 day strengths around 60 N/mm$^2$. Note that the grading of the sand is crucial to the ultimate strength of this grout.

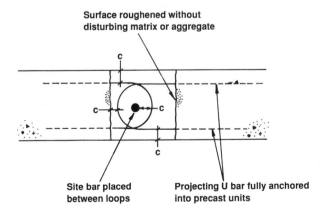

**Surface roughened without disturbing matrix or aggregate**

**Site bar placed between loops**

**Projecting U bar fully anchored into precast units**

c = h$_{agg}$ + 5 mm

**Fig. 7.17** Tension joints by lapping reinforcement in cast insitu infill.

**Table 7.3** Deformability $\lambda_o$ of different precast-insitu joints in tension (mm/N) $\times 10^{-5}$.

| | Type and diameter of bar (mm) | | | | | | | |
| --- | --- | --- | --- | --- | --- | --- | --- | --- |
| | Deformed bar | | | | Plain bar | | | |
| Type of joint | 6 | 8 | 10 | 16 | 6 | 8 | 10 | 16 |
| Straight bar in tension | 4.0 | 2.5 | 2.0 | 1.3 | 5.2 | 3.2 | 2.6 | 1.7 |
| 180° U-bar in tension | 3.2 | 2.0 | 1.6 | 1.0 | 4.2 | 2.6 | 2.1 | 1.4 |
| 90° hooked bar in tension | 6.4 | 4.0 | 3.2 | 2.1 | 8.4 | 5.2 | 4.2 | 2.8 |

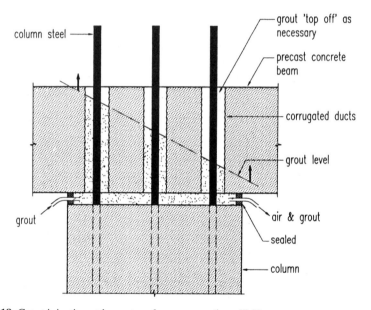

**Fig. 7.18** Grout injection at beam to column connections [7.9].

Grouting may be carried out by gravity pouring, but the annulus must be vented, or sufficiently large in diameter, e.g. 50 to 60 mm for a 25 mm bar, to prevent air pockets forming.

Bolting is used extensively to transfer tensile and shear forces. Anchorages such as bolts, threaded sockets, rails or captive nuts attached to the rear of plates are anchored in the precast units. Tolerances are provided using oversized or slotted holes in the connecting member. The tensile capacity of bolted connections should be governed by the yield strength of the bolt, as this gives a ductile failure. In most types of bolted joints tension is accompanied by shear. Shear capacities are governed by the local bearing strength of the concrete in contact with the shank of the threaded socket. Shear bolt failures are brittle and should be avoided. The PCI Manual [7.10] gives design data for bolting.

Welding is used to connect components through projecting bars, fully anchored steel plates or rolled steel sections, etc. The joint is made either directly between the projecting bars, as shown in Fig. 7.19(a), or indirectly using an intermediate bar or plate, as shown in Fig. 7.19(b).

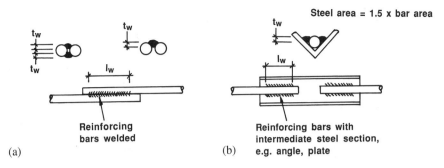

**Fig. 7.19** Tension joints by welding reinforcement. (a) Direct joint and (b) indirect joint.

Post-tensioning is used to resist tension and shear forces by the application of clamping forces across the joint. Cable ducts are inserted into the precast concrete components, or in the spaces around the components and, after erection, the cables are placed in the ducts and post-tensioned. Tensile capacities are computed from the state of stress in the post-tensioned components, and shear resistance is calculated using the shear friction hypothesis.

### 7.5.3 Shear joints

The most important statement to make is that shear resulting from flexural actions, for example in precast–insitu beams in bending, must be considered separately from shear derived from shear forces. The former was discussed in Section 6.7, whilst the latter concerns the transfer of direct shear forces between precast components through bond, friction, welds, shear keys and shear reinforcement, as shown in Fig. 7.20(a) with their associated failure modes. The graphs in Fig. 7.20(b) are plots of shear force vs. shear slip. The gradient of these curves gives shear stiffness which forms the basis of shear deflections in precast construction, i.e. shear deformation is the sum of joint slip and element deformation.

Shear friction should be used where the available shear interface is large, e.g. between precast floor units and insitu screed, and the possibility of delamination, e.g. due to vibrations, is small. No attempt should be made to deliberately roughen the surface texture of precast units beyond the usual machine slip-formed or extruded finish, or tamped finish. Large crevices harbour grease and debris, and provide possible sites for water to accumulate immediately before concreting commences (see Section 6.2.2). The shallow shear key provided by the normal trowelling procedures is not only structurally adequate in terms of interface stress, but has greater quality control. If no external normal force is present then reinforcement should be provided across the shear plane to ensure that the shear friction is mobilized, and the limiting ultimate shear stress is 0.23 N/mm$^2$, according to BS 8110, Part 1, clause 5.3.7[a] [7.6].

Deeper shear keys may be deliberately formed as shown in Fig. 7.20(c). The root depth should be at least 10 mm. The design is satisfied by following a few simple rules regarding geometry and cover distances. The angle $\alpha$ depends on the dimensions of the keyed surface. The resulting force $N$ has a magnitude $N = V \cot \alpha$ with $\alpha = 40°$ to $50°$, which must be carried either by reinforcement, of area $A_s$, crossing the interface, a

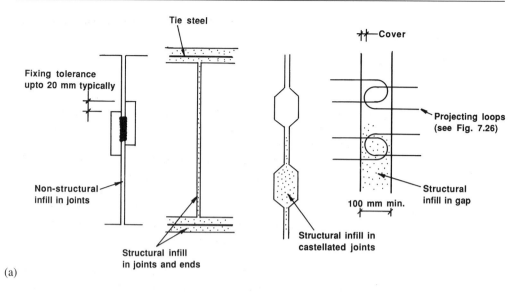

(a)

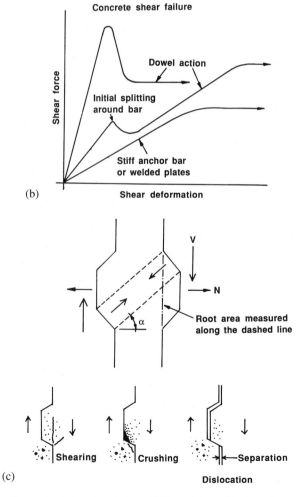

**Fig. 7.20** Design methods for shear joints. (a) Shear joints; (b) shear force vs slip behaviour and (c) deep keyed shear joint.

precompression $P$ from external sources (post-tensioning for example), or a combination of both such that:

$$A_s = \frac{V \cot \alpha - P}{0.87 f_y} \tag{7.5}$$

Sufficient reinforcement should be provided to control cracking, and the water content of the insitu infill should be low, e.g. about 160 kg/m$^3$, with a super-plasticizer being used to maintain medium workability.

Shear keys rely on mechanical interlock and the development of a confined diagonal compressive strut across the shear plane. A taper is usually provided to aid removal from formwork. This also assists in confining the concrete in the second direction. Failure is generally ductile due to the warnings given by concrete cracking in this way. However, as with shear friction methods, the interfaces should be prevented from moving apart, either explicitly by external forces, or implicitly by placing reinforcement according to Equation (7.5) across the shear plane. Providing that this is done, BS 8110, Part 1, clause 5.3.7[b] [7.6] states that no shear reinforcement is required if the ultimate shear stress is less than 1.3 N/mm$^2$, when calculated on the minimum root area. Where shear reinforcement is used the strength of the interface is based on the shear strength of the reinforcement alone, called 'dowel shear'. As shown in Fig. 7.21, it is not possible to add concrete-to-concrete frictional forces to dowel forces because when the latter is being mobilized the former is not, and vice versa. The shear capacity of a dowel is given by:

$$V_d = 0.6 f_y A_s \tag{7.6}$$

Shear stresses in welded joints may arise from direct shear or torsion. The usual structural models and material properties for the design of welds are used. Due consideration should be given to manufacturing and construction tolerances when specifying weld lengths, because the adjoining elements are captured in the precast components. This means adding 10 to 15 mm to the length of a weld. Figure 7.22 shows an example of the use of welded shear joint in the construction of a shear wall panel to column connection.

Shear deformations are the result of localized damage at the shear interface. Walraven and Mercx [7.11], Bljuger [7.12] and others have produced expressions for shear deformations in cracked concrete interfaces. The theory assumes that sliding movement between two faces is the result of matrix damage, which is related to the stiffness of the cement paste, and hence $E_c$ of the concrete. Also, the state of stress at the interface influences the damage in the paste, and therefore the normal stress $\sigma_n$, which is proportional to the area of reinforcement crossing the cracked plane, is also important.

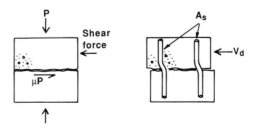

**Fig. 7.21** Shear friction hypothesis.

**Fig. 7.22** Shear jointing by site welding together steel plates in precast wall and column components.

Elastic shear deformabilities, expressed in units of mm displacement per kN applied shear force, are given in Table 7.4 [7.12].

Type I joints were also studied by the author using two precast blocks of equal strength, grade C25 and C50 for three diameters of helical strand – 9.3, 12.5 and 15.2 mm, shown in Fig. 7.23(a). The results given in Fig. 7.23(b) are in good agreement with the values obtained from Table 7.4. For example, experimental joint deformations are in the range 10 to $24 \times 10^{-3}$ mm/kN. Calculated values from Table 7.4 for the same geometry and material moduli are between 15 to $24 \times 10^{-3}$ mm/kN for grade C25 concrete, and between 12 to $19 \times 10^{-3}$ mm/kN for grade C50 concrete.

The results in Fig. 7.23(b) also show that the increase in dowel capacity is proportional to the cube root of the concrete strength ratio, i.e. $\sqrt[3]{50/25} = 1.26$. Because the failure mode is governed by the shear-tension capacity of the concrete it is hardly surprising to find that the effect of increasing diameter is small, i.e. less than 25 per cent for increases in cross-sectional areas of up to $2\frac{1}{2}$ times. This observation does not agree with the data given in Table 7.4.

Shear deformations in Type III joints are due to shear, bending and rotational deformations, idealized in Fig. 7.24. The shear stiffness is given by $k_s$ where:

**Table 7.4** Shear deformabilities (mm/kN).

| Description of joint | Diagram | Shear deformability (mm/kN) |
|---|---|---|
| Type I | Interface / Bar | $\dfrac{6}{\Phi}\left(\dfrac{1}{E_{c1}}+\dfrac{1}{E_{c2}}\right)$ |
| Anchor bar cutting through shear plane | Backing detail | $\dfrac{6}{\Phi E_{c1}}$ |
| | Welding / Bar | $\dfrac{1.5}{\Phi E_{c1}}$ |
| Type II | Precast element / Cast in situ concrete | $\dfrac{0.5}{E_c A_k}$ |
| Concrete element cutting through shear surface | Joint / Mortar layer / Floor slab | $5\times 10^{-3}$ |
| Type III Reinforced concrete flexural member connecting walls | Floor slab / Wall / Slab / Slab | To be determined as given below |

$\Phi$ = diameter of rebar (mm)
$E_{c1}$, $E_{c2}$ = elastic modulii of precast and insitu concrete, respectively (kN/mm$^2$)
$E_c$ = smaller of $E_{c1}$ and $E_{c2}$ (kN/mm$^2$)
$A_k$ = contact area at root of castellation (mm$^2$).

$$k_s = \frac{E_c bh}{3L} \qquad (7.7)$$

where $L$, $b$ and $h$ are the length, breadth and depth of the element. Shear deformations are not influenced by end conditions.

The flexural stiffness is given by $k_b$ where:

(a)

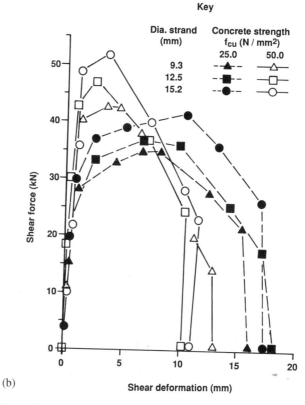

(b)

**Fig. 7.23** Shear joint tests by Bensalem at Nottingham University. (a) Experimental shear joint tests and (b) results from tests.

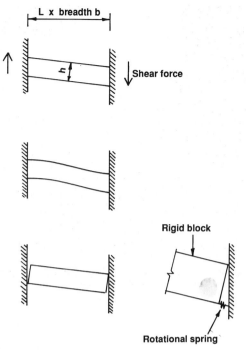

**Fig. 7.24** Components of shear deformations.

$$k_b = \frac{12 E_c I}{a_1 L^3} \qquad (7.8)$$

where $I$ is the second moment of area of the element in the plane of bending, $a_1$ is a factor; $a_1 = 1$ if both ends of the element are clamped, and $a_1 = 4$ if one end is clamped and the other end pinned.

The stiffness due to joint rotation is given by $k_j$ where:

$$k_j = \frac{b}{a_2 K L^2} \qquad (7.9)$$

where $a_2$ is a factor; $a_2 = 0.5$ if both ends of the element are clamped, and $a_2 = 1$ if one end is clamped and the other end pinned, and $K$ = bending deformability within the joint itself according to Table 7.5 [7.13] and Fig. 7.25.

### 7.5.4   Flexural and torsional joints

Flexural and torsional forces may be resolved as for a couple generating direct tension, direct compression and shear. The strain gradients across these joints are often large and so the materials making up the joint will behave differently under stresses of differing magnitude (rather in the same manner as the Brazilian and beam rupture tests give different tensile strengths). The main types of flexural joint are shown in Fig. 7.26(a) where a couple is generated between concrete in compression and steel in tension and shear. A

**Table 7.5** Bending deformability of an element clamped over a distance $L_s$.

| $L_s$ (mm) | 50 min | 100 | 150 | >200 |
|---|---|---|---|---|
| $K$ (radian/kNm/m) | 0.00020 | 0.00015 | 0.00012 | 0.00010 |

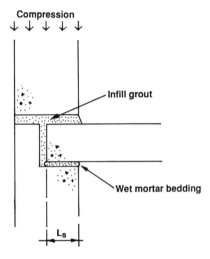

**Fig. 7.25** Definition of clamping and restraint factors in flexure and shear connections.

'cold' joint should not be left between the precast and insitu concrete in the compression zone. This is the term used to describe the placement of wet concrete against an ex-mould precast surface with no preparation or applied bonding agents.

It is also important that the concrete is adequately confined either by the use of binding links (see Section 10.2) or by the correct use of a bearing pad to prevent lateral bursting, as explained in Section 7.6. The main design criteria for the tension force is adequate anchorage, which is usually provided by casting a fully anchored steel plate or threaded socket, etc. into one or both adjoining members. Space is often at a premium in these joints and so it is recommended to use a large number of smaller components in the joint, rather than a small number of large components. The former option has the added protection of being able to yield and dissipate energy in case of overloading.

Empirical rules for the detailing of projecting loops are given in Fig. 7.26(b). The bearing stresses inside the loop are governed by BS 8110, clause 3.12.8.25 [7.6]. Semi-circular loops of internal radius $7 \times$ dia. may be used, otherwise a straight portion must be added to the loops making the splice zone larger. Lateral reinforcement of at least equal cross-sectional area to the loops should be provided across the loops.

## 7.6 BEARINGS AND BEARING STRESSES

### 7.6.1 Average bearing stresses

It is inevitable that some of the joints in precast frames will rely on one member bearing

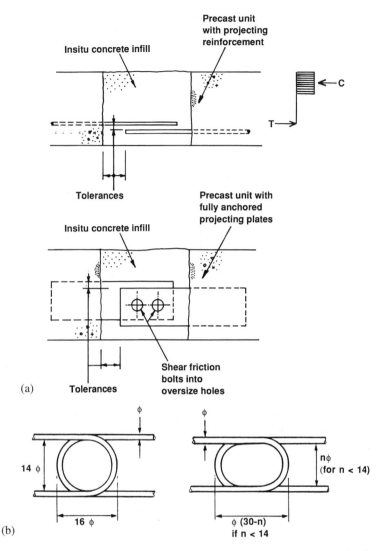

**Fig. 7.26** Flexural joint details. (a) Flexural jointing by welding or bolting projecting steel sections together to form a couple with the concrete in compression and (b) dimensions of loops for correct load transfer.

directly onto the cast surface of another member. Bearings are designed for structural strength and integrity, and BS 8110 recognizes the latter in stating (Part 1, clause 5.2.3) [7.6] that:

'... the integrity of a bearing is dependent on two essential safeguards: (a) an overlap of reinforcement in reinforced bearings, (b) a restraint against loss of bearing through movement ...'

and, it might be added, for a partial bearing due to accumulative manufacturing and site fixing tolerances. The main types of bearings, summarized in Fig. 7.27, are:

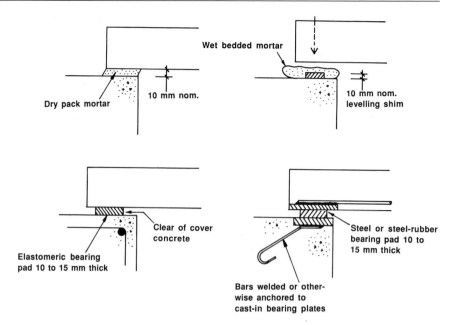

**Fig. 7.27**  Bearing types.

- dry bearing of precast to precast or precast to insitu concrete
- dry packed bearing where components are located on thin (3 to 10 mm thick) shims and the resulting small gap is filled using semi-dry sand/cement grout
- bedded bearing where components are positioned onto a prepared semi-wet sand/cement grout
- elastomeric or soft bearing using neoprene rubber or similar bearing pads
- extended bearings where the temporary bearing is small and reinforced insitu concrete is used to complete the connection
- steel bearing using steel plates or structural steel sections.

There is a broad division between precast components which are considered to be non-isolated and those which are isolated. Non-isolated components are connected to other components with a secondary means of load transfer which would sustain loads in the event of failure in the primary support. For example, hollow core flooring units which are grouted together would distribute shear forces to adjacent members in the event of a failure at the beam support and would be classed as non-isolated. On the other hand, a stairflight unit seated on to a dry corbel is an isolated component.

The design method considers the bearing stresses in both the abutting precast components and the sandwiched jointing material (if any) in between. Two values are determined in the plan dimension; bearing length is the greater dimension perpendicular to the direction of the span of the member, and bearing width. The effective bearing length is taken as the least of:

(1) the actual bearing length of the imposed member
(2) one half of (1) plus 100 mm
(3) 600 mm.

Conditions (2) and (3) are empirical values based on the acceptable degree of flatness possible in members exceeding about 1.0 m in the critical dimension.

The net bearing width, as defined in Fig. 7.28, is equal to the actual bearing width minus twice the spalling tolerance, called an ineffective bearing width. Note that the difference between the 'nominal' bearing width, as cited in design details, and the actual bearing width is equal to the sum of all inaccuracies. These equate to about 3 or 5 mm per metre distance between faces of precast or insitu supports, respectively. The minimum net bearing width is 40 mm for non-isolated components (or 60 mm for isolated), and the minimum nominal bearing width is about 70 mm. The ultimate bearing stress is given by:

$$f_b = \frac{\text{ultimate support reaction per member}}{\text{effective bearing length} \times \text{net bearing width}} \qquad (7.10)$$

These should not exceed the design ultimate bearing stresses, which are based on the cube crushing strength of the weakest of the two, or three, component materials (excluding masonry support), as follows (BS 8110, clause 5.2.3.4 [7.6]) for:

- dry bearings on concrete: $0.4 f_{cu}$
- bedded bearings on concrete or grout: $0.4 f_{cu}$ (bearing medium)
- elastomeric bearing (called flexible padding): between $0.4 f_{cu}$ and $0.6 f_{cu}$, use $0.5 f_{cu}$ or $f_c$ (bearing material)
- steel bearing of width $b_p$ cast into member or support and not exceeding 40 per cent of the concrete dimension $b$ (probably meaning bearing length): $0.8 f_{cu}$. Higher bearing stresses may be used only if proved by adequate testing. For wider bearing plates the allowable bearing stress $f_b$ is given as follows [7.14]:

$$f_b = \frac{1.5 f_{cu}}{1 + (2b_p / b)} \qquad (7.11)$$

This reduced stress is to cater for diagonal tension directly beneath the insert and

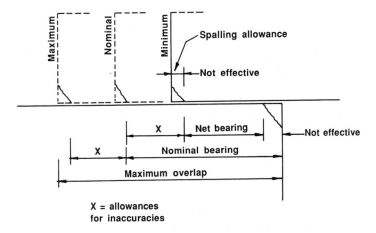

Fig. 7.28 Definitions of bearing lengths and tolerances.

close to the outside face of the column. Thus if $b_p/b > 0.44$, Equation (7.11) will govern.

In the case of a bearing plate $b_p \times h_p$ embedded in a section $b \times h$ the allowable bearing stress $f_b$ is given as follows [7.15]:

$$f_b = 0.6 f_{cu} \sqrt{\frac{bh}{b_p h_p}} < 2.0 f_{cu} \qquad (7.12)$$

The allowances for spalling are given in Tables 7.6 and 7.7.

### 7.6.2   Localized bearing stresses

In some cases localized bearing stresses will be much greater than the allowable ultimate values given above. Providing that the zones of localized high bearing are confined in the normal rc manner, these stresses may be treated separately and considerable enhancements in allowable values are permitted. If steel inserts or bearing plates have cast-in reinforcement welded to their backs, local bearing stresses up to $0.8 f_{cu}$ may be used. Test results [7.16] show that in columns, ultimate values equal to $1.0 f_{cu}$ were achieved using $H$ and $I$ section inserts. The IStructE manual [7.17] recommends:

'... where a local load represents a small part of the total loading, its influence can be neglected. It is difficult to provide clear guidance to cover all cases, but broadly, local effects can be ignored where the overall load does not exceed 80% of the

**Table 7.6**  At the supporting member.

| Method of support | Ineffective bearing (mm) |
|---|---|
| Steel | 0 |
| Concrete grade $\geq$ C30 | 15 |
| Concrete grade < C30 and masonry | 25 |
| Reinforced concrete less than 300 mm deep at outer edge | Nominal cover to reinforcement at outer edge* |
| Reinforced concrete where loop reinforcement at outer edge exceeds 12 mm diameter | Nominal cover plus inner radius of bent bar* |

\* Chamfers occurring in the above zones may be discounted.

**Table 7.7**  At the supported member.

| Reinforced at bearing | Ineffective bearing (mm) |
|---|---|
| Tendons or straight bars exposed at end of member | 0 |
| Loops not exceeding 12 mm diameter | 10, or end cover, whichever is the greater* |
| Loops exceeding 12 mm diameter | Nominal end cover plus inner radius of bent bar* |

\* Chamfers occurring in the above zones may be discounted.

allowable capacity, or where the local load does not represent more than 20% of the total load.'

Localized stresses may also result from the rotations of flexural members at simple bearings. Where rotations exceeding 0.01 radians are anticipated, suitable soft bearing membranes should be used to accommodate the rotation. A suitable material for 0.02 radians of rotation is 10 mm thick × 150 mm square neoprene pad. Otherwise, rotations will shift the line of action of loading onto a knife-edge near to the end of the spalling zone.

Finally, the combination of vertical load $V$ and significant horizontal loads $H$ can reduce the vertical bearing capacity by causing splitting or shearing in the supporting member. Where $H > V \tan 20°$ allowances should be made by providing sliding bearings, lateral reinforcement in the top of the supporting member or continuity reinforcement to prevent splitting in the supported member.

## 7.7 CONNECTIONS

### 7.7.1 Pinned connections

Pinned connections transfer purely shear forces, both for (dominant) gravity and uplifting forces. The connections lend themselves to simple detailing and construction, and may be formed in the simplest manner by element to element bearing. To increase bearing capacities, and reduce the localized spalling, steel inserts are frequently used by anchoring plates (or rolled structural sections) into the connecting elements. The connection is completed by surrounding the steel inserts with cementitious grout for fire and durability protection.

### 7.7.2 Moment resisting connections

Moment resisting connections are possible in many situations, in particular at foundations and between beams and columns. Moment resisting foundation design is dealt with in Section 7.14, but the idea of casting or bolting a steel plate to the end of one component and then connecting the plate to another component through waiting bars may be exploited elsewhere if space permits.

If the connection is to be structurally useful, either in relieving sagging moments in beams, or increasing the global frame stiffness, moments of resistances of at least 50 to 75 kN m are required for this purpose. If the moment is less than this then it is probably better to design the connection as pinned. A lever arm of about 150 to 250 mm is required between the compressive and tensile forces which generate the moment. Thus the couple will need to generate forces of around 300 kN. This is achieved using an area of concrete of about 20 000 mm$^2$ (grade C40 concrete assumed), and in the steel using 2 no. or 3 no. 16 to 20 mm diameter high tensile bars (grade 460 assumed).

If welding is used to provide the continuity of forces generous tolerances should be allowed in the positioning of the steel in the separate components to give the site operative the opportunity to place an intermediate bar between the bars and place a well-fashioned bead. Underhand welding should be avoided if possible, and the weld should be inspected afterwards.

Moment resisting connections should be proportioned such that ductile failures will occur and that the limiting strength of the connection is not governed by shear friction, short lengths of weld, plates embedded in thin sections, or other similar details which may lead to brittleness. Many of the principles behind these requirements have evolved through years of seismic research and development, and the common practice in the US, Japan and New Zealand is to design and construct moment resisting connections in the perimeter of the frame, where there is less size restrictions on beams and columns [7.18]. Large deep beams with ample space for this purpose are specified as ring frames, whilst the interior frames' connections are all pinned-jointed shallow beams. Figure 7.29 show details of a 457 mm square precast–insitu concrete beam and connection tested by Stanton *et al.* [7.1] which achieved a moment of resistance of 2100 kip-in (230 kN m). Versions similar to this have been studied by Park [7.18] and Pillai and Kirk [7.19].

Moment resisting connections may be formed between columns and precast walls or (sufficiently thick) cladding panels because the latter is usually capable of accommodating the transfer moment. However, caution must be exercised in assuming that the places where splices are made will attain their ultimate design limit, i.e. the receiving element may not be able to carry the connecting moment.

## 7.8    DESIGN OF SPECIFIC CONNECTIONS IN SKELETAL FRAMES

### 7.8.1    Floor slab-to-beam connections

Floor connections can be divided into two main categories:

- connections at supporting joints
- connections at non-supporting longitudinal joints.

In the latter, beams or walls spanning parallel with the floor slab are essential to the integrity of the precast frame and as such these joints are of equal importance to the main supporting joints.

Beam-to-slab connections using hollow core and double-tee floors are designed as simple supports despite the presence of reinforced insitu concrete strips cast in the ends of the units. Hollow core floors are usually laid dry directly onto the shelf provided by the boot of the beam, but neoprene bearing pads or (less frequently) wet bedding onto grout is also used in certain circumstances, e.g. double-tee floor slabs. Wet bedded bearings are sometimes used on refurbished beams with uneven surfaces.

### 7.8.2    Connections at supports

Although the connections at supporting joints are normally designed as simply supported, a degree of end restraint may be present due to the couple generated between the contact plane and tie back (rc or weld) as shown in Fig. 7.30. Moments of resistance of up to 25 kN m per metre run have been measured by Mahdi [7.20] using only 2 no. T 12 bars site-placed into the opened cores in the tops of the slabs. The magnitude of the restraint does not usually affect the flexural behaviour of the slab [7.21] because cracking in the top of the floor slab is often found only at the insitu-to-precast interface, and not in the

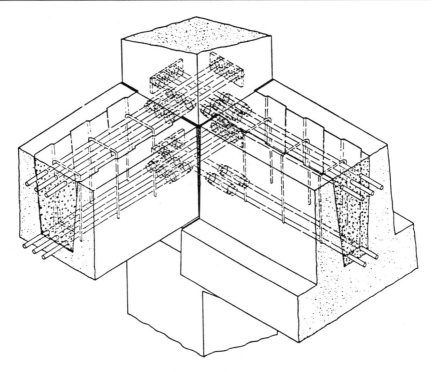

**Fig. 7.29** Moment resisting connections after Stanton [7.1].

slab itself. Longitudinal shrinkage of the floor slabs may also lead to cracking over the support, thus rendering the connection pinned in any case.

The objectives of design are to ensure transfer of vertical loading from the slab to the beam in both normal and abnormal (fire, accident, etc.) loading conditions. The connection has therefore to satisfy the requirements of load transfer, structural integrity and ductility. Some of the means of achieving this are shown in Fig. 7.31.

In hollow core construction, Fig. 7.31(a), openings are made in the top flanges of the units (during manufacture) to permit the placement of structural (grade C25 minimum) on site. Continuity of reinforcement is achieved either by direct anchorage between the precast beam and insitu strips, dowel action between loops, or between loops and other bars. The bars may be placed in the longitudinal gaps between the slabs providing that the geometry of the gap corresponds to the details in Fig. 5.9. Structural continuity, but not moment transfer, over internal beams is easily achieved as shown in Fig. 7.32(a)–(c). See the FIP Manual [7.21] for other possible details.

The length of embedment is taken as the greater of one anchorage bond length or the equivalent of the transfer length of the prestressing force in the precast unit. End hooks may be used to obtain the required length. The openings are wide enough to ensure good compaction using a small diameter poker vibrator. The infill usually penetrates the end of the opening a distance greater than is structurally required.

Hollow core units are laid directly onto dry precast beam seatings. The design ultimate bearing stress $0.4 f_{cu}$ is rarely critical. A nominal bearing length of 75 mm, resulting in a net length of 60 mm after spalling allowances have been deducted, gives rise to a bearing capacity in the region of 800 kN/m run. Rigid neoprene strips or wet mortar bedding,

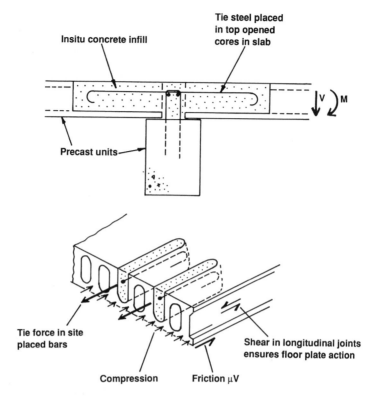

**Fig. 7.30** Spurious moment restraint at slab–beam connection.

which have been used in special circumstances, e.g. in refurbished buildings or on masonry bearings, ensure that a uniform bearing is made between them.

In double-tee floors a welded connection is made, Fig. 7.31(b). Double-tee slab bearings require greater consideration because it is vital that the loads should be equally shared between all four bearing points. A nominal bearing length of 150 mm minimum is recommended and the units should always be seated on rigid $100 \times 100$ mm neoprene (or similar) pads of about 8 to 10 mm thickness.

Design equations given in PCI Manuals [7.23] for ultimate bearing capacity take into consideration the dimensions $s$, $w$ and the coefficient of friction $\mu = T_u/V_u$ defined in Fig. 7.33. Using a static value for $\mu$ between dry elastomeric and concrete surfaces of 0.7 the analysis considers a potential shear failure crack inclined at 20° to 25° to the vertical free edge of the bearing. For typical double-tee geometry, the minimum bearing capacity (based on a design stress of 7 N/mm² for neoprene) is in the order of 170 kN/m. Again this is rarely critical. Bearing lengths of less than 100 mm are not recommended for double-tee units because there is a chance that misalignment of the pads could easily reduce the actual bearing length to (say) 70 mm (Fig. 7.34).

It is not advisable to attempt a flexural connection between double-tee units and support beams unless full anchorage of the force mechanisms providing the couple is guaranteed. The transfer moment is concentrated at the seating of the webs (usually at 1.2 m centres). Problems due to the combined effects of vertical and horizontal forces at the seating are difficult to avoid in such confined areas without the use of special anchorage devices.

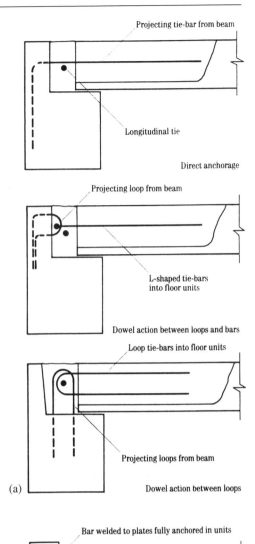

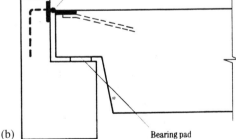

**Fig. 7.31** Floor slab to external beam connections [7.22]. (a) At hollow core slabs connections and (b) at double-tee slab connections.

Connections between flat plank flooring and supporting members present few problems providing a soft sand/cement mortar is used to bed 2.4 m wide units onto precast (particularly prestressed with camber) beams. A 75 mm nominal bearing is used. Continuity of reinforcement is provided by the mesh lapping with projecting reinforcement in the beams or walls.

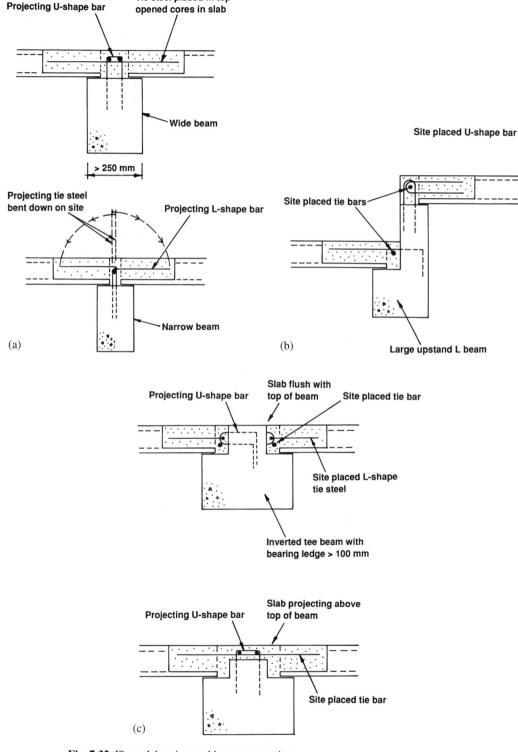

**Fig. 7.32** Floor slab to internal beam connections.

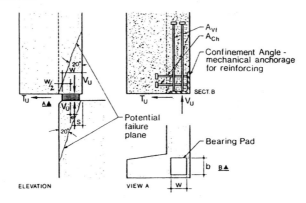

**Fig. 7.33** Shear friction hypothesis used to analyse edge bearings.

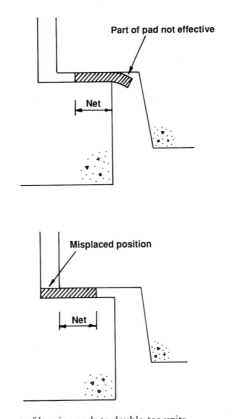

**Fig. 7.34** Misplacement of bearing pads to double-tee units.

### 7.8.3   Connections at longitudinal joints

These are provided between the edges of precast floor units and beams (or walls) running parallel with the floor. Their main function is to transfer horizontal shears, generated in the floor plate by diaphragm action (see Section 8.3), between floor slab and beam whilst allowing for the differential movement between adjacent precast components. Details of this joint are shown in Fig. 7.35(a)–(c).

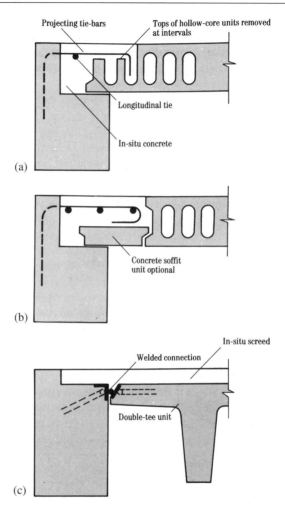

**Fig. 7.35** Floor slab to beam connections at longitudinal positions [7.22]. (a) Intermittent tie, (b) continuous insitu edge strip and (c) welded tie to double-tie.

In hollow core construction, recesses may be formed by removing part of the top flange. An insitu rc joint is formed at intervals, usually at 2.4 m centres, along the edge of the slab. Full depth insitu concrete strips may be preferred to ensure a positive shear key. The slab is sufficiently flexible to accommodate differential vertical movements due mainly to temperature and loading effects.

Typical longitudinal joints between double-tee and edge units are also shown in Fig. 7.35 where a welded connection (at 2.4 m intervals) is made between the units immediately after fixing. The strength of the weld is sufficient to maintain the position of the double-tee unit whilst the insitu concrete screed is hardening. The welded connection is designed to prevent separation of the reinforced screed – which is tied to the edge beam – from the flanges of the precast slab. The flanges of the slab are, in turn, reinforced (with mesh) to cater for flexural forces generated by the shear restraint of the weld to differential movement. Longitudinal cracking in double-tee units is very rare indeed.

It is not always necessary to use parallel edge beams with solid precast plank floorings because of the semi-monolithic form of the construction. However, if beams are used (to carry wall loads for example), an insitu connection between the beam and structural screed is made by lapping site-placed mesh to projecting bars from the edge beam.

### 7.8.4 Floor connections at load bearing walls – load bearing components

Horizontal joints in load bearing walls occur at floor and foundation levels. Primary forces in the joint are due to vertical compression from upper storey panels and horizontal shears from floor plate diaphragm effects. Secondary forces due to temperature, long-term shrinkage and creep, and spurious bending moments induced by end restraints are, by comparison, of minor importance. Connections at wall supports require careful detailing particularly if the floor units are supported within the breadth of the walls, and large wall loads are imposed (Fig. 7.36). Some hcus (e.g. 300 mm deep) may require strengthening to prevent web buckling by filling the voids to a depth coincident with the edges of the walls. Double-tee units may require rib end closure pieces to form a diaphragm.

There is a wide range of solutions available in forming these joints varying from an extensive use of insitu concrete and tie steel, Fig. 7.37(a), to welded connections made between fully anchored plates, Fig. 7.37(b). The former offers excellent structural integrity and ductility whereas the latter offers immediate stability to the precast frame. For these reasons a combination of both is sometimes used such that the bolt or weld design need only satisfy temporary stability conditions.

A general classification is made between *continuous* joints (e.g. using hollow core flooring) and *isolated* joints (e.g. using double-tee flooring where a connection is made at 1.2 m intervals). Typical details for continuous construction are shown in Figs 7.38 and 7.39(a) and (b) for internal and external use, respectively. It is important for the designer to allow for easy access to these joints, which is why the detail in Fig. 7.39(a)

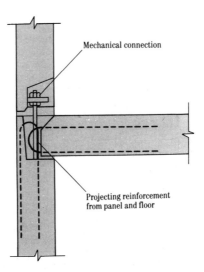

Mechanical connection

Projecting reinforcement
from panel and floor

**Fig. 7.36** Floor slab to wall joints [7.22].

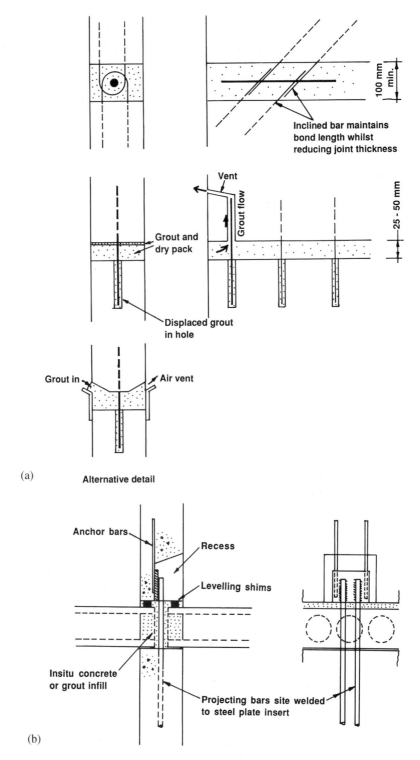

**Fig. 7.37** Load bearing wall to wall connections. (a) Using reinforced insitu joints and (b) using welded joints.

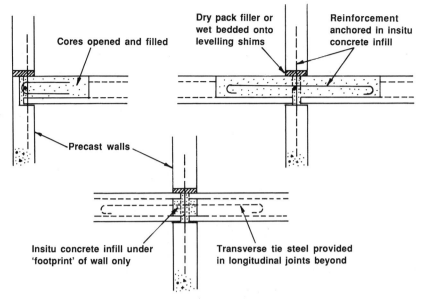

**Fig. 7.38** Continuous connections in floor slab to wall joints.

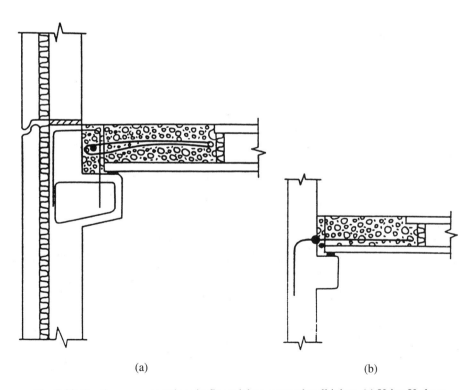

(a)                                                                (b)

**Fig. 7.39** Continuous connections in floor slab to external wall joints. (a) Using U-shape site bars and (b) using projecting L-shape bars.

using loose U-shape site bars is preferred to Fig. 7.39(b) where the projecting L-shape bars must be cast in the wall within an accuracy of about 50 mm.

Strength controlling factors for the correct transfer of vertical load are:

- the extent and compressive strength of confined insitu concrete
- the 'effective' width of shims and the use of steel levelling shims in the upper storey panel bearing
- the vertical splitting strength of the panel ends
- the strength of any mechanical connection.

A further subdivision of joints is where, for reasons of restricted access or bearing width, an extended bearing is formed as shown in Fig. 7.40. This is often referred to as an *open* joint, as opposed to the more frequently used *closed* joint. Differences in failure modes are due to the different stress-deformation characteristics of the insitu and precast concretes. Whilst crushing failure is theoretically possible, tensile splitting occurs first because transverse reinforcement is not present in most cores in hollow core slabs. Furthermore, to restrict the penetration of insitu concrete to the width of the wall bearing reduces the risk of introducing hogging restraint moments to floor slabs that are not suitably reinforced.

## 7.9   BEAM-TO-COLUMN, AND BEAM-TO-WALL CONNECTIONS

### 7.9.1   Definitions for different assemblies

The most fundamental feature in the design of multi-storey precast frames is the connection

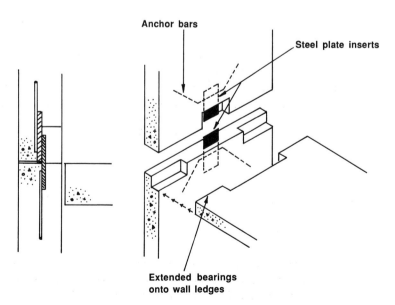

**Fig. 7.40** Intermittent connections and extended bearings in an 'open' floor slabs to wall joints.

between vertical and horizontal components. There is a broad subdivision in that either the vertical member is *continuous* (both in design and construction terms) and horizontal components are framed into it at various levels, or the vertical member is *discontinuous* (only in construction terms) and the horizontal components are either structurally continuous or separate across the junction. Although the former usually refers to skeletal column and beam construction and the latter to wall frames, there is an alternative approach, particularly in low-rise unbraced buildings, for single-storey columns to be erected floor-by-floor and an important insitu concrete connection made between precast components.

Fogarasi [7.24] gives a summary of many of the types of connections used in Europe, Asia and North America. Of the 20 or so types of connectors cited there is an even split between the use of concrete corbels with twin rib or rectangular beams, and cast-in steel inserts forming what many people refer to as the 'invisible corbel'.

### 7.9.2   Connections to continuous columns using hidden steel inserts

These joints are one of the commonest form of the beam-to-column connection. The structural mechanism, which is described in Fig. 7.41, is based on static strength, stiffness, load transfer into the connecting components, temporary stability and structural integrity. A wide range of joints, some of which are shown in Fig. 7.42(a)–(f), have evolved to satisfy these requirements. The basis of joint design is as follows (Fig. 7.41):

(1)  At A – to transfer the shear force carried by the reinforced (or uncracked prestressed) concrete at the ends of the beams into the connection. This usually takes the form of shear transfer from a combination of vertical shear links and/or bent up bars into a prefabricated steel section, called a *shear box*. The end of the beam is pocketed or recessed.

(2)  At B – to ensure adequate shear capacity in the plane of the physical discontinuity between beam and column (or other supporting member) taking into consideration ultimate and working values of static strength, cracking, durability and fire resistance. This usually takes the form of either a projecting (solid or hollow) steel section, or a gusseted angle or T-cleat bolted into anchored sockets.

(3)  At C – to transfer the compressive loads into the rc column. The effects of horizontal bursting forces, both above and below the connection in the case of eccentrically loaded columns, are taken care of using closely spaced links. Column anchorages are generally either fully anchored cast-in-sockets, or steel box or H-section cast-inserts.

### 7.9.3   Beam-to-column inserts at C

The options for this are as follows:

(1)  Direct frictional bearing between beam and column inserts with no positive mechanical action between the essential precast components. A non-structural top fixing, comprising either a bolted or welded cleat or plate, or an arrangement of reinforcement to provide torsional stability to the connection (Fig. 7.42(a)). Shear capacity is based on the net shear section of the projecting inserts.

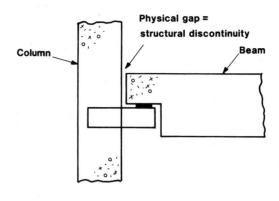

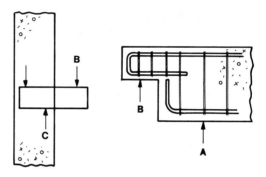

**Fig. 7.41** Basic principle for the design of beam-to-column connections using halving joints and cast-in steel inserts.

(2) As (1) with the addition of a *drop bolt* located between the top fixing and column insert and used to give immediate temporary stability to the connection and a positive mechanical tie between the precast components (Fig. 7.42(b)).

(3) A welded connection in which discrete weld preparations have been made to the adjoining beam and column inserts. Top fixings may be excluded because of the torsional stability provided by the weld (Fig. 7.42(c)). The vertical shear capacity is as (1) and excludes the contribution of the weld.

(4) A sliding plate fitted into a rectangular slot in the beam. A notch at the end of the plate fits over a lip at the bottom of a steel box cast into the column (Fig. 7.42(d)).

(5) A separate intermediate cleat. Typically these are rolled structural tee, rolled angle, fabricated plate angle or fabricated box gusseted for strength and drilled to receive a bolted connection to both beam and column components. Top fixings may be excluded because of the stability provided by the bolt (at least two) group (Figs 7.42(e) and 7.43). The shear capacity is based on the lesser of:

(a) the single shear capacity of the bolt (at least two) group to the column
(b) the capacity of the gusseted cleat (or box) designed as a solid *strut and tie*.

Side connections shown in Fig. 7.42(f) are also possible providing that the torsion moments are catered for in the beam.

The connection is designed for the least favourable position between contact surfaces taking into account the accumulation of frame and component tolerances. The gap between the precast components is concreted using a sand/cement (or concrete) grout containing a proprietary expanding agent. In some instances, particularly where the cover to the surface of the nearest steel insert exceeds about 50 mm, small diameter links are spot welded or otherwise attached to the inserts to form a small cage. The grout, which has a minimum design strength of 30 N/mm$^2$ (but frequently exceeds 50 N/mm$^2$ at seven days) provides up to a two hour fire protection and total durability protection to the connection.

## 7.10   COLUMN INSERT DESIGN

### 7.10.1   General considerations

A 'column insert' is the name used to describe a steel section which is embedded in to precast columns in order to transfer shear and axial forces, and sometimes bending and torsion moments to the column. There are many types of inserts including:

- universal column or beam (UC, UB)
- rolled channel, angle, or bent plate
- rolled rectangular or square hollow section (RHS, SHS)
- narrow plate
- threaded dowels or bolts in steel or plastic tubes
- bolts in cast-in steel sockets.

Column insert design may be conveniently subdivided into the following:

- insert type
- breadth of insert to column width ratio, $b_p/b$
- beam topography, i.e. number and orientation of beams at each column.

Insert type may be either the solid or tubular (Fig. 7.42(a)–(c)) or cast-in sections (Figs 7.42(d) and (e)). Inserts are classified as:

- 'wide sections', i.e. when the breadth of the bearing surface $b_p$ is in the range 75 mm $< b_p < 0.4\, b$.
- 'thin plates', which includes thin walled rolled sections with wall thickness less than 0.1 $b$, or 50 mm. In general, additional bearing surfaces are required in thin section connectors (see Section 7.10.5)
- 'broad sections', where the cover distance to the sides of the insert is small enough to cause concern over shear cracking, and as a consequence the permissible stresses in the joint are reduced.

This part of the beam–column connection has been the subject of considerable research and analysis [e.g. 4.12, 7.13, and 7.25 to 7.29] and is therefore designed with a greater degree of confidence than many other joints in the precast frame.

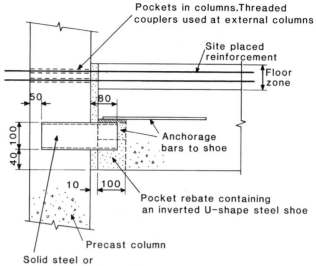

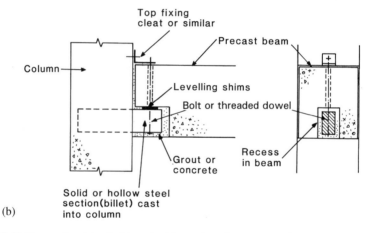

**Fig. 7.42** Types of cast-in steel inserts at beam-to-column connections.
(a) Using solid or hollow billet with top steel reinforcing bars
(b) Using solid or hollow sections with threaded dowel and top angle fixing
(c) Using solid billet with welded plate in beam
(d) Using open box and notched plate in beam
(e) Using rolled H-section and bolted on cleat
(f) Side billet connection (courtesy of Trent Concrete Ltd).

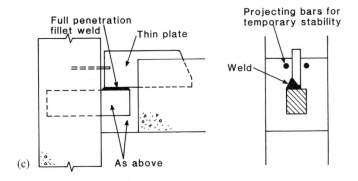

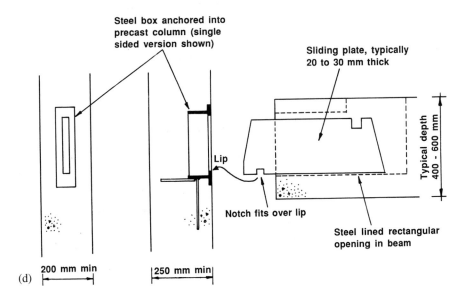

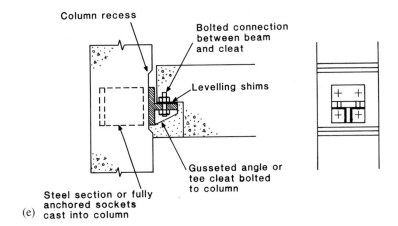

**Fig. 7.42** (*continued*)

(f)

**Fig. 7.42** (*continued*)

There are several aspects to embedded steel connections that require special attention, such as the need to minimize voids (and in particular lifting holes within 400 mm of the insert) under the embedded section, the need for fire protection, and to ensure that the concrete surrounding the embedded section is adequately confined by links. The most popular shapes for the insert are the SHS or RHS because of their:

- high shear to flexural capacity (greater than UBs)
- uniform shape (making it easy to weld additional plates and rebars to it)
- favourable torsional properties (in non-symmetrical situations)
- range of sizes available ($100 \times 50$ mm to $250 \times 150$ mm being the usual range).

**Fig. 7.43** Beam-to-column connection using stiffened cleat (courtesy of Composite Structures Ltd).

The shear span for any section is given by $a = M_u/V_u$, where $M_u = p_y S_{xx}$ and $V_u = 0.6 \, p_y \, 2 \, d \, t$. For SHSs and RHSs, $a = 90$ to $120$ mm, which is a convenient shear span for the majority of beam–column connectors (see Exercise 7.1 where the shear span is 114 mm). On the other hand, for UBs, $a = 150$ to $190$ mm, making it uneconomical.

The most economical section in terms of minimizing the size of the recesses in the beams is clearly the smaller thicker section. The minimum breadth of insert is taken as 75 mm. The minimum thickness for the steelwork is taken as 6 mm for rolled sections and 4 mm for box sections providing that the insert is sealed to the passage of air and moisture.

There are two ways of setting the insert in the column – either by casting directly into the column, or forming a hole in the column and placing the insert into the hole using either grout or resin-mortar. In the former, the insert must penetrate the sides of the mould, which is made good by plating afterwards. Unless this operation is well done it may cause uneven surfaces in the columns which are cast against the repaired mould afterwards. This does not happen if a hole is formed in the column.

The main problem here is that cracking may propagate around the corners of the hole whilst the column is being lifted from the mould, particularly if the breadth of the hole is more than one-third of the breadth of the column. For heavily loaded connections, in which the *working* stress underneath the insert is greater than $0.4 f_{cu}$ of the grout (calculated using a linear elastic analysis), an epoxy mortar should be considered. Apart from the obvious advantages in using a quick setting material of considerable compressive and tensile strength, epoxy based materials may be poured into a much smaller annulus than is possible with grout. This allows the hole in the column to be smaller with its attendant benefits in detailing and crack resistance. A note of caution should be included here, and the specialist supplier of epoxies should be consulted about the long-term durability of certain epoxy mortars. The manufacturers' details should be strictly followed with regard to mix proportions, ambient conditions and cleanliness.

In the wide sections, ultimate bearing stresses of $0.8 f_{cu}$ are used. For concentric loading, the ultimate bearing capacity is in the order of 800 kN (for a 300 mm square column with a 100 mm wide insert, allowing for 25 mm spalling in the cover concrete each side) and is rarely critical. The column connector is rated with respect to the design and ultimate load testing under eccentric loading. The IStructE Manual [7.17] proposes a method for determining the load and moment capacity of prismatic sections, and this has found favour with many design engineers.

The methods are adopted from the PCI Manuals using the design data given in BS 8110, Part 1, clause 5.3.5 [7.6] to replace those in the American standards. The method gives good agreement with the original PCI method and other more rigorous analyses. The equilibrium of forces and moments at the ultimate limit state is achieved by reinforcing (or otherwise guarding) against bursting, spalling and splitting, etc. Figure 7.44 shows typical details of a multi-sided connection. The general practice is to guarantee the confinement of concrete directly above and below the column insert using closely spaced links and to ensure the main longitudinal reinforcement is not interrupted by the insert.

The centre-line of the insert should not be less than $3b_p/2$ from the edge of the column. The insert must lie within the column reinforcement cage.

The beams reactions due to patch loading gives rise to non-symmetrically loaded connections. It is impossible for a column insert to be symmetrically loaded, and therefore the beam topography will always produce bending moments in the column. These may be as large as 150 kN m in extreme cases, but in general are in the order 20 to 60 kN m. Corner columns are subjected to biaxial bending, although it is usual for one of the beams at a corner to be lightly loaded due to the one-way spanning floor slabs. The most onerous design situation occurs in a single-sided insert.

### 7.10.2  Single-sided insert connections

The ultimate bearing capacity of the insert and the column into which it is cast is calculated as follows. Referring to Fig. 7.45(a), the bearing is assumed to be uniform over limited areas of the insert and, at ultimate capacity of the insert, to be limited to $0.8 f_{cu}$ or $f_b$ according to Equation (7.11). The line pressure consists of two components – one part to react the vertical load $V$, and a second part to produce a couple to react the bending moment. In reality the pressure distribution will be non-uniform [7.25], but for the purpose of design it may be taken as shown in the figure.

The following design method is adopted from the IStructE Manual on joints [7.17] and is similar to the design of shear boxes in beams in Section 4.3.11. The main differences here are:

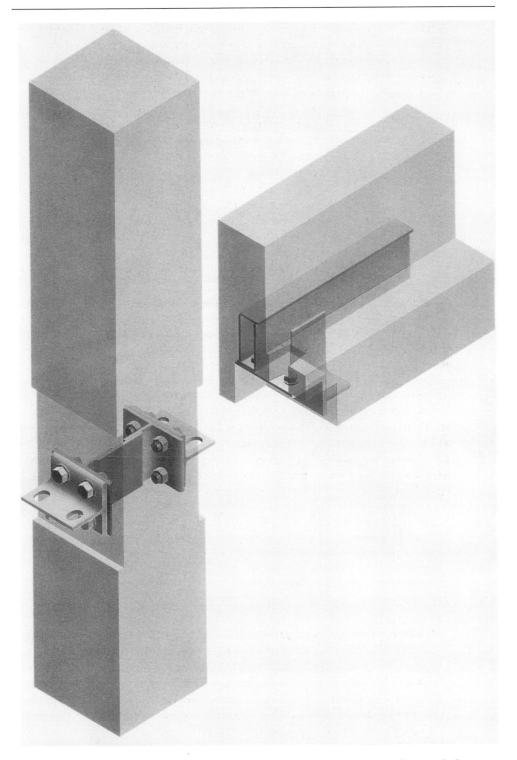

**Fig. 7.44** Multi-sided beam-to-column connection (courtesy of Trent Concrete Ltd).

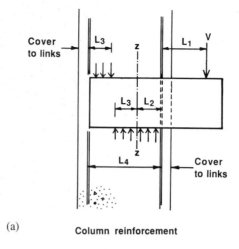

(a)            Column reinforcement

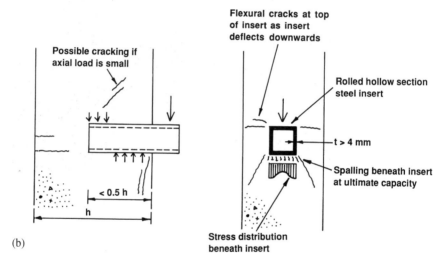

(b)

**Fig. 7.45** Column insert design and testing.
(a) General design of column insert
(b) Hollow steel section in deep column
(c) Web buckling in RHS column insert.

- that column insert lengths are restricted by the width of the column, typically 300 mm, and so therefore additional measures have to be taken
- punching shear beneath the insert is not a problem in columns
- cover concrete is ignored at the ultimate load. Full scale testing by the author found that surface spalling up to 15 mm from the face of the column occurs directly beneath steel inserts where $b_p/b < 0.5$.

The shear span $L_1$ must be taken assuming that the load $V$ acts at the centre line of the bearing area plus an allowance for tolerances. Assuming an ultimate bearing pressure of $0.8 f_{cu}$, calculate $L_2$ from the relationship:

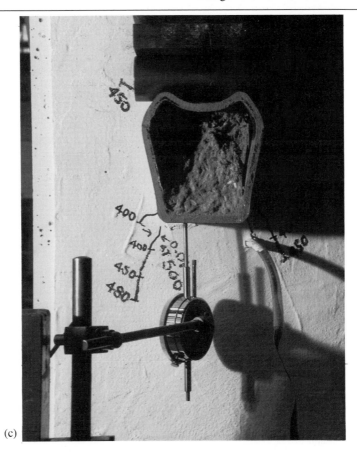

(c)

**Fig. 7.45** (*continued*)

$$V = 0.8 f_{cu} b_p L_2 \tag{7.13}$$

Zero shear occurs at the end of the stress zone $L_2$ and so the maximum bending moment at plane z–z given by:

$$M_{zz} = V (L_1 + 0.5L_2) \tag{7.14}$$

Since the moment inside the column at zz is:

$$M_{zz} = 0.8 f_{cu} b_p L_3 (L_4 - L_2 - L_3) \tag{7.15}$$

then $\quad L_3 (L_4 - L_2 - L_3) = L_2 (L_1 + 0.5L_2) \tag{7.16a}$

and $L_3$ can be found by solving a quadratic equation. Note that $L_4$ is exclusive of the cover concrete to the links: $L_4 = h - 2 \times$ cover. Check that $(L_2 + 2L_3) \leq 0.9 L_4$ so that the bearing surfaces do not overlap. If this relationship is not satisfied either (a) increase $b_p$ or provide additional reinforcement or steel plates, etc. welded to the insert (see Section 7.10.3).

For real roots in the solution of $L_3$, the expression:

$$(L_1 + L_2)^2 - L_1^2 < 0.5(L_4 - L_2)^2 \tag{7.16b}$$

must be satisfied.

In most practical cases the shear span $L_1$ for 300 mm wide columns is between 60 mm and 120 mm. The position of the neutral axis from the loaded face $X$ is given by:

$$X = L_4 + \text{cover} - 0.5(L_4 - L_2) \tag{7.17}$$

This works out at approximately $X = 0.7\,L_4$ from the face of the column. This is in good agreement with the experimental strain distribution results obtained by Marcakis and Mitchell [7.26] and shown in Fig. 7.46 where, for values of $L_1 = 102$ mm, $b = h = 203$ mm, $L_4 = 184$ mm, the neutral axis is found to be at $X = 0.77\,L_4$ from the face of the column.

Column reinforcement must not be cut or reduced at the position of the steel insert unless tests have shown that such reductions do not affect the capacity of the column. Displaced column links, called confinement links, should be grouped and placed at 50 mm centres above and below the insert (see Figs 7.47 and 7.48). The maximum compressive force occurs below the insert and is given by:

$$F = 0.8\,f_{\text{cu}}\,b_{\text{p}}\,(L_3 + L_2) \tag{7.18}$$

The horizontal bursting force is calculated from end block theory to give:

$$F_{\text{bst}} = \zeta\,F \tag{7.19}$$

where $\zeta$ is the bursting force coefficient referred to in Section 4.3.7. Thus the area of confinement steel $A_{\text{bst}}$ is given by:

$$A_{\text{bst}} = \frac{F_{\text{bst}}}{0.87 f_{\text{yv}}} \tag{7.20}$$

If the zone of pressure (i.e. $L_2 + L_3$) underneath the insert is small and located near to the front (i.e. nearest to the load) end of the insert, the area $A_{\text{bst}}$ should only be one leg of each of the confinement links. The links should be of the *closed* variety with proper anchorage. This is because the bursting forces will be present only in the front face of the column. Bursting forces will not affect the rear of the column until the pressure zone extends sufficiently far along the insert. Although there is a gradual increase in bursting forces in the column faces the following is recommended from some tests carried out by the author on asymmetrically loaded bearing plates:

If $L_2 + L_3 + \text{cover} < h/3$, all $A_{\text{bst}}$ to be provided by one leg of the links

If $h/3 < L_2 + L_3 + \text{cover} < h/2$, $\frac{2}{3}$ of $A_{\text{bst}}$ to be provided by one leg of the links

If $L_2 + L_3 + \text{cover} > h/2$, $\frac{1}{2} A_{\text{bst}}$ to be provided by one leg of the links.

Short-length billets (Fig. 7.45(b)) may fail by diagonal cracking above the billet and so confinement steel is placed above the insert as follows:

$$F_{\text{bst}} = \zeta\,0.8\,f_{\text{cu}}\,b_{\text{p}}\,L_3 \tag{7.21}$$

using Equation (7.20) to design the steel.

The maximum bending moment in the steel insert occurs at zero shear, at $L_2 = V/0.8\,f_{\text{cu}}\,b_{\text{p}}$ from end of stress block. The eccentricity $e'$ is given by:

$$e' = \left(\frac{L_2}{2}\right) + \text{cover} + \text{distance from face of column to reaction point} \tag{7.22}$$

Thus

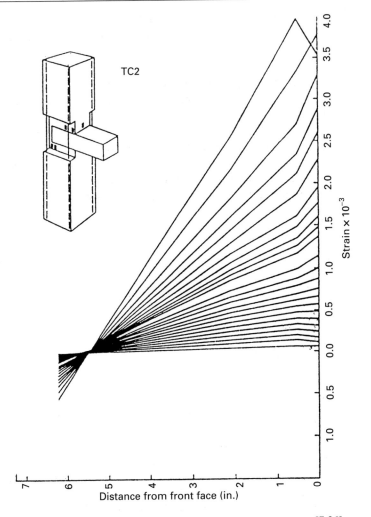

**Fig. 7.46** Strain distributions under steel billet in single-sided connector [7.26].

$$M_{zz} = V e' = V (L_1 + 0.5 L_2) \qquad (7.23)$$

Then the plastic modulus $S_{xx} > M_{zz}/p_y$ and the web area $> V/0.6 \, p_y$.

In the case of rectangular hollow sections, web buckling and bearing are avoided by filling the hollow with concrete or grout, either in the factory or during the grouting operation on site. Recent testing by the author found in $150 \times 100$ RHS sections designed (according to Fig. 7.42(b)) for 400 kN ultimate load, web buckling occurred at around 525 kN (Fig. 7.45(c)). Expanding agents for the infill may be used. Any projecting parts must be checked to ensure that they will not distort excessively nor buckle under the applied joint loads and rotations. Local spalling may result at the face of the columns under uneven loading created by distorted sections.

The foregoing analysis is highly simplified for the purpose of design, and conservative with respect to the real behaviour of the connection after construction. It is known that there are several inconsistencies in the assumptions used in the design manuals, some of

**Fig. 7.47** Confinement links near to cast-in rolled H-section.

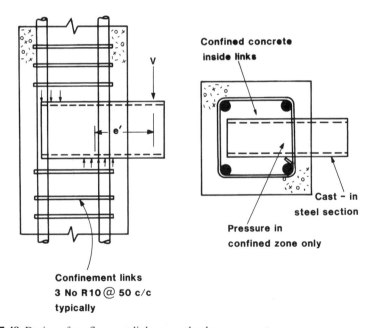

**Fig. 7.48** Design of confinement links around column connector.

which are explained in various papers on the strength of column inserts [e.g. 7.27, 7.28, 7.29]. For example, in Fig. 7.45(a)–(c) the position of the neutral axis depth is assumed constant, whereas it varies with the eccentricity and magnitude of loading, and is dependent on the effects of the relative confinement of concrete above and below the steel insert. It is also supposed that in the fully grouted assembly the beam has a restraining effect on bending of the insert, and that the insert is actually behaving not as a short cantilever, but as a stiff web. Holmes and Posner [7.29] also suggest that the calculation of bursting forces beneath inserts is more complicated than using a simple geometric relationship, and that the real effects of concrete shrinkage and compaction beneath the insert, and the presence of main longitudinal bars, must be considered.

Also, the effects of axial precompression, load eccentricity, shape and width of the insert and of additional reinforcement welded to the sides of the insert have been studied [7.26]. The most important results showed that for $b_p < b/2$ the effective width of the insert extended to the width of the confined column region as bounded by the links. Closely spaced links, i.e. at less than 75 mm spacing were necessary to ensure the connection failed in a ductile manner. Very high axial loads, e.g. 0.75 of the squash load, resulted in a decrease in ductility at failure. The insert is therefore under-designed with respect to the biaxial failure strength of concrete. Good design practice has in many cases responded to these findings.

## EXERCISE 7.2    Billet type single-sided column connector

Figure 7.49 shows the detail of a square hollow box type billet column connector used to support a rectangular precast beam. The maximum distance from the face of the column to the centre of the beam bearing is 60 mm. Given that the ultimate beam reaction is 150 kN, calculate the confinement reinforcement required below and above the billet.

Use $f_{cu} = 50$ N/mm$^2$; $f_{yv} = 460$ N/mm$^2$, cover to reinforcement = 35 mm, and grade 43 steel.

Calculate maximum line pressure under billet as the lesser of:

$$f_b = \frac{1.5 f_{cu}}{1 + (2b_p / b)} = \frac{1.5 \times 50}{1 + (2 \times 100 / 300)} = 45 \text{ N/mm}^2, \text{ or } f_b = 0.8 f_{cu} = 40 \text{ N/mm}^2.$$

Line pressure under billet $= 0.8 f_{cu} b_p = 40 \times 100 = 4000$ N/mm.

$$V = 4000 L_2$$

$$L_2 = \frac{150 \times 10^3}{4000} = 37.5 \text{ mm}$$

$$M_{zz} = 150 \left( 95 + 37.5 - \frac{37.5}{2} \right) = 17062 \text{ kN mm}$$

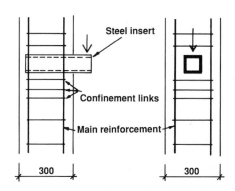

**Fig. 7.49** Details to Exercise 7.2.

Maximum length of billet = 300 − 35 = 265 mm.
Distance between end of billet and pressure at $L_2$ = 265 − 35 − $L_2$ = 192.5 mm.
Also $M_{zz}$ = 4000 $L_3$ (192.5 − $L_3$). Therefore, by solving quadratic equation $L_3$ = 26 mm.
Total vertical force below connector = 4000 (26 + 37.5) × $10^{-3}$ = 254 kN.
Total force above connector = 104 kN.
Bursting force coefficient (for $y_{po}/y_o$ = 0.33) = 0.22.
Therefore, $F_{bst}$ = 56 kN and 23 kN below and above connector.
Also $L_2$ + $L_3$ + cover = 99.5 mm < $h/3$, all $A_{bst}$ to be provided by one leg of the confinement links.

$$A_{bst} = \frac{56 \times 10^3}{0.87 \times 460} = 140 \text{ mm}^2$$

*Use 2 no. T 10 closed links below connector.*

$$A_{bst} = \frac{23 \times 10^3}{0.87 \times 460} = 58 \text{ mm}^2$$

*Use 1 no. T 10 closed link above connector.*

### Design of SHS insert

The maximum moment in the steel insert occurs at zero shear, at 150000/4000 = 37.5 mm from end of stress block.

Hence $e' = \dfrac{37.5}{2}$ + 35 cover + 60 = say 114 mm.

The maximum moment in the steel insert $M_{zz}$ = $V\,e'$ = 150 × 114 = 17 100 kN mm.
If $p_y$ for grade 43 steel = 275 N/mm$^2$ and $p_q$ = 165 N/mm$^2$

Then $S_{xx} > \dfrac{17100}{275} = 62.18 \text{ cm}^3$

and $2dt > \dfrac{150000}{165} = 909 \text{ mm}^2$

*Use 100 × 100 × 5 SHS ($S_{xx}$ = 67.1 cm$^3$ and 2dt = 1000 mm$^2$).*

### 7.10.3   Additional welded reinforcement to inserts

Where the insert lies close to the top of the column such that the restraining force $0.8\,f_{cu}\,b_p\,L_3$ cannot develop at its end, reinforcement should be welded to the sides of the rear (called remote) end of the insert and anchored by full bond development as shown in Fig. 7.50. The decision as to whether the top cover is adequate is made on practical grounds as well as assessing punching shear at the critical section. Generally, if the cover to the top of the insert is less than 150 mm then it is certain that a designer will not allow the insert to rely on holding down pressure $0.8\,f_{cu}\,b_p$, for the simple reason that it is likely that in such thin sections the limiting bearing stress would be about $0.3\,f_{cu}$.
Thus, referring to Fig. 7.50:

$$M_{zz} = 0.8\,f_{cu}\,b_p\,L_2\,(L_1 + 0.5\,L_2) \qquad (7.24)$$

$$M_{zz} = 0.8\,f_{cu}\,b_p\,L_3\,(d - \text{cover} - L_2 - 0.5\,L_3) \qquad (7.25)$$

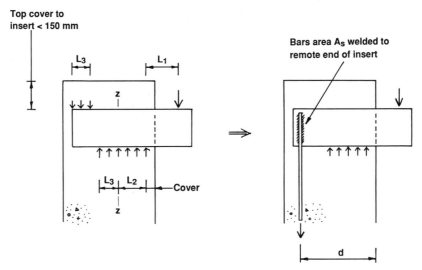

**Fig. 7.50** Additional tie back reinforcement welded to insert.

hence $L_3$ may be determined. Check $L_2 + L_3 < 0.7 \, L_4$ so that a couple may be generated. The steel required to replace the upthrust is:

$$A_s = \frac{0.8 f_{cu} b_p L_3}{0.87 f_y} \tag{7.26}$$

Where the top cover to the insert is greater than 200 mm, at least 3 no. closed links may be provided and the concrete well compacted, thus allowing the bearing pressures to develop.

   If the strength capacity of an insert is less than required due to overlapping pressure around the insert, one remedy is to add extra bars to both the front and rear of the insert. In calculating the strength of this it is assumed that the bars can act in tension or compression both above and below the insert, and can develop their ultimate yield strength by the action of bond alone. The bars are placed projecting an equal distance either side of the insert as shown in Fig. 7.51. Although a deformed bar is used in order to keep the anchorage bond lengths to a minimum, the ultimate stress in the bar is taken as the value used for mild steel reinforcement, i.e. 250 N/mm$^2$. In order to develop a full bond strength the bars should be positioned with a centroidal cover distance of at least 50 mm, and be enclosed within confining links of area not less than $0.5 \times A_s$, i.e. the bursting coefficient for rebars close to the edge of concrete is taken as 0.5. The weld can be deposited on both sides of the bar, i.e. effective thickness = 2 × throat thickness = 1.4 × weld leg. The weld design should be according to BS 5950, Part 1: clause 6.6, 1985 [7.30].

   The moment of resistance using this reinforcement increases from $M$, in the unreinforced case, to $M_r$. The corresponding vertical shear increases from $V$ to $V_r$. The equilibrium conditions are as follows. Referring to Fig. 7.51 and resolving vertically before the introduction of the bars $V = 0.8 f_{cu} b L_2$, and taking moments:

$$V (L_1 + 0.5 \, L_2) = 0.87 f_y A_s' (d - d') \tag{7.27}$$

and $A_s = A_s'$ because the vertical force $V$ is resisted by the concrete alone. When $V$ is

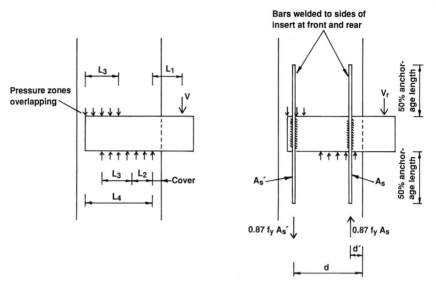

**Fig. 7.51** Additional reinforcement welded to insert.

increased to $V_r$ then:

$$V_r = 0.8 f_{cu} b L_2 + 0.87 f_y (A_s - A_s') \tag{7.28}$$

and

$$(V_r - V)(d + L_1 - \text{cover}) = 0.87 f_y A_s (d - d') \tag{7.29}$$

where $A_s > A_s'$ and $L_2$ is assumed to remain unchanged from the previous case. This may not be strictly true because of the influence of the welded reinforcement on the strain distribution in the cast-in section, but the assumption is sufficiently accurate not to warrant further detailed analysis. To ensure that the lever arm between the pressure zone beneath the insert and the bars welded to the rear end is sufficient, it is necessary to restrict $L_2 < 0.7 L_4$. See the experimental results in Fig. 7.46 [7.26].

### EXERCISE 7.3    Inserts close to the top of the column
A $300 \times 300$ mm column carries a single-sided asymmetrical ultimate load of 250 kN at a maximum distance from the face of the column of 60 mm. The top of the insert is located at only 100 mm from the top of the column. Design a suitable rectangular hollow section to be cast in to the column and suitable steel bars to be welded to the rear end of the insert.

Use $f_{cu} = 50$ N/mm$^2$, $f_y = 250$ N/mm$^2$, cover to reinforcement = 35 mm, and grade 43 steel.

### Solution
Assume $b_p = 100$ mm, the line pressure $= 0.8 \times 50 \times 100 = 4000$ N/mm.

$$L_2 = \frac{250 \times 10^3}{4000} = 62.5 \text{ mm}$$

$$M_{zz} = 250 \left( 95 + \frac{62.5}{2} \right) = 31\,575 \text{ kN mm}$$

If $p_y$ for grade 43 steel = 275 N/mm$^2$ and $p_q$ = 165 N/mm$^2$

Then $S_{xx} > \dfrac{31\,575}{275} = 115 \text{ cm}^3$

and $2dt > \dfrac{250\,000}{165} = 1515 \text{ mm}^2$

*Use 150 × 100 × 6.3 RHS ($S_{xx}$ = 148 cm$^3$ and $2dt$ = 1890 mm$^2$).*

Maximum length of billet = 300 − 35 = 265 mm.
Also $M_{zz}$ = 4000 $L_3$ (167.5 − $L_3$). Solving quadratic equation there are no real roots to $L_3$.
Thus use additional bars to rear of insert, at edge distance to bars:

$$= 35 + 8 \text{ links} + \text{say } 16 = 59 \text{ mm, say } 60 \text{ mm}$$

Then $d$ = 240 mm.
Using $M_{zz}$ = 0.8 $f_{cu}\, b_p\, L_3\, (d − \text{cover} − L_2 − 0.5\, L_3)$ = 4000 $L_3$ (142.5 − 0.5 $L_3$).
$L_3$ = 75 mm.
Total force in steel = 0.8 $f_{cu}\, b_p\, L_3$ = 301 kN.

$$A_s = \frac{301 \times 10^3}{0.87 \times 250} = 1384 \text{ mm}^2$$

*Use two no. T 32 bars (1610) at 45 mm cover welded to side of insert.*

Force in each bar = $\dfrac{301}{2}$ = 150.5 kN.

Weld length, say 8 mm continuous fillet weld (CFW) to both sides of the bar:

$$= \frac{150.5 \times 10^3}{1.4 \times 8 \times 215} = 63 \text{ mm} + 16 \text{ mm run out} = 79 \text{ mm} < 150 \text{ mm provided.}$$

*Use 8 mm fillet weld to side of insert × 150 mm long.*

### EXERCISE 7.4    Billet with additional reinforcement

Repeat Exercise 7.3 for a beam end reaction of 300 kN. Check the adequacy of the insert and provide additional reinforcement to the insert if necessary.

*Solution*

Assume $b_p$ = 100 mm.

$$L_2 = \frac{300 \times 10^3}{4000} = 75 \text{ mm}$$

$$M_{xx} = 300 \left( 95 + \frac{75}{2} \right) = 39\,750 \text{ kN mm}$$

If $p_y$ for grade 43 steel $= 275$ N/mm$^2$ and $p_q = 165$ N/mm$^2$

Then $S_{xx} > \dfrac{39\,750}{275} = 144.5$ cm$^3$

and $2dt > \dfrac{300\,000}{165} = 1818$ mm$^2$

*Use $150 \times 100 \times 6.3$ RHS ($S_{xx}$ 148 cm$^3$ and $2dt = 1890$ mm$^2$).*

Let $L_2 = 75$ mm as before.
Use edge distance to bars $= 35 + 8 + $ say $12 = 55$ mm, then $d - d' = 190$ mm.
$M_{zz} = 0.87\, f_y\, A_s\, (d - d')$ gives:

$$A_s = \frac{39\,750 \times 10^3}{0.87 \times 250 \times 190} = 961 \text{ mm}^2$$

*Use two no. T 25 bars (982) at 45 mm cover welded to side of insert.*

Force in each bar $= \dfrac{39\,750}{190 \times 2} = 104.6$ kN

Weld length, say 8 mm CFW along each side of the bar $= \dfrac{104.6 \times 10^3}{1.4 \times 8 \times 215} = 44$ mm $+ 16$ mm

run out $< 150$ mm provided.
*Use 8 mm fillet weld to side of insert $\times$ 150 mm long.*

### 7.10.4   Double-sided billet

Internal column connectors supporting a symmetrical arrangement of beams may be designed along similar lines as the single-sided connector in Section 7.10.1, by finding the equivalent point load. Patch loading is used where the maximum and minimum beam reactions are:

$V_{max} = 1.4 \times$ dead $+ 1.6 \times$ live load
$V_{min} = 1.0 \times$ dead.

The net overturning moment is obtained from the worst possible scenario when the construction tolerance $\Delta$ is added to the eccentricity of the greater load and deducted from the eccentricity of the smaller. Thus if the distance from the centroid of the column to the centre of the beam reaction is $e$, the net moment on the connector is:

$$M_{net} = M_{max} - M_{min} = V_{max}\,(e + \Delta) - V_{min}\,(e - \Delta) \tag{7.30}$$

The net eccentricity is given by:

$$e_{net} = \frac{M_{net}}{V_{max} + V_{min}} \tag{7.31}$$

The analysis may now proceed in the same manner as for the single-sided connection above. Note that $L_1$ may be negative if $e_{net} < h/2 - $ cover, such that the equation for overturning $M_{zz} = V\,(L_1 + L_2/2)$ is as follows:

$$M_{zz} = V\left[\left(e_{net} - \frac{h}{2} + \text{cover}\right) + \frac{L_2}{2}\right] \tag{7.32}$$

Only where the eccentricity $e$ is greater than a certain value do we need to consider the effects of bending. Referring to Fig. 7.52 we can adapt a type of 'middle-third' rule used in combined stress analysis in prismatic sections to find a limiting value for the eccentricity. Bending stresses occur where $e_{net}$ is given by:

$$e_{net} > \frac{L_4 - L_2}{2} \tag{7.33}$$

or

$$M > M_{max} - M_{min}$$
$$= \left(V_{max} + V_{min}\right)\left(\frac{h}{2} - \frac{V_{max} + V_{min}}{1.6 f_{cu} b_p} - \text{cover}\right) \tag{7.34}$$

where the maximum and minimum moments on either side of the connection are as given by Equation (7.30).

In these situations the pressure above the insert is very small (if not zero) and therefore no bursting steel is required. However, many precasters will always include at least one closed link, or even provide the *same* links above and below the insert to avoid the risk of the steel being misplaced or transposed. Sometimes when working on a very long column in a difficult environment it is easy to lose sight of which is the top of the column and which is the bottom!

A more realistic approach is to consider the steel billet supported by a deformable foundation, i.e. the concrete in the column. The distribution of the bearing pressure between the billet and the concrete depends on the relative stiffness of the two components and the position (or eccentricity) at which the load is applied to the billet. The analysis is akin to beams on elastic foundations subjected to end shear and end bending moment. For different permutations of $M$, $V$ and the relative stiffness between billet and concrete it is, in theory, possible for the billet to rise clear of the concrete below its middle, Fig. 7.53. Thus, in service (at which most of these connectors perform because of the large

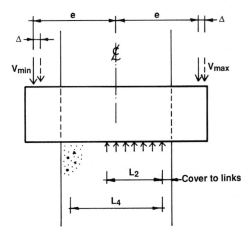

**Fig. 7.52** Double-sided billet design.

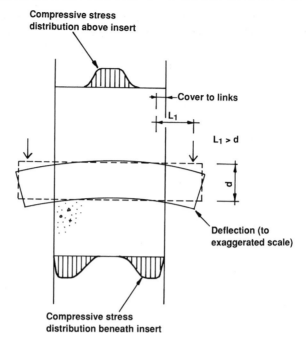

Compressive stress distribution above insert

Cover to links

$L_1$

$L_1 > d$

d

Deflection (to exaggerated scale)

Compressive stress distribution beneath insert

**Fig. 7.53** Pressure distributions under cast-in steel inserts.

factors of safety used in the industry) the bearing pressure distribution is as shown in Fig. 7.53.

At ultimate it is possible that localized stresses equal to the cube strength of the concrete may be present, but the design considers the overall effect of uniform pressure.

### EXERCISE 7.5    Double-sided steel billet design

A $300 \times 300$ mm column carries a double-sided symmetrical connector subjected to characteristic dead and live loads of 125 and 105 kN, respectively, at a distance from the face of the column of 55 mm. Using the appropriate loading combinations design a suitable rectangular hollow section to be cast into the column. Design the confinement reinforcement in the form of high tensile links in the vicinity of the connector.

Use grade 43a steel for the insert and grade C50 concrete. Cover to all reinforcement is 25 mm. The construction allowance for the position of the beams is 6 mm.

### Solution

Load Case 1: Patch load

| | | |
|---|---|---|
| Maximum beam reaction | $= (1.4 \times 125) + (1.6 \times 105)$ | $= 343$ kN |
| Maximum eccentricity | $= 55 + 6$ | $= 61$ mm |
| Minimum beam reaction | $= (1.0 \times 125)$ | $= 125$ kN |
| Minimum eccentricity | $= 55 - 6$ | $= 49$ mm |

Net maximum overturning moment

$$= [343 \times (150 + 61)] - [125 \times (150 + 49)] = 47\ 500\ \text{kN mm}$$

Net eccentricity = 47 500/(343 + 125) = 100 mm from centre line of column, i.e. 25 mm from the end of the insert. The effective reaction line is therefore within the column. Maximum length of insert in column $L_4$ = 300 – 2 × 25 cover = 250 mm.

Try breadth of insert $b_p = \dfrac{300}{3} = 100$ mm.

Bearing pressure under insert = $0.8\, f_{cu}\, b_p = 0.8 \times 50 \times 100 = 4000$ N/mm.

Then $\quad L_2 = \dfrac{(343+125)\times 10^3}{4000} = 117$ mm

$$M_{zz} = 468\left(\frac{117}{2} - 25\right) = 15\,678 \text{ kN mm}$$

Distance between far end of billet and pressure at $L_2$ = 250 – 117 = 133 mm.
Hence $M_{zz} = 4000\, L_3\,(133 - L_3)$. Therefore, by solving quadratic equation $L_3 = 44$ mm.
Check $L_2 + 2L_3 < 0.9\, L_4$. Then $117 + 88 < 0.9 \times 250$ (satisfactory).
Total vertical force below connector = $4000\,(117 + 44) \times 10^{-3} = 644$ kN.
Also $L_2 + L_3$ + cover = 186 mm > $h/2$, $A_{bst}$ to be provided by both legs of the confinement links.
Total force above connector = 176 kN.

### Load Case 2:  Maximum axial, no inaccuracies
Maximum beam reaction = $(1.4 \times 125) + (1.6 \times 105) = 343$ kN
and $e = 0$.

Then $\quad L_2 = \dfrac{(2 \times 343) \times 10^3}{4000} = 172$ mm < 250 mm available.

Total vertical force below connector = 686 kN.
Also $L_2 + L_3$ + cover > $h/2$, $A_{bst}$ to be provided by both legs of the confinement links.
Total vertical force above connector = 0.

### Load Case 3:  Maximum axial, with inaccuracies
Maximum beam reaction = 343 kN, and $e = 6$ mm. This is less than $e_{min} = 0.5\,(250 - 172) = 39$ mm from Equation (7.33), hence no bending stresses are present.
Total vertical force above and below connector as Case 2.

### Reinforcement design
Bursting force coefficient = $\left(\text{for } \dfrac{y_{po}}{y_o} = \dfrac{100}{300} = 0.33\right) = 0.22$.

Therefore, maximum $F_{bst} = 0.22 \times 686 = 150.9$ kN below connector.

$$A_{bst} = \frac{150.9 \times 10^3}{0.87 \times 460} = 377 \text{ mm}^2 \text{ divided by two legs} = 188 \text{ mm}^2.$$

*Use three T 10 (234) or two T 12 (226) closed links below connector.*

$$A_{bst} = \frac{0.22 \times 176 \times 10^3}{0.87 \times 460} = 97 \text{ mm}^2.$$

*Use two T 10 (157) or one T 12 (226) closed links above connector.*

### Design of RHS insert

The maximum moment in the steel insert occurs at zero shear, at $343\,000/4000 = 86$ mm from end of stress block.

Hence $e' = 86/2 + 25$ cover $+ 61 = 129$ mm.

$M_{zz} = V\,e' = 343 \times 129 = 44\,250$ kN mm.

If $p_y$ for grade 43 steel $= 275$ N/mm$^2$ and $p_q = 165$ N/mm$^2$

then $S_{xx} > 44\,250/275 = 161$ cm$^3$

and $2dt > 343\,000/165 = 2078$ mm$^2$.

*Use $150 \times 100 \times 8$ RHS ($S_{xx} = 183$ cm$^3$ and $2dt = 2400$ mm$^2$).*

The final arrangement of reinforcement is shown in Fig. 7.54.

### 7.10.5   Three- and four-way connections

Three- and four-way connections refer to the number of sides on which a column is loaded. Four-way connections are uncommon in precast structures (unlike steel structures where tertiary systems are used) but even where they are used, two of the reactions on opposite sides of the column are usually small and the connection may be conservatively designed as double-sided with a small eccentricity. In a multi-directional connection each leg may be designed as above, the welding between legs being carefully designed to transfer the loads. The problem is compounded if the inserts are not at right angles to one another.

The design of the welded joint between the two parts of the insert would also be based on bending and shear, and torsion if the inserts were not at right angles. It is recommended that the major load(s) be carried on the primary continuous insert, and the secondary load(s) on inserts welded to the primary insert (see Fig. 7.55). If the designer has a choice in the matter it is better that the primary insert is slightly deeper than the secondary

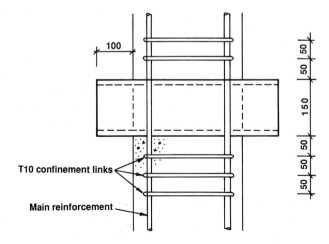

**Fig. 7.54** Details to Exercise 7.5.

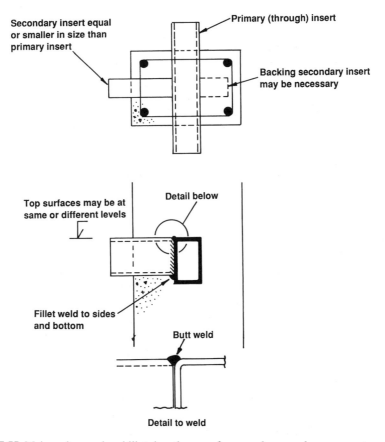

**Fig. 7.55** Main and secondary billets in a three- or four-way beam–column connector.

insert so that a fillet weld may be deposited around the full perimeter. If the inserts are equal depth a butt weld must be used across the top and bottom and the end of the secondary insert must be bevelled.

Three-way connections are more difficult to analyse because of the torsional behaviour caused by bi-axial loading. Little analytical or experimental work has been carried out on this type of connection. It is quite common for all three reactions to be large and therefore a conservative design approach would be to consider the single-sided part of the insert first and to transfer the uplift reaction into the double-sided insert. The design of the steel insert itself would be based on bending and shear as in the cases considered above.

**EXERCISE 7.6    Three-way billet connector design**

A 300 × 400 mm column carries a three-sided connector shown in Fig. 7.56(a). The inserts are at right angles to each other. The distance from the face of the column to the centre line of the bearing is 60 mm, exclusive of a construction tolerance of 6 mm. The characteristic beam end reactions (kN) to the three sides of the column are given in Table

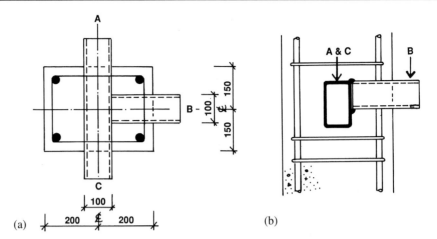

**Fig. 7.56** Details to Exercise 7.6. (a) Plan view of three-way connector and (b) section through primary insert.

**Table 7.8** Characteristic beam end reactions (kN).

| Position | Dead | Live |
|---|---|---|
| A | 70 | 45 |
| B | 40 | 15 |
| C | 85 | 45 |

7.8. Assume that the level of the top of the column connector is the same on all three sides.

Choose the most appropriate loading patterns and design suitable rectangular or square rolled hollow sections to be cast into the column. The size of the single arm insert may be smaller than the through insert.

Design the confinement reinforcement in the column in the form of high tensile steel links in the vicinity of the connector.

Use $f_{cu} = 50$ N/mm$^2$; $f_{yv} = 460$ N/mm$^2$, cover to reinforcement = 35 mm, and grade 43 steel and weld.

**Solution**

Refer to Fig. 7.56(b).

Assume $b_p = 100$ mm throughout, with a line pressure $= 0.8 \times 50 \times 100 = 4000$ N/mm.

**Load Case 1: Maximum axial**

Consider first the load at point B ($h = 400$ mm in this plane):

$V_B$ = maximum beam reaction = $(1.4 \times 40) + (1.6 \times 15) = 80$ kN

Then $\qquad L_{2B} = \dfrac{80 \times 10^3}{4000} = 20$ mm.

$$M_{zz} = 80 \times \left(35 + 66 + \frac{20}{2}\right) = 8880 \text{ kN mm.}$$

If $P_y = 275 \text{ N/mm}^2$ $p_q = 165 \text{ N/mm}^2$

Then $S_{xx} > \dfrac{8880}{275} = 32.3 \text{ cm}^3$

and $2dt > \dfrac{80 \times 10^3}{165} = 485 \text{ mm}^2.$

*Use 100 × 100 × 4 SHS ($S_{xx}$ = 54.9 cm$^3$ and 2dt = 800 mm$^2$).*
Distance from end of insert to pressure zone = 250 − 35 − 20 = 195 mm.

$$M_{zz} = 4000 \, L_3 \, (195 - L_3), \text{ hence } L_3 = 11 \text{ mm.}$$

Force beneath insert = $4000 \times (20 + 11) \times 10^{-3} = 124 \text{ kN.}$
Uplifting force above insert transferred to through insert = *44 kN.*
Consider maximum loads at A and C ($h$ = 300 mm in this plane):

$V_A$ = maximum beam reaction = $(1.4 \times 70) + (1.6 \times 45) = 170 \text{ kN}$
$V_C$ = maximum beam reaction = $(1.4 \times 85) + (1.6 \times 45) = 191 \text{ kN}$
$V_B = -44 \text{ kN from above.}$

Force beneath insert = 317 kN.

Then $\quad L_2 = \dfrac{317 \times 10^3}{4000} = 79 \text{ mm} < 230 \text{ mm available.}$

Eccentricity $\quad e_A = 150 + 60 - 6 = 204 \text{ mm}$
$\qquad\qquad\quad e_C = 150 + 60 + 6 = 216 \text{ mm}$
$\qquad\qquad\quad e_B = 0$

$M_{net} = (191 \times 216) - (170 \times 204) = 6576 \text{ kN mm}$

$e_{net} = \dfrac{6576}{317} = 21 \text{ mm} < \dfrac{(L_4 - L_2)}{2} = \dfrac{(230 - 79)}{2} = 75.5 \text{ mm}$

thus ignore eccentricity effects.

Total force beneath insert = 317 + 124 = *441 kN.*
Total force above connector = *0.*

## Load Case 2: Patch load to give the maximum overturning

Consider first the minimum reaction at point B, $V_B = 40 \text{ kN.}$
Then $L_{2B} = 10 \text{ mm.}$

$$M_{zz} = 40 \times \left(35 + 54 + \frac{10}{2}\right) = 3760 \text{ kN mm.}$$

Distance from end of insert to pressure zone = 250 − 35 − 10 = 205 mm.

$$M_{zz} = 4000 \, L_3 \, (205 - L_3), \text{ hence } L_3 = 5 \text{ mm.}$$

Force beneath insert = $4000 \times (10 + 5) \times 10^{-3} = 60 \text{ kN.}$
Uplifting force above insert transferred to through insert = *20 kN.*

Consider minimum load at A and maximum loads at C:

$V_A$ = minimum beam reaction = 70 kN

$V_C$ = maximum beam reaction = 191 kN

$V_B$ = –20 kN from above.

Force beneath insert = 241 kN.

Then $\quad L_2 = \dfrac{241 \times 10^3}{4000} = 60.3$ mm < 230 mm available.

Eccentricity as before: $e_A$ = 204 mm, $e_c$ = 216 mm, $e_B$ = 0.

$M_{net} = (191 \times 216) - (70 \times 204) = 26\,976$ kN mm.

$e_{net} = \dfrac{26\,976}{241} = 112$ mm $> \dfrac{(L_4 - L_2)}{2} = \dfrac{(230 - 60)}{2} = 85$ mm,

thus consider eccentricity effects.

$$M_{zz} = V\left[\left(e - \frac{h}{2} + \text{cover}\right) + \frac{L_2}{2}\right]$$

$$= 241 \times \left[(112 - 150 + 35) + \frac{60}{2}\right] = 241 \times 27 = 6507 \text{ kN mm.}$$

Distance from end of insert to pressure zone = 230 – 60 = 170 mm.

$M_{zz} = 4000\,L_3\,(170 - L_3)$, hence $L_3$ = 11 mm.

Force beneath insert = $4000 \times (60 + 11) \times 10^{-3}$ = *284 kN.*
Total force beneath insert = 284 + 60 = *344 kN* < Case 1 force.
Total force above connector = *44 kN.*

### Reinforcement design

Maximum value from Case 1, $F_{bst} = 0.22 \times 441 = 97$ kN below connector.

$$A_{bst} = \frac{97 \times 10^3}{0.87 \times 460} = 242 \text{ mm}^2 \text{ divided by two legs} = 121 \text{ mm}^2.$$

*Use two T 10 (157) closed links below connector.*

$$A_{bst} = \frac{0.22 \times 44 \times 10^3}{0.87 \times 460} = 24 \text{ mm}^2 \text{ from Case 2.}$$

*Use one T 10 (78) closed links above connector.*

### Design of RHS insert

The maximum moment in the steel insert occurs at zero shear at point C, at $191 \times 10^3/4000 = 48$ mm from end of stress block.

Hence $e' = \dfrac{48}{2} + 35$ cover $+ 60 + 6 = 125$ mm.

$M_{zz} = V e' = 191 \times 125 = 23\ 875$ kN mm.

If $p_y$ for grade 43 steel = 275 N/mm$^2$ and $p_q$ = 165 N/mm$^2$

then $S_{xx} > \dfrac{23\ 875}{275} = 86.8$ cm$^3$

and $2dt > \dfrac{191 \times 10^3}{165} = 1157$ mm$^2$.

*Use 150 × 100 × 5 RHS ($S_{xx}$ = 121 cm$^3$ and 2dt = 1500 mm$^2$).*

### Weld design between inserts

Let the top surface of the secondary insert be 10 mm lower than the primary insert to allow a fillet weld to be deposited.

From Load Case 1, the moment at the face of the primary insert

$$= [80 \times (150 + 66)] - 124\left(150 - 35 - \frac{31}{2}\right) = 4942 \text{ kN mm}.$$

The shear force = 124 − 80 = 44 kN.

Try 8 mm CFW.

$$S_{xx} = \text{(parallel axis theorem)} = (2 \times 100 \times 5.6 \times 53) + \left(\frac{2 \times 5.6 \times 111.2^2}{4}\right)$$

$$= 93983 \text{ mm}^3.$$

$M_r = 215 \times 93\ 983 \times 10^{-3} = 20\ 206$ kN mm > 4942 kN mm required.

$V_r = 215 \times 2 \times 5.6 \times 100 \times 10^{-3} = 240.8$ kN > 44 kN required.

*Use 8 mm CFW between primary and secondary inserts.*

## 7.10.6   Narrow plate column inserts

Narrow plate inserts are steel plates not thicker than one-tenth of the column width, or 50 mm, whichever is least, set vertically in the column to support a beam reaction as shown in Fig. 7.57. When loaded, such plates tend to produce a splitting effect in the columns, under the bearing surface, unless they are supported by transverse bearing plates or welded reinforcement.

The inserts are nearly always supplemented either by a wider bearing plate near to the loaded face of the column, or tensile anchorage reinforcement. In the latter, compressive bearing stresses are ignored in the thin plate and the total reaction, allowing for eccentricity of the imposed load, is carried in bond by reinforcement (which is additional to the overall compressive requirement of the column). The reinforcement is welded to the sides of the plate and by placing small diameter links (T 8 and 50 mm centres) around the bars the concrete immediately beneath the thin plate is highly confined. The compressive bond resistance of the bars is used if there is an insufficient tension anchorage bond length available above the connection, e.g. at roof level. Stirrups are used in the usual manner to prevent lateral buckling of these bars.

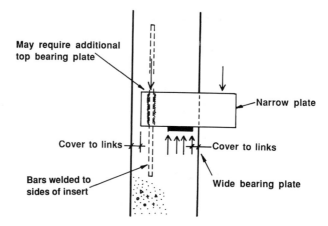

**Fig. 7.57** Narrow steel plate insert with additional reinforcement.

Narrow plate inserts with neither welded reinforcement nor transverse plates were tested by Holmes and Posner [7.31, 7.29], but the capacities obtained were very low, and only empirical relationships to column size and cube strengths were obtained. The capacities were affected by many variables such as shape of bearing surface, proximity of column reinforcement, and shrinkage stresses in the concrete, and it is therefore not recommended that such unreinforced narrow plates be used without careful testing.

To calculate the maximum shear and moments on the plate assume that load $V$ acts as described in Section 7.10.1. Calculate the force in the steel $F_{s2}$ from the relationship:

$$F_{s2} = V \times \frac{L_1}{L_2} \tag{7.35}$$

and calculate $F_{s1}$ from the relationship:

$$F_{s1} = F_{s2} + V \tag{7.36}$$

Provide reinforcement $A_{s1}$ and $A_{s2}$ to carry ultimate loads $F_{s1}$ and $F_{s2}$ through adequate bond and anchorage. Place additional column confinement links above and below the plate using bursting coefficient of 0.5 for point loads in steel bars. Note that where a plate occurs close to the top of a column, force $F_{s1}$ may be provided by reinforcement in compression, provided that sufficient links are arranged to restrain such reinforcement against buckling.

### 7.10.7   Cast-in sockets

The use of cast-in sockets as a means of transmitting vertical and horizontal forces to the column is not widely used. The design of inserts (and other fixings) in concrete is covered extensively in the CEB Report [7.32]. The pin anchor bar or welded heads attached to the rear end of the socket shown in Fig. 7.58 provides a full tensile anchorage to the socket. Vertical shear is provided by direct bearing under the barrel of the socket itself – typically 24 mm diameter by 100 mm long. The PCI Manual on cladding [7.10] gives extensive design guidance.

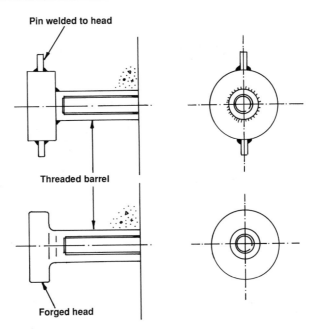

**Fig. 7.58**  Cast-in threaded sockets.

The spacing between cast-in sockets should not be less than two diameters of the outer barrel, or the size of the aggregate + 5 mm, whichever is the larger. Confinement links are calculated on the assumption that the compression beneath the sockets produces a knife-edge effect (even though the barrel is circular) radiating at 45° away from the sockets. If there are two sockets in the same horizontal plane (the usual case) the stresses overlap and create a uniform pressure zone, as shown in Fig. 7.59, such that $F_{bst} = \zeta\, V$ where $\zeta = 0.23$ and $V$ is the shear force in the sockets. The edge distance to a socket should be clear of any reinforcement and not less than the cover to the main bars plus 10 mm, or 75 mm whichever is the least.

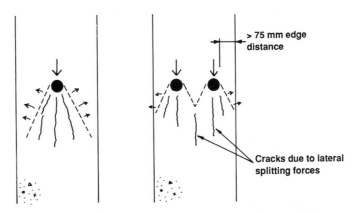

**Fig. 7.59**  Stress distributions in columns beneath cast-in sockets subjected to shear forces.

### 7.10.8 Bolts in sleeves

In this method steel or hard plastic sleeves are cast in the column, bound by the usual confinement links and main steel. Steel bolts or threaded rods are placed through the sleeves to receive seating cleats or similar attachments to the face of the column, as shown in Fig. 7.60. The structural mechanism is similar to cast-in sockets, but with the added complication that the bolt is not a tight fit in the sleeve. The resulting pressure distribution inside the sleeve is transferred directly to the concrete.

Full-scale testing by Mohamed [7.33] used M24 grade 8:8 threaded bolts supporting angle cleats to the faces of $300 \times 300$ mm precast columns to study the effects of the concentration of bolts on connector capacity. The number of bolts making the connection varied from one to four (i.e. two rows $\times 2$). Concrete strengths were varied from $61.9$ N/mm$^2$ to $30.5$ N/mm$^2$, and confinement reinforcement beneath the bolts was in the form of R 8 links at 50 mm centres. The ratio of the failure load for the multi-bolt connection to that of a single bolt was as given in Table 7.9.

The failure mode for the high strength concrete specimens was by flexural (due to the small connector eccentricity) and shear yielding of the bolts. Conversely the failure in

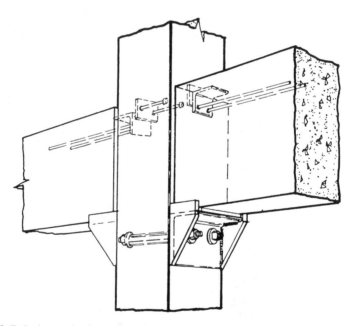

**Fig. 7.60** Bolts in cast-in sleeves in columns.

**Table 7.9** Ratio of failure load for multi-bolt connection to that of a single bolt.

| No. of bolts | Load capacity/Single bolt capacity |
| --- | --- |
| 2 | 1.7 |
| 3 | 2.6 |
| 4 | 2.8 |

the lower strength concrete was by large 'ovalization' of the sleeves and concrete splitting beneath the connector (see Figs 7.61 and 7.62).

## 7.11 CONNECTIONS TO COLUMNS ON CONCRETE LEDGES

### 7.11.1 Corbels

A corbel is a short cantilever projection from the face of a column (or wall) which supports a load bearing component on its upper horizontal ledge. Column corbels are not widely used in the UK compared with Continental Europe and North America. As in any type of frame it is necessary to minimize the structural zone by containing the connection within the overall depth of the beam. For reasons of appearance this calls for 'shallow' corbels, which are defined in Fig. 7.63 and designed as a short cantilever. Care is needed to ensure correct lapping and anchorage of reinforcement adjacent to contact surfaces, with full regard for fixing tolerances.

**Fig. 7.61** 'Ovalization' of bolt sleeves in shear connector tests [7.33].

**Fig. 7.62** Column splitting beneath four-bolt angle connectors [7.34].

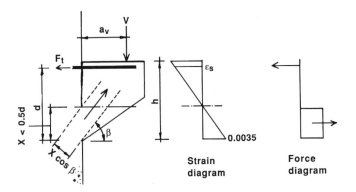

**Fig. 7.63** Shallow corbel definitions and design.

The usual beam connection is a single vertical dowel which is either a waiting bar cast into the corbel, or is site fixed into a hole. Care should be taken to ensure that the hole does not fill with water and freeze. The dowel diameter is between 16 and 25 mm typically, and the respective dowel hole is 35 to 50 mm diameter. Non-shrink grout, grade C25 (or C40 if it is being used on the site elsewhere), is poured into the hole from the top.

The design procedure is well documented in BS 8110, Part 1, clause 5.2.7 [7.6]. To qualify as a corbel in design, the distance $a_v$ from the load point to the face of the support should not be greater than $d$, the effective depth of the 'root' of the corbel. The depth of the face of the corbel should not be less than half the total depth. The width of a corbel must be less than about 500 mm, otherwise it qualifies as a continuous nib, where clause 5.2.8 is used.

In precast frames corbel design is carried out on an *ad hoc* basis to suit the specific requirements of the project. It is unusual to find a standard column corbel detail in production manuals. However, if used, the design and detailing is carried out with the usual attention to the prevention of spalling, cracking, etc. If a dowelled connection is made between the beam and corbel, horizontal hair-pin bars (typically R 8 diameter at 50 mm centres) are used to prevent bursting of the thin covering to the dowel hole and projecting dowel bar.

There are many complications in reinforcing multi-planar corbels, particularly if the level of the shoulder is at the same level in two directions. Congestion of the horizontal tie back may be relieved either by varying the height of the seating by at least 50 mm, or by casting in a special steel insert to transmit the tie forces as necessary.

A graphic illustration of the load transfer through a corbel is given by Bruggeling [7.34], reproduced in Fig. 7.64. The notation in this diagram refers to individual joint designs within the connection, and may be found in Bruggeling's book [7.35]. Extensive testing of corbels (over 200 tests) was carried out by Kriz and Raths [7.36] which showed the dependence on the shear span ratio $a_v/d$, and the percentage of horizontal reinforcement

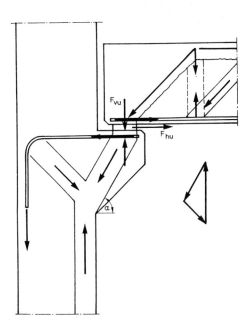

**Fig. 7.64** Shear force transfer between beam and column through half-joint and corbel [7.34].

on the ultimate load. The loads were applied through bearing pads of well-defined size and position, with no attempt to simulate concrete-to-concrete contact. Somerville [7.37] also gives design recommendations.

The finite element program named 'FIELDS' [7.38] has verified the behaviour of highly loaded rc shallow corbels assumed in design, as shown in Fig. 7.65. In all analyses the problem of bearing stresses in the bends of the reinforcement leads to small diameter bars being preferred, e.g. < 25 mm diameter. The disturbed region of the column, known as the *D-region*, may extend for a distance equal to about 1.5 $h$ above the table of the corbel and 1.5 $h \cos \beta$ below the root of the corbel. Additional links should be provided in the column in these regions.

### 7.11.1.1  *Shallow corbels*

Although BS 8110 [7.6] does not specifically refer to it as such, it is generally accepted that to qualify as a shallow corbel the distance $a_v < 0.6d$, plus some other criteria shown in Fig. 7.66. This figure shows three different ways of anchoring the main steel in the top corner of the corbel as follows:

- by welding the main steel to a bar across the front face of the corbel (Fig. 7.66(a))
- by contouring the bar to the profile of the corbel (Figs 7.66(b) and (d)).
- Using a horizontal U-shape bar in the top of the corbel (Fig. 7.66(c)).

Of the three options, (a) presents the most economical solution, but only if the diameter of the bar is less than about 16 mm.

Here the strut and tie model is used, which leads to a more favourable and accurate representation than by considering the corbel as a short cantilever in bending and shear. However it is first necessary to check that a shear failure will not occur by proportioning the depth of the corbel according to the following.

First, the ultimate shear stress $v = V/bd$ at the face of the column should not exceed $0.8\sqrt{f_{cu}}$. Shear reinforcement $A_{sv}$ is required if the applied shear stress $v$ is greater than the enhanced concrete stress $v_c (2d/a_v)$ such that:

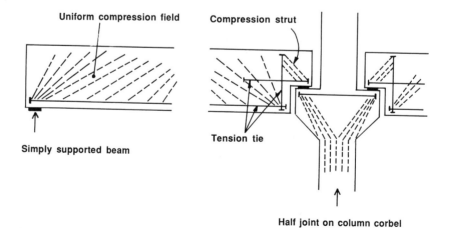

**Fig. 7.65**  Stress trajectories in a corbel from the program FIELDS [7.38].

$$A_{sv} = \frac{a_v b[v - v_c(2d/a_v)]}{0.87 f_{yv}} > \frac{0.4 a_v b}{0.87 f_{yv}} > 0.4\% \ bd \tag{7.37}$$

The enhancement factor $2d/a_v$ can often be as large as 3 or 4 in certain cases, and therefore the amount of shear reinforcement is surprisingly small.

A minimum of $A_{sv} = 0.5 A_s$ (where $A_s$ is obtained from Equation (7.42)) should be provided to ensure the confinement of the concrete in the compressive strut. Thus the minimum total area of steel $A_s + A_{sv} > 0.6\% \times$ concrete area. Horizontal links are placed in the upper two-thirds of the corbel. The bar diameter should not be more than 12 mm. The distance to the first link should not be more than 75 mm from the bearing surface. Zeller [7.39] measured failure shear stresses of $1.06 \sqrt{f_{cu}}$ to $1.12 \sqrt{f_{cu}}$ in rc corbels where $a_v/d = 0.5$ and the ratio of horizontal steel $\rho = 0.86\%$ was slightly greater than the value of 0.6% given above.

The strut and tie forces are calculated from the triangle of forces; it being assumed that a node is formed at the intersection of the main tension steel and a point directly beneath the centre line of the bearing. This latter point may not be strictly true as the effective bearing point may move further from the column when rotations (of up to 0.0001) take place. Thus before proceeding with the analysis it is wise to add up to 20 mm to $a_v$ in recognition of this fact. Lindberg and Keronen [7.40] have shown that the bearing point in a beam–column portal frame connection will move sideways when rotations take place in pinned connections.

The effective bearing area is confined to within the rc zones, i.e. any net dimensions are exclusive of the cover concrete $c$. If the breadth of the corbel is given by $b$, the bearing length is given by:

$$a_{eff} = \frac{V}{0.4 f_{cu}(b - 2c)} \tag{7.38}$$

where $V$ is the ultimate applied force.

The length of the seating ledge $a$, given by '$l$' in Fig. 7.66(b), is determined from the sum of the distance to the centre of the load $a_v$ plus half the bearing length $a_{eff}$, plus an edge distance allowance. For high tensile bars with an inner bend radius of $3\Phi$, the allowance is $5\Phi +$ cover to bar; for mild steel bars it is $4\Phi +$ cover. The overall depth $h$ of the corbel should be at least $1.5a$, and the depth of the front face of the corbel be $0.5h$. This means that the slope of the soffit of the corbel is not less than one in two. Hence the effective depth $d$ is given by:

$$d = 1.5 a_v + 0.75 a_{eff} + 7\Phi + 0.5c \tag{7.39}$$

Then:

$$\beta = \tan^{-1} \frac{d - 0.5X}{a_v} \tag{7.40}$$

Then:

$$F_t = \frac{V}{\tan\beta} \quad \text{and} \quad F_c = \frac{V}{\sin\beta} \tag{7.41}$$

The tensile tie capacity $F_{tR}$ is given by:

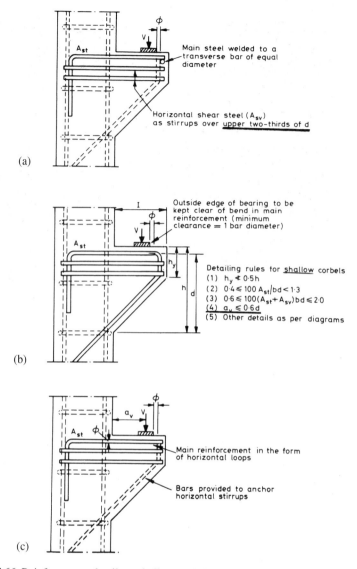

**Fig. 7.66**  Reinforcement details to shallow corbels
(a) Main steel welded to an edge bar [7.17]
(b) Main steel as a continuous bar [7.17]
(c) Main steel in the form of horizontal U bar [7.17]
(d) Corbel reinforced according to Figure 7.66(b).

$$F_{tR} = A_s \, 0.87 f_y \qquad\qquad (7.42)$$

The main tension reinforcement $A_s$ should not be less than 0.4 per cent of the section at the face of the column, and the steel should be fully anchored at the far face of the column. Figure 7.66(a)–(c) shows three different ways of reinforcing a shallow corbel.

The compressive capacity of the diagonal strut $F_{cR}$ is given by:

$$F_{cR} = 0.4 f_{cu} \, b \, X \cos \beta \qquad\qquad (7.43)$$

(d)

**Fig. 7.66** (*continued*)

To prevent the concrete strain from exceeding 0.002 at ultimate then $X < 0.5d$. Thus, Equation (7.40) may be written:

$$\tan\beta = \frac{0.75d}{a_v} \tag{7.44}$$

The required angle for $\beta$ is about 45° to 65° to the horizontal.

There are of course many combinations for the shape and size of the shallow corbel, but it is clear from Equations (7.38) to (7.40) that there is an optimum (or minimum) size for the corbel. By combining the above equations the maximum capacity of the corbel is given when:

$$d = \frac{a_v \tan\beta}{0.75} = \frac{V}{0.2 f_{cu} \ b \ \cos\beta \ \sin\beta}$$

$$V = \frac{0.2 f_{cu} b \sin^2\beta \left[ \dfrac{0.75 V}{0.4 f_{cu}(b - 2c)} + 7\Phi + 0.5c \right]}{\tan\beta - 1.125} \tag{7.45}$$

Hence $b$, $d$, $a_v$ and $a$ may be evaluated. The solution for $\beta$ is given by:

$$\tan\beta - K \sin^2\beta = 1.125$$

where $K$ is the dimensionless:

$$K = \frac{0.2 f_{cu} b \left[ \dfrac{0.75 V}{0.4 f_{cu}(b - 2c)} + 7\Phi + 0.5c \right]}{V} \tag{7.46}$$

The 'optimized design' considers the minimum compressive stress block at the root of the corbel. This gives rise to the largest possible value to β and hence the minimum horizontal tie reaction.

There is often a need to add horizontal U-shape steel bars directly beneath the bearing to cater for the horizontal friction force $\mu V$, particularly in the case of long prestressed beams, induced by creep shortening within the first six months or so. These bars are *additional* to the above. The top cover to the steel should be 25 to 30 mm (the exposed side cover may be greater). This frictional force can be eliminated if continuity steel is carried across the beam to corbel interface (e.g. stability ties), but this is not always possible at corners, or required in single-storey portal frames where corbels are often used. It is sometimes possible to provide a slip bearing such as a shear-deformable bearing pad, e.g. neoprene, which will prevent the frictional force being transferred to the corbel. However in order to make a positive site fixing beams are nearly always dowelled to the corbel.

One of the major drawbacks with corbels is that the mould must be specially shaped and built for each project (this does not happen with steel inserts where the cross-section of the mould is a continuous rectangle). To overcome this problem corbels have been designed and cast in a two-part process. In Part 1, fully anchored threaded couplers are cast into the face of the column at the level of the tension reinforcement to receive the corbel steel after the column has been stripped from the mould. A retarding agent is applied to the surface concrete of the column in the region of the corbel to expose the aggregate, but not to disturb it. In Part 2, further concrete is added to form the corbel a few days later.

### 7.11.1.2   Deep corbels

Deep corbels simulate inclined columns. To qualify as a deep corbel the distance $a_v$ should not be greater than $0.2d$. The outer edge of the corbel does not always have to be vertical but care should be taken at the top corner where a $25 \times 25$ mm splay should be detailed. Deep corbels are usually required because the capacity of the shallow corbel is insufficient. However the problem of local bursting stresses at the bearing surface is just as important as before, and so mild steel spreader plates are cast into the bearing surface. The tension steel $A_s$ is welded to the steel plate to form a positive anchorage.

The same design equations are used for deep and shallow corbels except that the shear stress in the compressive region should not exceed 1.3 N/mm$^2$ (BS 8110, Part 1, clause 5.3.7 [7.6]) and the compressive strut be restrained using horizontal bars as shown in Fig. 7.67. The IStructE Manual [7.17] suggests that apart from $A_s$, main strut reinforcement $A_{sc}$ should be provided such that:

$$A_{sc} = \frac{0.5\,b\,X\cos\beta}{100} \tag{7.47}$$

and horizontal links uniformly distributed over the full depth of the corbel of 0.4 per cent of the concrete area. The size of the bars should be at least a quarter of the diameter, with the spacing not exceeding 12 times the diameter of the main strut reinforcement.

### EXERCISE 7.7(a)   Shallow corbel design

A $300 \times 300$ mm column supports a precast concrete beam on a single-sided corbel. The

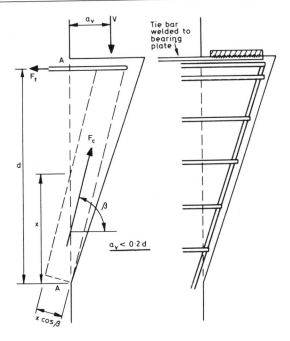

**Fig. 7.67** Deep corbel design [7.17].

nominal clear distance from the face of the column to the end of the beam is 15 mm. The beam is positioned on to a steel bearing plate on the centre line of the column. If the ultimate beam shear force is 200 kN, design a rc shallow corbel. The allowance for the rotation of the beam may be taken as 20 mm.

Use $f_{cu} = 50$ N/mm$^2$; $f_{yv} = 250$ N/mm$^2$; $f_y = 460$ N/mm$^2$, cover to reinforcement = 35 mm, except top cover to corbel = 25 mm.

**Solution**

Bearing plate area = $200 \times 10^3/0.6 \times 50 = 6667$ mm$^2$.

Use $100 \times 100$ mm × nominally 10 mm thick mild steel plate.

Distance to centre of bearing plate = gap 15 + half plate 50 + rotation allowance 20 = 85 mm.

Corbel length from face of column = 15 + half plate 100 + bend radius (say) 50 + cover 35 = 200 mm.

As a first approximation for the depth of the corbel, take a 45° load spread from the edge of the corbel to the root. Hence try $d = 200$ mm, and $h = 200 + $ cover 25 + radius 5 = 230 mm.

**Shear check**

$V < 0.8 \sqrt{f_{cu}} b \, d$

$$d > \frac{200 \times 10^3}{0.8 \times \sqrt{50} \times 300} = 118 \text{ mm} < 200 \text{ mm provided.}$$

Then if X = 0.5 $d$ in the limit, X = 100 mm.

$\beta = \tan^{-1} [200 - 0.5 \times 100]/85 = 60.5°$.

The compressive strut force $F_c = 200/\sin 60.5° = 229.9$ kN.
$F_{cR} = 0.4 \times 50 \times 300 \times 100 \times \cos 60.5° \times 10^{-3} = 295.5$ kN $> 229.9$ kN required.
Horizontal force $F_t = 200/\tan 60.5° = 113.1$ kN.

$$A_{s1} = \frac{113.1 \times 10^3}{0.87 \times 460} = 282 \text{ mm}^2.$$

Minimum $A_s = 0.4\% \times 300 \times 230 = 276$ mm$^2$.
Frictional force $= 0.4 \times 200 = 80$ kN.

$$A_{s2} = \frac{80 \times 10^3}{0.87 \times 460} = 200 \text{ mm}^2.$$

Total $A_s = 282 + 200 = 482$ mm$^2$.
*Use five T 12 (565) at 50 mm centres in top of corbel.*

### Shear reinforcement
$v = 200 \times 10^3/300 \times 200 = 3.33$ N/mm$^2$.
$100\, A_s/bd = 100 \times 565/300 \times 200 = 0.94$.
$v_c = 0.73 \times (40/25)^{1/3} = 0.86$ N/mm$^2$.
Enhancement factor $2\, d/a_v = 2 \times 200/85 = 4.7$.
$v_c = 4.7 \times 0.86 = 4.04$ N/mm$^2 > 3.33$ N/mm$^2$ required.
Provide minimum links $A_{sv} = 0.4 \times 85 \times 300/0.87 \times 250 = 47$ mm$^2$
or $A_{sv} = 0.5\, A_{s1} = 141$ mm$^2$.
*Use two R 8 (200) links at 50 mm centres.*
Minimum $A_s = 0.6\% \times 300 \times 230 = 415$ mm$^2 < 565$ mm$^2$ provided.

## EXERCISE 7.7(b)    Optimum shallow corbel design
Determine the minimum size of the corbel in Exercise 7.7(a) based on the optimum corbel design method. Calculate the area of tie steel required using this method and compare the answer with that in Exercise 7.7(a). Assume 12 mm diameter reinforcement and 35 mm cover to the steel.

### Solution

$$\text{Net bearing length } a_{\text{eff}} = \frac{200 \times 10^3}{0.4 \times 50 \times (300 - 70)} = 44 \text{ mm}.$$

$$\text{Then solving Equation (7.46) where } K = \frac{0.2 \times 50 \times 300\,[33 + (7 \times 12) + 35/2]}{200 \times 10^3} = 2.0175$$

gives $\beta = 72°$.
Hence $d = 227$ mm; $a_v = 55$ mm; $a = 173$ mm and $h = 260$ mm.
Then $F_t = V/\tan \beta + \mu V = 65 + 80 = 145$ kN.
$A_s = 362$ mm$^2$ compared with 482 mm$^2$ in Exercise 7.7(a). The reason for this reduction is because the optimized design considers the minimum compressive stress block at the root of the corbel. This gives rise to the largest possible value of $\beta$ (72° compared with 60.5° in Exercise 7.7(a)) and hence the minimum horizontal tie reaction.

### 7.11.2   Haunched columns

This design involves only the provision of an adequate bearing surface at the level of the soffit of the beam (Fig. 7.68). This type of joint is not widespread in multi-storey frames because of the increased dimension of the column at each floor level. The detail may be used in low rise unbraced structures where the increased column is beneficial in cantilever action, or in situations where the magnitude of the beam end reaction cannot practically or structurally be accommodated in a shallow corbel or steel insert, i.e. exceeding about 500 kN.

A bearing medium between the concrete surfaces of at least 10 mm thickness is highly recommended for many reasons:

- to ensure a uniform bearing pressure and ensure that the beam reaction is transferred to the column in the intended position
- to avoid eccentricity of load
- to prevent local spalling
- to accommodate tolerances, particularly in very long columns
- to allow beam rotations to take place.

The bearing pad may consist of a neoprene, or be a composite construction of two thin (3 mm thick) steel plates with neoprene (10 mm) sandwiched between. In all cases the edge of the bearing plate should not extend beyond a point connecting a 45° line to the edge of the top steel in the column, i.e. 2 × cover. Some precasters provide a 25 × 25 mm chamfer to ensure that the bearing pad does not extend to the edge of the column. Also the bearing pad should not project beyond the end of the beam where there is a clearance gap of 10 to 25 mm. Thus, referring to Fig. 7.69(a) the length $b_1$ of the bearing plate is given by:

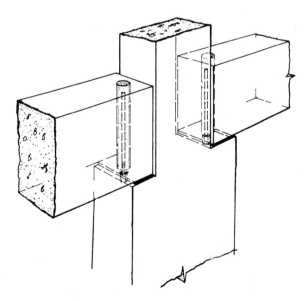

**Fig. 7.68**  Column haunch general arrangement.

$$b_1 = \text{shoulder} - (2 \times \text{cover}) - \text{clearance gap} \tag{7.48}$$

If the bearing pressure beneath the plate is greater than $0.6 f_{cu}$ additional plates must be cast into the column shoulder and, if necessary, additional bars welded to the underside of this plate. The designer has the choice of using a wide cast-in plate, say $b_p = b - (2 \times \text{cover})$, and accepting the lower bearing capacity given by Equation (7.11), or a narrower plate ($b_p < 0.44\, b$) and using $0.8 f_{cu}$. Referring to Fig. 7.69(b), if the line pressure beneath the cast-in plate is $f_b\, b_p$, then the pressure $f_c$ is given by:

$$f_c = \frac{V}{b_p b_1} < f_b \tag{7.49}$$

Adopting a $\theta = 20°$ load spread and using strut and tie action the horizontal force $V \tan \theta$ is resisted by bars welded to the underside of the cast-in plate and anchored by bond in

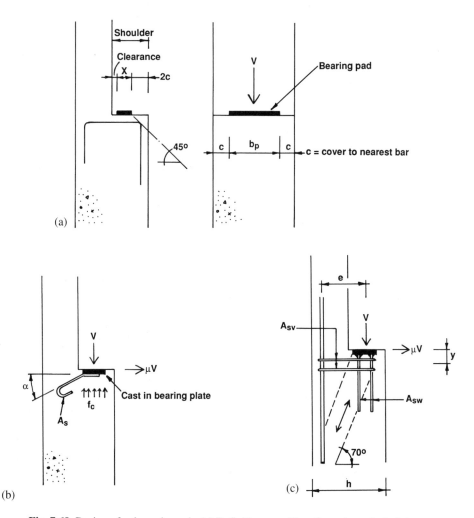

**Fig. 7.69**  Design of column haunch. (a) Definitions used in column haunch design; (b) column haunch using welded anchor bar and (c) column haunch using tie bars.

the column. As with the column insert design high tensile ribbed bars are used but the design stress is based on $f_y = 250$ N/mm$^2$.

Horizontal frictional forces $\mu V$ are treated in a similar manner as for corbels and beam nibs. Unless the horizontal force can be restrained by continuity reinforcement or some other type of positive tensile fixing, tie bars inclined at $\alpha$ to the horizontal are required beneath the bearing surface such that:

$$A_{sh} = \frac{V \tan\theta + \mu V}{\cos\alpha \; 0.87 f_y} \tag{7.50}$$

The thickness of the cast-in plate is determined from tie force considerations as normal, but this should not be less than 8 mm typically. Anti-bursting confinement links are designed and positioned according to Section 7.10.2.

Figure 7.69(c) shows an alternative method of transferring the horizontal forces into the column. A number of bars, usually four, are welded to the underside of the plate. Horizontal links $A_{sv}$ are placed around these bars such that the centre of the group of bars is at distance $y$ below the bearing surface ($y$ should not be more than about 100 mm). Assuming a $\theta = 20°$ load spread the compression in the shoulder will node at a distance $e/\tan\theta$ where $e$ is defined in Fig. 7.69(c). Then:

$$A_{sv} = \frac{V \tan\theta + \mu V \left[ \dfrac{y + e/\tan\theta}{e/\tan\theta} \right]}{0.87 f_{yv}} \tag{7.51}$$

The size of the vertical bars $A_{sw}$ welded to the plate and fully anchored into the column is given by:

$$A_{sw} = \frac{V \tan\theta + \mu V \left[ \dfrac{y + e/\tan\theta}{e/\tan\theta} \right]}{0.6 \times 0.87 f_{yw}} \tag{7.52}$$

A top fixing to the beam in the form of a bolted or weld cleat or plate, or by using continuity reinforcement across the column are specified in the same manner as for other beam–column connections.

The connection is essentially a pinned-joint, but it has the potential of developing a hogging moment capacity of considerable magnitude by the use of extended bearings and reinforced insitu concrete in composite action with the precast beam. Figure 7.70 describes the basic principles involved in this. Careful detailing and specific instructions to site are prepared to ensure composite action at the ultimate design load.

The main requirement is for a rigid bottom connection which does not rely on horizontal shear transfer to the column. If the beams are connected rigidly at the top, but not at the bottom, rotations will occur instead of bending moments. Spalling of concrete due to a lack of strength and rigidity at the bottom of the beam is shown in Fig. 7.71. See Chapter 8 in 'Prefabrication with Concrete' [7.15] for further details.

**EXERCISE 7.8    Haunched column design**

The single-sided column haunch connection shown in Fig. 7.72 is to be designed to

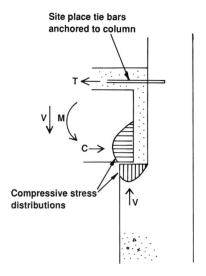

**Fig. 7.70** Principle of moment resisting connections at column haunches.

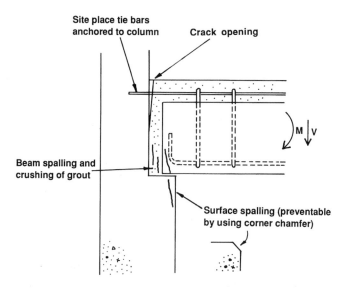

**Fig. 7.71** Spalling at support in moment resisting connection.

carry an ultimate beam reaction of 400 kN. Allow 15 mm clearance gap at the end of the beam. Design the reinforcement in the column, and specify any bearing and/or cast-in plates.

Use $f_{cu} = 40$ N/mm$^2$, $f_y = 460$ N/mm$^2$ and $f_{yw} = 250$ N/mm$^2$, cover to reinforcement = 25 mm, and grade 43 steel and welding electrodes.

**Solution**

Maximum bearing length $b_1 = 125 - $ gap $15 - 2 \times$ cover $= 60$ mm.

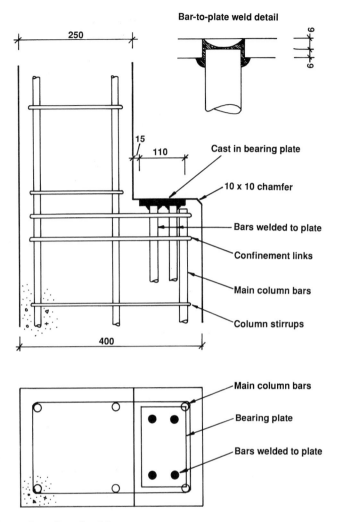

**Bar-to-plate weld detail**

250

15

110  Cast in bearing plate

10 x 10 chamfer

Bars welded to plate

Confinement links

Main column bars

Column stirrups

400

Main column bars

Bearing plate

Bars welded to plate

**Fig. 7.72**  Details to Exercise 7.8.

Maximum bearing width = $250 - 2 \times \text{cover} = 200$ mm.
Ultimate bearing pressure = $400 \times 10^3 / 60 \times 200 = 33.3$ N/mm$^2 > 0.6\,f_{\text{cu}}$.
Therefore try using cast-in bearing plate $170 \times 110$ mm in shoulder of column.

Maximum bearing pressure $f_{\text{b}} = \dfrac{1.5 \times 40}{1 + 2 \times 170/250} = 25.4$ N/mm$^2$.

Actual bearing pressure $= \dfrac{400 \times 10^3}{170 \times 110} = 21.4$ N/mm$^2 < f_{\text{b}}$.

Plate thickness $t = \dfrac{400 \times 10^3 \times \tan 20°}{170 \times 275} = 3.1$ mm$^2$.

but plate must be sufficiently thick to enable bars to be fillet welded through holes in the plate. Assume 6 mm weld (see detail in Fig. 7.72).
*Hence use 170 × 100 × 12 mild steel plate.*

### Vertical bars to plate

$$A_{sw} = \frac{400 \times 10^3 \times \tan 20°}{0.6 \times 0.87 \times 250} = 1115 \text{ mm}^2.$$

*Try four 20 mm diameter bars into 12 mm thick plate.*

### Bearing inside hole in plate

$P_b = 12 \times 20 \times 190 \times 10^{-3} = 45.6$ kN × 4 no. bars = 182.4 kN > 400 × tan 20°.
Axial force capacity of double-sided 6 mm fillet weld around 20 mm bar

$$= 2 \times 6/\sqrt{2} \times 215 \times 20 \, \pi \times 10^{-3} - 114.6 \text{ kN} \times 4 \text{ no. bars}$$
$$= 458 \text{ kN} > 400 \text{ kN compression.}$$

*Use four R 20 (1256) bars × 620 mm long welded to plate, 6 mm fillet weld to plate. Bars to penetrate hole in plate 6 mm.*

### Horizontal links

$$A_{sw} = \frac{400 \times 10^3 \times \tan 20°}{0.87 \times 460} = 363 \text{ mm}^2.$$

*Use two T 12 (452) links at 50 mm centres beneath plate.*

### Confinement links

Using $y_{po}/y_o = 170/250 = 0.68$, then $\zeta = 0.11$.

$$F_{bst} = 0.11 \times 400 = 44 \text{ kN.}$$

$$A_{bst} = \frac{44 \times 10^3}{0.87 \times 460} = 110 \text{ mm}^2 \text{ in one face only} < 226 \text{ mm}^2 \text{ provided by links}$$
above.

### 7.11.3 Connections to the tops of columns

There are many similarities in the principles involved in the design of this type of connection with those in wall-frame design, i.e. horizontal components are seated on the top of vertical members and an insitu concrete connection is made. The most common situation in beam and column construction occurs at the termination (usually at roof level) of a continuous column. The beam is seated and dowelled on to the column head, as shown in Figs 7.73(a) and (b), and 7.74, to form a simple support.

Bearing plates are provided between the components for the same reasons as given in Section 7.11.2. The size of the bearing pad should be at least 75 × 75 mm, or $h/3$ in larger columns. As before, the edge of the bearing plate should not extend into the cover concrete, and a 25 × 25 mm chamfer is provided so that the bearing pad does not extend to the edge of the column.

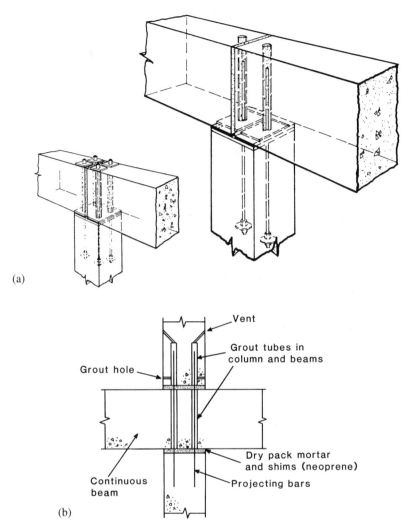

(a)

Vent

Grout tubes in
column and beams

Grout hole

Dry pack mortar
and shims (neoprene)

Continuous
beam

Projecting bars

(b)

**Fig. 7.73** Beam-to-column head connections. (a) Simply supported column head details
and (b) continuous beam-to-column connection.

The connection can transfer the vertical forces by providing confinement links according
to Section 4.3.10 and Table 4.10. The connection can also transfer applied horizontal
forces providing steel reinforcement is placed in the top of the column to prevent an
inclined crack forming as shown in Fig. 7.71. This steel is additional to the confinement
links.

Projecting column bars (or dowels) which are either cast in to the column at the factory
or grouted in holes in the top of the column on site, pass through circular corrugated
ducts (thin metal or plastic) cast in the beam. The inside diameter of the duct is two to
three times the diameter of the bars passing through, but not less than 50 mm. U-shaped
loops are placed around the dowel holes, typically R 8 at 50 mm centres. The strength of
these bars should be equal to the shear force in the dowels. The bars also provide the

**Fig. 7.74** Welded connection at column head (courtesy of Blatcon Ltd).

beam with lateral stability. The overturning moment $M = \Sigma V e$ from floor slabs reactions $V$ placed on one side of the beam at an eccentricity $e$ from the centre line of the beam, is resisted by a couple in the dowel. The dowel (area $A_s$ and section modulus $Z$) is designed to carry the bending moment $M = 0.87 f_y Z$ and the shear force $F = M/z = 0.6 f_y A_s$, where $z$ is taken as the lesser of the depth of the beam or length of projecting dowel.

Two methods are used when grouting the duct, depending on ease of access, the size of the annulus around the bar and the depth of the beam. If the beam depth is 300 mm or less and the annulus is more than about 15 mm the grout may be dispensed from above, with a back-stop filler placed around the outside of the joint to prevent grout loss. Otherwise for deeper beams pressure grouting from beneath the beam is recommended. The reason for concern is not so much to do with bond around the vertical bars, although this is important, but because of entrapped air voids in the bearing area.

A further application of the use of single-storey discontinuous columns is at a balcony connection where cantilever beams are designed continuously over a column head. As before, projecting reinforcement in the lower lift of column passes through sleeves in the beam and is fully grouted to ensure vertical continuity in the column. A full compression lap length is provided to the projecting bars. The bearing area is usually sufficient to enable a dry pack, e.g. plastic or steel shims, to be used with a solid structural expanding grout to form a joint of at least 20 mm thickness. There is no requirement for steel-to-steel bearings providing the design bearing stress does not exceed $0.6 f_{cu}$, and that the strength of the grout or dry-pack infill is at least similar, say within 20 per cent to that of the column.

The absence of a vertical joint at the face of the column ensures maximum efficiency in the use of precast concrete components subjected to shear and compression. The major disadvantage is that in jointing the column above and below each floor beam, two connections per floor level are made. The increased production and construction costs are in excess of the savings made in these simple connections.

## 7.12   BEAM-TO-BEAM CONNECTIONS

Where it is not possible to terminate a beam at a column, connections between secondary beams and primary beams may be made as shown in Figs 3.29 and 7.75. This connection requires special attention, particularly in the primary beam, where the combined effects of bending, shear, torsion and bearing stresses may cause problems within the shallow depth of the beam. Steel inserts are most widely used, particularly if the beam reaction is greater than about 100 kN. Park [7.41] shows a number of details for beam–beam connections used in New Zealand.

The IStructE Manual [7.17] makes a pertinent statement on this subject:

'These joints tend to be avoided … probably because their simplest forms are not normally appropriate to the client's requirements. In addition, pretensioned floors have excellent span potential so that the concept of main beams, secondary beams and slab can often be dispensed with and replaced by beam and slab.'

However, there are occasions, particularly surrounding medium-size voids in floor slabs, e.g. greater than about 2.5 m, where the floor slabs adjacent to the voids are too highly loaded to be supported without beams. The small beam span does not warrant adding extra columns to support the secondary beam, and so a beam–beam connection offers the most economic solution.

The design falls into three main categories which depend mainly on:

- the magnitude of the shear force in the secondary beam
- the difference in the depths, in particular the soffit level, of the adjoining beams
- the distance to the secondary beam from the end of the primary beam.

Let the primary beam depth be $h_1$ and secondary beam depth be $h_2$, and the breadth of the beams is $b_1$ and $b_2$, respectively.

(a) Type 1 – a simple pocket bearing may be made as shown in Fig. 7.75(a) where $h_1 - h_2 > 200$ mm, $b_1 > 250$ mm, and the end shear $V$ is given by:

$$V < 0.8\sqrt{f_{cu}}\, b_1\, (h_1 - h_2) \tag{7.53}$$

If the primary beam is *not* torsionally restrained the bearing may be recessed across a part of the primary beam, a distance the greater of 100 mm or $b_1/2$ in order to ensure that torsional stresses are minimized. If the primary beam is torsionally restrained by the floor slab for example, the distance may be the greater of 75 mm or $b_1/3$.

(b) Type 2 – a direct bearing on the top of the primary beam is made as shown in Fig. 7.75(b) if $h_1 - h_2 < 200$ mm, or the depth of the primary beam $h_1 < 200$ mm itself. The same bearing length as above is used.

(c) Type 3 – a side shear connector is made to a steel insert as shown in Fig. 7.75(c) if $h_1 - h_2 < 200$ mm.

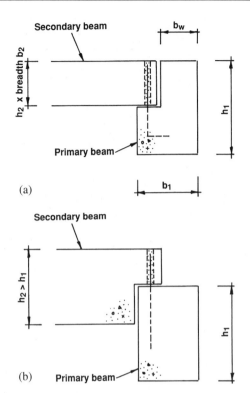

(a)

(b)

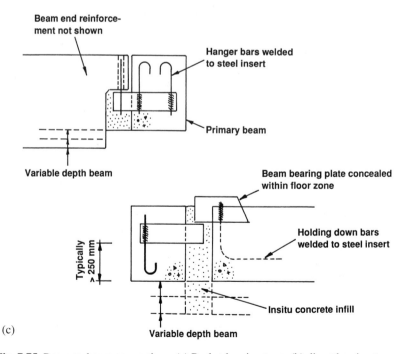

(c)

**Fig. 7.75** Beam-to-beam connections. (a) Pocket bearing type; (b) direct bearing type, (c) side shear box type.

(d)

**Fig. 7.75** (*continued*)  Beam-to-beam connections. (d) Making a beam-to-beam connection on site (courtesy of Trent Concrete Ltd).

In the recessed Type 1 connections, the shear force from the secondary beam is carried in to the main body of the primary beam in two ways. First, by a strut and tie model acting underneath the bearing in the same manner as for the L-shaped edge beam analysis in Section 4.3.5. The depth to the neutral axis $X$ from the bottom of the beam should not exceed 0.5 $d''$, where $d''$ is the effective depth to the tie steel in the bearing nib. The horizontal frictional force $\mu V$ is also considered as before. Figure 7.75(d) shows the construction of such a connection; interestingly the primary beam is cantilevered at the column head according to Fig. 7.73(b).

Second, by the resolution of forces in the longitudinal direction. Referring to Fig. 7.76(a) the beam geometry should be proportioned so that the line connecting the nodes in the diagonal compressive strut $C$ is inclined at $\beta = 45°$ to $55°$ to the horizontal. Then the force in $C$, and the diagonal–tension tie force $T$ (also inclined at $45°$ to the horizontal) is given by:

$$C = T = \frac{0.5\,V}{\sin^2 \beta} \tag{7.54}$$

The bottom horizontal tie force $H$ beneath the secondary beam bearing is given by:

$$H = 0.5\,V \cot \beta \tag{7.55}$$

A pair of hanger bars $A_s$ may therefore be provided as shown in Fig. 7.76(a) such that:

$$A_s = \frac{H}{0.87 f_y} \quad \text{or} \quad \frac{0.5\,V}{0.87 f_y} \tag{7.56}$$

whichever is greater. If the beam-to-beam connection is near to the end of the primary beam, i.e. closer than 2 $d$ from the end, where $d$ = effective depth of the primary beam, the compressive strut will be directed towards the end of the beam and will not induce the diagonal–tension force $T$.

Let the distances from the ends of the primary beam to the beam-to-beam connection be $x_1$ and $x_2$, where $x_2 > x_1$. Vertical shear links $A_{sv}$ should be provided immediately to the sides of the recess in the primary beam, in order to generate the maximum shear

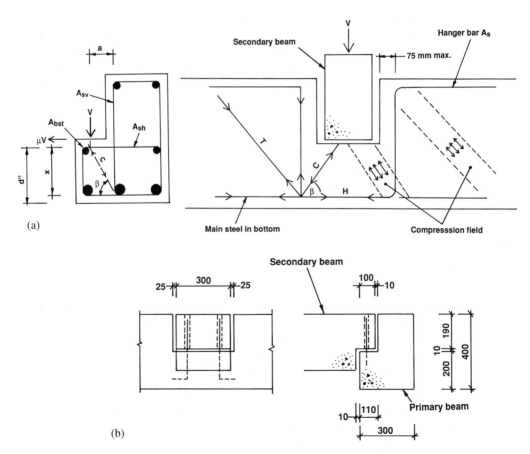

**Fig. 7.76** Practical beam-to-beam connections. (a) Reinforcement at primary beam; (b) recessed half-joint option in Exercise 7.9.

force $x_2 V/(x_2 + x_1)$ within the nodal distance equal to 0.75 $d$:

$$A_{sv} = \frac{x_2 V}{(x_2 + x_1) 0.87 f_{yv}} \tag{7.57}$$

Finally small lacer bars, T 8 or T 10, should be placed at the top corners of the tie steel in the boot.

The design methods used in the steel insert connections type 3 follow the same procedures as for beam end detailing in Section 4.3.12 and column insert detailing in Section 7.10. The major differences in the beam design is that the level of the bearing ledge is close to the top of the beam, and therefore a type of protruding Cazalay hanger is used. The protruding part is concealed within the floor slab zone. Narrow plates are the best option here as large anchor tie forces are possible if the depth of concrete beneath the plate is substantial. Tie bars or plate straps can be fully anchored into the beam, producing a ductile failure mode (see Fig. 9.2). A small bearing plate may be required at the remote end of the narrow plate.

The narrow plate is bevelled and fillet welded to the column insert. Two temporary stability studs are needed at the end of the plate to prevent the beam rotating on its knife-edge bearing.

## EXERCISE 7.9   Beam-to-beam connection design

A 300 × 300 mm secondary beam is supported at the mid-span position of a 400 × 300 mm wide primary beam. The ultimate shear force at the end of the secondary beam is 100 kN. The tops of the beams are level, and the maximum allowable protrusion for any kind of connector above the beams is 75 mm. It may be assumed that the beams are torsionally restrained.

Design a suitable beam-to-beam connection using two methods:

(a) recessed half joint with inclined bars
(b) cast-in steel inserts.

Use $f_{cu} = 40$ N/mm$^2$, $f_y = 460$ N/mm$^2$, $f_{yv} = 250$ N/mm$^2$, cover to exposed faces = 35 mm, cover to protected faces = 25 mm, and grade 43 steel and weld.

### Solution

**Method 1 Recessed half joint**
This is shown in Fig. 7.76(b).
Let site fixing tolerance = 10 mm to all sides, except 25 mm to the sides of the pocket.
Half-joint depth to primary beam = 200 mm, and breadth = 300 mm.
Half-joint depth to beam = 190 mm.

Bearing area without steel plates $= \dfrac{100 \times 10^3}{0.6 \times 40} = 4166$ mm$^2$.

Maximum bearing breadth = 300 − 2 × cover = 230 mm.
Bearing length = 4166/230 = 18 mm < minimum of 75 mm or $b/3$ = 100 mm.
*Design of strut and tie steel beneath bearing*
$d'' = 200 − 30 = 170$ mm and $x = 130$ mm.
$a = 30 +$ gap $10 + 100/2 = 90$ mm.
$\beta = \tan^{-1} 130/90 = 55.3° > 45°$ satisfactory.

Compressive strut force = $(0.5 \times 100)/\sin^2 55.3 = 74$ kN.
Then

$$X = \frac{74.0 \times 10^3}{0.4 \times 40 \times 300 \times \sin 55.3°} = 47 \text{ mm} < 0.5 \text{ d}''.$$

When the frictional force $\mu V = 0.7 \times 100 = 70$ kN acts, the tie force $H$ is

$$H = \frac{100}{\tan 55.3°} + 70 \times \frac{160}{130} = 155.4 \text{ kN}.$$

$$A_{sh} = \frac{155.4 \times 10^3}{0.87 \times 460} = 388 \text{ mm}^2.$$

*Use four T 12 (452) links in boot at 100 mm centres.*
Longitudinal hanger bars $A_s$ inclined at 45° to horizontal:

$$A_s = \frac{50 \times \sqrt{2} \times 10^3}{0.87 \times 460} = 176 \text{ mm}^2.$$

*Use two T 12 (226) inclined hanger bars in front face of primary beam.*
Shear links to side of recess

$$A_{sv} = \frac{50 \times 10^3}{0.87 \times 250} = 230 \text{ mm}^2.$$

*Use two R 10 (314) stirrups either side of recess.*
Splitting force directly beneath bearing point, using lowest value for $\zeta = 0.11$:

$$F_{bst} = 0.11 \times 100 = 11 \text{ kN}.$$

$$A_{bst} = 28 \text{ mm}^2.$$

*Use one R 8 (50) bar in top corner of boot.*
See Fig. 7.76(a) for final reinforcement details.

### Method 2 Cast-in steel inserts
This is shown in Fig. 7.76(c).
Choose narrow high-level beam plate, fillet welded to RHS or SHS insert.

### (1) Secondary beam connector design
The design requires an initial guess for the thickness of the narrow plate. Try 16 mm.
Bearing length for steel-to-steel contact:

$$= \frac{100 \times 10^3}{16 \times 190} = 33 \text{ mm} + \text{weld run-outs} = 60 \text{ mm}.$$

Eccentricity from centre of bearing to centre of tie bar:

$$= 30 + \text{gap } 10 + \text{cover } 25 + \text{link (say) } 10 + \text{radius (say) } 15 = 90 \text{ mm}.$$

Depth of narrow plate = 75 mm maximum protrusion plus embedded depth based on length of weld required to tie bars.
Referring to Fig. 7.76(d) tie force is approximately $T = 100 \times 380/290 = 131$ kN
Assuming weld size = 8 mm.

Weld length $= \dfrac{131 \times 10^3}{4 \times 0.7 \times 8 \times 215} = 28$ mm + weld run-outs = 44 mm, thus use 50 mm.

Total depth of narrow plate = 75 + 50 = 125 mm.

Narrow plate thickness $t = \dfrac{100 \times 10^3}{125 \times 165} = 4.85$ mm, based on shear < 16 mm

or

$$t = \dfrac{100 \times 10^3 \times 90 \times 4}{125^2 \times 275} = 8.38 \text{ mm based on bending} < 16 \text{ mm.}$$

*Use 125 × 16 mm narrow plate.*
Maximum moment in plate $M_{zz} = 100 \times 90 = 9000$ kN mm.
Line pressure beneath plate = $0.6\,f_{cu}\,t = 0.6 \times 40 \times 100 = 2400$ N/mm.
Try length of narrow plate $L_4 = 250$ mm.
Also $M_{zz} = 2400\,L_3\,(250 - 0.5\,L_3)$.
Hence $L_3 = 15.5$ mm; use 30 mm wide plate.
Then $C = 2400 \times 15.5 \times 10^{-3} = 37.2$ kN
and $T = 100 + 37.2 = 137.2$ kN.
Tie steel:

$$A_s = \dfrac{137.2 \times 10^3}{0.87 \times 250} = 630 \text{ mm}^2.$$

*Use two R 20 (628) bars welded to narrow plate.*
Weld length = 30 + 16 = 46 mm < 50 mm available.
*Use 8 mm CFW × 50 mm long-to-narrow plate.*
Bend radius to R 20 bars, $a_b = 16 + 20 = 36$ mm, force per bar = 137.2/2 = 68.6 kN.

$$r = \dfrac{68.6 \times 10^3 (1 + 1.11)}{2 \times 40 \times 20} = 90 \text{ mm.}$$

Horizontal shear in bar = $\mu V = 70$ kN is resisted by the shear in the tie bar, and friction–bond in the plate. Ignoring the latter, then

$$A_s = \dfrac{70 \times 10^3}{0.6 \times 0.87 \times 250} = 536 \text{ mm}^2 < 628 \text{ mm}^2 \text{ provided.}$$

**(2) Shear reinforcement design**
$b = 300$ mm, $d = 300 - 35 - 8 -$ say 8 = 249 mm, say 250 mm.

$$v = \dfrac{100 \times 10^3}{300 \times 250} = 1.33 \text{ N/mm}^2.$$

$100\,As/bd = 100 \times 628/300 \times 250 = 0.84.$
Hence, $v_c = 0.80$ N/mm$^2$.

$$\dfrac{A_{sv}}{s} = \dfrac{300 \times (1.33 - 0.8)}{0.87 \times 250} = 735 \text{ mm}^2/\text{m.}$$

*Use R 8 links at 125 mm centres (800) for a distance to the end of the narrow plate plus d.*

### (3) Design of wide plate

Average stress under wide plate $= \dfrac{37.2 \times 10^3}{100 \times 30} = 12.4 \text{ N/mm}^2$.

Overhang beyond edge of narrow plate $= \dfrac{(100 - 16)}{2} = 42$ mm.

Shear $V = 12.4 \times 30 \times 42 = 15.6$ kN.

Bending $M = \dfrac{12.4 \times 30 \times 42^2}{2} = 328$ kN mm.

Thickness $t = \dfrac{15.6 \times 10^3}{30 \times 165} = 3.15$ mm based on shear

or $t = \sqrt{\dfrac{328 \times 10^3 \times 4}{30 \times 275}} = 12.6$ mm based on bending.

*Use 100 × 30 × 16 wide plate.*

### (4) Design of primary beam steel insert and reinforcement
Distance to centre of secondary beam reaction from face of primary beam = 40 mm.
Because the lever arm is smaller than in column inserts, use SHS rather than RHS.
Try $b_p = 80$ mm.
Line pressure $= 0.8 \times 40 \times 80 = 2560$ N/mm.

$L_2 = \dfrac{100 \times 10^3}{2560} = 39$ mm.

$M_{zz} = 100 \times \left(40 + 35 + \dfrac{39}{2}\right) = 9450$ kN mm.

$S_{xx} > \dfrac{9450}{275} = 34.4 \text{ cm}^3$.

$2dt > \dfrac{100 \times 10^3}{165} = 606 \text{ mm}^2$.

*Use 80 × 80 × 5 mm SHS ($S_{xx} = 41.7 \text{ cm}^3$, $2dt = 800 \text{ mm}^2$).*
Maximum length of billet = 300 − 35 = 265 mm.
Thus use additional bars to rear of insert, at edge distance to bars = 35 + 8 links + say
8 = 51 mm, say 50 mm.
Then $d = 250$ mm.
Using $M_{zz} = 2560 \, L_3 \, (176 - 0.5 L_3)$.
$L_3 = 22.5$ mm.
Total force in steel $= 2560 \times 22.5 \times 10^{-3} = 57.3$ kN.

$A_s = \dfrac{57.3 \times 10^3}{0.87 \times 250} = 263 \text{ mm}^2$.

*Use two T 16 bars (402) at 45 mm cover welded to side of insert.* Bar anchorage length =
$32 \times (263/402) \times 16 = 335$ mm > 225 mm to bottom of beam. Hence provide hook to
give full anchorage.

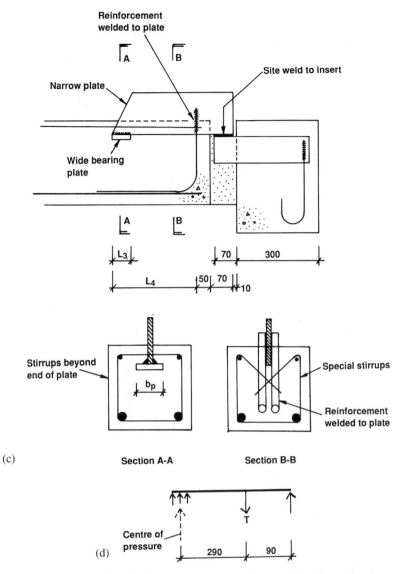

**Fig. 7.76** (*continued*) Practical beam-to-beam connections. (c) Cast-in steel option to Exercise 7.9 and (d) force equilibrium in steel insert in Exercise 7.9.

Assuming weld size = 8 mm,

Weld length $= \dfrac{57.3 \times 10^3}{4 \times 0.7 \times 8 \times 215} = 12$ mm + weld run outs = 28 mm < 80 mm available.

*Use 8 mm CFW × 80 mm long to side of insert.*
Confinement reinforcement beneath insert

$F = 100 + 57.3 = 157.3$ kN.

$$\frac{yv_{po}}{y_o} = \frac{80}{300} = 0.267. \text{ Hence } \zeta = 0.23.$$

$$F_{bst} = 0.23 \times 157.3 = 36.2 \text{ kN.}$$

$$A_{bst} = \frac{36.2 \times 10^3}{0.87 \times 460} = 90 \text{ mm}^2.$$

*Use two T 10 bars (157) at 50 and 100 mm below the front of the insert.*

## 7.13 COLUMN SPLICES

### 7.13.1 Types of splices

A column *splice* is the general term for a joint where a structural connection is made between a column and another precast component. The base member is usually a column, but it may also be a wall, structural cladding panel, beam, or in extreme circumstances a flooring unit. It does *not* include the connection to bases or other foundations.

Most precast concrete frame designers prefer to stagger the level of column splices to avoid forming a 'plane' of weakness, as shown in Fig. 7.77. The level of the first splices is usually shared between the third and fourth floors, except in five-storey frames where the splices (if used at all) are made at the second and third floors. The second splice is at two or three storeys above the first splice. Splices are located either at a floor level (within the structural floor zone) where they may be concealed in the floor finishes, or at a convenient working height, e.g. 1.0 m above floor level, near to the point of contraflexure in the frame where the bending moments are small. Column splices may only be used in a totally braced frame, i.e. it may not be unbraced in any direction, or in the braced part of a partially braced frame.

The choice of splice is often dictated more by site erection considerations than by structural strength. It is very important that the temporary stability of a structure is not placed at risk in using a connection which relies heavily on friction, wedging or other physical actions.

### 7.13.2 Column-to-column splices

Column-to-column splices are made either by coupling, welding or bolting mechanical connectors anchored into the separate precast components, or by the continuity of reinforcement through a grouted joint. The compressive capacity of the splice is made equal to that of the parent columns by confining the insitu concrete placed into the joint. Although the 'design' strengths of the precast and insitu concrete are equal, it is almost certain that the 'actual' strengths will differ – the precast being greater. (In the author's experience 28 day precast concrete cube strengths of 80 N/mm$^2$ are not unusual for columns where the 24 hour strength has been 40 N/mm$^2$.)

The effect of different concrete strengths on the compressive load capacity of columns and columns splices was studied by Chuan-Chien Shu and Hawkins [7.42]. The test variables were height of splice $v$, breadth of the column $b$, and cylinder strength of

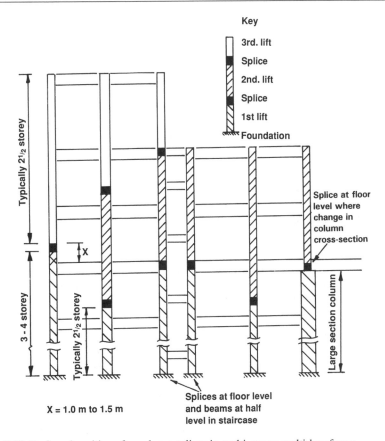

**Fig. 7.77** Preferred positions for column splices in multi-storey multi-bay frame.

the infill $f_{cy1}$. The effective (i.e. apparent) strength of the connection $f_c'$ was measured. In all cases the ratio $f_c'/f_{cy1}$ was greater than 1.0, but when the data for $f_{cy1}$ are converted to cube strengths the situation becomes non-conservative when $v/b > 1.5$. Thus providing that the height of the splice is not more than 1.5 times the minimum breadth of the column the connection will develop the design strength of the infill. The usual practice is for $v/b < 1$.

The essential features of design are to satisfy the requirements of ultimate strength and structural stability. This is achieved either by coupling, welding or bolting mechanical connectors anchored into the separate precast components or by the continuity of reinforcement through a grouted joint.

### 7.13.3   Coupled joint splice

The coupled joint shown in Fig. 7.78(a) provides a mechanical tie between the precast components which is capable of axial load and bending moment interaction. This connection requires absolute *precision* (to about ±3 mm) in placing projecting threaded

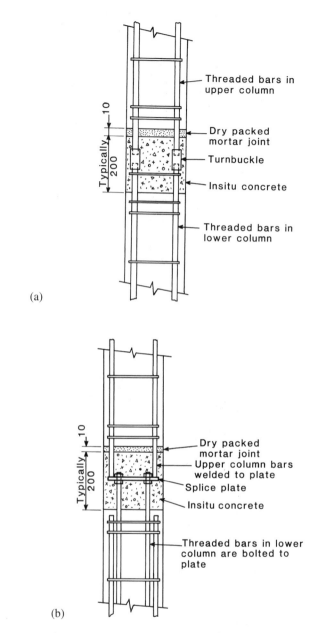

(a)

(b)

**Fig. 7.78** Types of splices. (a) Coupled joint, (b) welded plate.

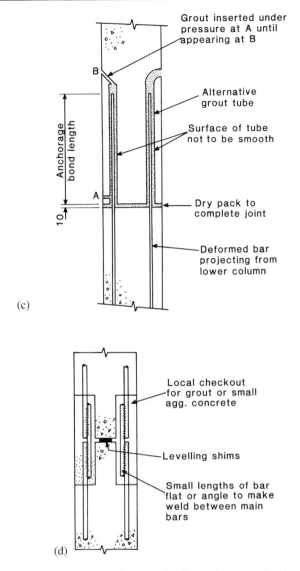

Grout inserted under
pressure at A until
appearing at B

B

Anchorage
bond length

Alternative
grout tube

Surface of tube
not to be smooth

A

Dry pack to
complete joint

10

Deformed bar
projecting from
lower column

(c)

(d)

Local checkout
for grout or small
agg. concrete

Levelling shims

Small lengths of bar
flat or angle to make
weld between main
bars

**Fig. 7.78** (*continued*). Types of splices. (c) Grouted sleeve, (d) welded lap joint.

bars and should only be used if the designer is satisfied that such tolerances can be achieved both in the factory and on site. The other main drawback is in connecting all couplers simultaneously.

The strength of the splice may have to be down-rated due to problems in the strain compatibility between the ordinarily reinforced precast components and the less ductile

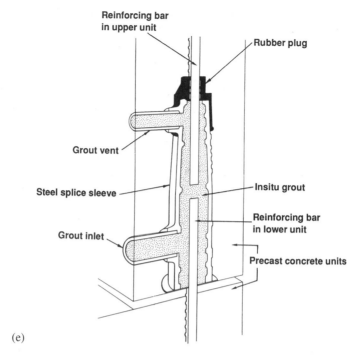

Reinforcing bar
in upper unit

Rubber plug

Grout vent

Steel splice sleeve

Insitu grout

Reinforcing bar
in lower unit

Grout inlet

Precast concrete units

(e)

**Fig. 7.78** (*continued*) (e) Grouted splice sleeve.

behaviour of the threaded coupler. Links designed to BS 8110 [7.6] are provided in the infilled region to provide stability to the compression reinforcement and to increase the strength of the infill by confinement. The projecting reinforcement in the upper and lower columns is threaded to opposite hands, or else the coupler is provided with opposite threads. A sufficient threaded length of bar within the coupler is assured by turning the coupler down to predetermined marks on the reinforcement.

The 28 day characteristic strength of the insitu infill is at least equal to the design strength of the column. A proprietary expanding agent is added to the cement/sand/6 mm aggregate mix to prevent shrinkage cracking between the different concretes. The height of the splice is generally less than 200 mm and is structurally adequate if the insitu connection is made in one pour. In dealing with larger volumes (greater than 0.05 m$^3$) or greater height (exceeding 300 mm) it is necessary to concrete in two stages by leaving a narrow 10 to 15 mm gap and dry packing with a 2:1 sand/cement mortar at a later date.

### 7.13.4   Welded plate splice

The splice, shown in Fig. 7.78(b), is formed by allowing projecting reinforcement from the (usually) lower column to be clamped (by a pair of nuts) to a plate which is welded to bars projecting from the upper column. The compressive strength of this arrangement is based on the strength of the infill concrete plus either the load capacity of the threaded bar/nut system, or the load capacity of the welded mild steel reinforcement. The strength of the infill should be equal to that of the precast column, but not more than 40 N/mm$^2$ unless testing has shown that higher infill strengths may be used. The flexural and shear strength of the plate is enhanced by the confinement of the insitu concrete. However, flexibility of the plate can cause problems in the temporary condition, particularly if long columns (up to three storeys) are left free-standing.

The strength of the splice is governed either by the strength of the plate in bending or shear across plane $x$–$x$ in Fig. 7.81, or by the strength of the bars. The upper column bars are mild steel because they are welded to the plate. The bars are placed into oversized holes such that there is an annulus of 4–5 mm around the bar to permit the placement of a full penetration fillet weld. Fillet welds made directly on to the surface of the plate are not acceptable. Small diameter (e.g. 3 mm) electrodes are used. The bar must penetrate the hole in the plate at a distance $l_x$ so that the strength of the weld is greater than the yield strength of the bar. A factor of safety of two is quite common to allow for variable workmanship. Thus, if the diameter of the bar is $\phi$, then:

$$\pi \, \phi \, l_x \, p_{weld} > 2 \, (0.87 \, f_y) \, A_s \tag{7.58}$$

If, for mild steel bar and grade 43 weld, $f_y/p_{weld} = 250/215$, then

$$l_x > 0.50 \, \phi \tag{7.59}$$

Thus the rules are:

- the splice plate should be at least equal to the 'designed' bar diameter
- the bar should penetrate the plate a distance to half its diameter.

Note that the actual bar used may be larger than the design value. In general the smallest size bar used is 16 mm diameter, with 25 mm being the most common for supporting two-storey columns.

The lower bars are threaded high tensile bars, with a reduced cross-sectional area measured at the root of the thread. For metric threads the effective diameter and strength of threaded reinforcing bars are as given in Table 7.10.

If the splice is concealed in the floor zone the size of the plate may be equal (or slightly greater if needed) to the size of the column. The cover to the threaded and welded bars may be equal to the main column reinforcement. However if the splice is made above floor level then the plate must be protected by concrete cover, typically 35 to 40 mm. Thus the cover to the splice bars is about 70 mm. This has the effect of reducing the strength of the splice by reducing the lever arm between the bars.

The centroidal distance $X$ between the bars is given by:

**Table 7.10** Metric thread data.

| Bar diameter | Area at root of thread | Tensile strength (kN) |
| --- | --- | --- |
| 16 | 157 | 62.8 |
| 20 | 245 | 98.0 |
| 25 | 353 | 141.3 |
| 32 | 561 | 224.5 |

$$X = \frac{h}{2} - \text{cover to bars} - \frac{\Phi}{2} \qquad (7.60)$$

where $\Phi$ = diameter of the threaded bar. The lever arm from the centre of this bar to plane $x$–$x$ is $X/\sqrt{2}$, and the effective breadth of the plate in plane $x$–$x$ may be approximated as $\sqrt{2}\ X$. In the temporary construction phase before the splice is grouted, the thickness of the plate is given by the greater of:

$$t^2 = \frac{2F_s}{p_y} \qquad (7.61)$$

$$t = \frac{F_s}{0.6p_y\ \sqrt{2X}} \qquad (7.62)$$

where $F_s$ is the force in the bar due to the self-weight of the column $W$ and the effects of overturning due to wind loads and lack of plumb. Wind pressures in the temporary stage cause moments $M$ at splices in the order of 5 to 6 kN m in a two-storey column, and about 10 kN m in three storeys. For combined bending and axial force, the force in the bars (four per column) is given by:

$$F_s = 0.25\left[\frac{M}{X} + W\right] \qquad (7.63)$$

This force is of course permanent and must form part of the overall calculation for the strength of the splice when completed. It is also vital that no signs of distress are seen in the splice during this stage. For these reasons the bars are often over-designed with a factor of safety of about two at ultimate.

After completion of the splice by grouting, etc. the load path through the connection by-passes the plate but not the bars. However the role of the plate is essential to the satisfactory performance of a splice by helping to confine the infill concrete. No confinement links are used here.

### 7.13.5   Grouted sleeve splice

One of the most popular (and the most economical) column splice detail is the grouted splice sleeve (Fig. 7.78(c)). Full scale tests [7.42] have shown that the axial load–bending moment interaction characteristics of this joint are equal to those of the parent column. Figure 7.79 summarizes the results. The joint possesses most of the advantages (confinement of concrete, thin dry packed joint, continuity of high tensile reinforcement, easy to manufacture and fix) and few of the disadvantages (fully compacted pressurized grout in the sleeves of the upper lift) associated with precast construction methods. Splices may be made in this way at virtually any level in the frame and are not restricted to column-to-column connections. The same provisions as to the size of the sleeve and the grouting method as given in Fig. 7.18 are applicable here.

The welded lap joint solution, Fig. 7.78(d), is not widely used because the height of unrestrained reinforcement, which is typically 500 to 700 mm in order to achieve the necessary weld length, cannot be justified. A single link can however be placed in the gap but this must then be made wider than is normally acceptable for dry-packed mortar.

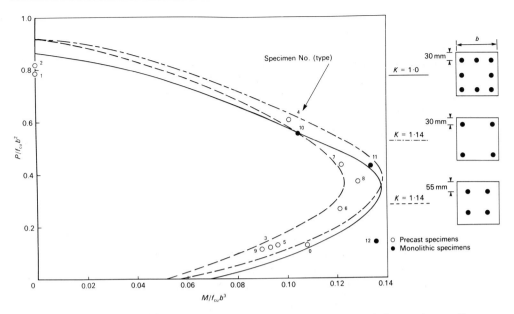

**Fig. 7.79** Moment–axial load interaction failure data for grouted sleeve column splice tests by Kuttab & Dougill [7.43].

The main disadvantages are associated with accuracy in manufacture and site erection, and that the upper column must be held rigidly whilst the welding is carried out. If the problems can be satisfactorily overcome, the splice may be considered structurally equal to a precast column reinforced with mild steel bars.

### 7.13.6    Grouted sleeve coupler splice

Reinforcing bars to be connected are inserted end to end into a steel sleeve and pressure grouted in the same manner as the grouted sleeve given in Section 7.13.5. The original splice sleeve was invented by A. A. Yee in the US and the proprietary system is now patented. The splice detail shown in Fig. 7.78(e) may be used for compression and tension, and is equally suited to horizontal as well as vertical (or inclined) splices.

The principle is simple enough in that the bond length of the bars may be reduced to about 8 × diameters because the grout inserted into the annulus between the bar is confined by the tapered and ribbed inside face of the sleeve. The annulus is approximately 8 mm minimum thickness, but special large opening sleeves may be used where site tolerances need to be greater than the standard practice. See Section 7.5.2 for the grouting specification. The sleeve contains a stop formed at its centre to ensure that each bar is embedded at the correct length. The bars may be of different diameter.

### 7.13.7    Steel shoe splices

Prefabricated steel shoes, Figs 7.80(a) and (b), are used where it may be necessary to

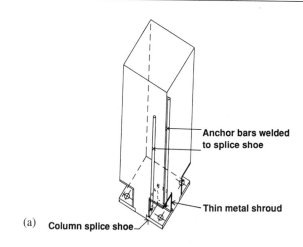

**Fig. 7.80** Column splice shoe details. (a) Column splice shoe tests by Bergström [7.44] and (b) ultimate failure condition at one corner of a column [7.44].

generate bending moment and tensile forces in splices. The so-called 'column shoe' may also be used at foundation connections. It is an attractive alternative to the welded splice plate in large cross-sections (greater than say 400 × 400 mm) where large plates may be wasteful. In all, four shoes are used, one at each corner of a rectangular column. Modified versions of the standard shoe are possible for non-rectangular columns. The connectors are expensive in terms of materials and manufacture, but compensate for this by providing a very rapid and structurally safe fixing on site, accommodating large tolerances. Positioning errors of up to 10 mm are possible by the use of cleverly designed eccentric hole plate washers.

Each consists of a thick (grade 50) steel plate, typically 12 to 40 mm thick and 100 to 150 mm square, joined to a thin plate metal shroud forming an 80 mm (approx.) open box cube, and 3 no. rebars in a triangular formation. The bars which are typically 16 to 40 mm in diameter provide the bond force to the concrete column. The base plate has a punched hole at its centre which receives the threaded coupler bars from the adjoining column (similar to the welded plate splice detail). The column shoe may be recessed in the column if the splice connection is exposed, otherwise the edge of the base plate is made flush with the column.

The tensile capacity of these connectors are always governed by the strength of the threaded portion of the coupler bar, and never by the bond strength of the rebars. Tensile strengths of up to 300 kN static force are possible.

Bergström [7.44] has carried out bending moment tests on (four) column shoe connections as shown in Fig. 7.80(a). The column size was 400 mm square. Applied moments of up to 230 kN m caused relative rotations between the foundation and the bottom of the column of up to 0.1 radians at ultimate, and 0.02 at the limit of proportionality. The average rotational stiffness of these connections was 12 750 kN m/rad, which is equal to approximately 0.8 times the uncracked flexural stiffness of a 4.0 m long column of the same size as used in the test. This value compares reasonably closely with the value of 1.0 according to BS 8110 [7.6], although the test result is on the unsafe side. Figure 7.80(b) shows the ultimate failure condition at one corner of a column.

## EXERCISE 7.10    Welded plate splice design

Calculate axial load capacity of the welded plate splice shown in Fig. 7.81, assuming that the bending moment is zero. Assume the edge distance to the holes in the plate = 20 mm.

Use $f_{cu}' = 40$ N/mm$^2$ (for the insitu), $f_y = 250$ N/mm$^2$ in the upper bars, $f_y = 460$ N/mm$^2$ in the lower threaded bars, 35 mm cover to the plate, grade 43 steel plate and weld.

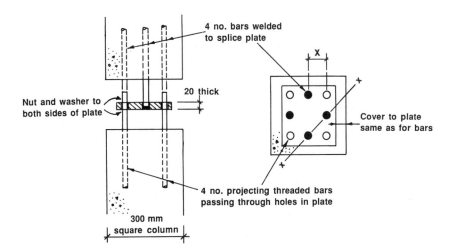

**Fig. 7.81** Details to Exercise 7.10.

### *Solution*

#### Stage 1: During construction

Assume 25 mm threaded bar, then $X = \dfrac{230}{2} - 20 - \dfrac{25}{2} = 82$ mm.

$b = \sqrt{2} \times 82 = 116$ mm.

Bending capacity $M_{R\ plate} = 275 \times 116 \times 20^2 \times 10^{-3}/4 = 3190$ kN mm.

$M_R = F_s X/\sqrt{2}$.

Hence $F_s = 3190/58 = 55$ kN.

Shear capacity $V_{R\ plate} = 165 \times 116 \times 20 \times 10^{-3} = 382.8$ kN.

Hence $F_s = V_R = 382.8$ kN > 55 kN bending capacity.

*Ultimate axial capacity of splice = 4 × 55 = 220 kN.*

Threaded bar $A_s = \dfrac{55 \times 10^3}{0.87 \times 460} = 137$ mm$^2$.

*Use T 16 bars threaded to M 16 (157).*

Welded bar $A_s = \dfrac{55 \times 10^3}{0.87 \times 250} = 252$ mm$^2$.

*Use R 20 welded bars (314).*

**(1) Weld capacity**

Use penetration $x = 0.5 \times 20 = 10$ mm.

Contact area $= \pi \times 20 \times 10 = 628$ mm$^2$.

Weld strength $= 628 \times 215 \times 10^{-3} = 135$ kN > 55 kN required.

#### Stage 2: After concreting

Splice designed as ordinary rc column (BS 8110, Part 1, equation 38 [7.6])

$N = 0.4 f_{cu}' A_c + 0.75 f_y A_{sc}$

Hence, for the welded bars

$$N = (0.4 \times 40 \times 88\,744) + (0.75 \times 250 \times 1256) \times 10^{-3} = 1655 \text{ kN}.$$

For the threaded bars

$$N = (0.4 \times 40 \times 89\,372) + (0.75 \times 460 \times 628) \times 10^{-3} = 1647 \text{ kN}.$$

Capacity of concrete in contact with plate, using $0.8 f_{cu}'$, and ignoring cover concrete

$$N = 0.8 \times 40 \times 230^2 \times 10^{-3} = 1693 \text{ kN}.$$

*Limiting ultimate axial capacity = 1647 kN.*

### EXERCISE 7.11    Grouted sleeve splice design

Calculate the ultimate moment of resistance of the column splice shown in Fig. 7.82, if the maximum and minimum axial force is 1350 kN and 950 kN, respectively. Determine the anchorage length of the projecting bars.

Use $f_{cu} = 30$ N/mm$^2$ (for the insitu), $f_y = 460$ N/mm$^2$.

### *Solution*

For minimum effective depth with splice bar touching inside face of sleeve:

$$d = 300 - \text{cover } 40 - (50 - 12.5) = 222 \text{ mm}$$

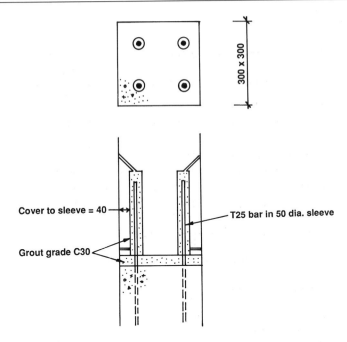

**Fig. 7.82** Details to Exercise 7.11.

$$\frac{d}{h} = \frac{222}{300} = 0.74$$

Therefore use BS 8110, Part 3, Column Chart 26 [7.6].

$$\frac{A_{sc}}{A_c} = \frac{1963 \times 100}{300 \times 300} = 2.2\%$$

for  $$\frac{N}{bh} = \frac{1350 \times 10^3}{300 \times 300} = 15.0$$

$$\frac{M}{bh^2} = 2.3 \ \textit{Therefore } M = 62 \ kNm.$$

for  $$\frac{N}{bh} = \frac{950 \times 10^3}{300 \times 300} = 10.55$$

$$\frac{M}{bh^2} > 2.3 \ \text{Therefore not critical.}$$

To determine bond lengths, check stresses in bars.
Axial stress = 15 N/mm$^2$.

Bending stress = $\dfrac{6M}{bh^2}$ = 13.8 N/mm$^2$.

Therefore all bars in compression.
Compressive bond length (for grade 30 concrete) = $29 \times 25 = 725$ mm.

### 7.13.8   Columns spliced onto beams or other precast components

Pin-jointed splices are often formed between columns and precast beams using any of the methods used in column-to-column placing. The danger is that in unbraced sway frames large torsional moments will be induced in the beam unless the joint is specifically designed to eliminate moment transfer. A pin-jointed version of the bolted plate splice is shown in Fig. 7.83 where two bolts are used in line to ensure that the connection actually behaves as a pin. Other similar welded or grouted details are easy to envisage.

Moment-resisting splices may be formed between columns and precast walls and cladding because the latter is usually capable of accommodating the transfer moment. However, caution must be exercised in assuming these places will attain their theoretical ultimate design limit. Full-scale testing [7.44] under in-plane loading found that where the faces of the precast components were flush there was extensive cracking and a tendency for the cover to spall on the face of the supporting member even though the connection was designed as pinned (see Figs 33 and 34 in the IStructE [7.17]).

Lightly loaded columns have successfully been spliced onto hollow core flooring units as shown in Fig. 7.84. Although this practice is not entirely endorsed by all precast concrete engineers, there are occasions where, to satisfy an architectural feature almost exclusively at roof level (e.g. Mansard roof), there is no other feasible or practical solution. The column base is assumed pinned because of flexibility in the slab. Holes for holding down bolts are pre-drilled during manufacture and the tops of the cores are opened to permit the placing of insitu concrete as appropriate. The dimensions of the position of the column and base plate are governed to a certain extent by the positions of the hollow cores in the finalized floor slab layout. The maximum point load normally considered is about 50 kN.

**Fig. 7.83** Pin-jointed base plate column-to-beam connection.

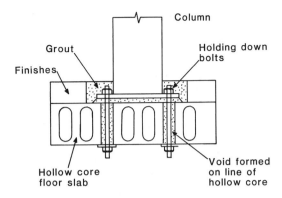

**Fig. 7.84**  Column-to-hollow core floor slab connection.

Column splices have (but rarely) been formed on double-tee units. The position of the column is restricted to being central over the webs of the flooring units. The joint which is made by welding or bolting to a base plate is considered to be a *last resort* in the correct application of precast concrete frame design. However lightweight steel columns have been spliced onto double-tee units to form a Mansard roof or plant room frame. The maximum point load normally considered is about 75 kN.

## 7.14   COLUMN BASE CONNECTIONS

Column connections to pad footings and other insitu (or precast) concrete foundations (e.g. retaining wall or ground beam) are of three main types:

(1)  grouted pocket
(2)  base plate; (a) greater or (b) smaller in plan dimension than the dimensions of the column
(3)  grouted sleeve.

Methods (1) and (2) are most commonly used. The base plate has the advantage that the column may be stabilized and plumbed vertical by adjusting the level of the nuts to the holding down bolts. This is particularly important when working in soft ground conditions where temporary propping may not provide adequate stability alone.

### 7.14.1   Columns in pockets

This is the most economical solution from a precasting point of view, but its use is restricted to situations where fairly large insitu concrete pad footings can easily be constructed. The basic precast column requires only additional links to resist bursting pressures generated by end bearing forces, and a chemical retarding agent to enable scabbling to expose the aggregate in the region of the pocket. In cases where the column reinforcement is in tension, the bars extending into the pocket must be fully anchored by bond, or other means. In order to reduce the depth of the pocket to a manageable size these bars may need to be hooked at their ends.

The insitu concrete foundation is cast using a tapered box shutter to form the pocket. The gap between the pocket and the column should be at least 75 mm at the top of the pocket. The foundation may also be used to support precast ground beams.

Vertical loads are transmitted to the foundation by skin friction and end bearing. Figure 7.85(a) shows the structural models for the two cases where the column is smooth and roughened. Exactly what constitutes a 'roughened' surface is never clearly defined in the literature, but we can assume that if the aggregate is exposed, but not damaged, then the surface may be classified as rough. It is possible to form shear keys in the sides of the pocket (according to the dimensions given in Fig. 7.20) and transfer the axial load by the action of shear wedging.

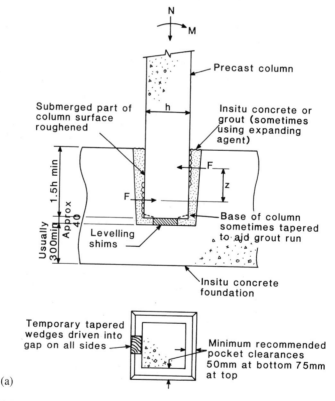

(a)

**Fig. 7.85** Pocket foundations. (a) Structural models in pocket foundations and (b) failure modes in pocket foundations.

If overturning moments are present half of the skin friction is conservatively ignored due to possible cracking in all of the faces of the precast/insitu boundary. Ultimate load design considers vertical load transfer by end bearing based on the strength of the gross cross-sectional area of the reinforced column and equal area of non-shrinkable sand/cement grout. The design strength of the infill grout is usually $f_{cu}' = 40$ N/mm$^2$ and the specification is the same as for the grout used in splices. There is a lack of analytical or experimental data on the real behaviour of pocketed connections, but this is most probably due to an almost total absence of failures. The only research on this topic has been limited to considering the prevention of concrete splitting in the sides of the pocket, particularly where the cover thickness is less than about 200 mm [7.45].

The depth of the pocket is governed by the bond length of the column reinforcement. This should be a full tension bond length (typically $32 \times$ diameter) if the column is designed at the balanced section. However it is most unlikely that this will be the case in multi-storey frames and so strictly speaking one should calculate the actual tensile stress in the bar and provide a bond length according to BS 8110, clause 3.12.8.3 [7.6]. In order to avoid using very deep pockets when using large diameter bars it will be necessary to provide a hook to the bottom of the bar. The minimum bond length, and hence pocket depth, should not be less than 12 bar diameters.

The failure mode may be by diagonal–tension shear across the corner of the pocket, as shown in Fig. 7.85(b), in which case links are provided around the top half of the pocket. Several small links, say T 8 to T 12, are preferred to larger bars in order to keep the corner bend radii small. Another mode of failure is crushing of the insitu concrete in the annulus. This is guarded against by using an ultimate stress of $0.4 f_{cu}'$ working over a width equal to the precast column only, i.e. ignoring the presence of the third dimension.

The pocket is usually tapered 5° to the vertical to ease the placement of grout in the annulus. This gives rise to a wedge force equal to $N \tan 5°$, where $N$ is the ultimate axial

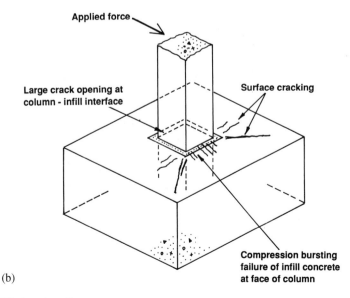

Applied force

Large crack opening at
column - infill interface

Surface cracking

Compression bursting
failure of infill concrete
at face of column

(b)

**Fig. 7.85** (*continued*)

load in the column. The precast column requires only additional links to resist bursting pressures generated by end bearing forces using $\zeta = 0.11$.

Horizontal shear and overturning moments dealt with as shown in Fig. 7.86(a) and (b). Compressive contact forces generate the vertical frictional resistance $\mu F$ (using $\mu = 0.7$), and a horizontal friction $\mu N$ underneath the bottom of the column. Horizontal forces $F$ in the contact region are distributed to the pad footing using horizontal links, and particular attention should be paid to this if the edge cover is less than the smaller dimension of the column. The recommended minimum depth of pocket $D$ is equal to 1.5 times the breadth of column, even though analysis may suggest values for

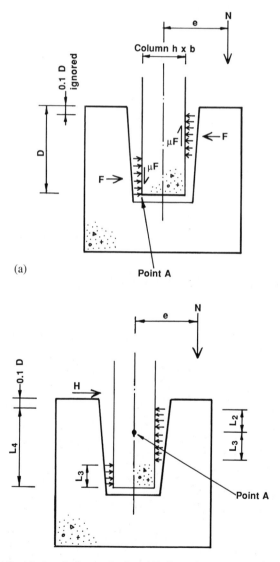

**Fig. 7.86** Pocket foundation design methods. (a) Insitu concrete pocket foundation design without shear and (b) insitu concrete pocket foundation design with shear.

$D < h$ for columns with small bending moments. The depth should not be less than $h$ because of the need to develop a diagonal compressive strut in the column to resist shear forces. Castellations to the sides of the column are frequently used in large cross-sections to reduce the penetration depth to $h$. The root depth is 40 mm minimum (see Fig. 7.87).

Bruggeling & Huyghe [7.15] propose that $D$ is related to the ratio of the moment $M$ and the axial force $N$ as follows:

If $M/N < 0.15\,h$, then $D > 1.2\,h$
If $M/N > 2.00\,h$, then $D > 2.0\,h$                 (7.64)

The text [7.15] does not refer to intermediate values but we can assume that they may be linearly interpolated.

Where the column reinforcement is at all in tension, the depth of the pocket should be governed by the tensile anchorage length of the bars. The column bars should extend for one bond length beyond a point 50 mm below the top of the pocket. This means that for high tensile 25 mm diameter bars in grade C50 concrete a bond length of 708 mm is required (BS 8110, Part 1, clause 3.12.8 [7.17]). The designer has the choice of either making the depth of the pocket = 50 + 708 = 758 mm + cover, or providing a bend or hook to the bars and making the pocket shallower. For example, the hook length for such a bar is 350 mm, therefore for the bar to be considered fully anchored the depth of the pocket $D = 50 + (708 - 350) +$ cover to bottom of column 35 = 443 mm. If the bars are in compression no such check is necessary.

The moment $M$ and axial force $N$ may be resolved into a single force acting at a distance $e = M/N$ from the centre line of the column. The moment is transferred from the column to the foundation by a set of diagonal compressive struts. The shear stress acting

**Fig. 7.87** Castellations in column sides for pocket foundations.

in the sides of the pocket is equal to the sum of the skin friction resisting part of the axial force (assuming a rectangular column) and that resulting from $\mu F$. The critical interface is where the overturning moment is attempting to lift the column upwards out of the pocket – it would be unlikely to puncture the bottom of the base of the pocket.

There are two design methods available, each of which assumes that a couple $Fz$ is generated between forces acting on opposite faces of the column. The analysis is for uniaxial bending only. There is no method for dealing with biaxial bending, but the method for dealing with biaxial bending in columns may be adopted here, i.e. an increased moment in the critical direction is considered as a uniaxial moment.

### 7.14.1.1   Method 1

In this method the force $F$ acts such that a couple $Fz$ is generated over a distance:

$$z = \frac{(D - 0.1D)}{2} = 0.45D$$

$$\text{or} \quad = \frac{(D - \text{cover})}{2} \tag{7.65}$$

whichever is greater. This is because the top $0.1\,D$ of the pocket is ignored within the cover zone, typically 50 to 60 mm. Thus, referring to Fig. 7.86(a) and taking moments about $A$:

$$Ne - \mu\, Fh - 0.45\, FD = 0 \tag{7.66}$$

$$F = \frac{Ne}{\mu h + 0.45D} < 0.4 f_{cu}'b(0.45D) \tag{7.67}$$

For the specific case of $D = 1.5\,h$, then:

$$F = \frac{Ne}{1.375h} < 0.27 f_{cu}'bh \tag{7.68}$$

where $f_{cu}'$ = strength of infill. Bruggeling & Huyghe [7.15] propose that $z = 0.75D$, but this seems to be rather excessive at the ultimate limit state. Hence simplifying Equation (7.66) for $D = 1.5h$ only, we get a limiting value for $e$:

$$\frac{e}{h} = \frac{0.375 f_{cu}'}{N/bh} \tag{7.69}$$

Similar equations may be derived for other values of $D/h$ and $f_{cu}'$, but the mode of action in shallow pockets will inevitably change from the above model where skin friction underneath the column will dominate. In reality the skin friction will also act over the sides of the column parallel to the direction of the moment, and so Equation (7.69) will be conservative in the presence of large axial forces.

### 7.14.1.2   Method 2

In this method the effects of shear are included (see Fig. 7.86(b)). An ultimate horizontal compressive line stress of $0.4 f_{cu}'\, b$ is taken in the insitu concrete infill across the breadth of the column. The reaction to $H$ is calculated from the relationship:

$$L_2 = \frac{H}{0.4 f_{cu}' b} \qquad (7.70)$$

where $L_2$ is measured from a point at $0.1\,D$ from the top of the pocket. Taking moments about the centre line of the column at the end of $L_2$ at point $A$:

$$N\,e + H\,(0.1\,D + 0.5L_2) = (\mu\,0.4\,f_{cu}'\,b\,h\,L_3) + [0.4\,f_{cu}'\,b\,L_3\,(L_4 - L_2 - L_3)] \qquad (7.71)$$

from $L_3$ which may be calculated with real roots to the quadratic only. Note $0.1\,D$ is replaced by the cover if this is smaller. The horizontal frictional force is only considered over the length $L_3$ because it is only mobilized in the presence of bending, and not shear.

The stresses in the opposite faces interface may not overlap (Section 7.10.2) such that $L_2 + 2\,L_3 < 0.9\,L_4$.

The size of the pad foundation is governed by two main factors:

- bearing pressures at service
- edge distance at top of pocket.

See Fig. 7.85. The first is determined using any statical method in order to avoid uplifting grouting pressures. The second is based on the shear stress resulting from the force $0.5F$ acting on the concrete at the top of the pocket. If the effective depth from the pocket to the edge of the foundation is $d_f$, then the condition is satisfied if:

$$0.2\,f_{cu}'\,b\,(L_3 + L_2) < v_c\,(L_3 + L_2)\,d_f \qquad (7.72)$$

where $v_c$ = shear stress of concrete = $0.8\sqrt{f_{cu}}$.

The reinforcement around the pocket must be capable of carrying the force $F$ plus the lateral force due to the taper in the pocket = $N \tan 5°$. Hence links $A_{sv}$ are provided in the top half of the pocket such that:

$$A_{sv} = \frac{[0.4 f_{cu}' b (L_3 + L_2)] + N \tan 5°}{0.87 f_{yv}} \qquad (7.73)$$

The steel is considered over two legs of the links.

Precast concrete pockets, Fig. 7.88, have not been successful from a commercial viewpoint. This is because of the dual effort of preparing a level insitu concrete pad on which precast foundation may sit, and afterwards bedding-in the pocket on to the pad. The precast pocket provides a pin-jointed foundation connection and may only be used in conditions of firm and level ground. This obviously rules out foundations on to pile caps, or on soft reclaimed land.

**EXERCISE 7.12**    **Column pocket design**

Design a column-to-pocket foundation connection required to support a $300 \times 300$ mm column subjected to an ultimate axial force of $N = 1000$ kN and a moment $M = 100$ kN m. Determine the minimum strength of the insitu infill.

Use $f_{cu} = 50$ N/mm$^2$ for the precast column, $f_{cu} = 25$ N/mm$^2$ for the foundation, and $f_y = 460$ N/mm$^2$ for the reinforcement. Cover to column bars = 35 mm, and to foundation bars = 50 mm.

**Fig. 7.88** Precast concrete pocket foundation.

### Solution

$e = 100 \times 10^3/1000 = 100$ mm $= h/3$, hence tension will develop in the column reinforcement.

Column design: $d/h = 300 - 35 - 8 -$ say $10/300 = 0.823$.

Use BS 8110, Part 3, Chart 47 [7.6].

$N/bh = 11.1$, and $M/bh^2 = 3.7$. Hence $A_{sc} = 0.85\%$ $bh = 765$ mm$^2$

*Use four T 16 (804) with R 6 links at 190 mm centres.*

Anchorage bond length for bars in tension $= 28.3 \times 16 = 453$ mm.

Hence $D = 50 + 453 +$ cover $35 = 538$ mm, say 550 mm.

Then $F = \dfrac{1000 \times 100}{0.7 \times 300 + 0.45 \times 550} = 218.6$ kN.

The limiting value for $F = 0.27 f_{cu}' \, b \, h$.

Hence $f_{cu}' = \dfrac{218 \times 10^3}{0.27 \times 300 \times 300} = 9$ N/mm$^2$.

*Use $f_{cu}' = 40$ N/mm$^2$.*

### Confinement steel in column

$F_{bst} = 0.11 \times 1000 = 110$ kN.

Links $A_{bst} = \dfrac{110 \times 10^3}{0.87 \times 460} = 274$ mm$^2$.

*Use four T 10 (314) links at 50 mm centres.*

## Reinforcement around pocket

Bars to be placed in upper half of pocket, i.e. to a depth of 275 mm.
Horizontal force induced by taper $= 1000 \tan 5° = 87.5$ kN.
Force across pocket $= F = 218.6$ kN.

Horizontal steel $A_{sv} = \dfrac{(87.5 + 218.6) \times 10^3}{0.87 \times 460} = 764$ mm$^2$

*Use four T 12 (905) links at 75 mm centres around pocket.*
Also provide vertical hanger bars to support confinement links, nominally 3 no. T 10 bars. See Fig. 7.89 for final details.


**EXERCISE 7.13     Moment resisting column pocket design**

A $400 \times 300$ mm column, with its major axis $x$–$x$, is subjected to the following ultimate loads and moments. Axial load $N = 500$ kN, bending moments $M_{x-x} = 250$ kN m and $M_{y-y} = 50$ kN m, and shear forces $H_{x-x} = 60$ kN. Calculate the minimum depth of pocket foundation, and the size of the pad foundation if the ground bearing pressure = 250 kN/m$^2$. Assume that the ultimate loads given above are based on $\gamma_f = 1.2$. Determine the reinforcement required in the column and the pocket.

Use $f_{cu} = 50$ N/mm$^2$ for the precast column, $f_{cu}' = 40$ N/mm$^2$ for the infill, $f_{cu} = 25$ N/mm$^2$ to the foundation, $f_{yv} = 460$ N/mm$^2$. Cover to column reinforcement = 35 mm, and cover to foundation reinforcement = 50 mm.

*Solution*

$h' = 400 - 35 - 8 - $ say $12 = 345$ mm.
$b' = 300 - 35 - 8 - $ say $12 = 245$ mm.

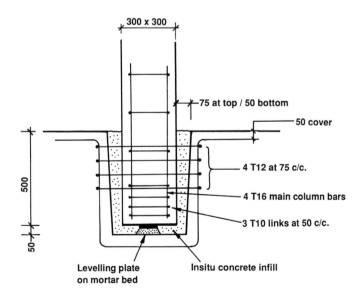

**Fig. 7.89** Details to Exercise 7.12.

$M_{x-x}/h' > M_{y-y}/b'$.

Also $N/f_{cu} b h = 0.083$, then $\beta = 0.9$ (BS 8110, Part 1, Table 3.24 [7.6])

Then $M_{x-x}' = 250 + \left(0.9 \times \dfrac{345}{245} \times 50\right) = 313.4$ kN m.

### Column reinforcement design

$d = 345$ mm.

$d/h = 0.862$. Use Chart 48 (BS 8110, Part 3 [7.6]).

Then $\dfrac{M_{x-x}'}{bh^2} = 6.5$ and $\dfrac{N}{bh} = 4.16$.

Then $A_{sc} = 3.3\% = 3960$ mm$^2$.

*Use 8 no. T 25 bars (3930) say satisfactory.*

Tension anchorage bond length $= 28.3 \times 25 = 708$ mm.

$e = 313.4/500 = 850$ mm, and $e/h = 1.57$.

Use $D = 2h = 800$ mm $> 708$ mm required for bar anchorage.

Line pressure in insitu infill $= 0.4 \times 40 \times 300 = 4800$ N/mm.

Then $L_2 = \dfrac{60 \times 10^3}{4800} = 12.5$ mm.

Moment about bottom of $L_2$ pressure $M = 313.4 + [60(50 + 6.25) \times 10^{-3}] = 316.8$ kN m.

Then $316.8 \times 10^6 = (0.7 \times 4800 \times 400 L_3) + [4800 L_3 (737.5 - L_3)]$.

Hence $L_3 = 69.7$ mm.

Horizontal force $F = 4800 \times 69.7 \times 10^{-3} = 334.6$ kN.

Force at top of pocket $= F + H_{x-x} = 4800 \times 82.2 \times 10^{-3} = 394.6$ kN.

Maximum shear stress at top of pocket $= 0.8 \sqrt{f_{cu}} = 0.8 \times \sqrt{25} = 4.00$ N/mm$^2$.

Then $d_f = \dfrac{394.6 \times 10^3}{2 \times 82.2 \times 4.00} = 600$ mm.

Total edge distance $= 600 + $ cover $50 = 650$ mm.

### Perimeter bursting steel in foundation around pocket

Bars to be placed in upper half of pocket, i.e. to a depth of 400 mm.

Horizontal force induced by taper $= 500 \tan 5° = 43.7$ kN.

Force across pocket $= F + H_{x-x} = 394.6$ kN.

Horizontal steel $A_{sv} = \dfrac{(43.7 + 394.6) \times 10^3}{0.87 \times 460} = 1095$ mm$^2$.

*Use five T 12 (1130) links at 70 mm centres around pocket.*

Also provide vertical hanger bars to support confinement links, nominally 3 no. T 10 bars.

### Base size

Allow 75 mm gap at top of pocket.

Then, breadth of base $B = 650 + 75 + 300 + 75 + 650 = 1750$ mm.

Depth of pocket $= 800 + 50 + 450 = 1300$ mm.

Bearing pressure due to self-weight of base $= 24 \times 1.300 = 31.2$ kN/m$^2$.

Net bearing pressure $= 250 - 31.2 = 218.8$ kN/m$^2$.

Length of base $H$ from the following:

Then

$$\frac{N}{BH} + \frac{6M_{xx}}{BH^2} + \frac{6M_{yy}}{HB^2} < 218.8 \text{ kN/m}^2$$

where $N = \dfrac{500}{1.2} = 416.7 \text{ kN.}$

$M_{xx} = \dfrac{250}{1.2} = 208.3 \text{ kN m.}$

$M_{yy} = \dfrac{50}{1.2} = 41.7 \text{ kN m.}$

From which $H = 2.68$ m, use 2.70 m.
Maximum bearing pressure $= +247.6 \text{ kN/m}^2$.
Minimum bearing pressure $= -8.8 \text{ kN/m}^2$ (small negative pressure satisfactory).

### Reinforcement in bottom of base

Taking moments across face of column, lever arm $= \dfrac{(2700 - 400)}{2} = 1150 \text{ mm.}$

Maximum pressure at face of column $= +138.4 \text{ kN/m}^2$.

$$M_{zz} = \left( \frac{138.4 \times 1.75 \times 1.15^2}{2} \right) + \left( \frac{109.2}{2} \times 1.75 \times \frac{2}{3} \times 1.15^2 \right) = 244.3 \text{ kN m.}$$

Then $M_{zz \text{ ult}} = 1.2 \times 244.3 = 293.2 \text{ kN m.}$

$B = 1750$ mm, $D = 1300 - 50$ cover $- 8$ (say) $= 1242$ mm.

$$K = \frac{293.2 \times 10^6}{25 \times 1750 \times 1242^2} = 0.0043 \qquad \frac{z}{D} = 0.95.$$

$$As = \frac{293.2 \times 10^6}{0.87 \times 460 \times 0.95 \times 1242} = 621 \text{ mm}^2.$$

*Use 5 no. T 16 at 330 c/c in bottom of base.*
*See Fig. 7.90 for final details.*

## 7.14.2  Columns on base plates

Base plates which are larger than the size of the columns are used where a moment connection is required. Figure 7.91(a) and (b) shows the structural mechanism for this type of connection, and Figs 7.92 and 7.93 show photographs of the base plate before and after casting in the column. The disruption to manufacture of the precast column may be considerable because the plate cannot be contained within the internal confines of the mould. On the other hand, base plates provide immediate stability when fixing the column on site, and the depth of the foundation is not excessive. The attitudes towards

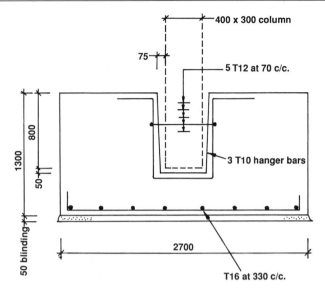

**Fig. 7.90** Details to Exercise 7.13.

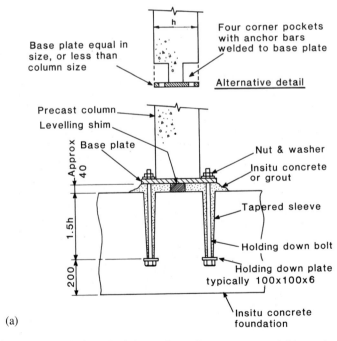

(a)

**Fig. 7.91** Base plate design principle. (a) General arrangement and (b) notation.

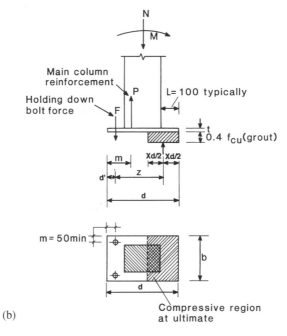

(b)

**Fig. 7.91** (*continued*)

**Fig. 7.92** Column base plate with starter bars welded to plate.

**Fig. 7.93** Column-to-foundation connection with extended base plate.

the choice in using base plates rather than pockets tend to be based more on production rather than structural decisions.

The following design method may be used to calculate the base plate thickness in the completed structure. Referring to Fig. 7.91(b), and resolving vertically. If $F > 0$:

$$F + N = 0.4 f_{cu}' \, b \, X \, d \tag{7.74}$$

where $X \, d$ = compressive stress block depth.
Taking moments about centre line of compressive stress block:

$$M = F (d - d' - 0.5 \, X \, d) + N (0.5 \, d - 0.5 \, X \, d) \tag{7.75}$$

also $M = N \, e$ such that:

$$\frac{N(e + 0.5d - d')}{0.4 f_{cu}' b d^2} = X\left(1 - \frac{d'}{d}\right) - 0.5X^2 \tag{7.76}$$

from which $X$ and $F$ may be calculated.
We can simplify by letting a stress factor

$$K = \frac{M'}{f_{cu}' b d^2} = \frac{N(e + 0.5d - d')}{f_{cu}' b d^2} \tag{7.77}$$

such that

$$X^2 - 2X\left(1 - \frac{d'}{d}\right) + 5K = 0 \tag{7.78}$$

If $X > N/0.4 f_{cu}' \, b \, d$, then $F$ is positive.

Assume $N$ number bolts each of root area $A_b$ and ultimate strength $f_{yb}$ to be providing the force $F$, then:

$$A_b = \frac{F}{N \times f_{yb}} \tag{7.79}$$

For grade 4:6 bolts use $f_{yb} = 195$ N/mm$^2$, and $f_{yb} = 450$ N/mm$^2$ for grade 8:8 bolts.

Except in extreme cases of very high bending and low axial load, the worst bending on the base plate occurs on the compression side. Use a base plate of thickness $t$ which is the greatest of:

$$t = \sqrt{0.8 f_{cu}' L^2 / p_y} \quad \text{(based on compression side)} \tag{7.80}$$

or

$$t = \sqrt{4 Fm / b p_y} \quad \text{(based on tension side)} \tag{7.81}$$

where $L$ = overhang of plate beyond column face, $m$ = distance from centre of bolts to centre of bars in column, and $p_y$ = yield strength of the plate. Steel grade 43 or 50 is used.

If $X < N/0.4 f_{cu} \, b \, d$, then $F$ is negative and Equation (7.75) is not valid. The analysis simplifies to the following:

$$X = 1 - \frac{2e}{d} \tag{7.82}$$

and

$$N = f_c \, b \, X \, d \tag{7.83}$$

because the grout is not fully stressed to $0.4 f_{cu}'$. Equation (7.80) is modified to:

$$t = \sqrt{2 f_c L^2 / p_y} \tag{7.84}$$

The column is sometimes cast without the base plate, for convenience of fitting it in an existing mould. After casting, the base plate is welded on to the protruding column bars, and the final section of concrete between the precast column and the base plate is cast. To avoid difficulties due to shrinkage, this can be, say, 50 mm of dry pack, or possibly a non-shrinking grout or concrete.

Reinforcement is fitted through holes in the base plate and fillet welded at both sides, Fig. 7.92. The design strength of the column is therefore determined using mild steel reinforcement which occasionally leads to cumbersome detailing. The bars are fitted through holes in the base plate and fillet welded according to the same details as given in Section 7.13.4 on splices. Additional links are provided close to the plate as is the practice at splices.

Vertical loads are easily distributed through the grouted infill beneath the plate (same specification for the grout as is used in splices). Overturning moments require a greater attention to detail both in the precast column and insitu foundation. Tensile forces must be transmitted by bond in the precast column, bending and shear in the base plate, tension in the cast-in holding-down bolt and bearing and shear in the foundation.

Holding-down bolts of grade 4:6 or 8:8 are used as appropriate. The length of the anchor bolt is typically 375 to 450 mm for 20 to 32 mm diameter bolts. The bearing area of the bolt head is increased by using a plate, nominally $100 \times 100 \times 8$ mm. The bottom of the bolt is a minimum of 100 mm above the reinforcement in the bottom of the footing. Confinement reinforcement (in the form of links) around the bolts is usually required, particularly where narrow beams and/or walls are used and where the edge distance is less than about 200 mm. The steel is designed on the principle of shear friction but should not be less than 4 no. R 8 links at 75 mm centres placed near to the top of the bolts. To err on the side of caution, anchor loops are usually provided around the bolts in order to achieve the full strength of the bolt if the horizontal edge distance is less than about 200 mm.

Larger compressive forces beneath plates which project beyond the column face in two directions cause biaxial bending in the plate. The maximum projection of the plate is therefore usually restricted to 100 mm, irrespective of size (see Fig. 7.93). The 100 mm is also a minimum practical limit for detailing and site erection purposes. The resulting thickness of plate and grade of concrete are given in Fig. 7.94.

Pinned jointed footings can be designed by decreasing the in-plane lever arm. Base plates using two bolts on one centre-line, or four bolts closely spaced also give the desired effect.

Base plates equal to or smaller than the column are used where a projection around the foot of the column is structurally or architecturally unacceptable, for example where the connection is made at a floor level and cannot be hidden in the finishing screed. The holding-down bolt group is located in line with the main column reinforcement. The base plate is set flush with the bottom of the precast column and small pockets, typically 100 mm cube (for access purposes), leave the plate exposed at each corner, as shown in Fig. 7.95 in the factory and Fig. 7.96 on site, or at each mid-side of the column. Short starter bars are welded to the base plate (as previously described) in various configurations. The plate is usually most critical in the temporary condition. In most cases, i.e. in columns less than 400 mm deep, these connections have a limited moment of resistance and are therefore considered pinned-jointed. Again, two bolts may be used to form a truly pin-jointed connection (although site workers prefer four bolts to assist with vertical alignment and frame stability).

An alternative to casting the holding-down bolts in to the foundation is to drill and fix expanding (or otherwise) bolts on site. The advantage in construction terms is obvious and it places less responsibility on to the contractor to locate the bolts accurately. However, the pull-out capacity of a drilled bolt is less than the capacity of a cast-in bolt of the same diameter. In some instances where the pull-out capacity is critical the cost of a drilled bolt may be greater. In other cases the smallest holding-down bolt used, typically 12 or 16 mm diameter, may be over-designed and a smaller drilled replacement may be the same price.

## EXERCISE 7.14  Column base plate design

Design a column base plate connection required to support a $300 \times 300$ mm column subjected to an ultimate axial force of $N = 1000$ kN and a moment $M = 100$ kN m.
Use $f_{cu}' = 40$ N/mm$^2$ for the grout, grade 43 base plate.

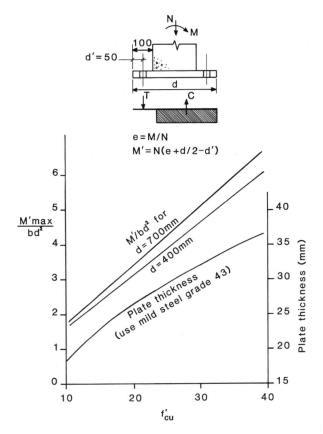

**Fig. 7.94** Column base plate design graph.

## Solution

Try $500 \times 500$ mm plate with $L = 100$ mm overhang and $d' = \dfrac{L}{2} = 50$ mm.

$$e = \frac{100 \times 10^3}{1000} = 100 \text{ mm.}$$

Then    $K = \dfrac{1000 \times 10^3 (100 + 250 - 50)}{40 \times 500 \times 500^2} = 0.06.$

$$X^2 - 2X\left(1 - \frac{50}{500}\right) + 5 \times 0.06 = 0$$

hence $X = 0.186$.

Because $X < N/0.4 f_{cu}' \, b \, d = 0.25$ then $F$ is negative.

Restart the analysis using $X = 1 - \dfrac{2e}{d} = 1 - 0.4 = 0.6$

**Fig. 7.95**  Column casting with flush base plate.

and $\qquad f_c = \dfrac{1000 \times 10^3}{500 \times 0.6 \times 500} = 6.66 \text{ N/mm}^2$

and $\qquad t = \sqrt{2 \times 6.66 \times \dfrac{100^2}{275}} = 22 \text{ mm}.$

*Use 500 × 500 × 25 mm mild steel base plate.*

### EXERCISE 7.15    Column base plate design

Repeat Exercise 7.14 using $N = 100$ kN and $M = 150$ kN.

Use $f_{cu} = 50 \text{ N/mm}^2$ for the precast column, $f_{cu}' = 40 \text{ N/mm}^2$ for the grout, and $f_y = 460 \text{ N/mm}^2$ for the reinforcement welded to the plate, cover to column bars = 35 mm, grade 43 base plate and grade 8:8 holding-down bolts.

### *Solution*

Try $500 \times 500$ mm plate with $L = 100$ mm overhang and $d' = \dfrac{L}{2} = 50$ mm.

**Fig. 7.96** Column-to-foundation connection with flush base plate.

$$e = \frac{150 \times 10^3}{100} = 1500 \text{ mm.}$$

Then    $K = \dfrac{100 \times 10^3 (1500 + 250 - 50)}{40 \times 500 \times 500^2} = 0.034$

hence $X = 0.1$.

Because $X < N/0.4 f_{cu}' \, b \, d = 0.025$ then $F$ is positive.

$$F = 0.4 f_{cu}' \, b \, X \, d - N = 400 - 100 = 300 \text{ kN.}$$

$$A_b = \frac{300 \times 10^3}{2 \times 450} = 333 \text{ mm}^2.$$

*Use two M 24 grade 8:8 holding-down bolts.*

$$t = \sqrt{\frac{0.8 \times 40 \times 100^2}{275}} = 34 \text{ mm.}$$

Before proceeding we need sizes of bars in column.

For $\dfrac{d}{h} = 0.80$, $\dfrac{N}{bh} = 1.11$, and $\dfrac{M}{bh^2} = 5.55$,

then from BS 8110, Part 3, Chart 47 [7.6], $A_{sc} = 0.04 \times 300^2 = 3600 \text{ mm}^2$.
*Use four no. T 32 plus 2 no. T 16 (3622) column bars.*
Then $m = 50 + \text{cover } 35 + \text{links say } 8 + 16 = 109 \text{ mm.}$

$$t = \sqrt{\frac{4 \times 300 \times 10^3 \times 109}{500 \times 275}} = 30.8 \text{ mm.}$$

(This is a well balanced design as the plate thicknesses for tension and compression are similar.)

*Use 500 × 500 × 35 mm mild steel base plate.*

### 7.14.3  Columns on grouted sleeves

The design of these joints is identical to the grouted sleeve splice described in Section 7.13.5. Full compression or tension anchorage lengths are provided in both the precast column and insitu foundation. High tensile deformed reinforcing bars are left protruding from the foundation, which is quite difficult to achieve with any accuracy (Fig. 7.97). This is the main disadvantage in using this connection. The tendency is for the bars to be touching the sides of the sleeves in the precast column thus preventing a full envelopment of grout around the bar. Also, the grouting cannot be inspected afterwards and so there is no guarantee that the bar is fully bonded.

However, notwithstanding these difficulties, because preventative measures can be taken, the joint may be considered in design as monolithic providing the bedding joint

**Fig. 7.97**  Projecting (starter bar) reinforcement at column foundation.

and grout sleeves are completely filled (by pressure grouting) to the satisfaction of the site engineer.

Work by Kuttab and Dougill [7.43], Somerville [7.45] and Korolev and Korolev [7.46] has shown axial force–moment interaction characteristics of the joint are equal to the column itself. An added bonus is that high tensile reinforcement is used throughout. The minimum internal diameter of the grout tube is 50 mm when using 25 mm diameter deformed bar. Nominal cover to the tube and the minimum distance between tubes should be at least 75 mm. Confinement links are provided in the usual manner using $\zeta = 0.11$.

# Chapter 8

# Designing for Horizontal Loading

*This chapter deals with the transfer of horizontal load from the floor plate at every floor level to the foundations. The stability of the structure and the methods of achieving it are discussed.*

## 8.1  INTRODUCTION

There is no doubt that precast multi-storey structures of up to about 50 m in height can be designed with economy, safety and excellent form. The main debate is the manner in which horizontal forces are transmitted through the components and their connections, particularly in very large precast structures where the connections become highly stressed. The obvious relationship between stability and details is evident from the photograph of The Bourse in Leeds, UK shown in Fig. 8.1. The tallest precast (wall frame) structure in Europe is the 54-storey building for Nationale Nederlanden in Rotterdam which consists of slip-formed insitu cores, precase load bearing walls and precast double-tee floor units.

This increasing awareness towards the structural integrity of structural connections in prefabricated construction has been where most of the recent research effort has been devoted. Most of the papers cited in this book are devoted either towards the structural integrity of the structure or of the connections. The details used to achieve robustness in a structure have a significant effect on the structural mechanisms by which horizontal forces are distributed, and vice versa.

Horizontal forces in precast structures derive mainly from wind loading and temperature gradients. In seismic zones earthquake loading is the predominant loading, but this is beyond the scope of this book. Secondary effects, such as linear changes in volume due to creep, tendon relaxation and shrinkage, etc., are mostly (but not totally) eliminated by precasting, where factory controlled moisture loss is more effective than on site. The low water content and the low water/cement ratio used in the predominantly horizontal components ensure that shrinkage effects are minimized. The practice in design offices is to restrict these secondary effects to individual joints between the precast components under consideration and not to allow them to accumulate over large areas of the structure.

With the exception of structural screeds, the use of large quantities of insitu concrete is confined to situations where shrinkage cracking and minimal creep strains can be tolerated and allowed for in design. In situations where this is not acceptable non-shrinkable concrete (or grout) is used.

Wind loading is in accordance with CP3, Part 2, Chapter 5 [8.1] and force coefficients for rectangular plan buildings are used for all building shapes. Although force and pressure coefficients are now available [8.2] for non-rectangular plan forms (L-, U-, H-shapes, etc.) and unclad skeletal (lattices), the information is not widely used at present. An ultimate horizontal loading of 1.5 per cent $G_k$ above each floor level replaces the ultimate wind force where appropriate. This takes into account the effects of lack of verticality in construction.

**Fig. 8.1** Construction of The Bourse, Leeds, UK (courtesy of Blatcon Ltd).

Two key factors need to be satisfied in design: horizontal load transfer by diaphragm action in the precast concrete floor plate (Sections 8.3 and 8.4), and stability in the vertical bracing elements (Sections 8.6 and 8.7). Diaphragm action may also be divided into two distinct avenues – behaviour with and without a structural screed. A general introduction to diaphragm action was given in Chapter 1, Section 1.4 and the reader should be familiar with this before proceeding.

## 8.2   DISTRIBUTION OF HORIZONTAL LOADING

The distribution of horizontal loading between stabilizing elements is determined from the position and stiffness of the bracing elements in the structure. The bracing elements may be either (in order of descending stiffness):

- precast components, such as walls, cores, or columns
- masonry infill walls
- moment resisting frames, e.g. continuous insitu framework
- steel cross diagonal ties

or some combination of each in the same structure. The relative stiffness of each bracing element determines the force carried by that element, and in turn defines the horizontal deflections in the floor plate.

Taking, for example, shear walls or cores as being the most popular type of bracing, and assuming that Young's modulus and shear modulus is constant for the materials used in the bracing, the stiffness of each unit is proportional to its second moment of area ($I$) in the uncracked condition (at the serviceability limit state). Thus the larger the $I$ value the greater will be the force required to cause horizontal displacement, and the greater will be the shear force in the floor slab adjacent to that wall or core. If the unit is more squat, i.e. its height to length ratio is small (less than about 2), shear deflections will govern and the response to loading will be a function of the plan cross-sectional area, rather than $I$.

Non-symmetrical walls (the vast majority) are considered for torsional stability, as shown in Fig. 8.2, although the disposition of columns is usually such that the shear centre of the system coincides fairly closely with the centre of the external wind pressure. However, for L-shape buildings shown in Fig. 8.3 where the eccentricity ratios $e_x/X$ and $e_y/Y$ exceed 0.15 the torsional equilibrium may increase individual floor diaphragm reactions by up to 50 per cent.

Any statical method may be used to determine these reactions, but a design method given by Pearce & Matthews [8.3] assumes that the floor plate is a rigid diaphragm and that the relative deflections of the bracing elements is proportional to the distance $a$ from the centroid of stiffness to the bracing. It also assumes that shear deflections are small compared with flexural deflections. Even though the distribution of forces will vary depending upon whether the response in a wall system is governed by flexure, shear or a combination of both, or if one part of the wall system reaches its elastic limit before another, only one parameter is used in the calculation of stiffness.

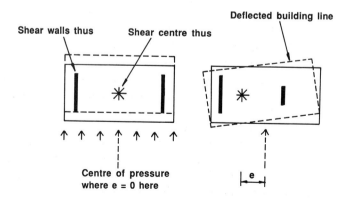

**Fig. 8.2** Definitions of floor plate and torsional stability in shear wall buildings.

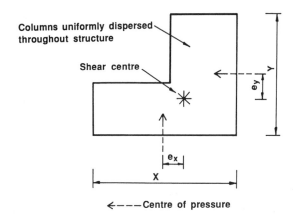

**Fig. 8.3** Eccentricity in shear wall systems.

In the following analysis the walls are assumed to be parallel with each other and to the direction of the load. Walls inclined to the direction of the load may be resolved into Cartesian components. Referring to Fig. 8.4 the cantilevered walls may be considered as linear springs of stiffness $8EI/L^3$, and the floor as a rigid beam. The floor plate reaction in each bracing element $H_1, H_2 \ldots H_n$ is given by the general expression:

$$H_n = \left[ \frac{I_n}{\Sigma I_i} + \frac{eI_n a_n}{\Sigma I_i a^2} \right] H \qquad (8.1)$$

where
$\quad H_n$ = reaction in wall $n$
$\quad I_n$ and $I_i$ = second moment of area in wall $n$ and in all walls
$\quad e$ = distance from the centroid of stiffness to the centre of pressure of externally applied wind load
$\quad a$ = distance from the centroid of stiffness of each bracing element
$\quad H$ = total reaction = total applied load.

Note that where there are only two walls the system is statically determinate.

If a wall is composed of cross walls, forming I-, T-, U- or L-configurations, as shown in Fig. 8.5(a), the inertia of the composite elements is used in place of the above providing the vertical joint at the intersections of the legs of the shape are capable of resisting the vertical interface shear force. If the walls are discrete components separated by columns (Fig. 8.5(b)) no interaction between the legs is considered.

In cases where 1.5 per cent $G_k$ is greater than the ultimate wind force, $e$ is taken as the distance from the centroid of stiffness to the centre of the dead load mass. This may be approximated from the summation of the centres of masses of the walls (external and internal) and floor slab at each level.

If the walls terminate at different floor levels the structural response of the building will be different above and below each floor level. The theory suggests that each change in the wall position will cause an immediate change in the response of the structure. In fact shear forces will develop in the floor plate and the changes will be gradual with the continuous walls taking a slightly larger share of the total force than Equation (8.1) gives.

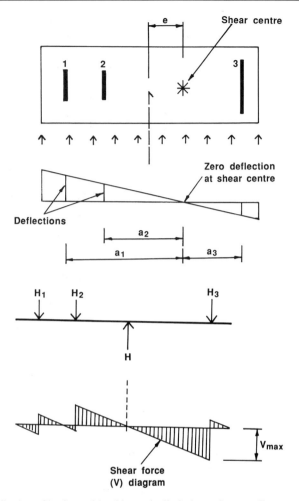

**Fig. 8.4** Distribution of horizontal load in statically indeterminate wall systems.

A similar situation occurs in buildings containing blocks of different height, as shown in Fig. 8.6. The reactions in the upper part of the taller block A must be transferred to the walls in the lower part of block A and to the walls in block B. In certain cases it is possible that the maximum wall reaction per floor will occur at an upper floor, possibly at the intersection of the two blocks, rather than at the foundation. However, the maximum wall shear and maximum overturning moment will always occur at the foundation.

This kind of problem is solved by dividing the elevation of the building into component rectangles, and using superposition to obtain the centroid of horizontal pressure above each floor level. There are two ways of subdividing the elevation – block by block or floor by floor. The latter is more convenient in the calculation of horizontal reactions because the eccentricity $e$ has a single value throughout. However, when using the method of superposition the overturning moment cannot be extracted from the analysis because the reactions differ between the walls in the various blocks in the upper and lower parts

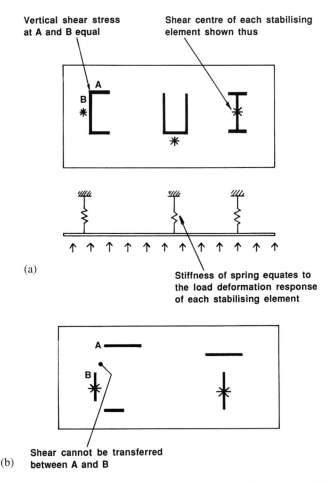

**Vertical shear stress at A and B equal**

**Shear centre of each stabilising element shown thus**

**Stiffness of spring equates to the load deformation response of each stabilising element**

(a)

**Shear cannot be transferred between A and B**

(b)

**Fig. 8.5** Distribution of horizontal load in two-dimensional wall systems. (a) Distribution in combined shear cores and (b) distribution in shear walls.

of the building. It is therefore necessary to determine the reactions at every floor level so that they may be summated at the foundation.

A specific problem occurs where one part of the building is very tall in relation to the remainder, for example in Fig. 8.7(a) where a one- or two-storey podium is connected to a multi-storey structure. At the level of the top of the podium the reactions in the walls in the tall building will suddenly be distributed through the floor plate as shear forces, moments and possibly torsion. It is better to structurally isolate the two parts of the building rather than risk possible floor plate cracking, as shown in Fig. 8.7(b).

Torsional effects in non-symmetrical shear wall systems may be balanced by walls at right angles (or near-right angles) to the direction of the load, as shown in Fig. 8.8. At least three walls are required, with at least two of the walls, called 'balancing walls', at right angles to the direction of the load. Provided there is shear continuity between the walls, any statical method may be used to determine the shear centre of the system, and the reactions in the balancing walls. If the walls are not connected to one another the

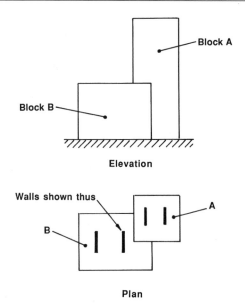

**Fig. 8.6**  Distribution of horizontal load in buildings of different height.

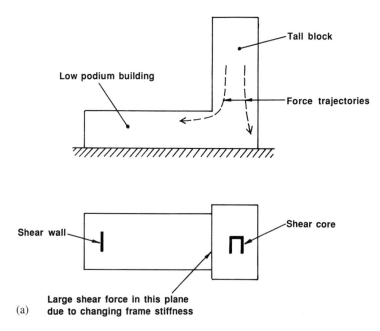

**Fig. 8.7**  Effects of different building heights. (a) Force trajectories between parts of building and (b) floor plate behaviour at narrow sections.

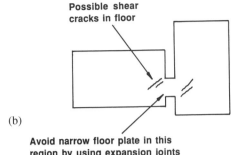

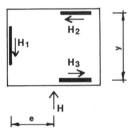

**Fig. 8.8** Shear system balanced by walls at right angles.

shear centre is taken at the centroid of the main wall parallel to the load. If this is a distance $e$ from the centre of pressure, then referring to Fig. 8.8:

$$H_1 = H, \text{ and } H_2 = -H_3 = \frac{He}{y} \tag{8.2}$$

Normally $y$ is at least 6 m. If there are more than two balancing walls the statical method given by Equation (8.1) may be used to determine the reactions in each of the balancing walls.

## EXERCISE 8.1   Distribution of horizontal loading

Calculate the percentage reactions in each of the shear walls in the building shown in Fig. 8.9. The walls are all of equal thickness and manufactured using the same concrete strength and amount of reinforcement.

### Solution

$I = L^3$ because $t, f_{cu}$ and $E$ are all constant.
Choose LHS as origin for $X = 0$, then referring to Table 8.1 calculate the wall stiffnesses.

$$\text{Shear centre distance} = \frac{2823}{77.9} = 36.24 \text{ m.}$$

Therefore $e = \dfrac{57.0}{2} - 36.24 = 7.74$ m.

The fraction of the total force in each of the walls is as follows:

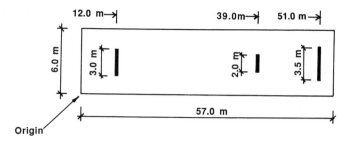

**Fig. 8.9** Details to Exercise 8.1.

**Table 8.1** Calculation for determining wall stiffnesses.

| Wall ref. | $I$ | $X$ | $IX$ | $a = x - \bar{x}$ | $Ia$ | $Ia^2$ |
|-----------|------|------|--------|--------------------|------|--------|
| A | 27 | 12 | 324 | 24.24 | 654 | 15 864 |
| B | 8 | 39 | 312 | 2.76 | 22 | 61 |
| C | 42.9 | 51 | 2 187 | 14.76 | 633 | 9 340 |
| Totals | 77.9 | | 2 823 | — | — | 25 265 |

- Wall A: $H_A = \left( \dfrac{27}{77.9} + \dfrac{7.74 \times 654}{25\ 265} \right) H = 0.550\, H = 55\%$

- Wall B: $H_B = \left( \dfrac{8}{77.9} - \dfrac{7.74 \times 22}{25\ 265} \right) H = 0.096\, H = 10\%$

- Wall C: $H_C = \left( \dfrac{42.9}{77.9} - \dfrac{7.74 \times 633}{25\ 265} \right) H = 0.354\, H = 35\%.$

Clearly the contribution of Wall B is small, but it may be significant in reducing diaphragm bending moments and shears as shown in Section 8.3.

**EXERCISE 8.2**

Repeat Exercise 8.1 omitting Wall B.

*Solution*

Taking moments about Wall A:

$$(51 - 12)\, H_C = \left( \dfrac{57.0}{2} - 12 \right) H$$

Then $H_C = 0.42\, H$, i.e. only an increase of six percentage points from Exercise 8.1. Also $H_A = 0.58\, H$, a three percentage point increase compared with Exercise 8.1. Thus, unless there are other requirements for the introduction of Wall B, this could be omitted from the shear system.

## EXERCISE 8.3

Repeat Exercise 8.1 omitting Wall C, and comment on the effectiveness of this new arrangement.

*Solution*

Taking moments about Wall A:

$$(39 - 12) H_B = \left(\frac{57.0}{2} - 12\right) H$$

Then $H_B = 0.61H$. This is a significant increase over the value $0.096\,H$ in Exercise 8.1, particularly for such a small wall. Because the two-wall system is independent of the size of the wall, the location of the wall should be consistent with its size, i.e. place larger walls near to the centroid of the pressure.

## EXERCISE 8.4   Distributions in buildings of different height

Calculate the horizontal forces in each of the shear walls in the y-direction for the braced structure shown in Fig. 8.10. Block A is four storeys tall, whilst Block B is six storeys and contains a $3 \times 3$ m lift and a 3 m high lift motor room. Each floor-to-floor level is 3.0 m.

Given that the design ultimate wind pressure is $0.8$ kN/m$^2$ and that the total characteristic uniformly distributed dead load is equivalent to $6$ kN/m$^2$ plan area, determine the maximum horizontal shear force, and the maximum average shear stress in the walls.

*Solution*

Using the block-by-block approach, and selecting the left-hand end of the building as zero datum, we have the following:

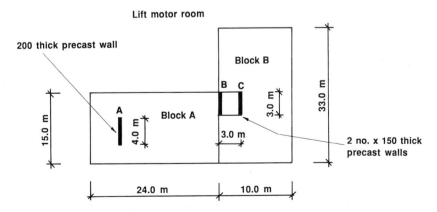

**Fig. 8.10** Details to Exercise 8.4.

## (1) Considering ultimate wind pressure

Total area of face of building in y-direction = $(24 \times 12) + (10 \times 18) + (3 \times 3) = 477$ m$^2$.

Design wind pressure = $0.8 \times 477 = 381.6$ kN.

Centroid of wind pressure $\bar{x} = \dfrac{[(288 \times 12) + (180 \times 29) + (9 \times 25.5)]}{477} = 18.67$ m from LHS.

## (2) Considering 1.5 per cent $G_k$

Total floor area = $(4 \times 24 \times 15) + (6 \times 10 \times 33) + (1 \times 9) = 3429$ m$^2$.

$$1.5\% \; G_k = \frac{1.5 \times 6 \times 3429}{100} = 308.6 \text{ kN} < \text{wind load.}$$

Continue analysis using wind load.

$I$ values in relative units = $t\,L^3$:  $I_A = 0.2 \times 4.0^3 = 12.8$ m$^4$

$\qquad\qquad\qquad\qquad\qquad I_B = I_C = 0.15 \times 3.0^3 = 4.05$ m$^4$.

Referring to Table 8.2, calculate the wall stiffnesses.

Shear centre distance = $\bar{x} = \dfrac{283.35}{20.9} = 13.55$ mm.

Therefore $e = 18.67 - 13.55 = 5.12$ m.

The fraction of the total force, and the design shear force in each of the walls, are as follows:

- Wall A: $H_A = \left( \dfrac{12.8}{20.9} - \dfrac{5.12 \times 96.64}{1904.6} \right) H = 0.353H = 134.7$ kN = maximum

- Wall B: $H_B = \left( \dfrac{4.05}{20.9} + \dfrac{5.12 \times 42.32}{1904.6} \right) H = 0.308H = 117.5$ kN

- Wall C: $H_C = \left( \dfrac{4.05}{20.9} + \dfrac{5.12 \times 54.47}{1904.6} \right) H = 0.342H = 130.5$ kN

where $H = 381.6$ kN.

The results show a well-balanced shear system with the shear forces in each wall varying over a small range.

The average maximum shear stress in each wall is given by $H/L\,t$ as follows:

- Wall A: $\tau_A = \left( \dfrac{134.7 \times 10^3}{200 \times 4000} \right) = 0.168$ N/mm$^2$

**Table 8.2** Calculation for determining wall stiffnesses.

| Wall ref. | $I$ | $X$ | $IX$ | $a = x - \bar{x}$ | $Ia$ | $Ia^2$ |
|---|---|---|---|---|---|---|
| A | 12.80 | 6 | 76.80 | 7.55 | 96.64 | 729.6 |
| B | 4.05 | 24 | 97.20 | 10.45 | 42.32 | 442.3 |
| C | 4.05 | 27 | 109.35 | 13.45 | 54.47 | 732.6 |
| Totals | 20.9 | | 283.35 | — | — | 1904.6 |

- Wall B: $\tau_B = \left(\dfrac{117.5 \times 10^3}{150 \times 3000}\right) = 0.261 \text{ N/mm}^2$

- Wall C: $\tau_C = \left(\dfrac{130.5 \times 10^3}{150 \times 3000}\right) = 0.29 \text{ N/mm}^2 = \text{maximum.}$

## 8.3  HORIZONTAL DIAPHRAGM ACTION IN PRECAST CONCRETE FLOORS WITHOUT STRUCTURAL SCREEDS

### 8.3.1  Background

The use of structural screeds in hollow core flooring systems has long been a controversial issue. The main emphasis of the debate has centred not around the vertical strength and stiffness characteristics of the floor, which is well documented, but on the horizontal behaviour of the floor acting as a plate or diaphragm as shown in Figs 8.11 and 8.12.

This debate has in part resulted from a lack of conclusive evidence of the real behaviour of the precast floor diaphragm, and a lack of understanding of the shear transfer mechanism between two precast units. Following the joint publication in the UK by the ICE and IStructE [8.4], which specified:

'In general it is advisable to use structural topping on precast floors so that the risk of cracking in the screed and finishes is minimised and the diaphragm action of the floor is ensured. This topping should include light fabric reinforcement.'

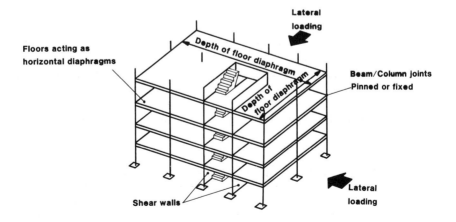

Diaphragm action of floors within a structure

**Fig. 8.11** Horizontal load transfer in hollow core floor slab.

**Fig. 8.12** Multi-bay hollow core floor prepared for insitu concrete ring beams to complete the diaphragm (courtesy of Reinforced Concrete Council).

There was conflicting opinion as to the use of an insitu rc structural screed to provide the necessary diaphragm action in the floor plate. Floor screeds of more than about 30 m length require contraction joints for anti-cracking due to shrinkage. This will create a plane of weakness where the shear can only be transferred through the precast slab. (New research is looking at the use of steel fibre rc for use in thinner (30 mm maximum) and larger area pours.) Staggered joints may be used in wider buildings, but this could be a crucial factor in certain situations. Structural screeds are known to be costly, time consuming and to counter the advantages inherent in a precast solution. However, their continued use has been based on the lack of understanding of an adequate shear transfer mechanism between adjacent hollow core floor units, which are joined together on site by placing rc infill at the ends of the precast units and insitu grout or small aggregate concrete between them.

The precast industry was faced with having to justify the use of a wholly precast hollow core floor diaphragm with no evidence of its actual behaviour. To add further confusion to the matter the Institution's Manual [8.4] was later revised to state that structural screeds were *not* necessary. In the meantime, however, the seeds of doubt had been sown and considerable resistance to the use of an unscreeded precast floor was encountered by the precast industry. A library computer search carried out found less than six directly related publications (e.g. [8.5 and 8.6]) and, prior to 1988, the only attempt to study the behaviour of a complete floor diaphragm using this type of proprietary unit was by Stroband & Kolpa (1/5 scale) [8.7], Sarja [8.8] and Svensson *et al.* [8.9]. Since then Stroband [8.10] has carried out some finite element work and Menengotto [8.11 and 8.12] some experimental studies on undulating edge profiles, to try to clarify how the precast diaphragm actually works.

Diaphragm action is not considered without a structural screed in floors comprising precast double-tee slabs where the maximum depth of flange is 50 to 75 mm. Although double-tee slabs are intermittently welded together longitudinally and at their ends to beams (see Section 5.3.2), which would provide the necessary shear and tensile forces in the plate, diaphragm action is not considered in large floors because of the inability to generate the compression or maintain a uniform centroidal axis across contiguous floor panels. Also, it would not be wise to rely entirely on welded connections for structural integrity.

Similarly, diaphragm action is not considered in unscreeded precast beam and block flooring for the obvious requirement of achieving shear transfer through the infill blocks. Diaphragm action is considered as being fully effective in the precast composite plank types of floors (Figs 5.1(d) and (e)), with no additional measures being taken other than to check the shear strength of the diaphragm at the critical positions, e.g. near to large voids, or narrow floor widths.

### 8.3.2   Details

The most important feature for ensuring horizontal diaphragm action in a hollow core floor slab is the edge profile, shown in Fig. 5.9. The edge is not made deliberately rough, but the drag of the casting machine on the semi-dry mix creates a surface roughness vital to diaphragm action. These units have edge profiles which permit the placement of insitu concrete in the longitudinal joints between adjacent units. Although different grades of insitu concrete are used in practice, the lowest strength is usually 25 N/mm$^2$ (see Table 4.3). Despite a slight roughening of the edges during the manufacturing process, these joints may be considered plain (i.e. uncastellated) and unreinforced.

It is also essential that the width of the gap between the units is large enough to permit the placement of properly compacted insitu concrete (or a sand/cement grout). Manufacturers have responded to this requirement in the last ten years by increasing the size of the former at the top edges of the units from about 10 to 20 mm (see Fig. 5.9). This improved profile is also beneficial in the lateral distribution of vertical shear forces, moments and deflections. The integrity of the floor system in the vertical plane ensures the success of the floor acting as a horizontal diaphragm.

Thus, the precast floor unit has a large in-plane stiffness, and the flexural deformations between the discrete precast slabs are small. The design of a diaphragm is therefore essentially a joint design problem as explained in Fig. 8.13. The essential features in the construction of a typical hollow core floor are:

- The precast units should be placed side by side such that there is no 'appreciable' gap between the units to allow shear-tension failure in the unreinforced insitu concrete. The size of this gap is not known, but its existence is thought to alter the shear failure mechanism by allowing diagonal tension to develop across gaps of more than 60 mm [8.13].
- Lateral tie steel, in the form of a continuous piece of 7-wire helical prestressing strand, or lapped high tensile deformed bar, is placed into a narrow (typically 50–80 mm wide) insitu concrete strip between the ends of the flooring units and supporting beams. This reinforcement, shown in Fig. 8.14, which is actually provided for stability purposes, provides the necessary tie forces required to prevent the precast slabs from moving apart sideways.

- L- or U-shaped shear friction bars are placed into insitu concreted slots formed in the broken out cores (sometimes referred to as 'milled slots') of the hollow core units, or placed into the longitudinal joints between the units. These bars also prevent longitudinal shrinkage movement, particularly if the precast slabs have been delivered to site at an early age, say three days. Shrinkage movements of 2–3 mm per 10 m long slabs have been seen where the slabs were seated but not tied to the supporting member.

### 8.3.3  Structural models for diaphragm action

Horizontal loads due to wind (or earthquakes) are transmitted to shear walls or moment

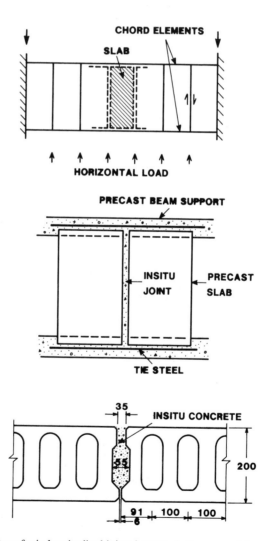

**Fig. 8.13**  Shear transfer in longitudinal joints between hollow core slabs.

resisting frames by considering the roof or floor, comprising individual precast concrete units, as a deep horizontal beam, truss or girder (Fig. 8.15). The structural walls or frames are supports for this analogous beam and the lateral loads are transmitted to these supports as reactions $H_n$ in Equations (8.1) and/or (8.2). The magnitude of the shear force at the wall may therefore be determined. The effective depth of the diaphragm is equal to the depth of the floor slab, but the part of the beam which is below the floor plate is ignored.

Various structural models have been proposed for the shear transfer mechanism, but the most critical situation is where the floors span parallel with the supporting shear walls. The web shear must also be transferred to the chord elements, i.e. the rc edge or spine beams spanning perpendicular to the floor units. This is commonly known as horizontal 'Vierendeel action' in which the Vierendeel frame components are simulated by insitu concrete strips peripheral to the precast slab units. In longer floors, where the length-to-breadth ratio $L/B$ is greater than about 5 or 6, the structural mechanism is probably closer to a deep beam than a Vierendeel girder. However, the problem is always one of shear transfer in the joints irrespective of the assumed model.

Equilibrating shear parallel with the supporting beam is determined using any statical method. In multi-bay floors, where the slabs are spanning parallel to the direction of the applied load, as in Fig. 8.16(a), the shear $V_h$ at interior support between span $L_1$ and $L_2$ is given by:

$$V_h = \frac{VS}{I} = \frac{6V(B-L_1)L_1}{B^3} \tag{8.3}$$

for $L_1 > L_3$, where $S$ and $I$ are the first and second moments of area at the interface.

Where the slabs are spanning perpendicular to the direction of the applied load,

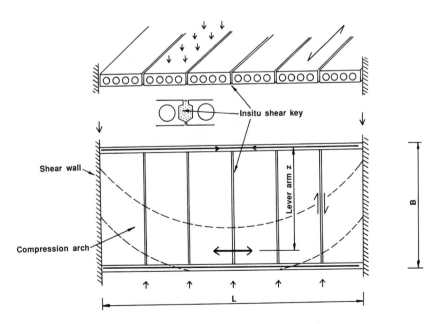

**Fig. 8.14** Peripheral tie steel is also used to ensure diaphragm action.

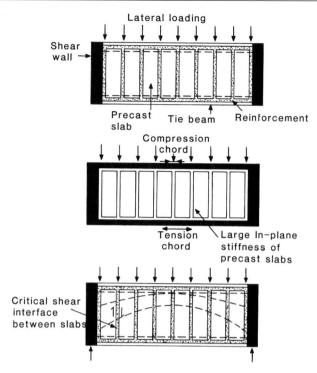

**Fig. 8.15** Structural models for floor diaphragm action.

Fig. 8.16(b), the maximum transverse shear between the slabs $V_h$ occurs at the neutral axis and is given by:

$$V_h = \frac{VS}{I} = 1.5\,V \tag{8.4}$$

The shear force $V_h$ is carried by dowel action using shear friction reinforcement, often called 'coupling bars', as indicated in Fig. 8.17. The purpose of these bars is to transfer shear, caused by small lateral rigid body translations in the slab, into the chord elements. A tie force is thus generated in the reinforcement in the beam and/or insitu perimeter strip. It is becoming increasingly popular to reinforce the insitu perimeter strip using 7-wire helical prestressing strand. The favourable mechanical properties of this type of strand and long lengths available on site make it an attractive alternative to high tensile bar.

The ties prevent the slabs from moving apart and simultaneously generate the clamping forces that create friction so essential to diaphragm action. In this manner, all the requirements specified in BS 8110, Part 1, clause 5.3.7(a) [8.14] are satisfied thus:

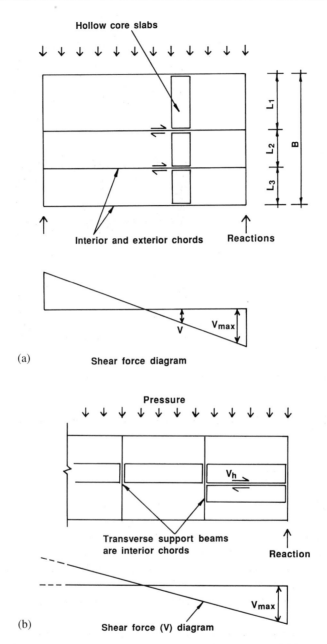

**Fig. 8.16** Shear forces along internal beams in floor diaphragm. (a) Slabs spanning parallel to wind loading and (b) slabs spanning perpendicular to wind loading.

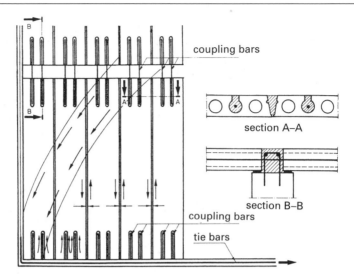

**Fig. 8.17** Tie bars cast into floor slots act as complementary shear friction reinforcement at the ends of the slabs [8.47].

> '... these should be restrained to prevent their moving apart. No reinforcement need be provided in or across the joint, and the sides of the units forming the joint may have a normal finish ... .'

An ultimate interface shear stress of 0.23 N/mm$^2$ is therefore used in design. The design clauses in BS 8110 are based on work carried out by Mast [8.15]. In computing the magnitude of the interface shear stress $\tau = V/BD$, the depth of the unit can only be taken to the depth of the insitu/precast interface, i.e. $D - 30$ mm in most types of hollow core slabs. The reason for this derives from observations made of the compaction of the concrete in the joint. It is found that grout loss occurs in the bottom of the joint and that the lower 10 to 15 mm is not grouted. Also, the lip at the bottom of the units prevents full penetration. The value also recognizes that differential camber will be present, further reducing the net contact depth.

The interfaces between the hcus and the insitu joint concrete are most likely to be the planes of cracking. Cracks occur due to the shrinkage of the insitu concrete, or by the restraint in the precast floor system. In practice, cracks up to 1 mm wide have been found, as shown in Fig. 8.18, and it is assumed that in the majority of practical cases cracks exist in the joints between precast floor units. This is obviously detrimental to the shear transfer behaviour of the floor slab and as such the shear must be transferred in cracked concrete. In this case the shear resistance is a combination of [8.16–8.18]:

- aggregate interlock in cracked concrete, by wedging action and shear friction
- dowel action through kinking and shear capacity
- axial restraint stiffness provided by reinforcement perpendicular to the cracks as shown in Figs 8.14 and 8.19.

If the shear force $V$ and the shear slip $\delta$ in Fig. 8.19 are plotted out, the relationship would be similar to the graph in Fig. 8.20(a). The gradient of the $V$–$\delta$ curve gives the

**Fig. 8.18** Interface shrinkage cracks between adjacent hollow core floor units. (The coin is a British penny measuring 20 mm diameter.)

stiffness $K_s$. Similarly as the shear force is applied to the cracked interface the two halves of the specimen would begin to move apart, given by $c_w$ (as this is in fact a crack width). The relationship between $\delta$ and $c_w$ is shown in Fig. 8.20(b).

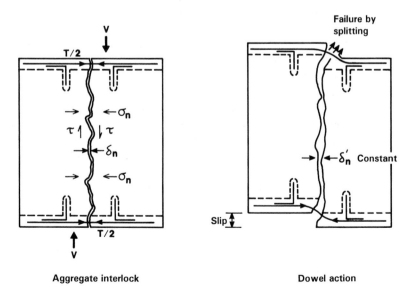

**Aggregate interlock**                    **Dowel action**

**Fig. 8.19** Shear transfer mechanisms of aggregate interlock and dowel action between discrete precast units.

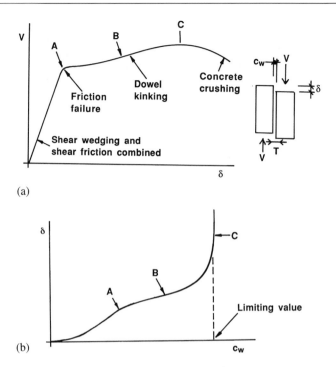

**Fig. 8.20** Idealized relationships in longitudinal slab joint. (a) Shear force vs. shear slip, (b) shear force vs. crack width and (c) shear force vs. transverse tie force.

Shear strength and stiffness is provided by aggregate interlock, and the structural integrity by dowel action of the reinforcing bars crossing the cracked interface. Aggregate interlock may be separated into two distinct phases, namely 'shear wedging' where the inclined surfaces either side of the crack are in contact, and 'shear friction' where the contact surfaces are being held in contact by the normal stress. Shear wedging provides a very high resistance, but cannot be relied upon for very long if the interface cracks are wide.

As the crack width increases, tensile forces $T$ are mobilized in the reinforcing bars, as shown in Fig. 8.20(c). To understand this a simple analogy is to slide one's knuckles over each other with an elastic band stretched around the fists. As the knuckles slide the elastic band tightens and the resistance to movement increases. This is the shear mechanism that is active in a hollow core floor diaphragm.

Cholewicki has developed a mathematical model to predict the strength and stiffness of the longitudinal joint between hcus based on shear friction and dowel action hypotheses [8.19]. Given this tool and a set of geometric and material properties it is possible to predict shear behaviour by calculating the stiffness $K_s$ and the strength $V$.

The floor plate is also subjected to horizontal bending where, as shown in Fig. 8.14, internal equilibrium is maintained by tension and compression chords. The tie force $T = M_h/z$ where $M_h$ is the applied diaphragm moment and $z$ the lever arm. $z$ depends on the aspect ratio for the floor, and on the magnitude of the bending moment. Maximum values for $z/B$ at the points of maximum bending are as follows [8.20]:

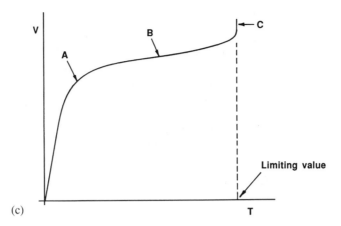

(c)

**Fig. 8.20** (*continued*)

| B/L | z/B |
|-----|-----|
| <0.5 | 0.9 |
| 0.5 < 1.0 | 0.8 |

Walraven [8.21] proposes a constant value for $z/B = 0.8$.

Where the aspect ratio $B/L > 1$ the behaviour will be closer to the strut and tied arch model than either the deep beam or truss models. The tie force is given by:

$$T = \frac{0.5V}{B/L} \qquad (8.5)$$

as a diagonal compressive strut develops across the floor.

### 8.3.4   Diaphragm reinforcement

The maximum horizontal bending moment $M_h$ is calculated from equilibrium of external wind pressures and the reactions from the shear walls obtained from Equation (8.1). Referring to Fig. 8.14 and given that the breadth of the diaphragm is $B$, the diaphragm reinforcement $A_{hdl}$ to be positioned in the chord elements over the tops of the beams at the ends of the floor slab is determined from the following expression:

$$A_{hdl} = \frac{M_h}{0.8\,B\,0.87\,f_y} \qquad (8.6)$$

where the term $0.8\,B$ is the lever arm between the compressive zone and the diaphragm steel. $A_{hd}$ should not exceed:

$$A_{hdl} = \frac{0.45\,f_{cu}{}'}{0.87\,f_y}\,0.4dB \qquad (8.7)$$

where $d$ is the depth of the precast floor slab $= D - 30$ mm, and $f_{cu}'$ is the strength of the insitu grout (or concrete) in the longitudinal gaps between the floor units.

It is necessary to collect the floor diaphragm tie steel $A_{hdl}$ in the chord elements. This may be achieved as follows:

- By utilizing the reinforcement already provided in the chord elements, such as edge beams, and by providing a positive non-slip tie between the beams. This tie may be continuous through the column, as shown in Fig. 7.42(a) for the billet connector, or may be provided as shown in Fig. 7.42(b) by using steel dowels in shear.
- By placing tie steel additional to the steel provided in the chord members. (This steel may also be used as part of the stability tie steel determined in Section 9.4.) Some designers prefer to pass the reinforcement through small holes in the column. Alternatively if the beam is wider than the column the bars are placed symmetrically on either side of the column.

Diaphragm reinforcement may be curtailed according to the usual rules governing lap lengths and anchorage. In many cases the point of maximum shear will coincide with minimum bending and therefore the full length of the slab $B$ may be used in computing the average value for $\tau$. However where the maximum moment and shears coincide the breadth of the diaphragm is reduced to $z$ to allow for the decay in shear stress in the compression zone. Thus:

$$\tau = \frac{V}{z(D-30)} \text{(N mm units only)} < 0.23 \text{ N/mm}^2 \tag{8.8}$$

If $\tau$ exceeds this value the shear force must be resisted entirely by shear reinforcement placed across the ends of the slabs as explained above. The steel area $A_{hd2}$ is given in BS 8110, Part 1, clause 5.3.7[d] [8.14] as follows:

$$A_{hd2} = \frac{V}{0.6 \times 0.87 f_y \tan a_f} \tag{8.9}$$

where $\tan a_f$ is the coefficient of friction as given in Table 5.3 of the code. Hollow core slabs are considered as being untreated and smooth, hence $\tan a_f = 0.7$. Thus:

$$A_{hd2} = \frac{2.74 V}{f_y} \tag{8.10}$$

Thus the total steel $A_{hd} = A_{hdl} + A_{hd2}$ may be determined. The minimum tie steel $A_{hdmin}$ (mm$^2$) is provided as a perimeter tie and given as follows:

$$A_{hdmin} = \frac{40\,000}{f_y} \text{ with } f_y \text{ in N/mm}^2 \tag{8.11}$$

## EXERCISE 8.5    Diaphragm reinforcement in floor slab

If the ultimate horizontal wind pressure acting on the floor slab used in Exercise 8.1 is 4 kN/m, calculate the maximum horizontal bending moment and shear force, and determine the floor diaphragm reinforcement. Check that the interface shear stress is no greater than the design value given in BS 8110 [8.14]. The overall depth of the precast floor slab is 150 mm. Choose a suitable position to curtail the chord reinforcement.

Use $f_{cu} = 25$ N/mm$^2$ for the insitu infill, and $f_y = 460$ N/mm$^2$.

**Solution**

Total wind force $W = 4 \times 57 = 228$ kN. Wall reactions from Exercise 8.1 enable the horizontal shear force and bending moment diagrams to be drawn as shown in Fig. 8.21.

*At Point A* where the moment is maximum and the shear is zero:

Flexural steel lever arm $z = 0.8 \times 6.0$ m $= 4.8$ m.

$$F_s = \frac{M_{max}}{z} = \frac{461}{4.8} = 96.04 \text{ kN}$$

$$A_{hd1} = \frac{96.04 \times 10^3}{0.87 \times 460} = 240 \text{ mm}^2$$

*Use 3 no. T 12 rebars continuous reinforcement.*
Check compressive stress in insitu concrete, where the depth of the floor slab = $D - 30$ mm.

$$F_c = 0.45 f_{cu} d 0.4 B = 0.45 \times 25 \times (150 - 30) \times 0.4 \times 6000 \times 10^{-3}$$
$$= 3240 \text{ kN} > 96.04 \text{ kN required. Therefore satisfactory.}$$

*At Point B* where the moment and the shear combine:

$$F_s = \frac{M_B}{z} = \frac{288}{4.8} = 60.0 \text{ kN}$$

$$A_{hd1} = \frac{60.0 \times 10^3}{0.87 \times 460} = 150 \text{ mm}^2 < \text{Point A.}$$

Check compressive stress in insitu concrete, where the depth of the floor slab = $D - 30$ mm and $z = 4800$ mm.

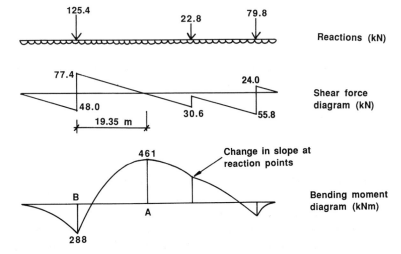

**Fig. 8.21** Details to Exercise 8.5.

$$F_c = 0.45 \times 25 \times (150 - 30) \times 0.4 \times 4800 \times 10^{-3} = 2592 \text{ kN} > 60 \text{ kN required.}$$

Therefore satisfactory.

Shear stress at $B$: $v_{ave} = \dfrac{77.4 \times 10^3}{120 \times 4800} = 0.134 \text{ N/mm}^2 < 0.23 \text{ N/mm}^2$ permissible,

hence $A_{hd2} = 0$.

*Use 2 no. T 12 bars.*

Chord reinforcement: $A_{ndmin} = \dfrac{40\,000}{460} = 87 \text{ mm}^2$.

*Use 1 T 12 bar (113) minimum.*

Hence steel may be curtailed when $M = 0.87 \times 113 \times 460 \times 4800 \times 10^{-6} = 216 \text{ kN m}$. The position may be obtained from the bending moment diagram. (It is quite likely that the designer would choose to continue the three bars along the full length of the building to avoid confusion and the possibility of misplacing the reinforcement.)

### 8.3.5   Design by testing

There has been extensive experimental work describing the shear transfer mechanism in cracked rc, and the formulation of some basic relationships between the dominant material and geometrical effects which are based mainly on experiments using small specimens [8.22, 8.23]. However, there is a lack of experimental data on the monotonic and/or cyclic behaviour of large scale units, and on the correlation between the small scale testing and the full size diaphragm.

Recent tests carried out by Davies, Elliott and Omar [8.24–8.26] on full scale 200 mm deep hcus, shown in Fig. 8.22, found that the attainable horizontal shear stress in the floor slab exceeded the permissible design stress for unreinforced uncastellated joints. The laboratory tests were carried out in a 'realistic environment' regarding materials specification, geometry and on-site practice.

The results have shown interface shear stresses are in excess of working load by a factor of at least 2.15, despite the presence of initial cracks in the interface up to 0.55 mm wide. The working load is calculated by multiplying the working stress by the net contact area in the longitudinal joint. The working stress is defined as the ultimate shear stress of 0.23 N/mm$^2$ divided by $\gamma_f$ of 1.4 for wind loading. Clamping forces normal to the precast units resulted in coefficients for the shear wedging and shear friction $T/V$ of at least 5.

It is also known that the behaviour of the interface depends on the roughness of the sides of the precast units. Physical measurements have been made on factory produced units to quantify this parameter [8.27]. The surface roughness was measured using an instrument which was placed on to the sides of slip formed units to measure roughness along a sampling length of 100 mm, as shown in the tracing in Fig. 8.23. The roughness factor is determined in accordance with BS 1134 [8.28]. This is represented by $R_a$, the arithmetical mean deviation of the edge profile (in mm), which is the average value of departure of the profile above and below the mean centre line throughout the sampling length. Values for $R_a$ in typical hollow core production are between 0.2 and 0.3 mm.

Tests were carried out on slabs selected with different surface textures, in a manner similar to Fig. 8.22. The specimens were grouted together in the longitudinal joint using

**Fig. 8.22** Experimental set up by Omar. (The precast floor slabs were manufactured by Bison Floors Ltd, UK.)

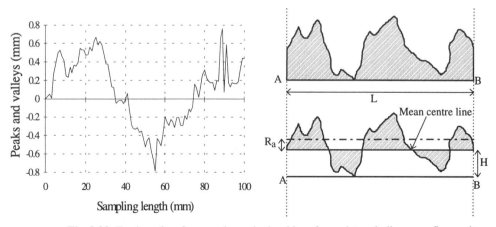

**Fig. 8.23** Tracing of surface roughness in the sides of proprietary hollow core floor units.

insitu concrete of 27 N/mm$^2$ cube strength and 10 mm aggregate. The behaviour was linear up to about $\frac{2}{3}$ of the ultimate load. The crack width at the ultimate shear load increased with the increasing roughness factor. This indicates that the interlocking mechanism is dependent on the texture of the adjacent surfaces. The peaks and valleys of the interfaces lock the opposite surfaces until a critical crack width is reached, usually at about 2.5 mm, after which a shear friction failure occurs. The resulting shear stress $\tau_{ua}$ (N/mm$^2$) may be expressed in terms of the roughness factor $R_a$ (mm) as follows:

$$\tau_{ua} = 0.22 + 0.207 \, R_a \tag{8.12}$$

The results in Fig. 8.24 show that for a smooth concrete surface with a roughness factor of zero, the extrapolated shear stress is 0.22 N/mm$^2$, i.e. just less than the BS 8110 design value of 0.23 N/mm$^2$ [8.14].

Designers also assume that the floor plate undergoes negligible shear deflection, i.e. the longitudinal shear stiffness is very large. In reality it is in the order of $K_s = 500$ kN/mm to 1000 kN/mm for 4 m long × 200 mm deep slabs [8.25]. Based on this experimental evidence the maximum horizontal shear deflection between these slabs at the design working shear stress of $0.23/1.4 = 0.16$ N/mm$^2$ would be approximately 0.2 mm. If this deflection was summated over a building length $L$ (m) a total shear deflection at the mid-point would be $0.05L$ (mm).

A further deleterious factor is that shear deflections of this nature are not recoverable. Figure 8.25(a) (from [8.25]) shows residual shear deformations of between 0.1 and 0.2 mm, equivalent to a shear strain of about 50 με after several completed ultimate load cycles. However when cycled at the working load the residual shear strain is less than 30 με (Fig. 8.25(b)). Dislocations and abrasion across the tops of the asperities in cracked concrete are responsible for these permanent deformations. The resulting degradation in stiffness is about 50 per cent of the initial monotonic value.

Actual stresses at the interfaces between slabs and walls depend on the geometry of the building and the number of walls, but typical ultimate values (in the region of 0.05 to 0.20 N/mm$^2$) are less than the permissible design ultimate stresses. A rigorous analysis on the hollow core floor diaphragm at the building shown in Fig. 8.26 and known as 'The Ark' (at Hammersmith, London) determined maximum serviceability values of about 0.13 N/mm$^2$ in the longitudinal joints [8.29].

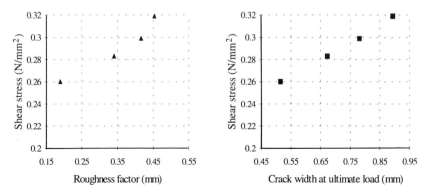

**Fig. 8.24** Effect of surface roughness and final crack width on interface shear stress capacity between hollow core units.

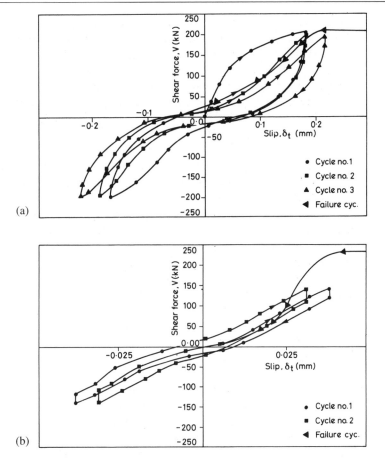

**Fig. 8.25** Results of cyclic load tests by Omar at Nottingham University. (a) Tested to ultimate load magnitude and (b) tested to working load magnitude. (Conversion factor: 200 kN force = 0.36 N/mm$^2$ stress.)

Sarja [8.8] tested a large floor slab, shown in Fig. 8.27, consisting of 5 no. × 1.2 m wide × 265 mm deep hollow core slabs. The slabs were tied together using an insitu ring beam containing either 2 no. 10 mm or 2 no. 16 mm diameter rebars. The average strengths of the materials were $f_{cu}$ = 21 to 24 N/mm$^2$ for the added concrete, and $f_y$ = 420 and 475 N/mm$^2$ for the 10 and 16 mm bars, respectively.

Shear tests were also carried out on pairs of slabs of equal size to the above. The tie steel in the ends of the slab was 2 no. 16 mm bars in one end and 1 no. 8 mm bar in the other end. No details are given in the report as to the effect of these differences.

The slabs were loaded in three-point bending over a span of 5.7 m (Tests 1–6) and pure shear (Tests 7 and 8). The resulting average shear stress and bending moment at failure are given in Table 8.3.

The results show that where the tie steel is 2 no. 10 mm diameter bars, and the tensile strength of these bars is only 64.4 kN, the lowest recorded shear stress is 0.18 N/mm$^2$. Omar [8.26] found that the tensile tie capacity of the steel should be at least 115 kN in order to generate shear stresses of 0.25 N/mm$^2$ or greater. Sarja's results confirm this;

**Fig. 8.26** Hollow core floor slabs used in The Ark, Hammersmith, London (courtesy of Bison Floors).

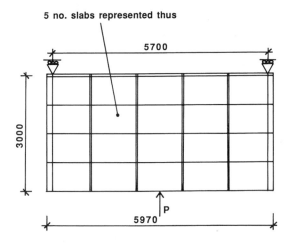

**Fig. 8.27** Test arrangement by Sarja [8.8].

when the tie steel force in Tests 5 and 6 was at least 175 kN the failure shear stress was 0.32 N/mm$^2$. The results for the pure shear tests are clearly greater than in the case of combined bending and shear.

### 8.3.6 Finite element analysis of the floor plate

Several attempts have been made by researchers to predict the behaviour of precast

**Table 8.3** Test results on hollow core floor diaphragm in bending and/or shear [8.8].

| Test ref. | Loading regime | Tie steel bars (mm) | Shear force (kN) | Average shear stress (N/mm$^2$) | Bending moment (kN m) |
|---|---|---|---|---|---|
| 1 | Three-point | 2 × 10 | 140 | 0.18 | 400 |
| 2 | bending | | 180 | 0.23 | 513 |
| 3 | | | 160 | 0.20 | 456 |
| 4 | | | 140 | 0.18 | 400 |
| 5 | Three-point | 2 × 16 | 247 | 0.32 | 705 |
| 6 | bending | | 287 | 0.37 | 819 |
| 7 | Pure shear | 1 × 8 | 406 | 0.52 | — |
| 8 | | | 382 | 0.49 | — |

diaphragms composed of discrete elements. Early attempts to simulate behaviour were hampered by the large number of variables to be studied and the results resembled no more than the displacements between solid rectangular blocks connected by springs. The introduction of finite elements into structural analysis in the late 1970s enabled the modelling of both the precast concrete units and the interface between them.

The pioneering work by Sarja in 1978 [8.8] showed the early potential for these techniques but unfortunately the input data relied heavily on known values, somewhat defeating the objective of the work. The numerical work modelled the experimental test slab shown in Fig. 8.27. The most interesting results from this work were the variations of the interface shear stress in one of the longitudinal joints. Figure 8.28 shows these distributions for three-point bend tests. The greatest values are close to the application point of the load. It is clear that these large stresses are due to compressive strut action

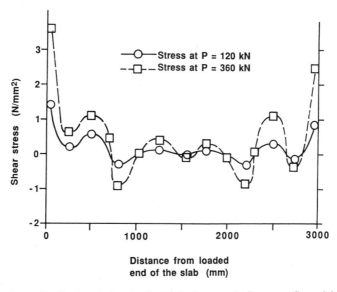

**Fig. 8.28** Stress distributions in longitudinal joint between hollow core floor slabs [8.8].

and some arching stresses are being interpreted as shear. The reduction in shear to zero at $\frac{2}{3}$ across the slab suggests the presence of flexural cracking, although why the shear stress increases again is unclear, but probably not important.

Recently de Roo & Straman at Delft University of Technology [8.30] used the finite element package DIANA to simulate all of the precast units, the longitudinal mortar joints, and transverse tie beams. Using non-linear interface slip surfaces to replicate the longitudinal shear–slip characteristics measured in tests, the relationship between the applied bending moment and the horizontal deflection of the diaphragm is shown in Fig. 8.29. The deflection at the working load was about 1 mm, and the span-to-deflection ratio at failure was 1/4000. The results have shown that the floor diaphragm failed due to yielding of the ties in the edge beam with a factor of safety of 3.13 with respect to the design strength. The compressive stresses were formed in a triangular distribution and concentrated across 2.5 m of the floor, giving a lever arm factor of $z/B = 0.97$. This suggests that the proposed design lever arm factor of 0.8 is very conservative.

The maximum compressive stress obtained in the modelling was 3.6 N/mm$^2$ and the maximum shear stress in the longitudinal joint was 0.4 N/mm$^2$. The final deformation pattern is shown in Fig. 8.30 where the critical longitudinal joint is not at the support but at the second interface between the slabs.

## 8.4 DIAPHRAGM ACTION IN COMPOSITE FLOORS WITH STRUCTURAL SCREEDS

Composite floor systems are designed on the basis that the precast flooring units provide

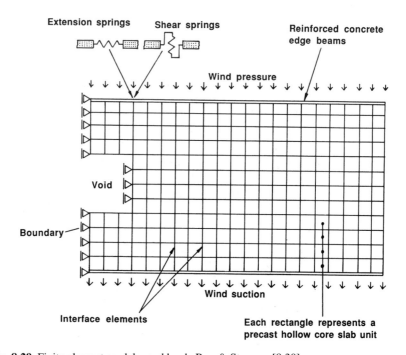

**Fig. 8.29** Finite element models used by de Roo & Straman [8.30].

restraint against lateral (in this case vertical) buckling in the relatively thin screed. In other words the precast floor is acting as permanent shuttering. The shear is carried entirely by the reinforced insitu concrete screed (the welded connections between double-tee units are ignored (see Fig. 1.27). The minimum thickness of structural screed is 40 mm. The design ultimate shear stress is taken as at least 0.45 N/mm$^2$ for grade C25 concrete (BS 8110, Part 1, Table 3.9 [8.14]) and the effective depth of the screed is measured at the crown (thinnest part) of the prestressed flooring unit.

Continuity of reinforcement in structural screeds is always extended to the shear walls or cores and it is safe to assume that the shear capacity of insitu diaphragms will not be the governing factor in the framing layout. Designers are careful not to allow large voids near to external shear walls, and to ensure that if an external wall adjacent to a prominent staircase is used then a sufficient length of floor plate is in physical contact with the wall.

## EXERCISE 8.6

Calculate the horizontal shear force capacity in the y-direction in the screeded double-tee floor slab shown in Fig 8.31 if the maximum thickness of screed is 75 mm. The precamber in the precast unit may be taken as span/400. Use $f_{cu} = 25$ N/mm$^2$.

## Solution

The maximum shear force occurs at a point of zero bending at the interface of the floor slab and shear wall. Maximum thickness of screed = 75 mm.
Area of mesh = $0.15\% \times 75 \times 1000 = 112.5$ mm$^2$/m.
*Use A142 mesh.*
Minimum depth of screed = $75 - (9000/400) = 52.5$ mm.
Minimum slab-to-slab contact length = 9 m.
Effective area of screed = $9000 \times 52.5 = 472 \times 10^3$ mm$^2$.

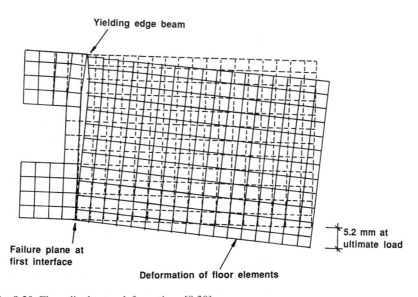

**Fig. 8.30** Floor diaphragm deformations [8.30].

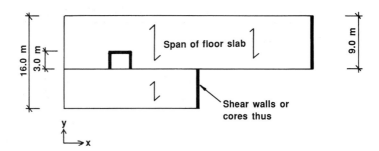

**Fig. 8.31** Details to Exercise 8.6.

Average shear stress for grade 25 concrete containing 0.15% reinforcement = 0.45 N/mm$^2$.
*Shear force = 212 kN.*

## 8.5 HORIZONTAL FORCES DUE TO VOLUMETRIC CHANGES IN PRECAST CONCRETE

The effect of shrinkage and temperature changes during the cement hydration time period is less significant in precast concrete than in cast insitu work. Early age deformations are unrestrained and therefore of little importance. In addition, precast connections are often designed such that volumetric changes do not generate large internal forces in the structure by assuming that once frictional forces are exceeded the components are allowed a very small free movement between one another before the insitu rc strips accommodate the tensile forces. Skeletal structures are also sufficiently flexible to accommodate such movements.

In calculating volumetric changes in precast concrete the two major obstacles in using standardized data intended for insitu concrete production are:

- the volume-to-surface area of many precast components is lower than normal, for example in double-tee units where the exposed surface area is large
- water content in many precast components is low, e.g. prestressed components such as hollow core slabs.

Hence measured values tend to be used in place of general data. Table 8.4 gives specific information on shrinkage, temperature and elastic shortening obtained from the PCI Manual [8.31]. The axial strain induced from volumetric changes in a range of precast concrete components is given in Table 8.4.

The data include elastic shortening strains in psc which equate to about 0.00035 (350 µε). Data collected by Bensalem [8.32] on the lateral (i.e. side-to-side) shrinkage in extruded hollow core slabs found that two-day old units shrank 0.2 mm per 600 mm width after several months exposure to moderate climatic conditions in the UK. The shrinkage strain is therefore $\varepsilon_{sh} = 0.00033$.

Horizontal forces and column bending moments due to drying shrinkage or temperature movement are shown in Fig. 8.32. The deformation is given by:

**Table 8.4** Approximate axial volume changes for unrestrained precast concrete elements.

| Volume-to-surface ratio (mm) | Axial strain | Typical components |
|---|---|---|
| 25 | 0.00062 | 300 deep double-tee slabs |
| 37 | 0.00067 | 700 deep double-tee slabs |
| 50 | 0.00069 | 150 deep hollow core slabs |
| 75 | 0.00086 | 250 deep hollow core slabs |
| 100 | 0.00086 | Narrow beams or columns (300 wide) |
| 125 | 0.00089 | Wide beams or columns (600 wide) |

$$\Delta = (\varepsilon_{sh} + \alpha t) L \tag{8.13}$$

where $\alpha = 10 \times 10^{-6}$ per °C

$t$ = temperature range.

The shrinkage strain $\varepsilon_{sh}$ may be either obtained from BS 8110, Part 2, Fig. 7.2 [8.14], with appropriate modification factors for low water content mixes (because these data are for water contents of 190 l/m$^3$), or obtained from Table 8.4.

Shrinkage strains $\varepsilon_{sh}$ are modified according to the volume of reinforcement in the cross-section $\rho = A_s/A_c$, and exposure parameter $K$ is as follows:

$$\varepsilon_{sh}' = \frac{\varepsilon_{sh}}{K\rho} \tag{8.14}$$

where $K = 25$ for indoor and 15 for outdoor exposure.

In a multi-storey structure the force is greatest at the first floor level because the foundation does not move. The maximum axial force (elastic analysis) is given by:

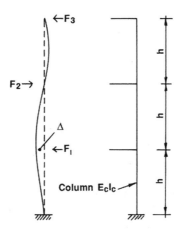

**Fig. 8.32** Horizontal forces and column bending moments due to drying shrinkage or temperature movement.

$$F_1 = \frac{10.31 E_c I_c \Delta}{h^3} \tag{8.15}$$

and $F_2 = 0.3900\, F_1$ and $F_3 = 0.0065\, F_1$ (8.16)

where $E_c$, $I_c$ and $h$ refer to the column.

The maximum bending moment is $M = 0.52\, F_1\, h.$ (8.17)

In multi-bay construction the forces accumulate in each successive bay. The connections must be designed to alleviate the forces to prevent tensile cracking in the connection and flexural cracking in the column.

**EXERCISE 8.7    Shrinkage and temperature-induced moments**

Determine the six-month drying shrinkage and temperature-induced horizontal deflections in the three-storey unbraced structure shown in Fig. 8.33, and calculate the resulting axial forces in the beams and the bending moment at the foundation at Column A. The floor slabs are simply supported on the beams and may be considered as being unconnected.

Use a temperature range of 20°C and indoor exposure conditions. The age at loading is 28 days. Use $f_{cu} = 40$ N/mm$^2$ and the water content = 190 l/m$^3$.

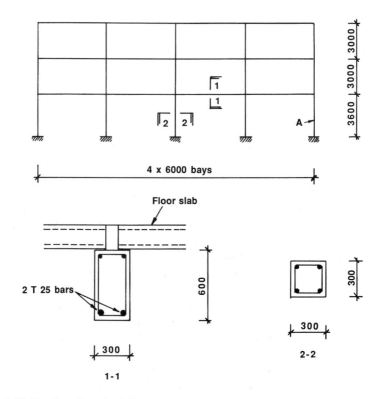

**Fig. 8.33** Details to Exercise 8.7.

*Solution*

**(1) Beam properties**:

$A = 180\,000$ mm$^2$, $u = 1500$ mm, effective depth of beam = 240 mm.

Steel ratio = $4 \times 490/180\,000 = 1.08\%$.

Modification factor for the reinforcement = $1 + (25 \times 0.0108) = 1.27$.

Shrinkage strain (estimated from BS 8110, Part 2, Fig. 7.2 [8.14]) = $125/1.27 = 98$ µε.

Free shrinkage per beam = $(\varepsilon_{sh}' + \alpha_t)L = [98 \times 10^{-6} + (10 \times 10^{-6} \times 10)] \times 6000$

$$= 1.2 \text{ mm} \times 2 \text{ beams}$$

$$= 2.4 \text{ mm (assume symmetry at centre column).}$$

**(2) Column properties**:

$I_c = 6.75 \times 10^8$ mm$^4$.

Short-term $E_c$ (BS 8110, Part 2, Table 7.2) = 28 kN/mm$^2$.

Thirty-year creep factor (estimated from BS 8110, Part 2, Fig. 7.1) = 2.5.

Proportion of long-term creep at six months = 60%.

Long-term $E_c = 28/0.6 \times 2.5 = 18.66$ kN/mm$^2$.

$$F_1 = \frac{10.31 \times 18.66 \times 6.75 \times 10^8 \times 2.4}{3600^3} = 6.68 \text{ kN}$$

$$M = 0.52 F_1 h = 0.52 \times 6.68 \times 3.6 = 12.5 \text{ kN m.}$$

## 8.6 VERTICAL LOAD TRANSFER

### 8.6.1 Introduction

The reactions from the horizontal floor diaphragm calculated in Section 8.2 are transferred to the stabilizing systems as point loads acting at the centroid of the area of the floor which is in contact with each bracing element, i.e. column, wall or core. The loads are reversible in nature and may not be coincident if floor levels differ in height.

In the split-level building example shown in Fig. 8.34(a) the columns on Line A near to the shear walls will be subjected to a horizontal shear force equal to the floor reaction in each shear wall. The columns on Line A are effectively loaded in the manner shown in Fig. 8.34(b), with point reactions at $x$ from each floor level. In the worst situation, the shear is maximum when $x$ is small, and (ignoring second order deflections) bending is maximum when $x = h/2$ as shown in the figure. Flexural and shear checks should be made on these columns, particularly as the axial compression may be small in gable end columns, where many shear walls are located. If necessary a deeper column should be used. The usual outcome is that a shear wall is positioned in the split-level area to eliminate the reactions completely.

This problem does not exist in unbraced structures because the horizontal loads are shared equally between all columns.

In a skeletal structure the bracing elements will be either:

- walls or cores, in the case of a braced structure
- columns, in the case of an unbraced structure, or
- both in the case of partially braced structures (refer to Sections 1.5 and 4.1).

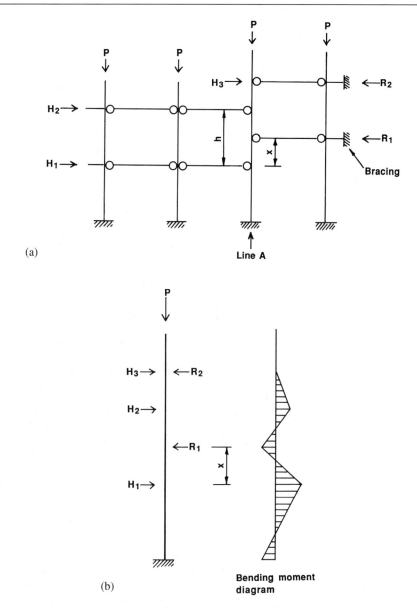

**Fig. 8.34** Design approach for split-level structures. (a) Horizontal load transfer in a split-level braced structure and (b) column bending moments in a split-level braced structure.

### 8.6.2   Unbraced structures

The stability of unbraced pin-jointed structures is provided entirely by columns designed as cantilevers for the full height of the structure (Fig. 1.20(b)). Figure 8.35 shows a photograph of a typical unbraced structure where partial restraints provided by moment-rotation or torsional stiff beam–column connections, deep external spandrel panels or internal brick/block walls have *all* been ignored in the design.

**Fig. 8.35** Spandrel beams in perimeter of frame.

The line of load application is at the centroid of the flooring system. The distribution of horizontal loading between columns is directly proportional to the second moment of area of the columns in the uncracked condition at the serviceability limit state divided by the bending height $h^3$. Non-symmetrical buildings (the vast majority) are considered for torsional stability, but the disposition of columns is usually such that the shear centre of the system coincides fairly closely with the centre of pressure. In most instances the columns will be equally loaded horizontally.

The maximum overturning moment in each column is $\Sigma H_i h_i$ where $H_i$ is the floor diaphragm reaction at each column, and $h_i$ is the effective height from a point 50 mm below the top of the foundation to the centroid of the floor plate $i$ (Fig. 8.36). The overturning moment is additive to the frame moments derived under column design. There is no moment distribution into the beams if the connections are pinned, and therefore the columns are designed using an effective length factor of 2.3.

The maximum height for an unbraced pin-jointed structure is about 10 m. Architectural restrictions on the sizes of columns and the magnitude of the moment-restraint required at the foundation are likely to be prohibitive. The moment carried by the columns is dependent on the degree of fixity between the column and the footing, and on the resistance

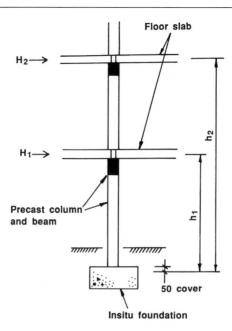

**Fig. 8.36** Definitions of lever arm for overturning moments in columns.

of the soil to footing rotation. BS 8110, Part 2 [8.14] gives the total stiffness of the foundation (i.e. column–footing–soil) equal to that of the column, and this is assumed in design. PCI manuals [8.31] are more explicit but the result is approximately the same.

Columns are manufactured in a single length and therefore the design of splices is not applicable. There are occasions however where a column is founded on to floor beams or slabs, Figs 7.83 and 7.84. This connection, which is designed as pin-jointed, does not contribute to overall stability (rather like a gable post in a warehouse).

Column reinforcement may be reduced at upper floor levels in accordance with the applied bending moment but, as in most precast components, increased fabrication costs (usually) outweigh savings in material.

**EXERCISE 8.8    Overturning moments in columns in unbraced structures**

Determine the maximum overturning moments in the columns shown in Fig. 8.37. The bay centres are 6 m. The columns, which have identical cross-sections, are founded in deep insitu concrete pockets with 50 mm cover to any reinforcement in the foundation. The storey heights are given to the centroid of the floor and roof slabs. Sway deflections are equal at roof level. The wind pressure may be taken as $0.7 \text{ kN/m}^2$.

*Solution*

The different floor spans have no effect on the overturning moments because the beam–column connections are pinned.

Add an extra 50 mm to bending heights of columns at foundation.

Bending height of Columns 1,2,3 = 10.0 + 0.050 = 10.05 m.

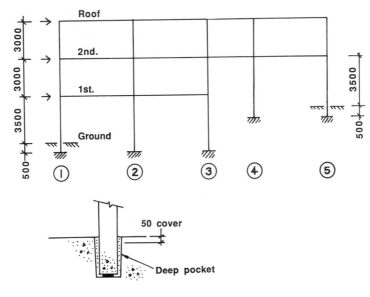

**Column foundation**

**Fig. 8.37**  Details to Exercise 8.8.

Bending height of Columns 4,5 = 7.0 + 0.050 = 7.05 m
Stiffness of Columns 1,2,3 = $EI/L_{1,2,3}$ = 0.099 $EI$
Stiffness of Columns 4,5 = $EI/L_{4,5}$ = 0.142 $EI$

$$\text{Proportion of horizontal force in Columns } 1,2,3 = \frac{3 \times 0.099\,EI}{3 \times 0.099\,EI + 2 \times 0.142\,EI}$$

$$= 0.513, \text{ or } 0.171 \text{ per column.}$$

Proportion of horizontal force in Columns 4,5 = 0.487, or 0.243 per column.
Wind pressure acting on three-storey face gives maximum condition:
Wind at roof*        = 0.7 × 6.0 × 1.5 = 6.3 kN acting at 10.05 m above foundation
Wind at second floor = 0.7 × 6.0 × 3.0 = 12.6 kN acting at 7.05 m above foundation
Wind at first floor   = 0.7 × 6.0 × 3.5 = 14.7 kN acting at 4.05 m above foundation.
* neglect any beam parapet in this exercise.
In Columns 1,2,3:

$$M = 0.171 \times [(6.3 \times 10.05) + (12.6 \times 7.05) + (14.7 \times 4.05)] = 36.2 \text{ kN m.}$$

In Columns 4,5:

$$M = 0.243 \times [(6.3 \times 7.05) + (12.6 \times 4.05)] = 23.2 \text{ kN m.}$$

*Thus, maximum moment = 36.2 kN m.*

### 8.6.3  Deep spandrel beams in unbraced structures

Deep beams, such as spandrel beams with a large upstand (say 750–1000 mm deep at least × about 150–200 mm wide), may be used to distribute sway moments into the frame by providing a vertical couple between the main connector, at the bottom of the beam, and a fixing at the top of the beam (see Figs 3.25 and 8.38). The principle is shown in Figure 8.39 where the stiffness and strength of the spandrel is considerably greater than in the column. For this to work the horizontal forces in the fixings in the spandrel beam must generate a moment given by:

$$M = \frac{\Sigma H_i h}{n} \text{ at the first floor}$$

and     $$M = \frac{H_i h}{n} \text{ at upper floors}$$       (8.18)

where $H_i$ = horizontal floor load, $h$ = storey height, and $n$ = number of columns in the row. The compressive and tensile forces in the couple are given by $F = M/z$, where $z$ = lever arm between the fixings. For this to be effective $z$ should be at least 600 mm.

The columns must be capable of resisting the same force by providing a tensile tie to the fixing and additional reinforcement surrounding it to carry the shear force, particularly in lightly loaded columns.

**Fig. 8.38** Welded joint at top of spandrel beam.

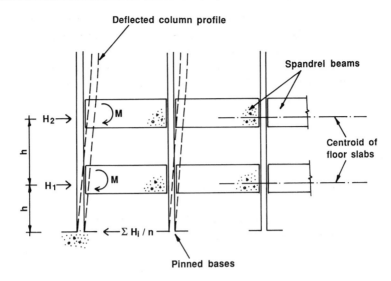

**Fig. 8.39** Spandrel beam restraint mechanism in unbraced structures.

A fully effective vertical shear resistance, one which does not rely on shear friction, must be provided between the spandrel beam and the column. Two options include projecting reinforcement grouted in, or castellated or roughened shear keys. A vertical interface shear resistance between the end of the spandrel and the column must be equal to the shear fixing force $F$. The shear force will be upwards in direction at one end of the spandrel beam, and unless the beam is very lightly loaded (imposed dead load only) this will not cause a problem. At the other end of the spandrel beam the force will be additive to the total shear force, and must be allowed for in the design.

The most effective method of force transfer is to stitch the two components using projecting loops (typically R 8 at 50–100 mm centres) and a vertical lacer bar connected into a recess. Welded joints are often made at the top of the beam which, as shown in Fig. 8.38, must be protected with grout. Bolted fixings usually require clearance holes which do not give positive connectivity unless friction grip bolts are used.

The columns are designed on the assumption that the spandrel is rigidly connected, both in terms of strength and stiffness, to it. Thus the normal rules for column effective length factors $\beta$ apply (see Section 4.4). A typical value for $\alpha_c = 0.2$ to 0.3; hence $\beta =$ approximately 1.1.

### 8.6.4  Braced structures

Braced structures offer the best solution to stability in multi-storey construction, irrespective of the number of storeys (Fig. 4.2(a)). Connection details and foundation design and construction are greatly simplified. Precast concrete wall units are inexpensive, have large in-plane stiffness and strength, are easy to erect and may be integrated with the structure using one of three methods:

- infill wall, either solid or hollow core
- cantilever wall, either solid or hollow core
- cantilever box.

Other methods of bracing are infill brick or blockwork walls, and steel cross-bracing.

As shown in Fig. 8.40, reproduced from the IStructE publication [8.33], floor diaphragm action occurs between vertical cores, composed of walls or a box. A centroidal line of load application from the floor plate to the core is once again assumed. The distribution of horizontal loading between shear walls or shear frames is dependent upon the following:

(1) In-plane deflection response

This is predominantly a flexural deflection in cantilever walls (although shear deflection may govern in 'long' walls, i.e. length to height ratio > 2), a shear deflection in infill walls and truss deflection in steel cross-bracing.

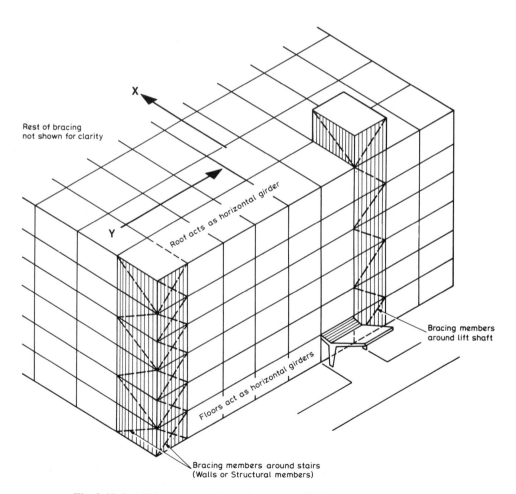

**Fig. 8.40** Stabilizing system in braced structures [8.33].

(2) Position

The structure must be balanced by disposing the walls according to their stiffness and in such a manner that the centre of pressure of horizontal loading lies between at least two of the 'larger' walls as shown in Fig. 8.2. Torsional effects, resulting from eccentricities, are statically determinate and modifications to the load distribution satisfy equilibrium in the direction of loading. If this cannot be satisfied lateral forces develop in the shear walls at right angles to the direction of loading, as explained in Section 8.2.

(3) Columns in the remainder of the structure

These may, if so desired, be designed to make a contribution to overall stability. This is rarely taken into account because:

(a) the contribution is usually small
(b) similar problems associated with moment-resisting bases, etc. in sway structure are reintroduced
(c) shear walls are usually over-designed because their size is often governed by lifting and pitching.

(4) Connections between walls, or between a wall and the foundation

Walls may be used in the structure for some other purposes (e.g. supporting half-landings) than for stability. They may be designed so as not to be moment or shear resisting. This may be achieved by under-reinforcing the connections and allowing load redistribution to take place away from the wall, whilst maintaining structural integrity and load bearing faculties.

(5) Expansion joints in the floor diaphragm

In general, precast structures exceeding 60 to 80 m in plan dimension are usually isolated, depending on the plan configuration.

The horizontal forces are transmitted through the structure in the manner shown in Fig. 8.41(a). It is not necessary for the walls to be located one above the other in the same

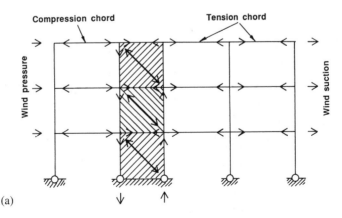

(a)

**Fig. 8.41** Horizontal load transfer in braced structures using coincident and non-coincident wall positions. (a) Walls in same vertical position and (b) walls in different vertical positions.

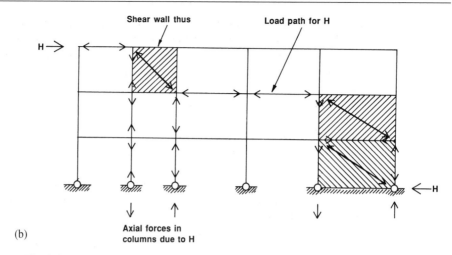

(b)

**Fig. 8.41** (*continued*)

vertical plane, Fig. 8.41(b), providing that the columns and beams surrounding the upper storey walls are designed to carry the vertical wall reaction. This figure shows the load path for the horizontal force *H*.

The moments in the columns are small, but the axial forces have to be considered, particularly in the lightly gravity-loaded cases where walls are positioned between 'gable' columns and uplift may occur, because gable columns support beams which do not carry floor slabs directly. On the other hand large compressive forces may occur in the columns adjacent to infill walls requiring a greater cross-section (or reinforcement) than in the remainder.

Columns adjacent to cantilever shear walls are assumed not to carry additional axial forces due to overturning moments, despite the presence of a continuous vertical shear key between the column and wall. At ultimate this shear key would probably break down and the shear wall would behave in isolation from the columns and beams.

### 8.6.5   Uni-directionally braced structures

Structures braced in one principle direction, and unbraced in the other are considered separately as far as the design of walls is concerned. This means that the effects of sway in the unbraced direction are ignored when the walls in the braced direction are designed. On the other hand the beneficial effects of lateral bracing is considered when the unbraced columns are designed and this often leads to rectangular section columns with the weaker axis in the plane of the bracing.

### 8.6.6   Partially braced structures

Partially braced structures are used in situations where stability walls are architecturally undesirable in the upper two (or maximum three) storeys (Fig. 4.2(c)). The structure is designed as fully braced up to a specified level, and unbraced thereafter. This may not always be the same level throughout the entire building and may alter in the different directions of stability. The columns are cantilevered above this level as in an unbraced structure, but because they are not founded at a rigid foundation, their behaviour is different to ordinary cantilever columns.

The advantages in using this system are many, e.g.:

- there are no bending moments at the foundation
- columns between ground and first floor, where a greater headroom is usually required, are braced
- column sizes and reinforcements in most one or two storey cantilevered structures are sometimes no greater than if the structure were fully braced, i.e. stability is being provided for 'free'
- column splices may be made in the braced part of the structure
- clear floor areas for open plan offices and staircases are punctuated only by columns.

An attractive application is the hybrid steel (or timber) roof/precast concrete structure, Fig. 8.42, where the height of the structure is too great for a wholly unbraced structure, and the upper arcade prohibits the use of stabilizing walls.

As explained in Section 4.4.4, columns in the unbraced part of the structure which are not in direct contact with the shear walls are designed as cantilevers with a maximum effective length factor of 2.35. Columns supporting raking steelwork which is connected at its lower end to a rigid (no-sway) part of the structure are designed as propped cantilevers. The shear forces at the base of the unbraced columns are carried in the floor plate to the stiffening elements in accordance with their stiffness and position.

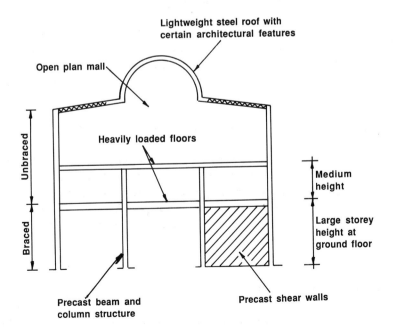

**Fig. 8.42** Hybrid steel–precast concrete partially braced structure.

However, bending moments resulting from sway in the unbraced part are carried over into the braced part of the structure, diminishing to zero with distance to the level of the floor plate below. The effective length factor for the columns in this region is 1.0.

It is equally plausible to use a partially braced structure with the lower floor (or two floors) unbraced, as shown in Fig. 8.43(a) and (b). The reason for doing this is to avoid shear walls at the ground floor. An open mall or car park may be required. However, there are few buildings today without service cores extending into the basement area and so it is unlikely that this option would be used.

The aforementioned rules for effective length factors are now reversed, except that it is possible to specify a moment resistant foundation which would give an effective length factor in the ground floor column (calculated using elastic stability functions) of 1.15.

### EXERCISE 8.9    Columns in a partially braced structure

Repeat Exercise 4.9 where the structure is braced in both directions using infill shear walls up to the second floor level. The building is 36 m in length × 12 m wide. The 300 mm square columns are at 6 m centres in both directions. The characteristic wind pressure is 1.0 kN/m$^2$.

The internal column will be designed. Assume that the horizontal force 1.5% $G_k$ is not critical in this exercise.

### Solution

Reference should also be made to Section 4.4.

In this exercise we are only concerned about the magnitude of the column bending moment at the second floor level in both directions. Determine column slenderness ratios and second order deflections according to Table 8.5.

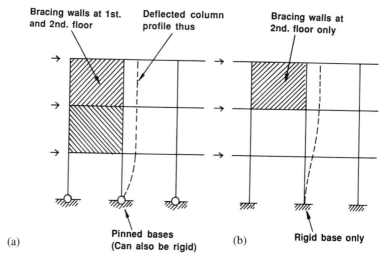

**Fig. 8.43** 'Inverted' partially braced structure. (a) Bracing from first floor upwards and (b) bracing from second floor upwards.

**Table 8.5** Column effective length factors.

| Column effective length factors | Slenderness ratios | $\beta_a$ | $a_u$ (mm) |
|---|---|---|---|
| Ground to first floor $\beta = 0.9$ (braced) | $10.8 < 15$ ∴ short | | |
| First to second floor $\beta = 1.0$ (braced) | $10.0 < 15$ ∴ short | | |
| Second to third $\beta = 2.35$ (unbraced) | $23.5 > 10$ ∴ slender | 0.276 | 82.8 K |
| Second to roof $\beta = 2.35$ (unbraced) | $47.0 > 10$ ∴ slender | 1.104 | 331.3 K |

**Load Case 1**

For maximum gravity loading without wind. From Exercise 4.9:

$N$ at third floor = beam reactions 310.4 + self-weight 9.1 = 319.5 kN.

$N$ at second floor = 319.5 + 590.4 + 9.1 = 919 kN.

$M_{min} = 0.05 \times 0.3 \times 919 = 13.9$ kN m (not critical, ignore further).

$M_{ecc}$ due to column eccentricity = 0.

$M_{add} = (319.5 \times 0.3313) + (599.5 \times 0.0828) = 155.5$ $K$ kN m.

Using BS 8110, Part 3, Chart 47 [8.14], $N/bh = 10.3$, hence $K = 1.0$.

Then $M/bh^2 = 5.75$.

$A_{sc} = 2.6\%$ $bh$.

**Load Case 2**

For patch gravity loading.

$N$ at third floor = 155.2 + 80 + 9.1 = 244.3 kN.

$N$ at second floor = 244.3 + 403.2 + 9.1 = 656.6 kN.

$M_{add} = (244.3 \times 0.3313) + (412.3 \times 0.0828) = 115.1$ $K$ kN m (but $K = 1.0$).

$M_{ecc}$ at second floor = $0.5 \times 43.35 = 21.7$ kN m.

Total $M = 21.7 + 115.1 = 136.8$ kN.

Then $N/bh = 7.29$ and $M/bh^2 = 5.066$.

$A_{sc} = 2.2\%$ $bh$ < Case 1.

**Load Case 3**

For combined gravity and wind loading in the x-direction.

Horizontal wind force at roof = $1.2 \times 1.0 \times 1.5 \times 12 = 21.6$ kN per 21 no. columns.

Horizontal wind force at third floor = $1.2 \times 1.0 \times 3.0 \times 12 = 43.2$ kN per 21 no. columns.

$$M_{wind} = \frac{(21.6 \times 6.0) + (43.2 \times 3.0)}{21} = 12.35 \text{ kN m}.$$

$N$ = at third floor = 256.8 + 7.8 = 264.6 kN.

$N$ at second floor = 264.6 + 475.2 + 7.8 = 747.6 kN.

$M_{ecc} = 0$.

$M_{add} = (264.6 \times 0.3313) + (483.0 \times 0.0828) = 127.65$ $K$ kN m (but $K = 1.0$).

Total $M = 12.35 + 0 + 127.65 = 140$ kN m.

Then $\dfrac{N}{bh} = 9.22$ and $\dfrac{M}{bh^2} = 5.185$.

$A_{sc} = 2.2\%$ $bh$ < Case 1.

**Load Case 4**

For combined gravity and wind loading in the y-direction.

Horizontal wind force in this direction is three times the above.
$M_{wind} = 3 \times 12.35 = 37.05$ kN m.
Total $M = 37.05 + 0 + 127.65 = 164.7$ kN m.

Then $\dfrac{N}{bh} = 9.22$ and $\dfrac{M}{bh^2} = 6.1$.

$A_{sc} = 2.85\%$ $bh >$ Case 1.
Provide $A_{sc} = 2.85\% \times 300 \times 300 = 2565$ mm$^2$.
*Use four T 32 bars (3220).*

## 8.7   METHODS OF BRACING STRUCTURES

Attempting to use precast concrete as much as possible, the most common methods are
(1) infill shear walls and (2) cantilever panel or hollow core walls. Figure 8.44 summarizes
the structural differences between the two types.

### 8.7.1   Infill shear walls

Unlike any other type of shear wall, infill shear walls rely on composite action with the
'unstable' pin-jointed column–beam structure for their strength and stiffness. Where an
infilled wall is built solidly, but not monolithically, into a flexible structure its resistance
to horizontal loading increases considerably due to composite action with the structure.
This is shown in the load response sequence in Fig. 8.45(a). Because the structure is (by

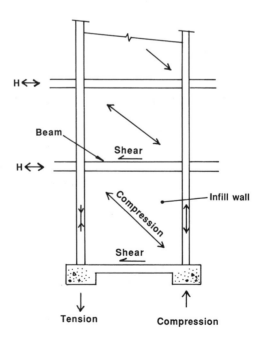

**Fig. 8.44** Behaviour of infill and cantilever shearwalls in pin-jointed frames.

design) flexible and the infill panel very stiff (large in-plane *EI*) there is a paradox in the fundamental use of infill structures. Theoretically the problem is similar to analysing stiff beams on elastic foundations [8.34], in that resistance to horizontal loading is affected by the amount of deformation and the interaction between the two media.

Optimum ultimate limit state design, i.e. collapse occurring in the structure and wall simultaneously, is very difficult to achieve because of the large number of variables used in the analysis. One of the most important of these is the quality of the vertical shear key connection, which for manufacturing purposes is usually unreinforced.

Most of the pioneering work on infilled frames – albeit using masonry infill – was carried out by Stafford-Smith & Carter [8.35], Mainstone [8.36] and Wood [8.37]. The design procedures suggested by these authors are widely used and respected in the UK. More recently Kwan and Liaum [8.38 and 8.39] have proposed a plastic theory to deal with ultimate collapse mechanisms and suggested that an optimum design can be attained using reinforced joints with finite interface shear strength. Kwan's analyses further support the UK approach.

Commentaries on the behaviour of infilled shear walls are numerous and therefore only the assumptions and methods used in current design will be raised here. Wright [8.40] produced a design method based on the principles developed by Stafford-Smith for masonry infill which were combined with Mainstone's work using micro-concrete. The design assumptions are:

(1) Ultimate horizontal forces are resisted by a compressive diagonal strut across the concrete infill wall. The effective width of the strut depends primarily on the relative

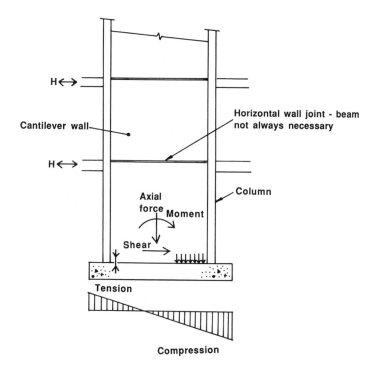

**Fig. 8.44** (*continued*)

stiffness of wall panel and structure, and on the geometry $h$ and $\theta$ as defined in Fig. 8.45(a), but also varies by reducing as the interface cracking load is exceeded.

(2) The tensile diagonal strength of the rc wall is ignored but the amount of reinforcement is sufficient to prevent excessive diagonal cracking and to maintain the intrinsic shape of the wall panel particularly at the corners.

(3) Slender wall panels are designed to BS 8110 [8.14] as slender braced *plain* concrete walls (reinforcement is usually less than 0.4 per cent) taking into account manufacturing and site inaccuracies and deflection-induced bending moments.

(4) The shear resistance at the horizontal interface between beam and wall panel is also based on shear in plain concrete walls (normal aggregate grade C25 minimum) using an ultimate average stress of 0.45 N/mm$^2$ or a value equal to 25 per cent of the vertical pre-compressive stress.

(5) The wall is not subjected to vertical frame loading, i.e. floor beams are assumed structurally isolated from the wall units even though the gap between them is grouted solidly.

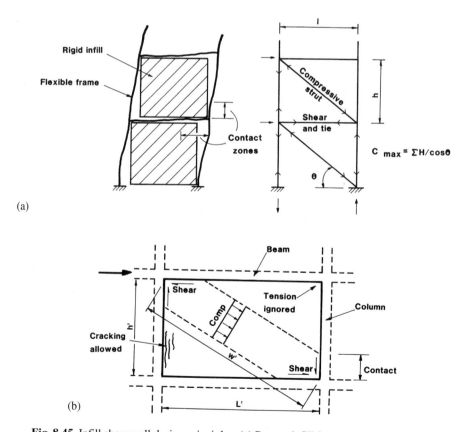

**Fig. 8.45** Infill shear wall design principles. (a) Precast infill frame strut analogy and (b) precast infill frame stresses.

(6) The wall is not subject to simultaneous in-plane and out-of-plane wind loading. Eddy currents in enclosed courtyards of I-shaped buildings may, in extreme circumstances, present mutually perpendicular wind pressure to parts of the structure.

Some analyses have been so refined that the infill frame can often be misrepresented as a simple pin-jointed truss where the walls, columns and beams are singularly replaced by diagonal struts, chord elements and ties, respectively. This analogy is inadequate because load transfer occurs over an extended region of the interface and the distribution of interaction over the contact length is semi-parabolic/triangular. The length of contact gives the effective width of the compressive strut and together with the appropriate reductions for slenderness and spurious eccentricity, the ultimate diagonal compressive strength of the wall is computed. If the permissible horizontal shear stress is not exceeded, load vs. infill panel size data may be presented as shown in Fig. 8.46. The derivation of these curves is given later. Note that the stiffness of the frame is based on 300 × 300 mm components throughout – a reasonable assumption.

The structural mechanism is as follows. On first application of a racking load there may be full composite action between the frame and wall if these are bonded together. At a comparatively early stage, however, cracks will develop between the two components, except in the vicinity of two of the corners where the infill panel will lock into the frame and there will be transmission of compressive forces into the concrete wall. At this stage, it is convenient to consider the concrete wall acting as a compression diagonal within the frame, the effective width of which depends on the relative stiffness of the two components and on the ratio of the height to the width of the panel. This action continues until the shear resistance is overcome and a crack, slightly inclined to the horizontal, is developed. Several more or less parallel cracks of this type may develop with further increase in

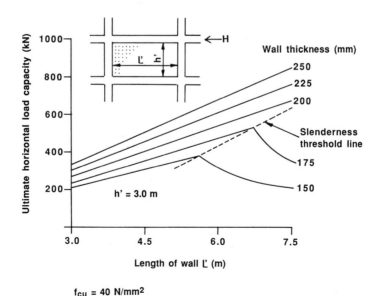

**Fig. 8.46** Precast concrete infill wall design graph.

load and failure may finally result from the loss of rigidity of the infill as a result of these cracks, or from local crushing or spalling in the region of the concentrated loads. In some cases the strength of the structure may be limited by the strength of the frame members or joints.

Essential problems in this approach are:

- to determine the contact lengths between frame and wall
- to find an effective width for the equivalent strut
- to establish the mode of failure and strength of the wall.

The contact length depends on the relative stiffness of the frame and infill, and on the geometry of the panel. Stafford-Smith first developed the analogy with a beam on an elastic foundation, by which the column of an unfilled frame under lateral load may be regarded as one half of a beam on an elastic foundation which, under a central concentrated load $P$, remains in contact with the foundation over a length known as the characteristic length $\alpha$ given by:

$$\frac{\alpha}{h} = \frac{\pi}{2\lambda h} \tag{8.19}$$

in which $\lambda h$ is a non-dimensional parameter expressing the relative stiffness of the frame and infill, where

$$\lambda = \sqrt[4]{\frac{E_i t \sin 2\theta}{4 E_c I h'}} \tag{8.20}$$

where

$E_i$ = infill modulus
$t$ = infill thickness
$I$ = minimum moment of inertia of beams or colums
$h'$ = height of infill
$E_c$ = concrete frame modulus
$\theta$ = slope of infill.

An 'equivalent' strut analogy is thus developed from the determination of $\alpha$. This depends on aspect ratio $h'/L'$. Failure occurs either by local crushing at the corner, diagonal splitting due to excessive compression, or shear failure. Figures 8.47 and 8.48 (reproduced from [8.35]) are used to determined the failure loads appropriate to these conditions. Resulting ultimate design stresses are assessed against the following:

$$\frac{0.67 f_{cu}}{\gamma_m} \quad \text{for concrete in compression } (\gamma_m = 2.2)$$

$$\frac{f_k}{\gamma_m} \quad \text{for brickwork in compression } (\gamma_m = 3.5)*$$

$$\frac{f_v}{\gamma_{mv}} \quad \text{for brickwork in shear } (\gamma_{mv} = 2.5)* \tag{8.21}$$

* Note that in [8.35] the compressive and shear strength are given as $f_c$ and $f_{bs}$, respectively.

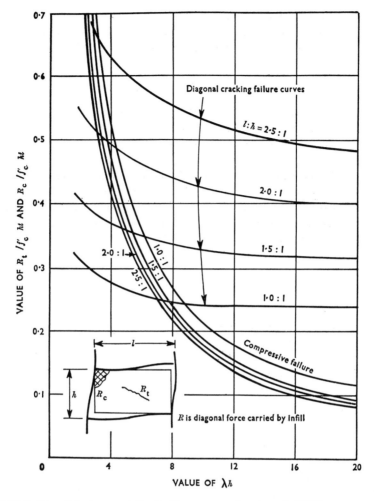

**Fig. 8.47**  Infill wall design graph for limiting compressive strength [8.35].

Mainstone's method considered an upper bound solution by restricting the width of an equivalent strut to 0.1 $w'$, where $w'$ = diagonal length of infill, but where the effects of slenderness must be considered. Vertical column reactions in compression or tension are carried to the column as appropriate.

Design methods are given in Sections 8.7.2 and 8.7.3 for concrete and brick infill walls, respectively.

### 8.7.2   Design methods for infill concrete walls

Concrete walls are considered to be plain walls, according to BS 8110, Part 1, clause 3.12.5 [8.14], because the minimum area of reinforcement is provided only for lifting purposes. The wall is built in on all sides and is therefore braced. The ultimate horizontal force is resisted by a diagonal compressive strut across the infill wall. Referring to

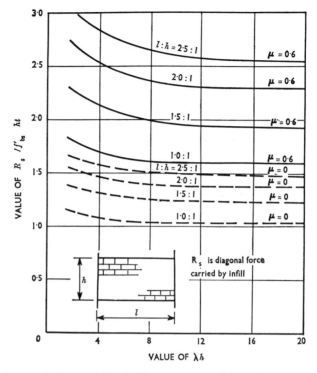

**Fig. 8.48** Infill wall design graph for limiting shear strength [8.35].

Fig. 8.45(a) the strength of the strut $R_v$ is given by:

$$0.1 = \frac{R_v}{0.3 f_{cu} w' t} \tag{8.22}$$

where $f_{cu}$ = compressive cube strength of infill wall. The horizontal resistance is given by:

$$H_v = R_v \cos \theta = 0.03 f_{cu} L' t \tag{8.23}$$

Where infill slenderness ratio $w'/t > 12$, Equation (8.23) is modified in accordance with BS 8110, Part 1, clause 3.9.4.16. As the wall is built in on two sides at the corner, the effective length of the wall is taken as $L_e = 0.75w'$. The design is as follows:

$$\frac{R_v}{0.1w'} \leq (t - 1.2e_x - 2e_{add})0.3 f_{cu} \tag{8.24}$$

where

$$e_x = 0.05t, \text{ and } e_{add} = \frac{L_e^2}{2500 \, t} \tag{8.25}$$

from which

$$\frac{R_v}{0.1w't} = \left[ 0.94 - \frac{1}{1250}\left(\frac{L_e}{t}\right)^2 \right] 0.3 f_{cu} \tag{8.26}$$

with a limit on the slenderness ratio

$$\frac{L_e}{t} = \frac{0.75w'}{t} < 30 \tag{8.27}$$

Values of $H_v/L't$ are plotted graphically in Fig. 8.46 for a specified aspect wall ratio.

At corners, Stafford-Smith shows that the length of contact is $0.5\ L'$ along the beam and $\alpha = \pi/2\lambda$ up the column, where $\lambda =$ stiffness factor as before. In consideration of horizontal shear forces, BS 8110, Part 1, clause 5.3.7 recommends an interface shear stress of $0.45$ N/mm$^2$ without castellations or other shear key aids because the joint is in compression as well as shear. Hence:

$$R_v \sin \theta = 0.45\ \alpha t \tag{8.28}$$

$$R_v \cos \theta = 0.45\ L' t \tag{8.29}$$

Residual horizontal shear may be taken through added interface reinforcement passing through beams and grouted into holes in the wall or otherwise fixed to wall panels. Excess vertical shear is carried by the beam–column connector.

Bearing under the corners of the wall is checked against clause 5.2.3.4 where $f_{cu} =$ weakest concrete applicable:

$$R_v \sin \theta \le 0.6\, f_{cu}\ L't \tag{8.30}$$

### EXERCISE 8.10    Concrete infill wall design

The three-storey building shown in Fig. 8.49 is 50 m long $\times$ 6 m wide and is braced at each end using infill concrete walls. Determine the minimum thickness of the wall at ground floor if the characteristic wind pressure $w_k$ is 0.8 kN/m$^2$.
Use $f_{cu} = 40$ N/mm$^2$ for the wall and $f_{cu} = 50$ N/mm$^2$ for the frame.
$L = 6000$ mm, $L' = 5700$ mm.
$h = 4000$ mm, $h' = 4000 - 100 - 400 = 3500$ mm.

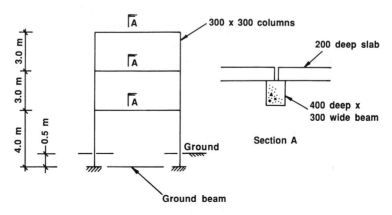

**Fig. 8.49** Details to Exercise 8.10.

$\theta = 31.55°$, $w' = \sqrt{5700^2 + 3500^2} = 6690$ mm.

$I = 300 \times 300^3/12 = 6.75 \times 10^8$ mm$^4$.

Slenderness ratio $= 0.75 \times 6690/t = 5017/t < 30$, hence $t > 167$ mm. Try $t = 170$ mm.

$E_{\text{wall}} = 28$ kN/mm$^2$, $E_{\text{frame}} = 30$ kN/mm$^2$.

Design ultimate wind load per end $H = \gamma_f w_k\, A = 1.4 \times 0.80 \times 25 \times (1.5 + 3.0 + 3.25) = 217$ kN.

Diagonal force $= 217/\cos 31.55° = 255$ kN.

Diagonal strength capacity $R_v$ is given by: $\dfrac{R_v}{0.1 w' t} = \left[0.94 - \dfrac{1}{1250}\left(\dfrac{L_e}{t}\right)^2\right] 0.3 f_{cu}$

where $L_e = 5017$ mm.

Thus, $R_v = 0.1 \times 6690\, t\left[0.94 - \dfrac{20\,136}{t^2}\right] \times 0.3 \times 40 \times 10^{-3} = 255$ kN.

Solving the quadratic gives $t = 164.2$ mm $< 170$ mm.

Check interface strength, using relative stiffness parameter $\lambda$ to give contact length against column $\alpha$ using $\dfrac{\alpha}{h} = \dfrac{\pi}{2\lambda h}$.

$$\lambda = \sqrt[4]{\dfrac{28 \times 170 \times 0.891}{4 \times 30 \times 6.75 \times 10^8 \times 3500}} = 1.968 \times 10^{-3}$$

$$\dfrac{\alpha}{h} = \dfrac{\pi}{2\lambda h} = \dfrac{\pi}{2 \times 1.968 \times 10^{-3} \times 4000} = 0.2$$

Contact length to column $= 800$ mm.

Now

$$R_v \sin\theta = 0.45\,\alpha\,t$$

$$R_v = \dfrac{0.45 \times 800 \times 170 \times 10^{-3}}{0.523} = 117 \text{ kN} < 255 \text{ kN}$$

Residual force $= (255 - 117) \sin 31.55° = 72.2$ kN is carried to ground beam and does not transfer to column.

Also shear capacity along horizontal plane is given by:

$$R_v \cos\theta = 0.45\,L't$$

$$R_v = \dfrac{0.45 \times 5700 \times 170.5 \times 10^{-3}}{0.852} = 512 \text{ kN} > 217 \text{ kN design force, therefore}$$

adequate.

*Use 170 mm thick concrete wall.*

## EXERCISE 8.11    Slender precast infill wall strength check

Calculate the horizontal shear ultimate capacity of the 175 mm thick precast concrete shear wall shown in Fig. 8.50, and determine the minimum strength of the precast concrete. The wall is built into a precast concrete frame on all sides.

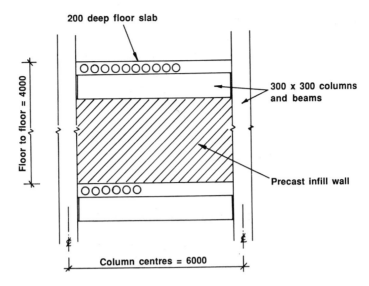

**Fig. 8.50** Details to Exercise 8.11.

*Solution*

$$I_{\text{frame}} = \frac{300 \times 300^3}{12} = 6.75 \times 10^8 \text{ mm}^4.$$

$$\theta = 31.55°.$$

Diagonal length $w' = \sqrt{5700^2 + 3500^2} = 6690$ mm.

Slenderness ratio $= \dfrac{0.75 \times 6690}{175} = 28.7 < 30$, but because $> 12$, reduction factors apply.

Diagonal strength is given by $\dfrac{R_v}{0.1 w' t} = \left(0.94 - \dfrac{28.7^2}{1250}\right) 0.3 f_{\text{cu}}.$

$R_v = 9.871 f_{\text{cu}}$ kN (with $f_{\text{cu}}$ in N/mm$^2$).
$H_v = R_v \cos 31.55° = 8.412 f_{\text{cu}}$ kN.
Shear capacity on horizontal bedding joint:

$$H_v = 0.45 \, L' t = 0.45 \times 5700 \times 175 \times 10^{-3} = 448.9 \text{ kN}.$$

Then if 448.9 kN $= 8.412 f_{\text{cu}}$, we get $f_{\text{cu}} = 53.4$ *N/mm*$^2$, for a limiting shear capacity $=$ 448.9 kN.

### 8.7.3   Design method for brickwork infill panels

The ultimate horizontal forces are resisted by a diagonal compressive strut across the infill. The criteria are based on a stiffness factor $\lambda$ as before. The method is the same except:

- a factor $k$ replaces the constant 0.1 for the width of the diagonal strut
- the crushing strength of the concrete is replaced with $f_k$, the strength of brickwork in compression (see BS 5628, Part 1 [8.41])
- a further criterion is the local crushing resistance of the brickwork at the corners, obtained from Fig. 8.47
- Young's modulus for brick = $450 f_k$ (N/mm² units)
- the horizontal shear limit is obtained from Fig. 8.48 with $\mu = 0.6$ for solid (or unperforated) bricks, because vertical compressive stress acts at the same time as the shear.

Hence

$$R_{vs} = f_s h't \times \text{value from the design graph} \tag{8.31}$$

where $f_s$ is from BS 5628, Part 1, clauses 25 and 27.4.

Therefore $f_s = \dfrac{0.35}{2.5} = 0.14\,\text{N/mm}^2$.

Hence

$$R_{vs} \leq 0.14\,h't \times \text{value from graph in Fig. 8.48.} \tag{8.32}$$

Compression limit obtained from Fig. 8.47 (compressive failure curves):

$$R_{vc} = f_k h't \times \text{value from graph} \tag{8.33}$$

where $f_k$ is from BS 5628, Part 1 based on the strength of mortar from Table 1 and clause 27.3 of this code. The bricks should have a minimum crushing strength of 21 N/mm².

### EXERCISE 8.12    Brickwork infill wall

The building in Exercise 8.10 is braced at each end between the first and second floor using 215 mm thick infill brick walls. Check the adequacy of these walls using 21 N/mm² strength brick in $1:\frac{1}{4}:3$ mortar. $\gamma_m = 3.5$.

### Solution

$L' = 5700$ mm, $h = 3000$ mm, $h' = 3000 - 200 - 400 = 2400$ mm.
$\theta = 22.83°$, $w' = 6185$ mm.
$t_{wall} = 215$ mm, $f_k = 12$ N/mm², $E_i = 5.4$ kN/mm².

$$\lambda = \sqrt[4]{\frac{5.4 \times 215 \times 0.715}{4 \times 30 \times 6.75 \times 10^8 \times 2400}} = 1.438 \times 10^{-3}$$

$\lambda h = 4.31$, and $\dfrac{L'}{h'} = 2.38$.

$$\frac{R_v}{(f_k / \gamma_m)h't} = 0.41 \text{ for compression failure, and 0.58 for diagonal splitting failure}$$

(subject to graphical approximations). Thus critical value is lowest:

$$R_v = 0.41 \times (12.0/3.5) \times 2400 \times 215 \times 10^{-3} = 725 \text{ kN}$$

$$\frac{R_s}{(f_v/\gamma_{mv})h't} = 2.75 \text{ for shear failure using } \mu = 0.6.$$

$$R_s = 2.75 \times \left(\frac{0.35}{2.5}\right) \times 2400 \times 215 \times 10^{-3} = 198.7 \text{ kN}.$$

$$H = 198.7 \cos 22.83° = 183 \text{ kN}.$$

Design ultimate wind load at second floor
$$H = \gamma_f w_k A = 1.4 \times 0.80 \times (37.5 + 75)\,126 \text{ kN} < 183 \text{ kN capacity.}$$

Also design $R = \dfrac{126}{\cos 22.83°} = 137 \text{ kN.}$

Check interface strength, using relative stiffness parameter $\lambda$ to give contact length against

column $\alpha$ using $\dfrac{\alpha}{h} = \dfrac{\pi}{2\lambda h}$ :

$$\frac{\alpha}{h} = \frac{\pi}{2\lambda h} = \frac{\pi}{2 \times 1.438 \times 10^{-3} \times 3000} = 0.364.$$

Contact length to column = 1092 mm.

$$R_v \sin \theta = 0.14 \, \alpha t$$

$$R_v = \frac{0.14 \times 1092 \times 215 \times 10^{-3}}{0.387} = 84.9 \text{ kN} < 137 \text{ kN.}$$

Therefore residual force = $(137 - 84.9) \sin 22.83° = 20.2 \text{ kN}$ is carried to first floor beam.

### 8.7.4  Infill walls without beam-framing elements

The function of horizontal beams in the framing action in infill wall design may be replaced by using dowels between upper and lower walls as shown in Figs 8.51 and 8.52. The units are economical to manufacture in shallow steel or timber moulds, Fig. 8.53. The dowels serve two functions by providing horizontal shear forces as well as holding the walls together by developing tension across the interface. The diagonal strut analogy for infill walls is replaced by shear panel concepts where both horizontal and the complementary vertical shear forces are resisted by mechanical means. The vertical shear resistance between the wall and columns may be provided either by site welding together cast-in plates, or by using projecting dowels grouted into a keyed joint along the full interface.

This method of construction, although more expensive than the infill method, has the advantage that fairly large voids may be formed in the walls, providing that there remains sufficient contact between the precast components to allow panel action to take place. The wall panels are also relatively inexpensive to manufacture, requiring simple rebar cages usually consisting of two sheets of A142 or A193 mesh. The units shown in Fig. 8.53 also included projecting dowels, dowel holes and cast-in plates.

**Fig. 8.51** Construction of precast shear walls.

### 8.7.5   Use of slip-formed hollow core walls as infill walls

Slip-formed hcus, made in a similar manner to hollow core slabs, may be used as infill walls as shown in Figs 8.54 and 8.55(a) and (b). The latter shows two alternative forms of construction. The method is not used extensively in some countries, and hardly at all in the UK, but it should not be dismissed as an alternative form of bracing even though the thickness of the wall will inevitably be greater than for a solid wall of equal strength.

Slip-formed hollow core walls are notoriously difficult to handle in the vertical position owing to an absence of cast-in lifting points. Friction clamps provide the best method of pitching and lifting, but the safety of these devises must be guaranteed at all times.

The wall units must be symmetrical and lightly prestressed. In the first method (Fig. 8.55(a)) the hollow cores lay in a horizontal plane and the ends of the cores are filled with insitu concrete (6 mm aggregate) for a distance of about 100 mm generally, and 500 to 600 mm in at least two cores per unit. Note the similarity with the hollow core slab arrangement. The edges of the units are profiled to give a 'tongue and groove' joint, which helps to locate the walls one above another. Wall units are bedded onto wet mortar, with steel or plastic shims used to obtain the correct levels. Some form of temporary

**Fig. 8.52** Partially complete construction of precast shear walls.

stability is required, usually in the form of sturdy clamps placed around the columns at the corners of each wall unit. This type of construction relies on columns for temporary stability.

The design method is an extension of the solid infill wall design except that there are four additional considerations:

- the reduced cross-sectional area and inertia of the hollow core wall unit compared to a solid wall of equal thickness
- the horizontal shear force in the longitudinal joint between the units
- the shear key to the column
- crushing of the webs in the corners of the wall panel.

The diagonal strut model is shown in Fig. 8.55(a). Because the ends of the wall units are filled with grout the contact width is equal to the full width of the wall. However, if Equation (8.19) is to be used to determine the contact length $\alpha$, the relative stiffness of the hollow core wall must be modified to take into account the voids. The wall must also be checked for diagonal compressive capacity, and for the effects of slenderness

**Fig. 8.53** Wall panels in stock yard.

**Fig. 8.54** Hollow core shear wall under construction (in Far East).

as in the design of the solid wall in Section 8.7.1. Thus an effective thickness of wall $t_{ef}$ is found by equating $EI_{hcw}$ of an hcu $EI$ of a solid wall of equal thickness. Assuming that $E$ is constant, and the breadth of the wall is $b$ then the conversion is made as follows:

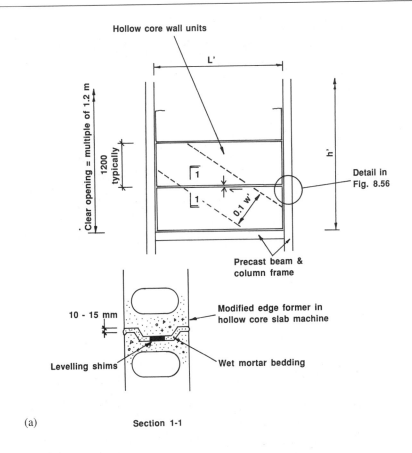

(a)    Section 1-1

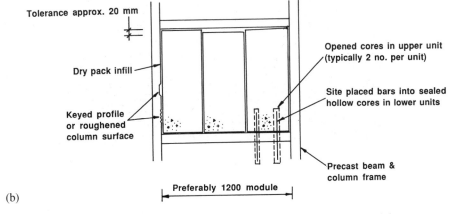

(b)

**Fig. 8.55** Hollow core units used as infill shear walls. (a) Hollow cores and joints horizontal and (b) hollow cores and joints vertical.

$$t_{ef} = \sqrt[3]{\frac{12\,I_{hcw}}{b}} \tag{8.34}$$

The effect of this reduction will have a compound effect on strength, especially if the wall becomes slender. The compressive strength of the wall is based on the total thickness of the two flanges, i.e. the webs between the hollow cores are ignored. As a result the overall thickness of a hollow core wall is likely to be greater than an equivalent solid wall. However, as the thickness of the greater proportion of solid walls in use is governed not by ultimate design, but by lifting in the factory, the differences may not always be so great.

The second criteria concerns the development of sufficient vertical compression across the joint to enable the joint to be considered as a compression joint according to BS 8110, Part 1, clause 5.3.7[b] [8.14], for which the ultimate shear stress is 0.45 N/mm². The code does not specify the degree of compression that must exist across the joint, but it must be active in all design situations. The minimum compressive stress due to self-weight gives an ultimate compressive stress of about 0.04 N/mm², and is clearly insignificant. However, it is not possible for a horizontal shear force $H$ to exist in the joint unless it is accompanied by a diagonal force $R$, where $R = H/\cos\theta$. The vertical component of $R = R \sin\theta$. The force $R$ acts across a diagonal of width $0.1 \times$ diagonal length of wall $= 0.1\,w'$ and therefore compresses the joint over a distance $x = 0.1\,w'/\sin\theta$.

However, because the wall unit has a large in-plane stiffness, $EI_{xx}$, it is known that diagonal forces acting across a hollow core diaphragm exert uniform pressures and displacements over a considerable length of the interface. Omar [8.26] found differential crack openings of less than 0.05 mm in a 4 m long hollow core slab when subjected to shear, and several investigators have measured greater shear capacities between hcus when a diagonal force is acting across the joint.

Thus the compressive stress may be determined and, if necessary, added to the self-weight pressure of 0.04 N/mm². The ratio of the vertical to the horizontal stress will depend on $\theta$, but for $\theta = 35°$ (a common value) it is between 0.70 and 0.75. This is equal to the required coefficient of friction in a plain surface (BS 8110, Part 1, Table 5.3).

In all cases the wall units must be prevented from moving apart and shear-connected to the columns. The simplest method is to cast projecting U- or L-shaped tie bars in the columns and fill the mating opened hollow core with expansive concrete, as shown in Fig. 8.56. The projection from the column should be a full anchorage length, typically 500 mm for a 12 mm bar. Mild steel bars are often cast against the inside of the mould and broken out as soon as the column is demoulded.

The area of steel required may be calculated as follows. The vertical shear force is complementary to the horizontal force and equal to $H \tan\theta$. If the vertical shear stress exceeds 0.23 N/mm² (this is used because only a part of the vertical joint is under compression and because the column is flexible), reinforcement should be provided to resist the entire shear force. The area of shear steel projecting from the column is given by:

$$A_s = \frac{H \tan\theta}{0.6 \times 0.87 f_y} \tag{8.35}$$

This steel may be distributed uniformly over the storey height. At least two cores per wall unit should be opened. The opened core should be filled using a concrete grade C40 minimum.

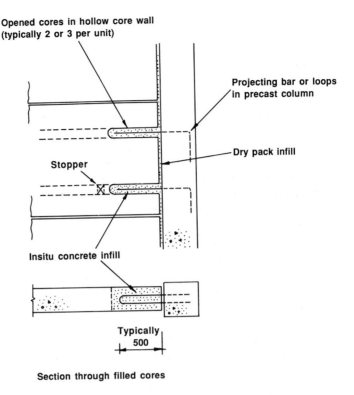

**Fig. 8.56** Shear connectors between infill hollow core wall and columns.

Local crushing of the webs in the corners of the wall panel may occur in flexible frames where the contact zones α are small. This may easily be prevented by filling the first one or two cores with concrete in the factory for a distance of about 600 mm.

In the second method, Fig. 8.55(b), the wall–wall joints are vertical. The main differences with the former method are:

- the joint is rebated and filled (inside a shutter) in the same way as floor slabs
- the interface shear stresses are the result of complementary shear forces trying to rack the panel in a lozenging effect.

The largest force occurs in the vertical joint nearest to the centre of the panel. The joints are all subjected to a normal compression $R \cos \theta$ which is equal to the shear force $H$. Thus a limiting interface shear stress of 0.45 N/mm$^2$ is used in the design. The surface of the column in contact with the wall is either roughened by removing laitance and revealing the coarse aggregate, or castellated to form a shear key. The effects on slenderness and strength capacity of the hcu are the same as in the former method.

The horizontal joint between the top of the units and the floor beam is made by dowelling the hollow cores through holes (preferably slots) in the beam. The tolerance required in this joint to take care of variations in the height of the hcus is at least 20 mm. The figure has exaggerated the effect. The dowels are grouted into two to three sealed hollow cores per unit prior to the upper floor units being placed. The joint is completed

by pouring insitu concrete through preformed openings in the upper wall units. Expansive cements are used throughout.

Hollow core walls are reinforced longitudinally using a symmetrical arrangement of prestressing wires to produce a uniform prestress of around 2 N/mm². This is sufficient for lifting purposes. If the walls are exposed to lateral wind loading, they should be designed to span horizontally between columns or vertically between beams. It is unlikely that this will be a governing factor in the design of the units.

**EXERCISE 8.13    Hollow core shear wall design**

Calculate the horizontal shear capacity of a 200 mm thick hollow core shear wall for the geometry and loading shown in Fig. 8.57.

Use $f_{cu} = 50$ N/mm² for the precast wall, $f_{cu} = 40$ N/mm² for the insitu infill, and $f_y = 250$ N/mm² for the loops in the joints.

*Solution*

Assumed geometric properties of hollow core wall per 1200 mm wide unit (actual values from manufacturers' literature) are:

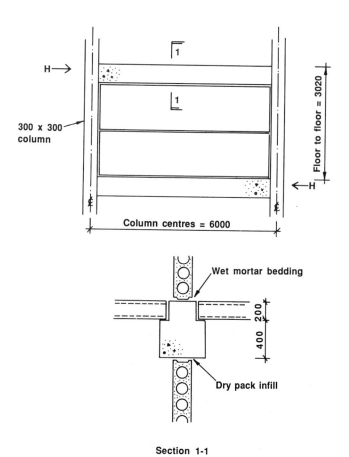

**Section 1-1**

**Fig. 8.57** Details to Exercise 8.13.

$t = 200$ mm, $I = 650 \times 10^6$ mm$^4$, flange thicknesses $= 25$ mm.
$L' = 5700$ mm, $h' = 2420$ mm to give 20 mm tolerance, $\theta = 23.0°$, $w' = 6192$ mm.
For the purpose of slenderness the effective thickness of the hcu is:

$$t_{ef} = \sqrt[3]{\frac{12 \times 650 \times 10^6}{1200}} = 186 \text{ mm.}$$

Slenderness ratio $= 0.75 \times 6192/186 = 25 < 30$, but because $> 12$, reduction factors apply.
For the purpose of strength the effective thickness is $t_{ef} = 50$ mm.
Diagonal strength is given by

$$\frac{R_v}{0.1 w' t_{ef}} = \left(0.94 - \frac{25.0^2}{1250}\right) 0.3 \times 50.$$

$R_v = 204.3$ kN.
$H_v = R_v \cos 23° = 188.1$ kN.
Shear capacity on horizontal bedding joints using actual thickness of 200 mm:

$$H_v = 0.45 \, L't = 0.45 \times 5700 \times 200 \times 10^{-3} = 513 \text{ kN} > \text{compression capacity.}$$

*Then horizontal shear capacity* $H_{max} = 188.1$ *kN.*
Vertical shear force $V = H \tan \theta = 188.1 \times \tan 23° = 79.8$ kN to be carried by projecting
loops in shear.

$$A_s = \frac{79.8 \times 10^3}{0.6 \times 0.87 \times 250} = 612 \text{ mm}^2.$$

*Provide four R 10 (628) loops at equal spaces.*


### 8.7.6 Cantilever shear walls and shear boxes

Cantilever walls and boxes are generally very large physically because they are most
often provided as stabilizing elements around stairwells and liftshafts. The individual
precast components are tied together vertically using site placed continuity reinforcement
or other mechanical connections such as welded plates or bars.

Various failure modes are considered; shear slip failure would occur where $L/h$ is
greater than (approximately) 2.5 to 3, as shown in Fig. 8.58(a). Flexural tension failure
occurs particularly if the wall is lightly loaded axially, Fig. 8.58(b), and flexural
compression if the wall is heavily loaded, Fig. 8.58(c). The wall panels themselves are
designed as rc (not plain) walls in the conventional rc manner.

The slenderness of the wall, out of plane eccentric loading and constructional tolerances
are taken into consideration. Particular emphasis is given to the reinforcing of corners –
particularly to avoid cracking at the early lifting stage – and around windows and doors,
etc. Figure 8.59 shows typical details for these regions. Large fenestrations, openings for
doors or services, etc. are readily included by considering alternative load paths for the
vertical and horizontal forces. The positions of the joints may have to be amended if
particularly large voids fall on the proposed joint lines.

Complete storey height units require only horizontal connections between them. Shear
forces and overturning moments in the wall system generate regions of compression,

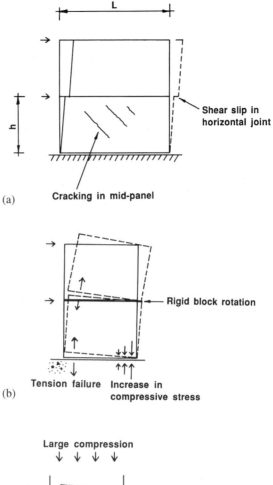

(a)

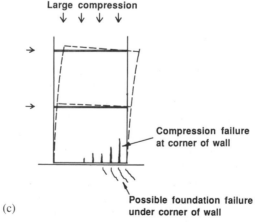

(b)

(c)

**Fig. 8.58** Deformations in wall panels. (a) Shear failure in long wall; (b) flexural tension due to large overturning moment and (c) compression failure due to combined axial forces and moments.

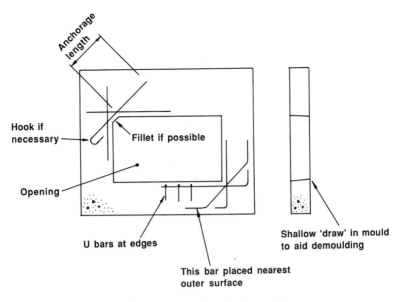

**Fig. 8.59** Reinforcement details at large openings in shear walls.

shear and tension. Fully anchored reinforcement (or other mechanical devices such as welded or bolted plates, or wall shoes, etc.) is used to transfer the tensile forces between units in addition to maintaining the pre-compression across the faces of the joint. The joint is assumed to be cracked only in flexure and the shear friction theory is used to justify taking the full length of the wall into consideration in the determination of the average ultimate shear stress. The limiting value is $0.45 \text{ N/mm}^2$. In reality the shear resistance of the concrete under high pre-compression is usually sufficient to resist to imposed shear.

Units assembled from individual flat panels are connected on site using vertical joints whose main function is to resist shear forces resulting from the cantilever action of the wall. The resistance to sliding depends on the size and shape of the joint in addition to the amount of reinforcement projecting from the adjoining panels. Typically, the joint is at least 60 mm across the root (10–15 mm at the outside edges). Larger root gaps are occasionally necessary, although full scale experimental work [8.13] has shown a reduction of 44 per cent in the ultimate shear strength when the root gap is increased to 120 mm. Vertical and castellated joints have been studied fairly extensively [8.42–8.45] and the ultimate design stress of $1.3 \text{ N/mm}^2$ given in BS 8110 [8.14] has been shown to be conservative providing sufficient reinforcement necessary to develop aggregate interlock by shear friction is provided.

Mechanical connections are favoured by some engineers in that they provide immediate stability to the walls. Fully anchored plates are used to provide the site welder with easily accessible down-hand welding.

### 8.7.7   Hollow core cantilever shear walls

The hollow core walls described in this section are wet-cast, rc panels with two or three preformed rectangular cores. The walls are used with the cores placed vertically and the design is in no way connected with the slip-formed infill walls discussed in Section 8.7.5.

The arrangement of these walls is shown in Figs 3.35 and 8.60. The walls must be placed between columns, even though they do not rely on composite action with the columns. The design section in BS 8110, Part 1, clause 3.9.3.6.3 is not very informative and so therefore design methods for columns are adopted with due consideration for transverse (out-of-plane) bending.

Hollow core wall elements are manufactured using grade C40 to C50 concrete and contain typically 0.2 per cent reinforcement in each face. The hollow core is about 75 to 100 mm wide, depending on the overall thickness of the wall, and there are usually three or four cores in the length of the unit, see Fig. 3.35. Typical sizes of wall are $b = 200$ to 250 mm, core size $75 \times 1200$ mm, $H =$ half or full storey height, and $L = 3$ to 5 m. The core is formed using a sacrificial polystyrene (or similar) former. The inside surface therefore provides a good shear key with the site placed concrete. A rebate is provided in the sides of the unit to facilitate a keyed and grouted interface with the column.

Site placed reinforcement includes starter bars cast in the foundation (the length of bar projecting is $1.0 \times$ tension anchorage length) to which additional bars are lapped

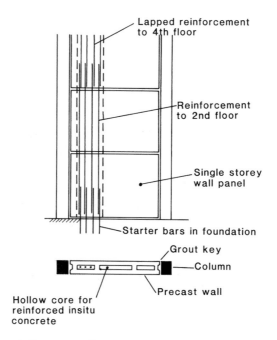

**Fig. 8.60** Cantilever hollow core walls.

to provide the holding down (tension) force $F_s$. Insitu concrete (usually grade C40) is placed in the hollow cores to provide the compressive force $F_c$. The reinforcement is provided along the full length of the wall, with the larger bars, typically 16 mm diameter high tensile bars, nearest to the sides, or furthest from the neutral axis. The centre-most cores of the wall may be reinforced using 10 or 12 mm diameter bars. The spacing of the bars is typically between 100 and 250 mm depending on the strength requirements.

The design is carried out at the ultimate limit state as follows. Referring to Fig. 8.61, the overturning moment $M$ and axial compression $N$ are calculated as before. Ignoring the compressive forces in the bars in the compression zone, the tensile force in the reinforcing bars $F_s$ and the compressive force in the concrete $F_c$ are given by:

$$F_s = 0.87 \, n A_s f_y \tag{8.36}$$

$$F_c = 0.45 \, f_{cu} b \, 0.9 \, X \tag{8.37}$$

where $n$ = number of bars in tension zone, i.e. in the zone $2(d - X)$ at a spacing $s$, where $d$ is the distance to the centroid of the steel, and $X$ is the depth to the neutral axis. The design value of $0.45 \, f_{cu}$ is used because the flexural stresses dominate compared with the smaller axial stress, and so BS 8110's idealized rectangular stress block devised for beam design is appropriate here [8.14]. The main difference between beam design and cantilever wall design is that the $d$ and $n$ will vary depending on the relative magnitudes of $N$ and M. Then:

$$N = F_c - F_s = 0.405 f_{cu} b X - \frac{1.74(d - x) A_s f_y}{s} \tag{8.38}$$

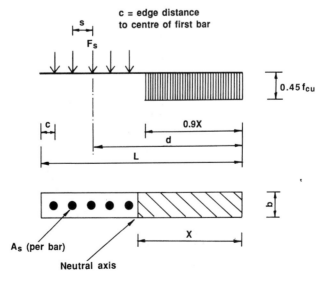

**Fig. 8.61** Design principles for cantilever shear walls.

$$M = 0.405 f_{cu} bX(d - 0.45X) - N(d - 0.5L) \text{ or}$$

$$= \frac{1.74(d - X)A_s f_y (d - 0.45X)}{s} + N(0.5L - 0.45X) \tag{8.39}$$

depending on whether the steel or concrete is critical. From these equations $X$ and $d$ may be determined for given geometry, materials and loads. The minimum length of wall is given by:

$$L = d + (d - X) + \text{cover to first bar (usually 50 to 60 mm)} \tag{8.40}$$

If this length is greater than the available length either the value of $b$, $f_{cu}$ or $A_s/s$ must be increased until it is less than the available length. The usual option is to decrease the bar spacing $s$ so that the same size of bar may be used in all walls. However, if $s$ becomes ridiculously small, say less than 75 mm, the size of bar should be increased. It is quite feasible to double up on the bars by placing them in pairs providing that the bond lengths are increased accordingly.

### EXERCISE 8.14    Hollow core cantilever wall design

Redesign the walls in Exercise 8.10 using hollow core cantilever walls 2.7 m long × 200 mm thick, i.e. the wall is to be placed between 300 mm square columns at 3.0 m centres. Assume that the flooring is supported by other beams and that the axial load on the wall is due only to self-weight.

Use $f_{cu} = 40$ N/mm$^2$ for the precast concrete, and 12 mm diameter high tensile bars. Assume cover to centre of first bar = 50 mm.

### Solution

Try $s = 150$ mm, $A_s = 113$ mm$^2$.

$N = [0.9 \times 0.2 \times 2.7 \times (4.0 + 3.0 + 3.0)] \times 24 = 117$ kN.

$M = 1.4 \times 0.8 \times 25 \times [(1.5 \times 10.050) + (3.0 \times 7.050) + (3.25 \times 4.050)] = 1383$ kN m.

Then $N = F_c - F_s = 0.405 f_{cu} bX - 1.74(d - X)A_s \dfrac{f_y}{s}$

$\qquad = 0.405 \times 40 \times 200X - 1.74(d - X)113 \times \dfrac{460}{150}$

$\qquad N = 3240X - 603(d - X) \tag{E.1}$

Checking first the compression capacity:

$M = 0.405 f_{cu} bX (d - 0.45X)$

$M = 3240 dX - 1458X^2 \tag{E.2}$

From Equation (E.1), $d = 6.373X - 194 \tag{E.3}$

Combining Equations (E.2) and (E.3) gives $X = 285$ mm and $d = 1624$ mm. The ratio $X/d = 0.175 < 0.5$ preventing any danger of over reinforcement. Thus $h = d + (d - X) + c = 3013$ mm $> 2700$ mm assumed. Therefore increase $A_s$ or decrease $s$ to increase capacity. Try $A_s = 113$ mm$^2$ and $s = 115$ mm. Repeat above to give $X = 318.7$ mm and $d = 1483$ mm. Thus $h = d + (d - X) + c = 2697$ mm $< 2700$ mm assumed.

Checking tensile capacity $M = 1.74\ (d - X)\ A_s f_y\ (d - 0.45X)/s + N(0.5L - 0.45X) = 1367$ kN m, we find this less than the required moment 1383 kN m. It is therefore necessary to decrease the spacing of the bars further, and to obtain values for $d$ and $X$ using Equation (E.1) and the equation $d + (d - X) = 2650$ mm.

Try $A_s = 113$ mm$^2$ and $s = 110$ mm.

Then $d = 1490$ mm and $X = 330$ mm, from which $M = 1420$ kN m $> 1383$ kN m.

*Use T 12 at 110 mm centres.*

## EXERCISE 8.15   Hollow core cantilever wall design

A hollow core cantilever wall 2.70 m long × 200 mm thick is to be reinforced using T 12 bars at 150 centres. Determine the maximum possible moment of resistance.

Use $f_{cu} = 40$ N/mm$^2$ and $f_y = 460$ N/mm$^2$. Assume cover to first bar is 50 mm.

### Solution

For maximum $M$, let $N = 0$.

Then $0.405\ f_{cu}bX = 1.74(d - X)f_y \dfrac{A_s}{s}$.

$A_s = 113$ mm$^2$ per bar; $s = 150$ mm; $b = 200$ mm.

Then:

$$3240\ X = 603(d - X)$$
$$d = 6.373\ X \tag{E.4}$$

From geometry:

$$d + (d - X) = 2700 - 50 = 2650 \text{ mm}$$
$$d = 1325 + 0.5\ X \tag{E.5}$$

Combining Equations (E.4) and (E.5) gives $X = 226$ mm and $d = 1438$ mm.

Then:

$M = 3240\ X\ (d - 0.45X) = 978$ kN m based on concrete.

or $M = 603\ (d - X)(d - 0.45X) = 977$ kN m based on steel (the two values should in fact be equal).

*Thus, M = 977 kN m when N = 0.*

## 8.7.8   Solid cantilever shear walls

Solid rc shear walls rely entirely on the connections in the horizontal and vertical planes. The horizontal sway displacement of these types of walls is a function of the shear-stiffness $K$ of both the wall panel and the joint. The latter is the most flexible (and weakest) part, possibly 1/100th of the former in terms of stiffness [8.46]. Two reasons for this large difference is that the precast wall must be strong enough to resist lifting forces at the factory within 12 hours of casting and have considerable stiffness and strength in service.

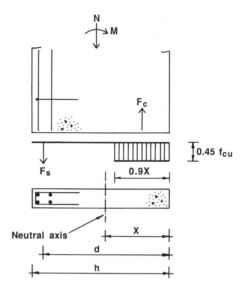

**Fig. 8.62** Design principles for solid cantilever shear walls.

As with the hollow core cantilever wall the design is carried out at the ultimate limit state with the moment $M$ and axial compression $N$ calculated as before. Referring to Fig. 8.62, the idealized stress block given in BS 8110 [8.14] is used to give the following forces in the concrete $F_c$ and reinforcing bars $F_s$:

$$F_c = 0.45f_{cu}b\,0.9X \tag{8.41}$$

where $X$ = distance to neutral axis.

$$F_s = 0.87A_sf_y \tag{8.42}$$

Then:

$$N = F_c - F_s = 0.405f_{cu}bX - 0.87A_sf_y \tag{8.43}$$

$$M = 0.405f_{cu}bX(d - 0.45X) - N(d - 0.5h)$$

or

$$= 0.87\,A_sf_y\,(d - 0.45X) + N(0.5h - 0.45X) \tag{8.44}$$

depending on whether the steel or concrete is critical. Here $h$ = length of wall, and $d$ is the effective depth to the centroid of the steel bars in tension. This point should be coincident with the centroid of the group of holding down bolts.

The side and end cover to these bars, and the clear horizontal distance between them, is very important and should not be less than 50 mm. This means that if two rows of bars are chosen the minimum thickness of the wall should be 200 mm. If one row of bars is used the thickness wall may be 150 mm – the minimum thickness for a precast wall.

The steel bar(s) at the bottom of the wall are often welded to wall splices or shoes and are therefore mild steel. The bar(s) at the top of the wall are often threaded, projecting through the splice plate. These bars are high tensile. To ensure the complete transfer of force, the vertical bars must be surrounded by horizontal steel on either side of the vertical bars. This horizontal bar is usually U-shaped with a minimum area of 0.4% × area of wall. Additional bars, typically 2 or 3 R 10 at 50 mm centres, are added at the very top and bottom as extra precaution against debonding.

Mechanical connectors between walls may be made using splice plates or the so-called 'wall shoe', Fig. 8.63. Designs vary according to load capacity and the breadth of wall available. In general the cost of such connectors increases as the breadth reduces.

The principle of the wall shoe is similar to the column shoe (Fig. 7.80) in that a punched hole plate is welded to a shroud box and rebars. The hole in the plate receives a threaded bar in the adjoining wall (or foundation). Sizes and capacities vary from 140 × 80 × 80 mm boxes with 16 mm diameters bars capable of resisting tensile forces of about 75 kN, to 250 × 150 × 150 mm with 32 mm bars having a capacity of at least 250 kN. As with the column shoe the design strength is based on static strength of the threaded bar, and not the strength of the welded box or the bond strength of the rebars anchored in the wall.

Base plates may be used to connect the wall at the foundation with the necessary modification to the design as the compression plate and tension plate are separate. If the length of the wall = $h$, then the length of the base plate must be equal to the length of the

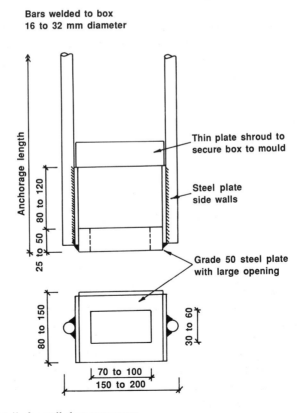

**Fig. 8.63** Details for wall shoe connector.

compression stress block $Xh$. Note that in the design of base plates, the compressive stress is limited to $0.4 f_{cu}'$. The method given in Section 7.14.2 may be used with the following modification to Equation (7.76):

$$\frac{N(e + 0.5h - Xh/2)}{0.4 f_{cu}'bh^2} = X\left(1 - \frac{Xh}{2h}\right) - 0.5 X^2 = X - X^2 \tag{8.45}$$

because the distance to the centre of the bolts in tension is equal to $Xh/2$.

We can simplify by letting a new stress factor

$$K = \frac{M'}{f_{cu}'bh^2} = \frac{N(e + 0.5h)}{f_{cu}'bh^2} \tag{8.46}$$

such that

$$X^2 - X\left(1 + \frac{N}{0.8 f_{cu}'bh}\right) + 2.5 K = 0 \tag{8.47}$$

where $b$ = breadth of wall and $h$ = length of wall in plane of bending. (Note: $K$ in Equation (8.46) is not the same as $K$ in Equation (7.77).) The magnitude of $N/0.8 f_{cu}bh$ will be small (typically 0.005) in comparison with all other values, and so Equation (8.47) may be approximated to:

$$X^2 - X + 2.5 K = 0 \tag{8.48}$$

without loss in accuracy.

The holding-down force in the bolts is given by Equation (7.74), and the bolt size by Equation (7.79). Usually there will be four bars per plate, two on either side of the wall.

The length of the base plate should be greater than $Xh$ by at least 50 mm to allow for edge spalling beneath the plate. It is quite common for the size of the plate to be governed by the physical size of the bolts and the spacing between them – more for the contractor's benefit of casting them into the foundation than for the precaster's needs. The edge distance to the bolts should be at least 50 mm, and the centre distance between them at least 75 mm for bolts up to 25 mm diameter, and 100 mm for larger diameters.

As a final check on the arithmetic, the sum of the forces in the bars and underneath the base plate should be equal to $M/z$, where $z = (d - Xd)/2$.

For the base plates, splices and/or wall shoes to be structurally effective the reinforcement in the wall must be gathered near to the connector to observe the strength of the shoe. The recommended spacing between the bars is 200 mm vertically and horizontally. The walls may be lightly symmetrically, partially prestressed (approximately <4 N/mm²) to assist in pitching the more slender walls. Prestressed walls would also be reinforced as for ordinary rc walls. It is therefore necessary to design the connections first before proceeding with the reinforcement design.

## EXERCISE 8.16    Solid cantilever wall design

Design a solid rc cantilever shear wall 2700 mm long × 200 mm thick. Given the ultimate axial force in the wall is 150 kN, and the overturning moment is 1400 kN m at the foundation, design a steel plate connection. If the ultimate axial force and moment at the first floor are 100 kN and 580 kN m, design a steel splice connection which is flush with the sides of the wall.

Use $f_{cu} = f_{cu}' = 40$ N/mm², $f_y = 460$ N/mm², $f_y = 250$ N/mm² for bars welded to plates, grade 43 steel plate and weld. Cover to bars in wall = 35 mm.

**Solution**

Design of base plate

Let $b_{plate} = 200 + 2 \times 100 = 400$ mm.

$d = 2700$ mm.

$e = 1400 \times 10^3/150 = 9333$ mm.

Then $K = \dfrac{150 \times 10^3 (9333 + 1350)}{40 \times 400 \times 2700^2} = 13.74 \times 10^{-3}$.

Hence $X = 0.0356$, and $Xh = 96$ mm.

Then $F_c = 0.4 \times 40 \times 400 \times 96 \times 10^{-3} = 615$ kN.

Bolt force $F = 615 - 150 = 465$ kN.

*Use 6 no. M 20 grade 8:8 holding-down bolts (660 kN), 3 no. each side of wall at 75 mm centres.*

Plate length $> 96 + 50 = 144$ mm from calculations, or $50 + 75 + 75 + 50 = 250$ mm to allow access for fixing bolts.

Actual compressive stress beneath plate $= \dfrac{615 \times 10^3}{400 \times 250} = 6.15$ N/mm².

Plate thickness given by Equation (7.84) and/or Equation (7.81) using $L = 100$ mm, $m = $ say 110 mm, $b = 250$ mm, and $F = \dfrac{465}{2} = 232.5$ kN.

$$t = \frac{\sqrt{2 \times 6.15 \times 100^2}}{275} = 21.1 \text{ mm based on compression beneath plate}$$

or

$$t = \frac{\sqrt{4 \times 232.5 \times 10^3 \times 110}}{250 \times 275} = 38.6 \text{ mm based on tension in bolts.}$$

*Use 2 no. $\times 250 \times 400 \times 40$ mm mild steel plates.*

**Reinforcement in wall**

Place the reinforcing bars such that the effective depth $d = 2700 - \dfrac{250}{2} = 2575$ mm.

Using $N = 0.405 f_{cu}bX - 0.87 A_s f_y = 3240 X - 217.5 A_s$.

Then $X = 46.3 + 0.067 A_s$.

Also $M = 0.87 A_s f_y (d - 0.45X) - N(0.5h - 0.45X)$.

$1400 \times 10^6 = 217.5 A_s (2575 - 21 - 0.03A_s) + [150 \times 10^3 (1350 - 21 - 0.03A_s)]$

Hence $X = 196.3$ mm, which coincides closely with the size of the base plate.

$A_s = 2239$ mm².

*Use 6 no. R 25 bars (2950) at 75 mm centres, 3 each face.*

These bars should be aligned with the positions of the holding-down bolts.

High tensile bars to be lapped with the above. Repeat the above using $f_y = 460$ N/mm$^2$.

Then $X = 46.3 + 0.1235A_s$.

$1400 \times 10^6 = 400.2A_s \, (2575 - 21 - 0.056A_s) + [150 \times 10^3 (1350 - 21 - 0.056A_s)]$

$A_s = 1217$ mm$^2$.

*Use 6 no. T 20 bars (1890) lapped to R 25 bars.*

Vertical steel in remainder of wall = 0.4% = 800 mm$^2$/m per both faces.

Distance between main bars = $2700 - 200 - 200 = 2300$ mm.

*Use T 10 at 190 mm centres (413).*

Horizontal steel in wall = 0.25% = 500 mm$^2$/m per both faces.

*Use T 8 at 200 mm centres (252), plus T 8 at 200 mm centres, U-bars at end of wall.*

**Wall splice at first floor**

Referring to Fig. 8.64, ensure that the effective depth to the bars in the splice is maintained at $d = 2575$ mm.

Design mild steel bars in upper wall welded to splice plate:

Using $N = 0.405 \, f_{cu}bX - 0.87 \, A_sf_y = 3240 \, X - 217.5 \, A_s$.

Then $X = 30.9 + 0.067 \, A_s$.

Also $M = 0.87 \, A_sf_y \, (d - 0.45X) + N \, (0.5h - 0.45X)$.

$580 \times 10^6 = 217.5 \, A_s \, [2575 - 14 - 0.03 \, A_s] + [100 \times 10^3 \, (1350 - 14 - 0.03 \, A_s)]$.

$A_s = 814$ mm$^2$.

*Use two R 16 bars (804) in between the projecting threaded bars.*

High tensile threaded bars in lower wall. Repeat the above using $f_y = 460$ N/mm$^2$.

Then $X = 30.9 + 0.1235 \, A_s$.

$580 \times 10^6 = 400.2 \, A_s \, (2575 - 14 - 0.056 \, A_s) + [100 \times 10^3 \, (1350 - 14 - 0.056 \, A_s)]$.

$A_s = 442$ mm$^2$.

*Use four T 16 bars threaded to M 16.*

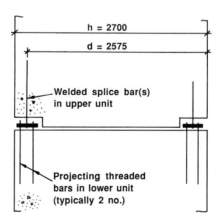

**Fig. 8.64** Details to Exercise 8.16.

### Splice plate design

Referring to Fig. 8.65, size of plate = 250 mm long × 200 mm wide.
Upper bars welded to plate to be located in mid-sides of plate. Lower threaded bars to be located at corners of plate. Refer to Section 7.13.4 for design method.
Edge distance to bars = cover 35 + link 8 + radius 8 = 51 mm, use 60 mm.

Distance between bar centres $= \dfrac{(250-120)}{2} = 65$ mm.

Force in bars in splice plate = $0.87 \times 460 \times 442 \times 10^{-3}/4 = 44.2$ kN.
Lever arm = 34 mm.
Moment in plate = $44.2 \times 34 = 1503$ kN mm.
Effective breadth = 76.3 mm.

Then $t = \sqrt{\dfrac{4 \times 1503 \times 10^3}{76.3 \times 275}} = 16.9$ mm.

Shear in plate = 44.2 kN.

Then $t = \dfrac{44.2 \times 10^3}{76.3 \times 165} = 3.5$ mm.

*Use 250 × 200 × 20 mm splice plate.*

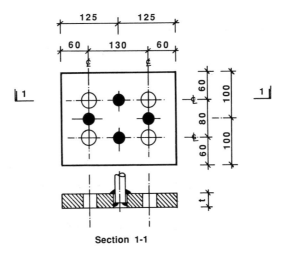

Section 1-1

**Fig. 8.65** Details to Exercise 8.16.

# Chapter 9

# Structural Integrity and the Design for Accidental Loading

*This chapter describes how a precast structure is designed against possible progressive collapse, and looks at the way connections are designed so that alternative load paths are available in the event of an accident.*

## 9.1  PRECAST FRAME INTEGRITY – THE VITAL ISSUE

Precast concrete frames are scrutinized by checking authorities more for structural stability, robustness and integrity than for the performance of the individual reinforced and prestressed concrete components and connections. This means that there is more concern for the way in which precast members interact with each other, than for the structural performance of an individual beam, slab or column. The main question is whether the frame will react to forces as an entity or, as in the case of a Meccano set, will the removal of one bolt cause the whole structure to collapse? Codes of practice refer to this topic as 'robustness' and require that:

'… situations should be avoided where damage to small areas of a structure or failure of single elements may lead to collapse of major parts of the structure.' [9.1].

The popular term for this is 'structural integrity' and the design aspects involve all the items dealt with in this book, namely component design, connections, diaphragm action and structural stability. These items cannot be dealt with in isolation and, contrary to popular opinion, precast design is an integrated process.

Frame stability is a crucial issue in multi-storey precast construction and adequate stiffness, strength, and the provisions to guard against the possibility of a collapse which is disproportionate to its cause must be provided. Tie forces between the precast components must be mobilized in the event of accidental damage or abnormal loading, and alternative means of transferring loads to the foundations must be sought if a component is rendered unserviceable. In all cases the precast floor must act as a rigid, yet ductile, horizontal diaphragm.

## 9.2  DUCTILE FRAME DESIGN

In precast work the design of ductile structures is entirely dependent on the stiffness and ductility of the connections. The stiffness is the rate of change of load vs. deformation, i.e. the tangent gradient $K$ in the idealized plot shown in Fig. 9.1. Ductility is measured in terms of a ductility factor $\Phi$, the ratio between the deformation at the ultimate load $\Delta_u$ and the deformation at the yield point $\Delta_y$ in Fig. 9.1. These are simple definitions and the two characteristics are easy to understand in terms of a single element or joint. However, their effect on the behaviour of three-dimensional precast concrete structures is complex.

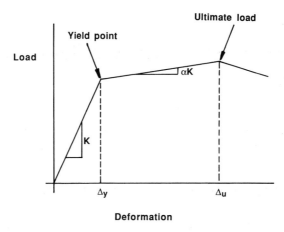

**Fig. 9.1** Idealized deformation behaviour in non-linear mediums.

The flexural stiffness of many of the beam–column connections in use is between $0.1\,K_b$ and $1.3\,K_b$ where $K_b = 4EI/L$. This is the flexural stiffness of a prismatic beam of length $L$ to which the connector is attached [9.2]. The ductility factor $\Phi$ for these connections varies from about 1.5 to more than 6. Figure 9.2 (a) and (b) shows examples of moment vs. connection rotation data obtained from the full scale beam–column–hollow core slab assembly shown in Fig. 9.3. The relative beam–column rotations were measured at each side of the column with the positive moment inducing hogging at the column face. Initial cracking occurred at the face of the column at approximately 5 mrad. but the connection responded to increased moment at four times this value.

Similarly, the stiffness of the half-joint connection tested (by the author) as shown in Fig. 4.26 is only about twice the shear stiffness of the end of a beam subjected to an end shear force ($1.2\,VL/AG$). This information is hardly surprising as the connections are designed as pinned joints with large shear capacities. The results of these and many other tests show that the designer is able to design the connections in order to control frame behaviour.

Precast frame design is not based on 'capacity' design procedures where the ductility of a component or connection is considered implicity, and a value of $\Phi$ is aimed for in the way that strength and stiffness is. However, if a connection is found to be ductile, this is usually the result of features being introduced to the design for reasons other than making the particular connection fail in a ductile manner.

For example, consider the beam–column connection in Fig. 9.4. The steel billet has been designed to fail in bending, rather than by shear, crushing or buckling. The column has also been designed such that the confinement links will yield and enable the concrete underneath the billet to fail gradually (more like the compressive failure of a cylinder than a cube). The beam half-joint has been designed so that bursting and shear failures are controlled by strategically placed and fully anchored reinforcement. If any one of these three elements was to fail prematurely and catastrophically the connector load vs. deformation behaviour might be as shown in Curve A in Fig. 9.5. If the connector was so designed that all three elements were to fail simultaneously the deformation might be as Curve B, with the changing stiffness in the 'grey' area reflecting the

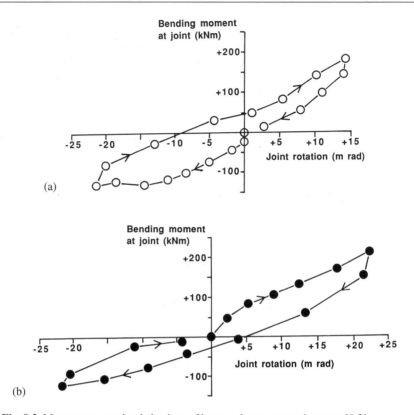

**Fig. 9.2** Moment vs. rotation behaviour of beam column connection tests [9.2].

redistribution of forces from one zone to another. However, the presence of the continuity tie steel passing through the floor slab and column will increase the ductility of the entire connection by allowing more than one of the constituent parts of the connection to yield without total failure. The results would be Curve C.

The design of components and connections should be such that undesirable modes of failure are prevented. These include:

- shear-tension failures in incorrectly or unreinforced joints subjected to building movements, etc., Fig. 9.6
- buckling of compression reinforcement in site-made connections where confinement links have not been specified or omitted because of fixing difficulties, Fig. 9.7
- anchorage failures in tie steel because inadequate length of the projecting reinforcement and inadequate bearing lengths, Fig. 9.8
- large plastic rotations in connections that have been designed as pinned-joints but have been constructed so that bending moments and torque have been transferred into components which have not been designed to resist the resulting forces, e.g. Fig. 9.9 for torsion and Fig. 7.5 for flexure.

The relationships between the stiffness and ductility of individual components/connections and the behaviour of the total structure is a complex non-linear three-dimensional problem

**Fig. 9.3** Test assembly used in flexural connection tests [9.2]. Column and beam size = 300 × 300 mm. Slab depth = 200 mm (Roth type). Tie steel across top of beam = 2 no. T 25 ribbed bars. Precast and insitu concrete strengths = 50 N/mm$^2$ and 25 N/mm$^2$, respectively.

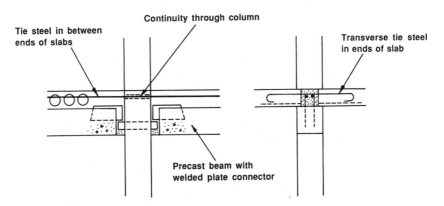

**Fig. 9.4** Principle of ductile beam–column connections because of continuous tie steel in the insitu infill strip over the beam.

with which research institutions are only just beginning to grapple [9.5]. Most of the work in this respect is being carried out on steel and insitu concrete frames where the fundamental behaviour of the materials is already understood. The work on precast skeletal frames is less advanced because the integrity and robustness of these structures involve an even larger number of parameters, including workmanship – the most difficult aspect to quantify.

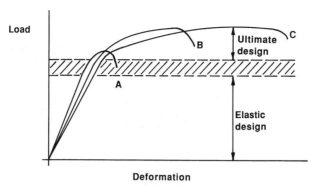

**Fig. 9.5** Different failure criterion in connections.

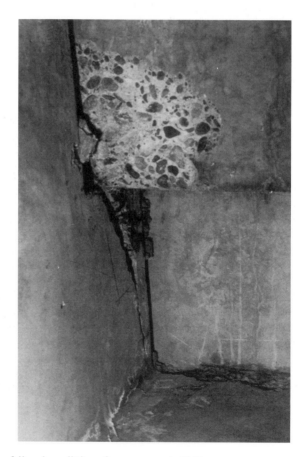

**Fig. 9.6** Shear failure by splitting of cover concrete [9.3].

As a consequence the industry has been forced to adopt design rules based on semi-empirical work carried out in the 1970s on model tests. The idea of using steel ties to maintain integrity is an acceptable one providing that the workmanship on site is compatible with the designer's good intentions.

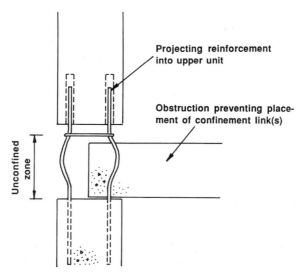

**Fig. 9.7**  Buckling in unconfined compression bars.

**Fig. 9.8**  Ultimate failure of displaced hollow core slab by Engström [9.4].

## 9.3   BACKGROUND TO THE PRESENT REQUIREMENTS

The term 'progressive collapse' was first used in the UK following the partial collapse of a precast concrete wall frame at Ronan Point, London in 1968 (see Fig. 9.10). A gas explosion in a corner room on the eighteenth floor blew out one of the external walls,

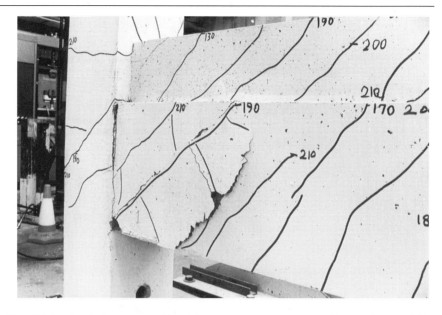

**Fig. 9.9** Torsional moments transferred into columns because continuity steel is provided in the floor slab.

and because of inadequate structural continuity between the wall and floor elements, the removal of one wall element was sufficient to cause the total collapse of about 23 m$^2$ of floor area per storey over the height of the entire building.

Immediately following the accident several contracts were suspended for a time until the results of structural investigations and testing at Imperial College, London and the UK Building Research Station were known. The investigations found deficiencies in the manner in which the precast components were tied to one another, and poor detailing and unsatisfactory workmanship were blamed for the disproportionate amount of damage. However, the most crucial factor was that there was no suitable design information to guide the designer towards a robust solution.

Today, three alternative methods are used to design for accidental damage. Referring to Fig. 9.11 (a)–(c) these are:

(a) protected members
(b) alternative load paths
(c) use of ties.

The methods may be adopted separately or in combination in different parts of the same structure. It is not permitted to superimpose the mechanical effects of the three methods, for example one cannot say that a member is partially protected and partially tied – it has to be either fully protected or fully tied.

Methods (a) and (b) are classified as 'direct' methods because an appreciation of the severity and possible location of any accidental damage is known or assumed. The structure can be designed to withstand the occurrence of primary damage without an uncontrolled escalation of the damage. The key words here are 'uncontrolled escalation'. In method

**Fig. 9.10** Progressive collapse of part of Ronan Point (courtesy of the Brick Development Association).

(c), in the event of accidental loading, a notional tie force is capable of being activated at every location in the structure. Because the severity and possible location of the loads are not known, the method is classified as 'indirect'.

Following the Ronan Point collapse, tests were carried out [9.6] in which horizontal pressures, due to controlled gas explosions inside chambers, were monitored and measured. Pressures of 15.8, 16.9, 18.5 and 22.7 kPa caused structural damage without collapse. A pressure of 24.9 kPa caused considerable damage with an outer wall showing a characteristic 'roof shape' fracture-like crack pattern, but without collapse. From these (and other) test results, the 5th Amendment to the UK Building Regulations in 1978 [9.7] specified that any element on which the stability of the structure is to depend, following an accident, should be able to withstand a design ultimate pressure of 34 kPa, to which no partial safety factors should be applied. Such a member, including its connections to adjacent members, is called a 'protected member', or (using BS 8110, Part 2, clause 2.6 terminology [9.1]) a 'key element'.

These members should be designed, constructed and protected as necessary to prevent removal by accident. All other structural components that provide support vital to the

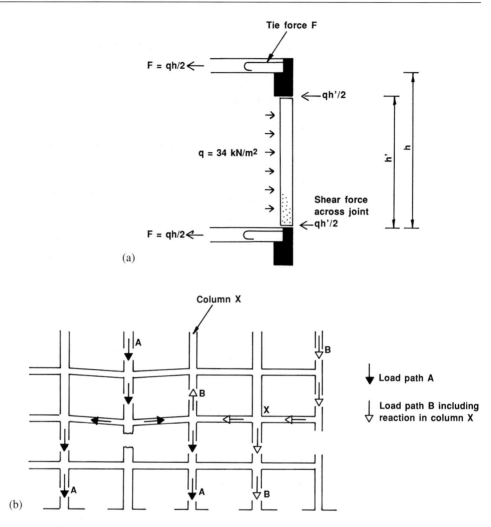

**Fig. 9.11** Alternative means of protection against progressive collapse. (a) Protected members; (b) alternative load paths and (c) use of ties.

stability of a key element should also be considered a key element. Few precast structures have incorporated protected members, although certain non-load bearing precast cladding panels have been designed as such in high risk buildings.

An alternative direct design method, called 'alternative path design' specifies walls, beams and columns, or parts thereof, which are deemed to have failed and the remaining structure is analysed based on these criteria. The components in the remaining structure are called 'bridging elements'. At each floor level in turn (including basement floors), every vertical load bearing member, except for key elements, is sacrificed and the design should be such that collapse of a *significant* part of the structure does not result, as shown in Fig. 9.12. Note that BS 8110 does not quantify the permissible area of collapse. This method is not widely used in the UK because of the implicit necessity to satisfy the

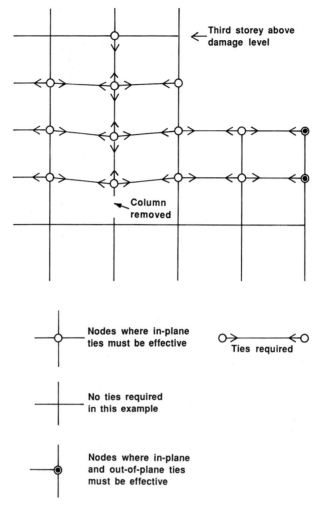

(c)

**Fig. 9.11** (*continued*)

requirements of BS 8110 [9.1] which permits the indirect method of using a fully tied solution.

The indirect approach to the control of accidental damage does *not* depend on the representation of the force vectors likely to be encountered, or on the assessment of the effects of limited damage. The design is therefore based on empirical rules to provide minimum levels of strength, continuity and ductility. The first stage in securing stability is a satisfactory structural layout of slabs, beams, walls and columns where the fewest possible number of components rely upon others for safety. Figure 9.13 shows the kind of layout to avoid where, for various reasons, columns are not permitted at certain locations. The load in the slab at point $X$ must pass through no less than six connections to reach the ground, i.e.:

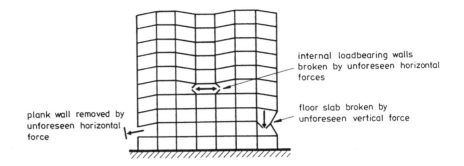

internal loadbearing walls broken by unforeseen horizontal forces

floor slab broken by unforeseen vertical force

plank wall removed by unforeseen horizontal force

**Fig. 9.12** Containment of damage and debris loading in multi-storey structures [9.16].

(1) slab bearing
(2) beam-to-beam connection
(3) beam-to-stub column
(4) stub column-to-beam
(5) beam-to-beam
(6) beam-to-column.

Failure at one point (Point (6)) would affect six structural components. Similarly at Point Y the load must be carried through several connections, including a cantilever beam end connection.

Continuity of tie steel placed in the slabs to span over beams or walls, and for beams either around the sides of, or through sleeves in columns as shown in Fig. 9.14 afford the highest degree of protection. Continuity around corners is also important, Fig. 9.15, where specially profiled hollow core slabs are specified to enable the correct bend radius of 500 mm to be provided in the tie strands. If catenary action is assumed as shown in Fig. 9.16, allowance should be made for the horizontal reactions necessary for equilibrium.

To proceed with the design: BS 8110, Part 1, clauses 2.2.2.2 and 3.12.3 set out the requirements as to when and how the ties should be used, and this *de facto* avoids the requirement to design and construct key and bridging elements. Interaction between elements is obtained by the use of horizontal floor and vertical column and wall ties positioned as shown in Fig. 9.17. These are as follows:

- horizontal internal and peripheral ties, which must also be anchored to vertical load bearing elements
- vertical ties.

Horizontal ties are further divided into floor and beam ties (Fig. 9.18):

- floor ties, to provide continuity between floor slabs, or between floor slabs and beams
- internal and peripheral beam ties, to provide continuity between main support beams
- gable peripheral beam ties, to provide continuity between lines of main support beams.

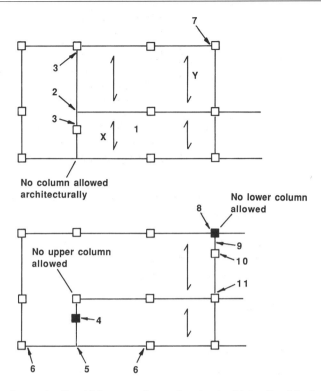

**Fig. 9.13** Floor layout details which may endanger the structural integrity of the building in case of an accident.

## 9.4  THE FULLY TIED SOLUTION

### 9.4.1  Horizontal ties

The specific requirements relating to ties in precast concrete structures are given in BS 8110, Part 1, clause 5.1.8 [9.1]. These are satisfied either by using individual continuous ties provided explicitly for this purpose in insitu concrete strips, or using ties partly in the insitu concrete and partly in the precast components.

The structural model is as follows. In the event of the complete loss of a supporting column or beam at floor level $n$, the floor plate at levels $n$ and $(n + 1)$ must resist total collapse by acting in catenary as shown in Fig. 9.16. At the moment the accident occurs an alternative load path for the floor beams which were previously supported by the damaged member may not be immediately available. If a column is removed the tie forces over the beams must be mobilized. If a beam is lost then it is the floor ties which do the work. With increasing deformation there will be a considerable amount of moment distribution away from the critical region. This will be followed by large plastic hinge rotations at the column connections which will lead to membrane action in the floor plate. A new equilibrium state will develop as shown in Fig. 9.16. The column which was directly above the damaged unit carries the beam end reactions of the beams at level

**Fig. 9.14** Sleeves in columns to receive continuity tie steel.

**Fig. 9.15** Use of helical strand as stability tie steel at external corners.

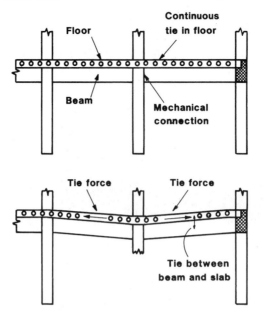

**Fig. 9.16** Catenary action between precast members.

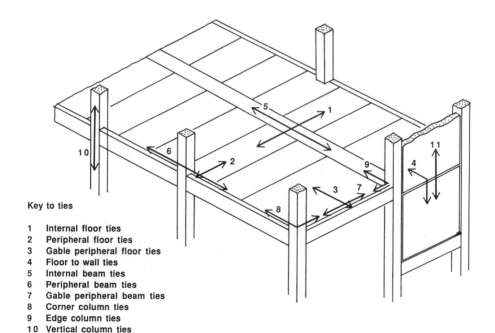

Key to ties

1  Internal floor ties
2  Peripheral floor ties
3  Gable peripheral floor ties
4  Floor to wall ties
5  Internal beam ties
6  Peripheral beam ties
7  Gable peripheral beam ties
8  Corner column ties
9  Edge column ties
10  Vertical column ties
11  Vertical wall ties

**Fig. 9.17** Types of ties in skeletal frames.

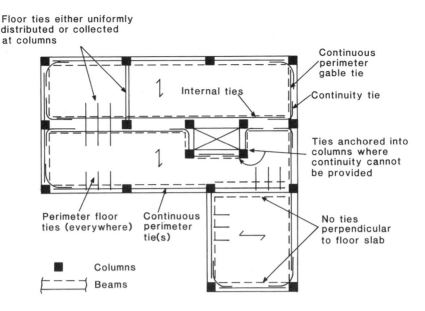

**Fig. 9.18** Definitions of floor ties [9.17].

$n$ as a tie in vertical suspension. The column must therefore act as a tie according to Section 9.4.5 to prevent debris loading on the floor below.

This then is a so-called 'collapse mechanism' which may be highly complex in buildings of irregular form. The problem is also three-dimensional and non-isotropic depending upon the orientation of the floor slabs, walls and beams. A floor plate comprising structurally screeded double-tee units will be particularly difficult to analyse, and no attempt to do so has yet been made.

Tie bars are either a high tensile deformed bar using a design strength of 460 N/mm$^2$ ($\gamma_m = 1.0$), or (increasingly popular) a helical prestressing strand using a design strength of 1580 N/mm$^2$, Fig. 9.15. (Note: ten per cent less than $f_{pu}$.) The strand is laid unstressed, but stretched tightly. The favourable mechanical properties of strand and the long lengths available on site make this an attractive alternative to the high tensile bar.

A tie is considered continuous if it is correctly lapped or spliced according to BS 8110, Part 1, clause 5.3.4. The details given for column splices in Section 7.13.2 may be appropriate here. The lap length for a deformed bar is usually taken as 40 diameters (according to BS 8110, Part 1, Table 3.29 for grade C25 concrete). The lapping length for the strand is based on the transmission lengths given in BS 8110, Part 1, clause 4.10.3. Using $f_{cu} = 25$ N/mm$^2$ in place of $f_{ci}$ and $K_t = 360$ for 7-wire drawn strand, a length of $72 \times$ diameter is found. Omar [9.8] carried out pull-out tests on a number of 12.5 mm diameter helical strands embedded over a length of 1200 mm (94 diameter) into 100 mm square prisms of grade C25 concrete. A maximum stress of 1790 N/mm$^2$ was measured in the strand prior to fracturing, with zero slip being recorded at the rear end of the strand. Thus a lap length of 1200 mm is usually specified.

The position of the lap is staggered to avoid concentration bars. The tie, of diameter $\phi$, must be embedded in insitu concrete of at least $2(\phi + h_{agg} + 5$ mm) wide. In most instances this means that the insitu must be at least 50 mm wide (as explained in beam design in Section 4.3.4). The partial safety factor for steel reinforcement may be taken as 1.0 (BS 8110, Part 1, clause 2.2.4.2). The partial safety factor for concrete in flexure is 1.3 but moments of resistance are not determined in this kind of situation.

Ties passing through precast columns or walls are fed through oversized sleeves (usually two to three times the diameter of the tie bar(s)) and later concreted in. Problems occur with this detail at external corners where the tie bars change direction, of up to 90°. These sleeves are impossible to form. Also, the position of the sleeve must be known in advance and correctly positioned in the mould. This can often cause problems of two kinds. First, many manufacturers prefer not to puncture the side of the mould unnecessarily. Second, too many slots and sleeves can weaken columns and lead to small diagonal cracking at the corners of the sleeve.

To overcome these problems precast contractors prefer to run tie steel past the face of the column, rather than through it. This means that the width of the beam, over which the tie bars are placed, must be greater than the width of the column around which they will pass. Otherwise part of the floor slab has to be broken out to allow the ties to be bent and cranked around the corners of the column.

Continuity around corners is achieved either by concreting the tie into a 45° splayed corner or using threaded couplers or anchored plates, as shown in Fig. 9.19. Continuity is not usually provided between the flights and landings in precast staircases because these units are not essential to the stability of the frame. However, Fig. 4.48(b) shows how this might be achieved if, in the very rare event, continuity of ties could only be placed in the floor landing.

The main problem with using gathered tie steel in this manner occurs at perimeter re-entrant corners where, as shown in Fig. 9.20, the resultant force from the perimeter tie steel is pulling outwards into unconfined space. One cannot rely on the shear capacity of the column to restrain this force. Two alternatives are possible, as shown in Fig. 9.21:

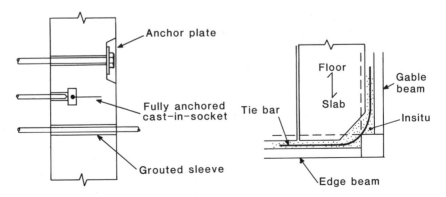

**Fig. 9.19** Ties anchored using cast-in sockets or bearing plates.

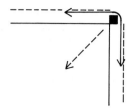

**Fig. 9.20** Stability ties at re-entrant corners.

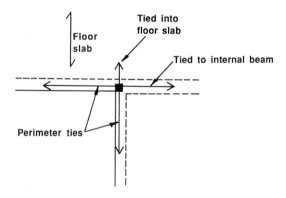

**Fig. 9.21** Continuity of ties at re-entrant corners.

(1) The tie steel continues through or past the side of the column as though it were an edge column. This is only possible if an internal beam is located in line with the edge beam, which is not always the case as shown in Fig. 9.22. In many buildings, particularly offices, a lift shaft or stair-well is located at re-entrant corners. This may cause additional tying problems because of the lack of space to place ties in walls, etc. Cast-in couplers (or similar) may be used to anchor the tie steel to the walls.

(2) A structural screed containing a steel fabric may be used in this localized area only, but this will probably cause embarrassment to the structural zone in the remainder of the floor. If the floor spans in this part of the building can be proportioned so that a shallower precast floor or precast beam is used in this area, a topping screed will not increase the overall floor depth. A 50 mm reduction is all that is required.

### 9.4.2   Calculation of tie forces

The basic horizontal tie force $F_t$ is given in BS 8110, Part 1, clause 3.12.3 [9.1], by the lesser of:

$$F_t = (20 + 4n) \text{ or } 60 \text{ kN/m width} \tag{9.1}$$

where $n$ is the number of storeys including basements. $F_t$ is considered as an ultimate value and is *not* subjected to the further partial safety factor of $\gamma_f = 1.05$ given in clause 2.4.3.2.

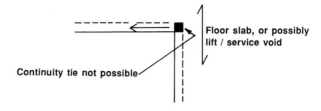

**Fig. 9.22** Difficulties in positioning ties at re-entrant corners.

If the total characteristic dead $(g_k)$ + live $(q_k)$ loading is greater than 7.5 kN/m² and/or the distances $(l_r)$ between the columns or walls in the direction of the tie is greater than 5 m, the force is modified as the greater of:

$$F_t' = \left[ \frac{(g_k + q_k)}{7.5} \right] \times \left[ \frac{l_r}{5} \right] F_t \qquad (9.2)$$

or      $F_t' = 1.0\, F_t$ \hfill (9.3)

The catenary tie forces are calculated by simple statics, but the assumptions implicit in their formulations are rather arbitrary. The precast reinforced or prestressed members are designed so that they can generate the necessary tie forces, and so it is only continuity tie forces between the units which must be provided insitu.

Internal floor ties parallel with the span of the flooring are either distributed evenly using short lengths of tie steel anchored by bond into the opened cores of the hollow core floor units, or grouped in full depth insitu strips at positions coincident with columns. The FIP manual [9.9] on hollow core floors shows a large range of alternative details from which Fig. 9.23(a) has been taken.

In the first case, the mechanism relies on an adequate pull-out force generated by tie steel cast into insitu concreted hollow cores (see Section 9.5). A typical configuration is to use 12 mm diameter high tensile bar or 12.5 mm diameter helical strand at 600 mm centres, i.e. two bars per 1200 mm wide slab. For greater quantities it is better to decrease the bar spacing rather than to increase the size or number of bars. At no time should two bars be placed in one core. If high tensile bars are cast into beams and arranged so that they are to be manually bent into the slots in the slab, the maximum size of bar that can easily be handled is 12 mm diameter. Providing that the floor slabs are tied to edge beams or tied across internal beams the tie force across the floor is provided in the precast floor unit itself. Adequate reinforcement will be provided in all but the least reinforced units using the minimum area of high tensile steel of 0.13 per cent $A_c$.

Ties along gable edge beams are placed into the broken out cores of slabs at intervals varying between 1.2 and 2 m. As shown in Fig. 9.23(b) the success of the ties relies on adequate anchorage of small loops cast into insitu concrete which itself is locked into the bottom of the hollow core. Generous openings (say 300 mm long) should be made in the floor slabs to ensure that any projecting tie steel in the beam may be lapped without damage to the slab or tie bar. Here the tie force is $F_t$.

The magnitude of the tie forces $F_t'$ between precast hollow core slabs is as given by Equation (9.2) or (9.3). The area of steel is:

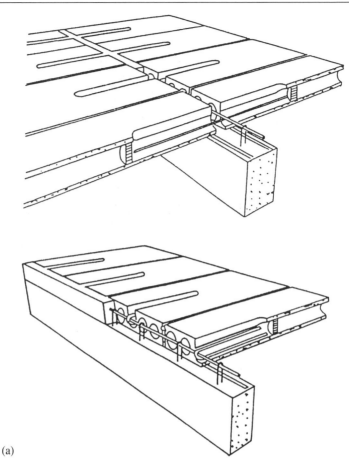

(a)

**Fig. 9.23** Continuity for floor ties in hollow core slabs. (a) Parallel with the floor slab [9.9], and (b) perpendicular to the floor slab.

$$A_s = F_t'/\gamma_m f_y \qquad (9.4)$$

where $\gamma_m = 1.0$ for reinforcement (clause 2.4.4.2).

Continuity between beams is provided across the column line by calculating the magnitude of the internal and peripheral tie force $F_t'$ according to Equation (9.2) or (9.3). Here the tie force in the beam $F_t'$ is the summation of all the internal tie forces across the span $L$ of the floor, i.e.:

$$F_t' \text{ (beam)} = F_t'L \qquad (9.5)$$

If the spans of the floor slabs on either side of the beam are different, $L_1$ and $L_2$, the summation of half of each span, i.e. $0.5 (L_1 + L_2)$, is used to replace $L$ in Equation (9.5). In an edge beam $L_2 = 0$. Where the floor slabs to one side of the beam are spanning parallel with the beam, and on the other side are spanning on to the beam with a span $L_1$, the beam tie force is taken as:

$$F_t' \text{ (beam)} = 1.0 \, F_t' + 0.5L_1 \, F_t' \qquad (9.6)$$

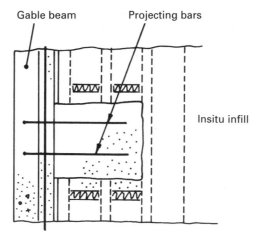

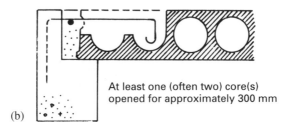

(b)

**Fig. 9.23** (*continued*)

Where the floor slabs are spanning parallel with the span of an edge beam, a nominal tie force $1.0\ F_t'$ need only be provided. The area of steel is calculated according to Equation (9.4).

The internal tie bars should be distributed *equally* either side of the centre line of the beam, be separated by a distance of at least 15 mm to ensure adequate bond (assuming 10 mm size aggregate), and be positioned underneath projecting loops from the beam. Peripheral tie bars should be similarly positioned and, according to BS 8110, Part 1, clause 3.12.3.5, be located within 1.2 m of the edge of the building. This is not usually a problem except in the case of cantilevers where there is no beam nearer than 1.2 m from the end of the cantilever. In this case the tie must be located in a special edge beam which is cast for the sole purpose of providing the peripheral tie.

The success of the tie beams depends largely on detailing. Figure 9.24 illustrates the concept; two tie bars are fixed on site and pass underneath the projecting reinforcement loops. The hooked bars, which must pass underneath, are fully anchored into the hollow cores of the slab. Finally, the tie bars pass through small sleeves preformed in the columns, or pass by the side of the column. Figure 9.25 shows how continuity may be satisfied at external positions.

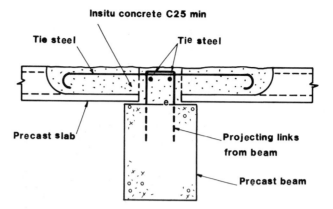

**Fig. 9.24** Stability ties at internal positions.

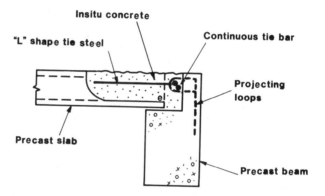

**Fig. 9.25** Stability ties at external positions.

### 9.4.3   Horizontal ties to columns

Each external column is tied horizontally into the structure at each floor and roof level with ties capable of developing a force equal to the greater of (a) or (b) as follows (BS 8110, Part 1, clause 3.12.3.6 [9.1]):

(a)  the lesser of 2.0 $F_t$ or $0.4h\,F_t$ where $h$ is the storey height in metres
(b)  three per cent of the vertical ultimate axial force carried by the column at that level using $\gamma_f = 1.05$.

Corner columns should be tied into the structure in two mutually perpendicular directions with the above force.

Ties may be provided at columns in one of two ways. If there is no positive tie force between the beam and column, e.g. the beam–column connector in Fig. 7.42(a) where only a frictional bearing is provided, loose tie steel should be placed on site and be fully anchored as shown in Figs 9.14 and 9.26. The net cross-section of any threaded bar (to the root of the thread) is used in calculating the area of the tie bar. The column

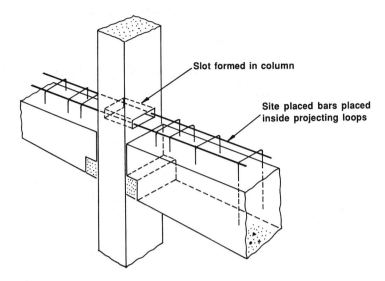

**Fig. 9.26** Continuity for beam ties through columns.

should be checked to ensure that the tie force can be resisted, and all anchor plates, cast-in sockets, etc. should be designed accordingly. Resin anchored reinforcement may also be used, but this involves the use of proprietary materials and techniques. The manufacturer's specification with regard to the size of hole and insertion of the resin should be followed.

The second method is to design the beam–connector so that the horizontal shear capacity may be used to provide the necessary tie force from the column to the beam. This imposes additional axial forces in the beam which must be dealt with as explained in Section 4.3.10. There must be a positive no-slip horizontal connection between the beam and column. This is guaranteed by surrounding any mechanical connectors, such as bolts or dowels, with insitu concrete. The site workmanship should be especially supervised in these situations because evidence of poor compaction can easily be covered over.

**EXERCISE 9.1**

Calculate the horizontal tie requirements in the floors of the four-storey building shown in Fig. 9.27 where the floor-to-floor storey height = 3.2 m. All floors and roof slabs are 200 mm deep hcus. The characteristic superimposed dead and live floor loads are 1.6 kN/m$^2$ and 5.0 kN/m$^2$ respectively. The characteristic superimposed dead and live roof loads are 2.0 kN/m$^2$ and 1.5 kN/m$^2$, respectively. Assume that the building façade dead load is 12 kN/m. Assume that the beam–column connector has no horizontal tie capacity.

Use $f_{cu}' = 25$ N/mm$^2$ for the insitu infill concrete $f_y = 460$ N/mm$^2$.

*Solution*

$F = 20 + (4 \times 4) = 36$ kN/m.

$g_k$ = self-weight of slab (Table 3.1) = 3.0 kN/m$^2$, plus superimposed dead load = 1.6 kN/m$^2$.

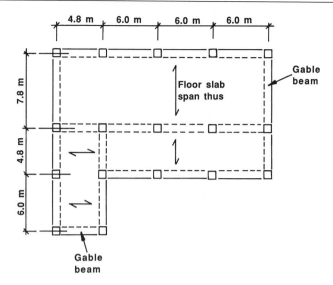

**Fig. 9.27** Details to Exercise 9.1.

Total $g_k$ = 4.6 kN/m².
$q_k$ = 5.0 kN/m².
Then $(g_k + q_k)/7.5 = 1.28$.
Thus, if $l_r < 5$ m then $F_t' = 36 \times 1.28 = 46.08$ kN/m, and if $l_r > 5$ m then $F_t' = 46.08 \times l_r/5$.

### Floor ties

These are placed in slots at ends of hollow core slabs. Preferable distance between ties = 600 mm, i.e. two per 1.2 m wide units.

**(1) Span = 7.8 m.**

$$F'_t = \frac{46.08 \times 7.8}{5} = 71.9 \text{ kN/m run.}$$

$$A_s = \frac{71.9 \times 10^3}{460} = 156 \text{ mm}^2/\text{m.}$$

*Use T 12 ties at 600 mm centres (188).*
Anchorage length = $40 \times 12 = 480$ mm.
Tie force in each bar = $0.6 \times 71.9 = 43.2$ kN must be resisted by shear in the projecting loops in the edge beams. Then size of projecting bar is given by:

$$A_s = \frac{43.2 \times 10^3}{0.6 \times 460} = 156 \text{ mm}^2.$$

*Use two  T 10 projecting loops in pairs (157).*

**(2) Span = 4.8 m < 5 m.**
$F'_t = 46.08 \times 1 = 46.08$ kN/m run, and $A_s = 100$ mm²/m.
*Use T 10 ties at 600 mm centres (131).*

Anchorage length = 400 mm.

Tie force in each bar = $0.6 \times 46.1 = 27.7$ kN, and $A_s = 100$ mm$^2$·

*Use two T 8 projecting loops in pairs (100).*

**(3) Gable ends**

$F'_t = 36 \times 1 = 36$ kN/m run, irrespective of span, and $A_s = 78$ mm$^2$/m.

*Use T 10 ties at 1000 mm centres (78).*

Anchorage length = 400 mm provided by hook end.

Tie force in each bar = 36 kN, and $A_s = 130$ mm$^2$.

*Use two T 10 bars in pairs (157).*

### Peripheral and internal beam ties

**(1) Peripheral ties to edge beams**

(a) 6.0 m span beams supporting floors spanning 7.8 m

$$F'_t = \frac{46.08 \times 6}{5} \times \frac{7.8}{2} = 215.6 \text{ kN.}$$

$$A_s = \frac{215.6 \times 10^3}{460} = 468 \text{ mm}^2.$$

*Use two T 20 edge ties (628).*

(b) 6.0 m span beams supporting floors spanning 4.8 m

$$F'_t = \frac{46.08 \times 6}{5} \times \frac{4.8}{2} = 132.7 \text{ kN, and } A_s = 288 \text{ mm}^2.$$

*Use two T 16 edge ties (402).*

(c) 7.8 m span gable beams

$$F'_t = \frac{46.08 \times 7.8}{5} = 71.9 \text{ kN, and } A_s = 156 \text{ mm}^2.$$

*Use one T 16 gable edge tie (201).*

**(2) Internal ties**

(a) 6.0 m span beams supporting floors spanning 7.8 m and 4.8 m

$$F'_t = \frac{46.08 \times 6}{5} \times \frac{(7.8 + 4.8)}{2} = 348.4 \text{ kN, and } A_s = 757 \text{ mm}^2.$$

*Use two T 25 internal ties (982).*

(b) 4.8 m span beams supporting floors spanning 7.8 m only

$$F'_t = \frac{(46.08 \times 7.8)}{2} + (46.08 \times 1) = 225.8 \text{ kN, and } A_s = 490 \text{ mm}^2.$$

*Use two T 20 internal ties (628).*

**(3) Horizontal column ties at first floor level**

(a) *Edge columns at 6 m centres* carry 23.4 m$^2$ slab

Axial load at roof = $6.5 \times 23.4 = 152$ kN.

Axial load at floors = $9.6 \times 23.4 + 12 \times 6 = 297$ kN.

Tie force = greater of:

- lesser of $2 \times 36 = 72$ kN, or $0.4 \times 3.2 \times 36 = 46.1$ kN.
- $0.03 \times 1.05 \times [152 + (2 \times 297) + \text{self-weight } 21] = 24.2$ kN.

Use $F_t = 46.1$ kN.

$$A_s = \frac{46.1 \times 10^3}{460} = 100 \text{ mm}^2.$$

*Use one T 12 edge column tie (113).*
(b) *Internal columns at 6 m centres* carry 46.8 m² slab.
Axial load at roof = $6.5 \times 46.8 = 304$ kN.
Axial load at floors = $9.6 \times 46.8 = 450$ kN.
Tie force = greater of:

- lesser of $2 \times 36 = 72$ kN, or $0.4 \times 3.2 \times 36 = 46.1$ kN.
- $0.03 \times 1.05 \times [304 + (2 \times 450) + \text{self-weight } 21] = 38.6$ kN.

Use $F_t = 46.1$ kN as above.
*Use one T 12 internal column tie (113).*

*Note*: If the beam ties are not tied to the columns in any way, the column ties must be used. However, if the beam ties pass through sleeves in the column, and are enclosed within the main column reinforcement cage, then only the T 10 gable tie would need to be increased to T 12 to cater for the column tie force.

### 9.4.4 Ties at balconies

Where cantilevered floor or roof slabs are used to form balconies, there is a major problem in placing floor ties over the support beam. Figure 5.25 illustrates the problem. The simplest answer is to use a structural screed and provide the tie steel in the mesh, but where this is not desirable the designer is faced with having to provide a tie within the depth of the floor slab as shown in Fig. 9.28. Openings are formed in the sides of the floor slabs through which site bar(s) are placed. The site bars must pass beneath the top steel in the hollow cores of the floor slab (Fig. 5.26) and insitu concrete introduced in through slots cut in the top of the slab. The method is not 100 per cent satisfactory because of the problems in feeding the tie bars into the slabs. Also in the case of an accident to the supporting beam the floor slabs would hang in catenary as shown in Fig. 9.29, but the beam would no longer be integral.

### 9.4.5 Vertical ties

The reason for vertical ties is given above where, as shown in Fig. 9.11(c), the column is in vertical suspension. BS 8110, Part 1 [9.1] calls for continuous vertical ties in all buildings (the provision for ties in five or more storeys was deleted in 1993) capable of resisting the tensile force given in clause 3.12.3.7. The vertical tie force capacity is

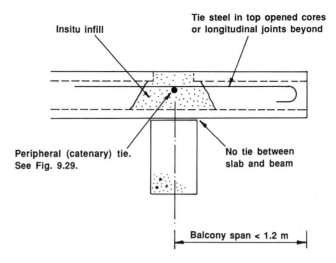

**Fig. 9.28** Perimeter floor ties across balcony slabs.

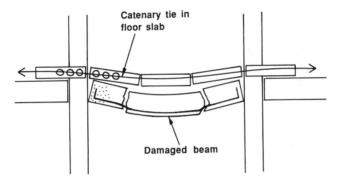

**Fig. 9.29** Catenary action in balcony slabs without beam–slab tie.

calculated from the *summation* of the ultimate beam reactions $N$ only at the floor (or roof) immediately above where the tie is designed, not the total load from every floor above the tie. This means the reinforcement provided in the column, and in any splice within that column, should be at least:

$$A_s = \frac{N}{f_y} \tag{9.7}$$

The loading on the floor may be taken as the characteristic dead plus one-third of the superimposed live (unless the building is being used for storage where the full super load is taken). The partial safety factor for all gravity loads $\gamma_f$ may be taken as 1.05.

The tensile capacity of any column splice should also be able to resist the force $N$. In the case of grouted sleeve splices the anchorage bond length for the reinforcement may be calculated using BS 8110, Part 1, clause 3.12.8.3 where the bond stress obtained from Table 3.28 is without the partial safety factor of 1.4. The weld lengths in the welded splice are also determined without the partial safety factor.

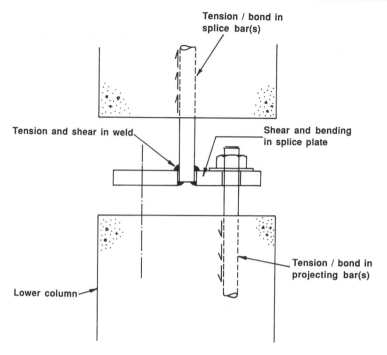

**Fig. 9.30** Stability vertical tie in welded plate column splice.

In the welded plate splice the thickness of the plate is determined by considering the bare steel sections alone, i.e. the insitu concrete is ignored. The root thread area is used to determine the tensile capacity of the projecting bars, and the pull out and anchorage bond capacity of the bars welded to the plate gives the tensile capacity of the upper bars. The strength of the plate is checked in bending and shear as shown in Fig. 9.30. The effects of 2-D plate bending and prying forces at the nut are ignored.

**EXERCISE 9.2**

Check the vertical tie requirements in the column and welded plate splice used in Exercise 7.10 and shown in Fig. 7.81. The characteristic column dead and imposed loads at each of the floor levels are 340 kN and 250 kN, respectively. The column concrete is grade C50.

**Solution**

Column reinforcement:

$$N = (1.05 \times 340) + \frac{(1.05 \times 250)}{3} = 445 \text{ kN.}$$

$$A_s = \frac{445 \times 10^3}{460} = 966 \text{ mm}^2.$$

*Provide four T 20 column bars (1260).*

### Column splice
Use mild steel welded reinforcement to upper column and high tensile threaded bar to lower column.

### Upper column bars

$$A_s = \frac{445 \times 10^3}{250} = 1780 \text{ mm}^2.$$

*Provide 4 no. R 25 column bars (1963).*
Bond length to column.
Force in each bar = 111.25 kN.
Bond stress for grade C50 concrete for mild steel in tension $= 0.28\sqrt{50} \times 1.4 = 2.77$ N/mm$^2$.
Bond length $= (111.25 \times 10^3)/(2.77 \times 25 \pi) = 512$ mm.

### Lower column bars
$A_s = 966$ mm$^2$.
*Provide four T 20 column bars with M 20 thread (980 mm$^2$ root area).*
Bond length $= (111.25 \times 10^3)/(4.95 \times 20\pi) = 358$ mm, use 360 mm.

### Plate design
Ignore the insitu concrete infill surrounding the plate and bars.
Plate size $= 300 - 35 - 35 = 230$ mm square.
Eccentricity of upper to lower bars = 82 mm.
Moment in plate $= 111.25 \times 82/\sqrt{2} = 6452$ kN mm.
Shear in plate = 111.25 kN.
Effective width of plate $b = 82 \sqrt{2} = 116$ mm.
Then, modulus $S_{xx} = b\,t^2/4$ and shear area $= bt$.
Using $p_y = 275$ N/mm$^2$, then $t > 28.5$ mm.
Using $p_q = 165$ N/mm$^2$, then $t > 5.8$ mm.
*Use 230 × 230 × 32 plate.*

## 9.5  CATENARY SYSTEMS IN PRECAST CONSTRUCTION

There remains a popular belief that under certain loading conditions precast concrete frames cannot offer the ductility inherent in monolithic construction because of a lack of continuity in the joints. There is no evidence to support this argument as a routine inspection of joint details, specifications and modern construction practice would show. To support these claims the following is a summary of some of the experimental work carried out to test the integrity of precast connections and components.

Various attempts to quantify the floor membrane forces present in precast catenary systems have been made by Regan [9.10], Burnett *et al.* [9.11, 9.12], UK Building Research Establishment (BRE) [9.13], Odgard [9.14] and Engström [9.4 and 9.15]. In all cases the focus of attention has been on the development of a catenary system using hollow core floor slabs without structural screeds. The objectives were to show (or not as is the case) that short lengths of tie steel (either rebar or helical strand) embedded in the ends of the precast slabs would satisfy the new state of equilibrium shown in Fig. 9.16, for example, without collapse.

The most appropriate work to the skeletal structure is by Burnett [9.12]. Full scale experimental testing consisted of hollow core slabs tied across their supports using short lengths of 7-wire helical strand in the longitudinal joints between adjacent units. Although it was not possible to achieve the new state of equilibrium it was possible to achieve catenary action at loads of up to 80 per cent of the slab failure load, which suggested that by increasing the anchorage length and the amount of continuity reinforcement full catenary action might be possible. The modes of failure at the centre joint were either by concrete splitting forces above the tie steel or by rupture in the ties, and at the outer joints by hogging bending cracks in the tops of the hollow core slabs.

Test carried out at the BRE [9.13] on precast floors and beams found that by the correct disposition of continuity reinforcement over a missing support, deformations up to 20 per cent of the adjacent single span could be achieved without collapse occurring. In some tests, failure occurred when reinforcement broke out through the soffit of the precast units rather than in the insitu joints. It is debatable whether this is good or bad news.

Engström [9.4] has studied ductility of tie connections in precast floors, and rotation mechanisms in cantilevering wall frames. This is in fact an inverse catenary system where the outside support is removed and hogging moments and tie forces develop in the top of the floor plate. The situation is particularly critical in the 'long floor' frame layout where there are no transverse beams (Fig. 1.29(a)) and the cantilevering floor system becomes the primary bough.

The hollow core specimens were tested as shown in Figs 9.31 and 9.32 which would effectively simulate the catenary action at point X in Fig. 9.11(b). The tie steel consisted either of grade S400 high tensile bar, grade S260 mild steel bar, or grade S1690 prestressing wire. The insitu concrete cube strengths varied from 24.6 to 33.3 N/mm². As expected, initial failures occurred abruptly due to flexural tension failure in the interface between the precast slab and insitu infill, after which the slabs were connected by the tie steel alone. The important result, shown in Fig. 9.33, is that the full tensile capacities of the bars were mobilized when the rotation of the slabs (given by $\phi$ in Fig. 9.33) was at least 0.1 to 0.2 radians. Generally, ultimate failure occurred at $\phi = 0.3$ radians and the forces

**Fig. 9.31** Arrangement of slab continuity tests by Engström [9.4].

**Fig. 9.32** Initial cracking at precast hollow core–insitu infill [9.4].

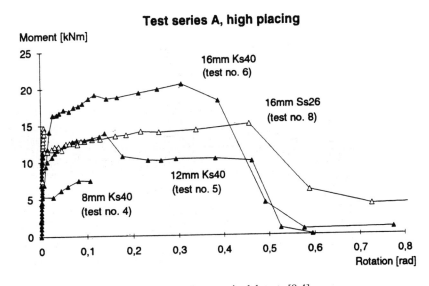

**Fig. 9.33** Moment–rotation data from Engström's slab tests [9.4].

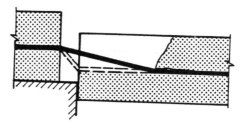

**Fig. 9.34** Displaced support mechanism [9.4].

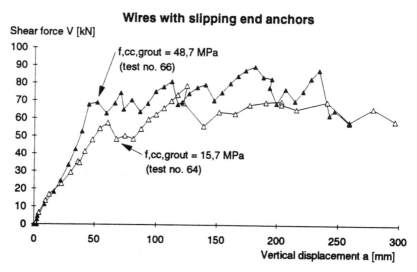

**Fig. 9.35** Shear force vs. vertical deflection in Engström's tests [9.4].

in the continuity bars were at least 60.5 kN per T 12 bar. This is equal to the upper limit of 60 kN for stability ties (Equation (9.1)).

Thus, the tests show that adequate tensile forces can be mobilized in the tie steel without other modes of failure occurring first.

Engström also tested the behaviour of hollow core floor slabs displaced off their bearings by the large sideways movement of roller supports (Fig. 9.34). High tensile bar and prestressing wires were cast into the top opened cores of the slabs and when the slabs were displaced from their bearings the failure mode was by top-splitting of the concrete above the tie steel as shown in Fig. 9.8. The prestressing wires (2 no. 6 mm diameter per core) generated tie forces of 63 kN, i.e. two-thirds of the uniaxial yield strength, before splitting occurred. The vertical shear force vs. displacement behaviour is shown in Fig. 9.35 where maximum vertical displacements of 250 mm were possible. See also Fig. 9.8. The minimum value for the high tensile steel was 151 kN for 1 no. T 16 bar, i.e. 1.45 times the yield strength.

The conclusions from this work which are relevant to the design of stability floor ties according to BS 8110 [9.1] is that providing that the tie steel is anchored 750 mm into the hollow cores and is correctly positioned with at least 30 mm top cover, the steel will generate tie forces greater than 60 kN before failure occurs by splitting of cover concrete.

# Chapter 10

# Site Practice and Temporary Stability

*Temporary stability is an important part of design. This chapter explains that in certain situations the design of the frame is affected by the construction method. Certain components are more critical in the temporary condition than when the structure is complete.*

## 10.1 THE EFFECTS OF CONSTRUCTION TECHNIQUES ON DESIGN

The design for temporary stability, lifting of precast concrete components and site construction methods in an integral part of the overall design procedure. Careful instructions must be relayed from the design office to the site regarding the sequence of erection and, vice versa, the design engineer should be aware of the contractor's chosen method of construction. In the majority of cases precast design engineers are seconded to site to learn of the construction methods and see for themselves the practical methods of frame erection. This exchange of personnel and information, and the permanency of site-fixing gangs in the precast industry, is one of the major factors contributing towards the successful completion of structurally sound and undamaged frames.

All accredited precast frame producers will supply on request written statements of the principles of site erection, methods of making structural joints and materials specifications. These are in general accordance with the requirements of BS 8110, Part 1, clauses 5.3, 6.11.3 and 6.14 [10.1] with regard to:

- critical jointing details, materials and methods
- critical dimensions allowing for manufacturing and site tolerances
- temporary propping or fixings
- rules for advancing the construction ahead of completed and maturing sections of the building.

What is not always clear to the outside observer is where and how these criteria are met in the design, and the logic behind many of the decisions taken by precast designers is not always obvious. The two main aspects involved in the design for the temporary condition are:

- individual frame components and joints
- overall frame stability.

Checks are made to ensure that the young precast concrete components are handled correctly, both in the factory and on site, to prevent cracking, spalling and premature debonding, etc. This is particularly important in components such as plain stocky walls or lightly loaded columns where the serviceability requirements for reinforcement are small, or near to the sharp corners at openings where stress concentrations gather. In these situations, additional reinforcement is provided to satisfy lifting and pitching criteria and because the steel is not always designed according to the usual codes, but often by rule of thumb, these details may seem strange to the unaccustomed eye.

The components must be capable of supporting their own weight, often within 18 to 24 hours of age. Unlike cast insitu concrete units which only ever resist forces in the intended directions, precast components must also be designed to resist stresses of the type which are only possible during handling, pitching and during the construction stage. These may occur within five days of casting. For example, the long narrow spandrel beam of the type shown in Fig. 3.25 would be lifted from the mould, and may even be delivered to site in a flat horizontal manner with the weaker axis in the plane of bending as shown in Fig. 10.1. The handling reinforcement would be provided in the front face of the beam, some of which will inevitably lie close to the neutral axis of the beam when it is in the upright position. Thus, reinforcement has been provided solely for the purpose of lifting.

Once in position in the structure the spandrel beam would possibly require some temporary lateral or torsional bracing to prevent cracking whilst the flooring was being

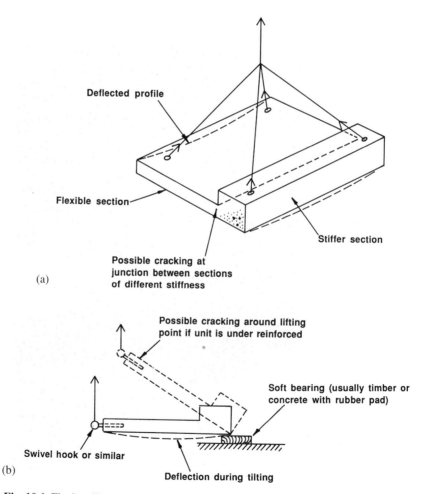

**Fig. 10.1** Flat handling of deep and narrow spandrel beams.

laid. Because the element may be subjected to horizontal wind loading before the flooring is fully tied to the beam, the element would have to be designed to resist the wind load by flexure about the weak axis. Horizontal shear forces must be resisted by the end connectors, which means that certain types of beam–column connectors are not suitable for this purpose.

The second most important issue is that load paths through the partially completed structure may be different to that in the final form. An example of this is where floor slabs are placed on one side only of an internal beam, as shown in Fig. 10.2. The connection is checked to resist the torsion and if found to be inadequate props are placed to prevent the beam rotations from damaging the connector. A more extensive discussion on this subject is given in the following sections.

Practical connection design is a very important aspect, and some examples of impractical connections were shown in Fig. 7.3. Although there are a large number of connectors which may be used successfully in one situation and unsuccessfully in another, there are a number of basic 'rules' which lead to practical design and which, in turn, affect the design of the structure:

**Fig. 10.2** Non-symmetrical loading of slabs on beams.

(1) Components should have a sufficiently wide bearing to assist the site operatives in positioning the component without excessive barring. The seating area should be designed with tolerances and not, as shown in Fig. 10.3, with a captive groove or restricted position.

(2) The seating should be designed so that the impact resistance associated with setting of precast units is at least twice the self-weight $W$ of the unit. This figure should be increased to $3W$ in the case of large 2-D units where the setting down forces are not equally distributed between all fixing points.

(3) The seating capacity of temporary bearings, or extended bearings, should be similarly designed so that the cumulative effects of tolerances, shrinkage and movement resulting from accidental construction damage (e.g. crane impact causing a column to flex) do not cause failure of the seating as shown in Fig. 10.4(a) and (b).

(4) Allow tolerance to at least one end of an element, and avoid captive dimensions which may cause a lack of fit in the connection.

(5) Avoid designing components or connections which rely on accumulative dimensions without tolerances in each individual element, e.g. vertical wall panels, or the layout of 1197 mm wide hollow core floor slabs which should be at 1200 mm centres.

## 10.2   DESIGNING FOR PITCHING AND LIFTING

### 10.2.1   Early lifting strengths

Concrete strengths in a precasting works are governed by the 12 to 24 hour strength. This

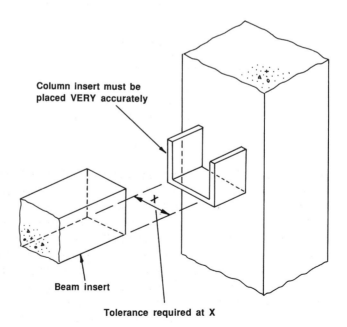

**Fig. 10.3** Restricted fixings.

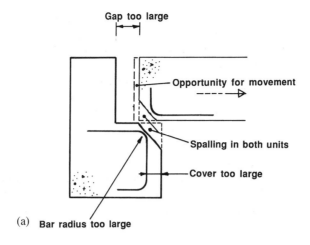

Gap too large

Opportunity for movement

Spalling in both units

Cover too large

(a)  Bar radius too large

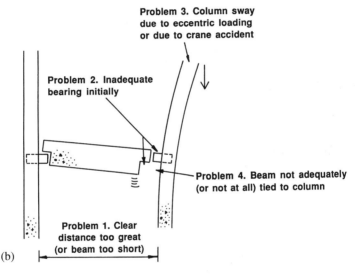

Problem 3. Column sway
due to eccentric loading
or due to crane accident

Problem 2. Inadequate
bearing initially

Problem 4. Beam not adequately
(or not at all) tied to column

Problem 1. Clear
distance too great
(or beam too short)

(b)

**Fig. 10.4** Problems due to inadequate precautions against bearing failure. (a) At slab-to-beam bearings and (b) at beam-to-column connections.

is usually 20 to 25 N/mm$^2$ for reinforced work and about 40 N/mm$^2$ for prestressed concrete; values that enable the units to be lifted from the casting zone and moved to a more convenient location for maturing ready for site. The maturity period can be as little as four or five days and so it is important that the internal microstructure of the young concrete unit is subjected only to strains *well* below the tensile cracking limit. This may involve doubly reinforcing beams and staircase units which would otherwise have only a single bottom layer of flexural reinforcement and, in the case of beams, nominal top steel.

Steam curing helps not only in accelerating the strength gain but gives an improved homogeneous microstructure to the concrete. Steamed units may be lifted at 8 hours, in the case of dry cast concrete, and 12 hours for wet casting. The use of surface retarders or colouring pigments is not detrimental to the strength of components.

## 10.2.2 Lifting points

Lifting points are selected so that the flexural and shear reinforcements provided in service are, as far as possible, fully utilized in lifting. The lifting points are also chosen to minimize deflections. Bending moments and shear forces are calculated in the usual manner. They are based on a loading which includes a 25 per cent mould suction plus a 25 per cent impact allowance. It may be necessary to increase the suction to 50 per cent in the case of units in contact with large surface areas, such as wall panels, spandrel beams, cladding units. The sides of the mould should be lifted clear of the unit so that it leaves the mould in a vertical ascent.

Lifting points are usually at $\frac{1}{4}$ to $\frac{1}{6}$ of span from either end of the units, although this varies depending on the type of unit. Prestressed units are lifted closer to the ends than reinforced units. The optimum situation for equal sagging and bending moment is where the lifting point is at $0.208\,L$ from the ends of a unit length $L$. Thus $0.2\,L$ is used (Fig. 10.5(a)).

Floor slab units, staircases, landings and flat wall units greater than about 750 mm wide require four-point lifting. Theoretically it should be possible to lift these units from two points but the imbalance becomes more critical in wider units with higher centroidal positions. Very long columns and beams (approximately > 13 m) of slender section (length/depth > 50 to 55) may also require four-point bending because of unacceptable flexure stresses and defections in a two-point lift. Spreader beams are used as shown in Fig. 10.5(b). The line of the crane hook must coincide with the gravitational centroid of the unit.

Temporary bracing across narrow legs or upstands in U- or L-shaped sections may be a more suitable alternative to excessive reinforcement or thicker sections as shown in Fig. 10.6(a). Spreader beams may also be used to keep the lateral reaction force as small as possible, Fig. 10.6(b).

Ductile reinforcement (for lifting hooks) is designed using service (not ultimate) stresses, roughly $0.6\,f_y$ together with an ultimate load factor of 1.4 (exclusive of above allowances). By way of comparison, The Chain Tester's Handbook [10.2] suggests a factor of safety of 2.0 on steel design stresses, i.e. $0.87\,f_y/2 = 0.43\,f_y$ giving approximately the same result. Additional reinforcement may be required around the lifting points to prevent cracking due to radial bond stresses. This is particularly critical near to corners where the cover distance to the lifting point is less than 50 mm. Small links (typically R6 or R8) are positioned to resist a lateral force equal to $0.2 \times$ lifting reaction.

Lifting loops and sockets are used to raise a unit through 90° to the vertical (or to turn a unit for finishing work). Not only is the socket fully anchored in the unit, but the concrete directly above the socket in the flat lift position is reinforced to transfer the socket reaction into the body of the unit as shown in Fig. 10.7. A standard hook or bend may be used to reduce the anchorage length.

Most proprietary devices are designed to eliminate the need for additional steel (except for turning) by using a special shape shaft, conical, bulbous, or otherwise, which produces wedging forces around the shaft. These help to confine the concrete radially and reduce the need for additional confinement steel around the shaft. Figure 10.8 shows some of the more common types of swivel devices.

Most precast components are delivered to site horizontally (although some large wall units greater than 2.5 m wide are dispatched near to the vertical on A-frames). Columns are lifted vertical using a pitching point at 0.3 times the height ($L$) of the column from

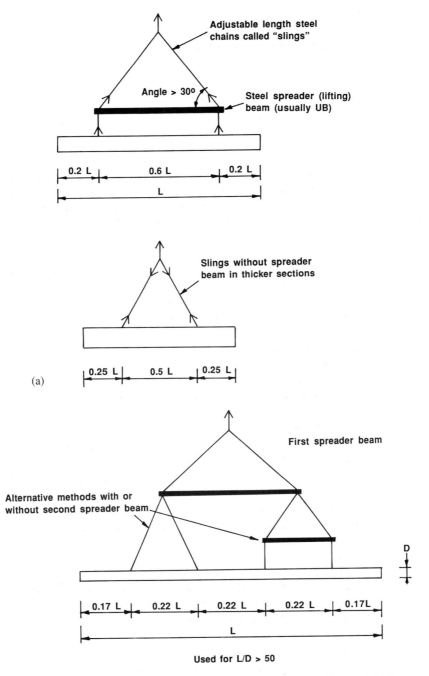

**Adjustable length steel chains called "slings"**

**Angle > 30°**

**Steel spreader (lifting) beam (usually UB)**

| 0.2 L | 0.6 L | 0.2 L |

L

**Slings without spreader beam in thicker sections**

| 0.25 L | 0.5 L | 0.25 L |

(a)

**First spreader beam**

**Alternative methods with or without second spreader beam**

D

| 0.17 L | 0.22 L | 0.22 L | 0.22 L | 0.17L |

L

(b)  **Used for L/D > 50**

**Fig. 10.5** Flat handling of long components. (a) Two-point lifting positions and (b) four-point lifting positions.

the top (Fig. 10.9). The maximum reaction in the crane's chains is 0.72 $wL$, where $w$ is the self-weight plus 25 per cent impact allowance. The maximum bending moment is 0.045 $wL^2$. Dual pitching using sliding chains may be used for very long columns, say

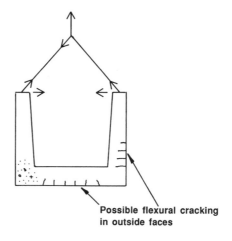

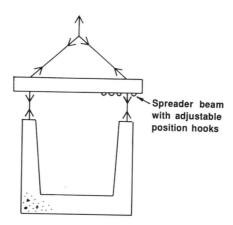

**Fig. 10.6** Handling methods in thin-walled components. (a) Incorrect procedure for lifting units and (b) correct procedure using a lifting beam.

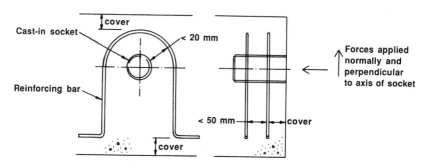

**Fig. 10.7** Hanger reinforcement around lifting holes or sockets.

>15 m, as shown in Fig. 10.10. The pitching points are placed at $0.16\,L$ and $0.60\,L$ from the top end of the column and the maximum bending moment is $0.013\,wL^2$, i.e. a threefold reduction in moment compared to a single pitch point.

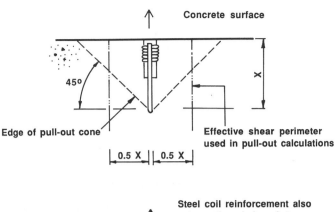

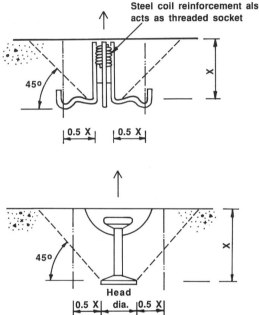

**Fig. 10.8** Types of lifting devices.

It is important that the pitching point is located at the centre of gravity of columns to avoid tilting during the fixing operation. Although this has virtually no implication on the design of columns, verticality in placement ensures that all the reinforcement provided for the temporary condition is equal stressed, e.g. anti-bursting bars at the foot of the column.

The age at which most columns are pitched (7 to 14 days) has enabled the concrete to mature to its design strength, and in many cases more than this. Using a single pitching point, the maximum vertical deflection in a 12 m long 300 × 300 mm column is about 10 mm. Maximum tensile strains are less than the tensile cracking strain for grade C50 concrete, 25 m long columns have been pitched using two points only but the I-cross-section columns were axially prestressed to enable this to happen.

Walls and flat panels are pitched from the top. The maximum bending moment is $0.125\ wL^2$, where $w$ is the self-weight plus 25 per cent impact allowance, and this may

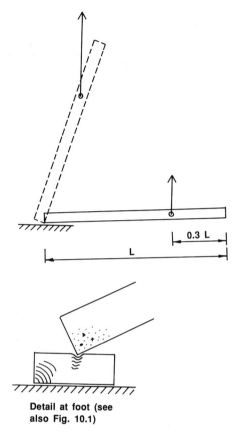

**Fig. 10.9** Single point pitching.

lead to the most onerous design situation. Lifting points are carefully located vertically over the centre of gravity of the unit, although this may prove impossible in some units. It is occasionally desirable to allow the unit to tilt slightly (no more than about 5°) inwards at the top or bottom to assist in the fixing. Special lifting devices have evolved to ensure that the concrete is not damaged during the pitching operation.

Hollow core floor slabs are treated as a special case because the manufacturing process does not facilitate lifting hooks, and prestressing cambers (caused by hogging moments) must be considered when the lifting points are not close to the ends, about 400 mm from the ends of the unit. The design of these slabs is governed only by handling when the length is greater than the cut-off points shown in Fig. 5.2.

Slabs are hoisted on site either by using slings, chains or specially designed clamps (Figs 10.11 and 10.12). The unit should be level during lifting so that an even contact is made with the supporting member. When slings or chains are used, the horizontal component of the reaction at the top of the unit creates a transverse bending moment in the unit. The tendency is for the force to try to collapse or to 'fold' the unit as shown in Fig. 10.13. Although the design of the unit is such that this does not happen, if large areas of the slab have been removed for, say, site placed top steel at a cantilever, the problem may arise. Preventive measures include leaving small bridges of unbroken slab

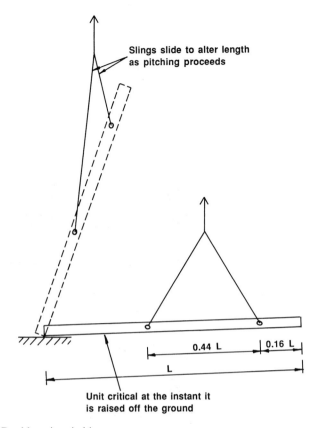

Slings slide to alter length as pitching proceeds

0.44 L

0.16 L

L

Unit critical at the instant it is raised off the ground

**Fig. 10.10**  Double point pitching.

between the slots, usually 100 mm long at about 400 mm centres, or placing a temporary strut (acting as a spreader beam) across the chains to resist the horizontal forces.

Double-tee units are hoisted using four lifting points, each over the line of the webs (Fig. 10.14). Reinforcement details at these points may be obtained from the manufacturer, but typically comprise a large inverted U- or V-shape high tensile reinforcement or helical strands embedded to the bottom of the web. Small bars are place horizontally across the top of the lifting hook to prevent splitting cracks. Spreader beams are used on units of length greater than about 15 m to avoid the horizontal component of the force in the low inclination chains from causing splitting stresses in front of the lifting hook. Ideally the inclination of the chains should not be less than about $\theta = 35°$ to the horizontal. Low inclinations of $\theta$ also lead to greater forces in the chains themselves, given by $V/\cos\theta$, where $V$ is the self-weight reaction at that point.

Precast planks are hoisted at four points, but the lifting criterion is not usually critical because the planks are used in relatively short lengths compared with other types of flooring.

Staircase units require two sets of chains for site fixing – one fixed and one adjustable, in order to maintain the correct inclination as shown in Fig. 10.15. The site lifting criteria is less onerous than at the factory because the same lifting points are used in both situations. This is because the unit must be simultaneously seated level at both ends to avoid the

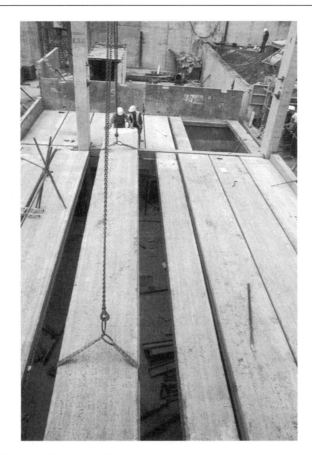

**Fig. 10.11** Hoisting 15 m long hollow core floor units.

inclined edges from digging into the bearing ledge and causing horizontal frictional forces between the members.

**EXERCISE 10.1**

Check the lifting requirement for a 12 m long × 300 mm square column containing 4 no. T 20 bars which is lifted from the mould using only two points. Use $f_{cu}' = 15$ N/mm², high tensile reinforcement and 30 mm cover to 8 mm diameter links.

*Solution*

Lifting points to be at 1/5 points = 2.40 m from ends.
Self-weight = 0.3 × 0.3 × 24 = 2.16 kN/m plus 50 per cent suction and impact for a narrow mould = 3.24 kN/m.
Ultimate moment $M = 1.4 \times 3.24 \times 2.40^2/2 = 13.06$ kN m.
Shear $V = 1.4 \times 3.24 \times 2.4 = 10.9$ kN.

**Flexure**

$b = 300$ mm; $d = 300 - 30 - 8 -$ say 12 = 250 mm.
$K = 0.046$; hence $A_s = 138$ mm² < 628 mm² provided; thus lifting is not critical.
Shear $v = 0.15$ N/mm², therefore not critical.

**Fig. 10.12**  Lifting clamps for hoisting hollow core floor units.

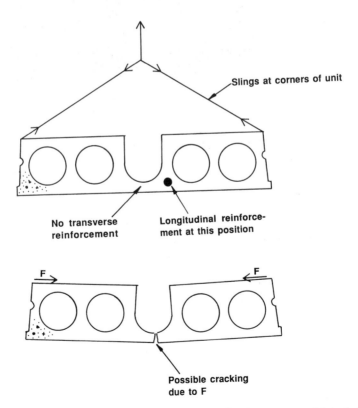

**Fig. 10.13**  Problems encountered in lifting where top flanges are removed from hollow core units.

**Fig. 10.14** Hoisting double-tee units (courtesy of Blatcon Ltd).

**EXERCISE 10.2**

Design the lifting reinforcement for a 600 mm deep double-tee slab weighing 12 tonnes, assuming that the angle of the 4 no. lifting chains is 60° to the horizontal. Allow 50 per cent suction and impact allowance at the mould and a factor of safety of 2 to the strength of the lifting hooks. Use ductile mild steel reinforcement $f_y$ = 250 N/mm$^2$, and grade C35 at the initial lifting. Assume cover to bottom of lifting hook = 50 mm.

**Solution**

Referring to Fig. 10.16, use V-shape lifting bar with the inclined bar at 45° to the horizontal.

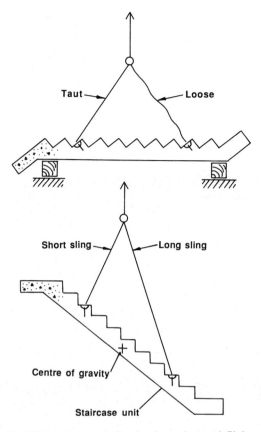

**Fig. 10.15** Adjustable lifting slings on inclined units such as stairflights.

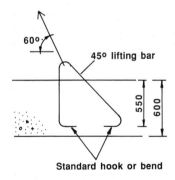

**Fig. 10.16** Details to Exercise 10.2.

Service stress in bar $= 0.6\,f_y = 0.6 \times 250 = 150$ N/mm$^2$.
Weight of unit $= 12$ tonnes $= 117.72$ kN.
Maximum vertical reaction $= (117.72 \times 1.50)/(4 \sin 60°) = 51$ kN.
Shear area $A$ of lift bar based on $f_q = 0.6 \times 150 = 90$ N/mm$^2$ in double shear.

$$A = \frac{51 \times 10^3}{2 \times 90} = 283 \text{ mm}^2 \times \text{factor of safety (FOS) of } 2 = 566 \text{ mm}^2.$$

Axial forces in each leg of hook (by resolution of forces) = 36.04 kN in the inclined leg and 18.02 kN in the vertical leg.

Maximum tensile area $A_s = (36.04 \times 10^3)/150 = 240 \text{ mm}^2 \times \text{FOS of } 2 = 480 \text{ mm}^2$.

*Use one R 32 bar (804).*

Tensile anchorage bond length in grade C35 concrete = 33 dia. × 480/840 = 632 mm > (600 − 50) = 550 mm available depth; provided 550 mm anchorage plus standard end bend.

### 10.2.3   Handling

Loading arrangements on delivery vehicles should be such that the components are not subjected to forces and stress which have not been catered for in the design. The components should be loaded evenly so that the distribution of weight is uniform on each component. Components should have semi-soft (e.g. softwood) bearers placed along their length. Where components are stacked one above the other, the bearers should line over each other. The bearer should be large enough to keep the components vertically separated and strong enough to resist dynamic loads such that the concrete components do not make contact with each other during transit.

When stacking units on the ground on site, the guidelines will be similar to the above. The ground should be firm such that no differential settlement may take place and cause spurious forces and stress in the component (see Fig. 10.17). The height to which components may be stacked is about 1.5 m – the usual criteria is for an operative being able to man-handle the component (crane assisted) whilst standing on firm ground. Figure 10.18 shows the result of not abiding by the rules!

The type of lifting equipment (crane, hoist, etc.) should be compatible with the geometry of the structure, weight and size of components, the nature of ground and site access. Consideration should be given to the maximum weight and radius of lift from the agreed area for lifting so that the lifting equipment is able to lift and place the components smoothly and without jerking.

Components which are to be inclined by pitching should be first placed on firm ground, and pitched in a separate operation (Fig. 10.9). A special (steel, concrete or timber) shoe may be required when pitching heavy and thin units because it is not possible to reinforce the units for the very large point loads which develop on uneven surfaces. It is important that the point in contact with the ground does not slide, slew or move sideways. This could disturb the stability of the lifting device and initiate a collapse.

### 10.2.4   Cracks

Small hairline cracks are often seen near to the lifting points of long, slender components, such as columns or beams. The decision as to whether these cracks are structurally detrimental in the final building must rest with the project designer, but the following may be taken as general guidance for cracks less than 0.3 mm wide:

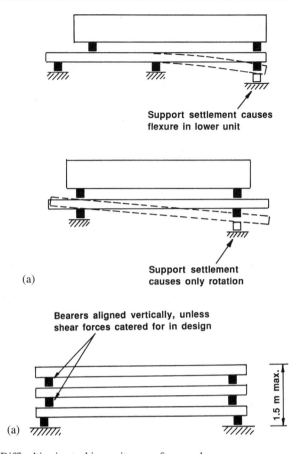

**Fig. 10.17** Difficulties in stacking units on soft ground.

(1) If the crack forms in a region which is permanently in compression: no action.

(2) If the crack closes completely to the naked eye when the unit is placed in its permanent location which is not in an area of tension: no action.

(3) If the crack width is visible to the naked eye when the unit is placed in its permanent location, which is an area of tension: measure the crack width using a graduated microscope before and after the construction of the components carried by the said unit. Then:

    (a) If the crack width does not increase: no action.

    (b) If the crack width increases but is no greater than 0.3 mm, check the serviceability limit state of deflection and crack widths based on the flexurally cracked section properties.

    (c) If the crack width increases to greater than 0.3 mm, refer to the designer to check both the serviceability calculations and the durability conditions. If necessary carry out structural repairs to the crack, but if this is not acceptable then consider replacing the unit or carrying out load testing to determine its flexural stiffness. Ultrasonic pulse velocity testing may be used to search for

**Fig. 10.18** Damaged prestressed slabs due to careless stacking.

internal cracks in concrete, but the results must be treated with extreme caution if reinforcement is present between the measuring points.

Calculations to determine the onset of flexural cracking may be carried out by dividing the bending moment by the gross section modulus (ignoring the area of reinforcement if desired), and checking that this does not exceed about $0.37 \sqrt{f_{cu}'}$. Although this value is greater than $0.24\sqrt{f_{cu}}$ given in the code, it is a more realistic value to take. The lifting strength may be conservatively taken as $f_{cu}' = 20$ N/mm$^2$. Hence the critical flexural cracking stress is about 1.6 N/mm$^2$.

Crack widths may be determined from BS 8110, Part 2, clause 3.8.3 [10.1] where, using the usual notation for serviceability calculations, gives:

$$\text{crack width} = \frac{3a_{cr}\varepsilon_m}{1 + 2\left(\dfrac{a_{cr} - c_{min}}{h - x}\right)} \tag{10.1}$$

where $$\varepsilon_m = \frac{M(a - x)}{E_c I_c} - \frac{b(h - x)(a' - x)}{3E_s A_s(d - x)} \tag{10.2}$$

It is unlikely that the tensile stress in the reinforcement will exceed $0.8 f_y$ during lifting. The term $\varepsilon_m$ is calculated using the modular ratio method.

Shear cracks during handling are uncommon, but torsional cracks are possible in grade C20 concrete where the torsional shear stress is about 1 N/mm$^2$. According to Adlparvar [10.3] torsional cracking in asymmetrically loaded L-shape beams commences when $v_t > 0.2 \sqrt{f_{cu}}$ (for $f_{cu} = 60 - 70$ N/mm$^2$). In non-symmetrical edge beams of the type shown in Fig. 3.25, where the ends of the beam are restrained in the temporary condition against rotation, torsional shear stresses due to the self-weight of the beam plus impact allowances may result in helical cracking near to the end supports (Fig. 10.19(a)). The problem here is created by the temporary restraint, because the beam may have been designed on the basis that floor loads would equalize the torque $Fe$, even though the reactions $F_1$ and $F_2$ may be different as shown in Fig. 10.19(b). The solution is to prop the beam at two or three points before the crane is released. Care must be taken at upper storey levels to ensure that all props line up vertically otherwise the prop reaction may cause twisting of the component at the lower levels.

### EXERCISE 10.3

(1) Check the column in Exercise 10.1 for flexural cracks. Use the code of practice value for concrete tensile stress of 0.24 $\sqrt{f_{cu}}$ as the limiting criteria for cracking.
(2) If the limiting tensile stress is exceeded, calculate the crack widths in the soffit of the section.

### Solution

(1) Young's modulus for grade C15 concrete is not given in BS 8110 (grade C20 is the lowest). However, by extrapolation to C15 one may use the value of $E_c = 23$ kN/mm$^2$ without too much worry. Alternatively, the equation $E_c = 9.1 f_{cu}^{0.33}$ gives 22.2 kN/mm$^2$.

Therefore modular ratio

$$m = \frac{200}{23} = 8.7.$$

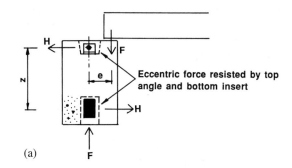

(a)

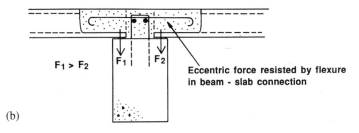

(b)

**Fig. 10.19** Non-symmetrical floor loading on beams. (a) Temporary restraint and (b) permanent restraint.

Then uncracked section inertia

$$I_u = \left(\frac{300 \times 300^3}{12}\right) + (7.7 \times 2 \times 628 \times 100^2) = 772 \times 10^6 \text{ mm}^4.$$

Section modulus

$$Z_u = \frac{772 \times 10^6}{150} = 5.147 \times 10^6 \text{ mm}^3.$$

Then flexural stress (using $M = 13.06/1.4 = 9.33$ kN m)

$$\frac{M}{Z_u} = \frac{9.33 \times 10^6}{5.147 \times 10^6} = 1.81 \text{ N/mm}^2.$$

> limiting flexural stress = $0.24 \sqrt{15} = 0.93$ N/mm$^2$.

(2) Thus, calculate by referring to Kong and Evans' design graphs [10.4]. If

$$\frac{mA_s}{bd} = \frac{8.7 \times 628}{300 \times 250} = 0.073$$

then

$$\frac{x}{d} = 0.32 \text{ and } I_c = 0.047 \ bd^3.$$

Then

$$x = 80 \text{ mm and } I_c = 220 \times 10^6 \text{ mm}^4$$

and

$$Z_c = \frac{220 \times 10^6}{(300 - 80)} = 1.0 \times 10^6 \text{ mm}^3.$$

Then flexural strain in the soffit

$$\frac{M}{E_c Z_c} = \frac{9.33 \times 10^6}{23\,000 \times 1.0 \times 10^6} = 406 \ \mu\varepsilon.$$

Then

$$\varepsilon_m = 406 \ \mu\varepsilon - \frac{300 \times 220 \times 220}{3 \times 200 \times 10^3 \times 628 \times 170} = 180 \ \mu\varepsilon.$$

To determine crack width in soffit over the longitudinal bar

$$a_{cr} = c_{min} = 38 \text{ mm to main longitudinal bar}$$

then crack width = $3 \times 38 \times 180 \times 10^{-6} = 0.02$ mm.

## 10.3    TEMPORARY FRAME STABILITY

### 10.3.1    Propping

In some cases temporary propping will be required to stabilize a component prior to permanent stability being achieved. This will or may affect:

- columns and 2-D walls vertically
- beams, staircases and floor slabs horizontally.

Where propping is required the ends of the props should rest against head-plates and sole-plates to spread point loads (Fig. 10.20 [10.5]). The component should be reinforced to take the reaction(s) from the prop (both as an imposed load and a reaction), and special cast-in fixings should be provided if there is a chance that shear or tensile forces will develop.

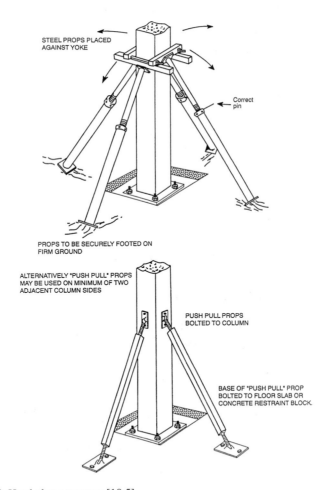

**Fig. 10.20**  Head plates to props [10.5].

The notion of propping refers not only to the use of adjustable length props, but includes any form of bracing in which a reactive force is developed. For example, timber wedges driven in between precast components at the foundation may create additional stresses for which the component should be designed. Great care should be taken when using hydraulically driven jacks to manoeuvre components into position as large and unforeseen reactions, which may be present elsewhere, could lead to over-stressing. This is unlikely to happen with manually powered equipment.

Whenever jacks are used they should be correctly stabilized to prevent sideways forces developing. If jacking takes place from a lower level floor, the precast contractor should be advised as to the maximum permitted load, and its position, to be carried by the floor below. The use of tow ropes, chain blocks, etc. to pull components into position should be correctly specified in several ways. The permitted positions at which ropes may be attached to components (in particular to any projecting, fixing or reinforcing bars) should be clearly defined. If the ropes are tied to cast-in fixings the fixing must have adequate anchorage and strength.

### 10.3.2 The effect of erection sequence

The sequence of erection is controlled by many factors, the more significant ones being crane accessibility, structural form (plane shape, unbraced/braced frame) and positions of stability walls.

Unbraced frames are inherently stable after completion of the moment connection at the foundation, but this does not mean that temporary stability may be ignored. A fully tied first floor slab must be assured before temporary props are removed from the columns, and the verticality of the columns is checked upon the completion of each floor level. Where more than one column is structurally tied into the frame side-sway can only occur by simultaneous lateral movement of all columns. Because absolute rigidity cannot be achieved, the function of any bracing is to limit side-sway to within reasonable limits. The maximum permitted value is 0.05 times the width of the column, and this is well below the usual site fixing tolerances.

In braced frames, which rely on stabilizing components, construction proceeds in an area that contains at least one stabilizing element in the two orthogonal directions, and work proceeds outwards and upwards from this point. Temporary steel cross-bracing or diagonal propping is used if the geometry of the frame makes this sequence of fixing impossible. There is a degree of redundancy here in that most 'pinned' bases have semi-rigid stiffness (the exceptions are columns founded on retaining walls or torsionally flexible insitu ground beams) and this may be called upon to assist in the temporary condition up to the level of the first column-to-column splice.

Two types of loading are considered at this stage: wind loading and eccentric loading due to lack of plumb or overhanging components. The effects of wind currents on rectangular blocks, plates, etc. have been reported by the Engineering Science Data Unit (ESDU) [e.g. 10.6 and 10.7] and by Cook on space frames [10.8]. Amplification factors on the basic wind pressure of up to 1.8 are possible on a partially clad building. However, these forces are not usually critical unless the frame is partially clad at the *same* level as the floor plate is incomplete. In this case the line of columns on the windward and leeward external faces are checked against forces calculated using local wind pressure coefficients.

Eccentric loading due to non-symmetry is more easily calculated and allowed for. The main modes of possible failures are summarized in Fig. 10.21. The divergent mode, Fig. 10.21(a) is possible where particularly large concrete panels are fixed out of sequence, or too early with respect to the maturity of the structural frame. This may occur in cold weather where insitu concreting is not permitted and the contractor is anxious to continue with the precast construction. These situations are always anticipated and temporary tie backs (e.g. using *turfors*) are used. The alternative is the sway mode in Fig. 10.21(b) where rotations between the beam and slab, and between the column and beam, are both

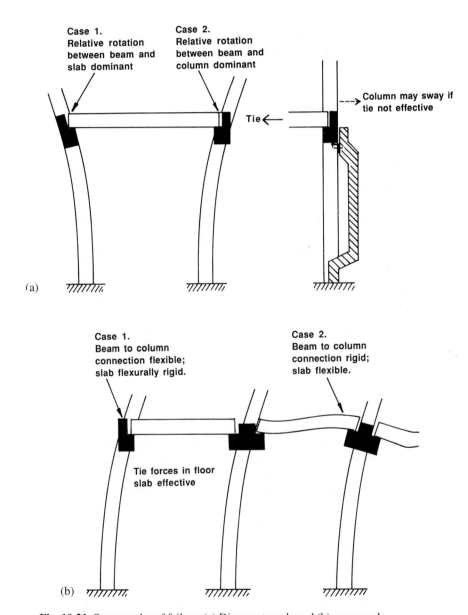

**Fig. 10.21** Sway modes of failure. (a) Divergent mode and (b) sway mode.

possible in the temporary condition. Props are used to prevent side-sway until the joints are grouted.

Lack of plumb is largely eliminated by very strict fixing tolerances, although frame movements due to temperature, shrinkage and self-weight loading will inevitably cause deviations in the order of ±10 mm for a 60 m long frame. Precast columns are surprisingly flexible to straightening and it cannot be emphasized too strongly that, unlike steel structures, precast frames are vertically aligned at every floor level before the insitu concrete tie beams are cast. 'Push-pull' props with tension as well as compression capacity are used extensively in this exercise and fully anchored threaded sockets, brackets, etc. are supplied in the relevant precast units.

### 10.3.3   Special consideration for braced frames

The sequence of construction should ensure that floor diaphragm action must be allowed to develop between the bracing frames or walls, particularly if the skeletal structure is to be clad, in part or in whole, around the perimeter, or internally. Figure 10.22 shows the correct sequence. Point A lies on a critical plane and should be dealt with earlier than later in the fixing sequence. Floor slabs should not be omitted from the floor near to bracing frames or shear walls.

Each floor slab level should be grouted such that the grout has time to mature before fixing of the floor (or roof) slab at the *second next* upper level commences. The usual maturity time is three to five days in temperate climates (10–20°C) or one or two days in warm humid climates (25–30°C). The tie connections between adjacent floor slabs, beams, columns and walls should be fixed according to the contract drawing. Cover distances and lap length should be checked, and the threaded penetration length into cast-in-sockets marked before the threaded portion is mated.

The bracing frames, shear cores or shear walls, etc. should be constructed at least one storey ahead of beam, floor and staircase erection at any floor level. If the strength and

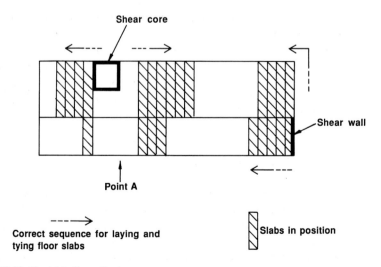

**Fig. 10.22**  Unstable floor diaphragm.

stiffness of the wall relies wholly on composite action with the skeletal frame through an insitu grouted interface, the grout should be given time to mature before construction of the components at the floor level above the wall commences.

The wall should be temporarily propped or secured to the stabilized frame, perhaps using one of the methods shown in Fig. 10.23(a)–(d). Temporary supports should remain in position until the insitu grout has matured and the wall is supported or tied into the frame (Fig. 10.24). This may cause a delay in construction if the wall lies on the critical construction path. If the wall-to-frame connection is totally provided by bolting or welding there should be no delay to subsequent construction.

The construction of column splices should be carried out exactly according to the specification, particularly with reference to the sequence of fixing upper columns and the mixing and placement of insitu concrete, grout, etc. The lower column should be fully tied into the skeletal frame to the floor level below the level of splice. No splices should be made to columns that are not connected to a fully tied and braced part of the structure more than one storey level below the level of the connection.

The upper column should be propped for all types of column splices – even if the permanent connection between columns is made immediately, e.g. bolted, coupled, or welded (Fig. 7.78). The props should remain in position until the splice connection has developed full strength and the upper column is tied to the skeletal frame at the next floor level above the splice. If the next floor level is staggered the column should be tied into the next floor level only, providing the column is connected to ties in two mutually perpendicular directions.

Temporary stability is considerably enhanced if the level of the column-to-column splices is staggered in alternate rows of columns, Fig. 7.77. This avoids a plane of weakness in the permanent structure and enables construction to proceed with at least half the number of columns continuous above a critical horizontal plane. The axial force and bending moment capacity of some of the smallest column splices (for a $300 \times 300$ mm column) is sufficient to allow for about 15 mm of sway at the top of a two-storey column, a value which is unlikely to be exceeded in practice.

The floor slab is a convenient platform on which to work and therefore columns surrounding lift shafts, atriums, etc. are particularly difficult to deal with. Special devices have been developed to stabilize these components but it is also thanks to the judgement of the erection foreman that temporary stability is guaranteed.

Frame erection continues no more than two storeys ahead of a fully tied framework. This may be restrictive in tall buildings of a small plan area, e.g. $300$ m$^2$, where the curing period for the insitu concrete infill is less than the fixing time for a complete floor. Consequently, bracing props are left in position longer than normal to enable the precast frame erection to proceed.

### 10.3.4   Special considerations for unbraced frames

Unbraced frames are stabilized by cantilever action of the columns, or by frame action between the columns and floor plate, or the columns and beams. Therefore the sequence of construction is less critical because each column is capable of carrying horizontal loads, in both the temporary and permanent situations. The columns should be adequately propped and column–frame connections completed and allowed to mature according to the same restrictions as given in Section 10.3.3.

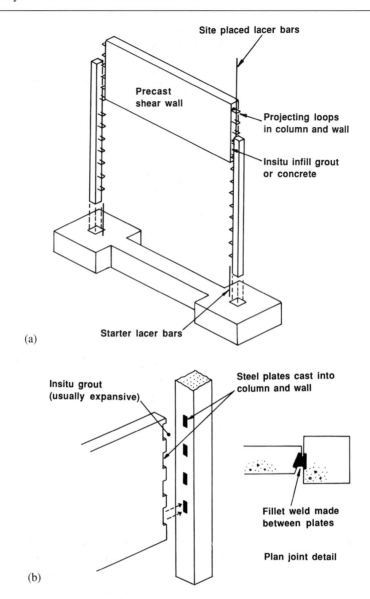

**Fig. 10.23** Wall-to-frame connections. (a) Lapped reinforcement; (b) welded plates; (c) temporary angle cleats and (d) temporary column clamps.

The principle of concreting floor slabs no more than two floors behind the level of floor construction is also applicable in unbraced frames. The importance of this lies not so much in horizontal stability, because the columns can act as cantilevers by design, but in the safety and integrity of the floor slab in cases of accidentally dropping precast concrete components (or plant) onto the floor slab. Experience has shown that ungrouted slabs can be dislodged from their seating by the impact loading from a slab unit dropped from the next upper floor level, whereas grouted slabs (more than three days maturity of grout) may survive the impact loading from two such slabs.

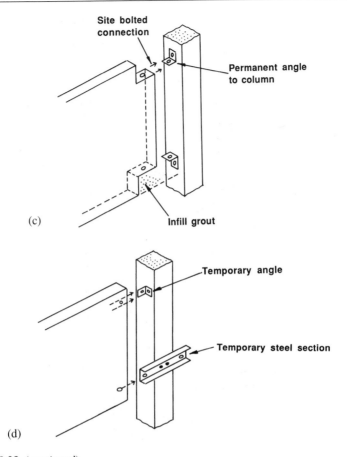

**Fig. 10.23** (*continued*)

### 10.3.5   Temporary loads

Imposed loads due to the combined storage of materials, storage of precast components and construction traffic and plant should be assessed, and maximum allowances given to the various parts of the framework. Strict guidance rules should be made available so that the precast contractor is aware of the limitations.

Where a precast component has been designed to act compositely with another medium, consideration must be given to the non-composite strength and stiffness of the component. Working drawings should carry this information in the form of allowable loads per unit area, or per individual component, irrespective of the location of the imposed load.

Wherever possible the stacking of precast components should be at ground level. Where this is not possible or practical, e.g. on confined sites, the weight and position of the component should be determined so as not to violate the above guidelines. The positions of stacked components should not impede the insitu site work necessary to achieve permanent stability.

In some instances damage will occur to the waiting components or to the components in the frame due to impact. This may be the result of negligence on behalf of the precast

**Fig. 10.24** Wall and column props in position.

contractor or by oversight on behalf of the designer. Similar structural damage caused by following trades should likewise be reported and assessed. In some cases structural remedial work will be necessary or the complete removal (and possible replacement) of the damaged part(s) of the structure may be required.

## 10.4   ON-SITE CONNECTIONS

### 10.4.1   Effect of fixing types

The most important factors in practical connector design are:

- accessibility
- temporary stability
- tolerances.

Strength criteria may be satisfied simultaneously with the above, but unsuccessful connector designs have failed because of a lack of understanding in erection sequences and the cumulative manufacturing and site dimension deviations. The connection must allow for precast components to be lowered unhindered onto their supporting member and for adequate access to the connection thereafter. In difficult situations a slewed approach is sometimes permissible, but in general this is avoided.

Structural joints between precast components require small quantities of materials. Small aggregate concrete, cementitious mortars, epoxy mortars and adhesives must be carefully specified and accurately located and compacted. The IStructE Manual [10.9] is used by designers for compliance with proposed jointing details.

There is a wide range of fixings and connections in a precast structure, many of which are used to perform a unique function. The precast contractor should be made aware of the importance, or not, of each fixing type, and of the permitted deviations. The designer should ensure that the fixing(s) performs its intended function, either by strict compliance with the manufacturer's details, or full scale testing. In the former, the influence of the position of the fixing in the component, and the measures taken to ensure the full safe working capacity of the fixing should be observed. Construction tolerances should be stated on the drawings.

Fixings should be clearly referenced and coded to avoid error. Where shims or packs are used to aid construction methods, the designer should ensure that they are capable of sustaining the necessary load, and provide the degree of flexibility required. The load bearing components should be designed for the effect of temporary point loads resulting from the use of small shims (particularly steel shims). The precast contractor should be informed of the permitted maximum load capacity through a shim.

The strength of bolted or welded connections should be assessed for both the temporary and permanent condition such that no element is over-stressed at any time. The effect of torsion and shear stresses induced by non-symmetrical loading (e.g. slabs onto one side of a beam) which would not normally occur in service should be catered for, e.g. Fig. 10.2.

The contribution to strength and stiffness from insitu concreted connections should be assessed with respect to the time taken for the concrete to mature. Although adequate strength may be achieved in compression the bond resistance between rebar/dowel and the concrete may take longer to fully develop. The connections should be protected from rapid drying or rain penetration.

### 10.4.2   Strength and maturity of connections

Immediate strength is achieved by bolting or welding although for the joint to achieve full service strength, insitu concrete may be required. In this situation the strength of the partially complete joint is checked for temporary loading and, if necessary, the construction sequence is interrupted and work is carried out elsewhere until an adequate strength is achieved. Temporary propping is *not* always required during this operation. Insitu concrete and grouts are specifically designed for early strengths and low shrinkage characteristics. Strengths around the 10 N/mm$^2$ mark are typical for temporary loading and this is usually achieved within two days.

The most important joints are:

- column splices, where the joint must achieve adequate compressive strength before upper storey beams and slabs are positioned

- floor slabs, where insitu concrete infill is used to complete the floor diaphragm to transfer horizontal loads to stabilizing components
- shear walls, etc., which must be integral with the frame at least one floor behind construction.

These are dealt with in Section 10.5.


## 10.5   ERECTION PROCEDURE

### 10.5.1   Site preparation

The General Contractor (GC) should prepare well consolidated and level ground around the full perimeter of the frame and within the interior of the frame. Any service pipes or similar must be protected from crane and vehicle impositions. The extent of this working ground depends on the type of crane being used by the precast frame erector, but should not be less than about 6 to 7 m wide, as measured from the edge of the frame. The working ground level should preferably not be more than 0.75 m above or below the finished ground floor slab level.

A typical precast erection gang consists of four or five persons. The foreman is responsible for the works and will liaise, in the absence of a resident engineer, with the GC and the personnel at the precasting company. The foreman also orders the deliveries of precast components, and is often the most skilled person in the team for fixing and directing the actions of the crane. Two other fixers assist the foreman, and two labourers carry out the duties after the precast components are fixed, e.g. shuttering, grouting, making good, etc.

In most instances the GC is responsible for the construction of the foundations and substructure in general (except where precast foundations are used). Insitu concrete foundations should be cast at least 14 days before commencement of the frame erection. The strength of the insitu concrete to the foundations should be at least 10 to 15 N/mm$^2$ before frame erection commences, unless the precast designer specifies any special requirements. A levelling pad is prepared by the precast contractor at least three days before frame erection commences. If this is not possible then 150 mm square solid steel packs, or sandwich steel-clay tile packs may be used. The clay tile should not be more than 10 mm thick and the steel plate not less than 3 mm thick.

The precast contractor should prepare sound bases for the column props. Where props are used on either side of the column the foundation for each prop need only be required to carry a compressive force. A timber or concrete spreader pad might be required if the ground conditions are poor. Where only one prop is permitted in any one direction the foundation pad must be able to resist both compression and uplift. The precast designer should inform the GC of the size and position of the pad. The pad must be levelled horizontally and located centrally on the column line.


### 10.5.2   Erection of precast superstructure

Figure 10.25 shows the fundamental erection sequence [10.10]. The important feature of this sequence is the time at which the temporary props are removed. The following sections discuss this sequence in detail.

### 10.5.2.1 Columns

Columns should be lifted off the delivery vehicle using at least two lifting points, and correctly stacked and pitched according to Section 10.2. The column may be inspected for any cracks when it is at an inclination of about 5° to the horizontal. Any cracks should be marked, measured and reported to the designer, whereupon any necessary corrective action will be taken.

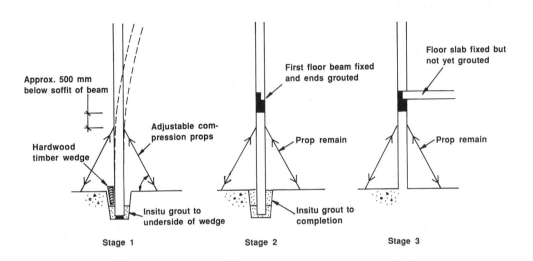

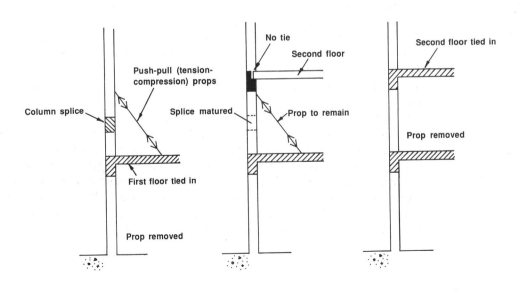

**Fig. 10.25** Precast frame erection sequence [10.10].

At the ground level connection, the column is lowered on to the foundation on the pre-levelled plate. If a pocket foundation is used, timber wedges (two per face) are forced down into the gap between the column and foundation and the column is propped at about 2.5 to 3 m above foundation level. If the column is placed over projecting reinforcement to form the grouted sleeve connection props are placed as above, together with horizontal wedges to prevent the bottom of the column moving sideways. If holding-down bolts are used the base plate is secured by fully tightening the nuts to *every* bolt, and no wedges are needed.

The props should be of adequate compressive strength and subtend an angle of approximately 45° to 60° to the horizontal. The typical prop rating is 3 to 4 tonnes compression at 4 m extended length. The action of forces from the props should provide stability in two mutually perpendicular directions, and be coincident with the centroid of the column and the reaction from the wedges. This may be achieved using four compression props, or two tension–compression props, or a combination of both. The props should remain in position until all the columns supported by the said props are fully tied to the first floor slab. It is not allowed to remove the props from one particular column on the premiss that it is itself tied into the frame. Where the column is asymmetrically loaded the props should remain in position until such time that the column is tied in at *two* floor levels. This is to provide the necessary horizontal prop reaction force (Fig. 10.26).

In situations of bad weather, high winds or a disrupted fixing sequence, the precast contractor may wish to leave the props in position until the first and second floor are completed. Similarly, very tall columns, exceeding 15 m in height × 300 mm depth may have to be propped at the first floor level so that no more than 9 to 10 m height of column is unpropped whilst fixing commences at the second floor level.

Columns founded in pockets are manoeuvred into position and aligned vertically using the tapered wedges and props, respectively. The wedges should be at least 250 mm long and tapered at approximately 10°. It may be necessary to insert two wedges if the gap is particularly wide. Sturdy seasoned hardwood should be used because the wedges are driven downwards in order to move the column into position. Crow bars (approximately 1 m long × 40 mm diameter) are also used to manoeuvre columns on base plates or projecting reinforcement. Optical instruments, such as two theodolites reading azimuth angles in two mutually perpendicular directions, are suitable (Fig. 10.27). Plumb lines may be used for shorter (less than 10 m) columns, but are best suited to calm weather. When the column is correctly positioned the annulus between the column and foundation should be filled to the level of the wedges or 250 mm depth, whichever is the greater, using a concrete grade C40 containing 6 or 10 mm size aggregate and an expanding agent. The slump should be 50 to 100 mm. A poker vibrator should be used to compact the concrete which is placed in pockets. The wedges may be removed a minimum of 24 hours later, or when the compressive cube strength of the insitu concrete reaches 10 N/mm$^2$. The foundation is then filled to the top using the same mix as before.

Columns founded on base plates assume a more immediate fixity, although a structural connection is not considered until insitu grout underneath the plate has achieved a reasonable strength.

Further vertical alignment of the column is possible after the base has been concreted. Columns are sufficiently flexible to allow horizontal movements of not more than 10 mm over a 3 m storey height for 300 × 300 mm sections. Realignment of columns from upper

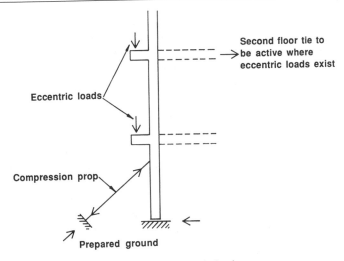

**Fig. 10.26** Stability of column subjected to eccentric load.

**Fig. 10.27** Vertical alignment of columns using optical methods.

**Fig. 10.28**  Straightening of upper columns using ropes and wrenches.

storeys may impose forces on the supporting framework which have not been allowed for in the design, and should not be attempted without first referring to the designer. In all cases compressive cube strengths should be obtained for the concrete used at foundations according to BS 1881 [10.11].

It may be necessary during construction to adjust some of the columns vertically. This may be due to column curvature (unsymmetrical curing, temperature variations, etc.), or due to the forces imposed on the columns during frame erection to the upper floor levels. Figures 10.27 and 10.28 show wire ropes attached to these six-storey columns at the second floor level. It is necessary to tension the rope by 'pulling' the column using a ratcheted wrench (e.g. Turfor or Monkey wrench). Great care must be taken because it is surprising how flexible tall concrete columns are! The column is secured by props at the floor level as shown in Fig. 10.29.

Where a threaded connection is made to the end of a projecting dowel, the threaded part of the dowel must be undamaged, and the protective covering placed over the thread at the factory is still in place. Grout splashes should be removed from the thread. As before, the annulus around the dowel should be filled using a pourable grout before the holding-down plate (or washer) is placed over the end of the dowel hole. The entire length of the nut must be in *complete* contact with the threaded dowel.

**Fig. 10.29** 'Push-pull' props for temporary column stability (courtesy of F.C. Precast Ltd).

No attempt should be made to manoeuvre precast components into position using the nut and dowel as leverage. If the element requires realignment it should be lifted by the crane or raised off its bearing using jacks and moved in the correct manner.

### 10.5.2.2    *Beams*

The rules governing the lifting of beams are the same as for columns, except that beams may, and should, be lifted directly on to the framework. The positions of the lifting points and lifting devices (spreader beams, for example) should not interfere with the connections or fixing methods.

Fixing procedures for beams depend largely on the type of connector used. A broad distinction can be made between those connectors which provide an immediate restraint to shear and torsional forces (e.g. bolted cleats, billet with top fixing) and those which do not (dowelled corbel, billet with insitu tie top fixing). In the latter, temporary propping may be used because the beam is unstable until the insitu grouted connection has matured. In all cases the shear capacity of the connection is not affected by the sequence of erection and is not dependent on the strength of the insitu concrete infill.

The most onerous situation is the asymmetrically loaded spine beam shown in cross-section in Fig. 10.19(a). The eccentricity of the loading is about $e = 270$ mm and this results in a torsional moment of about 7 to 8 kNm in a typical frame. This must be resisted by the connector or nullified by propping. With the exception of the welded plate type connection where propping is the preferred method, all connectors are capable of generating this restraint. Similar problems occur in some of the wider L-shaped edge beams and in special cantilevered sections, but in general these are easily overcome by the addition of an extra top fixing cleat or by ad hoc site instructions to prop.

Beams of non-symmetrical shape, deep beams (exceeding 900 mm in depth) and spandrel beams are provided with a second fixing to improve temporary stability. The fixing may remain exposed, in which case stainless steel is used, or be covered in the general grouting operations. The operation is straightforward and has little effect on design.

Non-symmetrical perimeter components create torsional moments in their connections, and hence horizontal deflections in columns. This may lead to further deflections due to the $P$-$\Delta$ effect as subsequent floor loads are added. If column rotations (which will inevitably) lead to beam or slab rotations it is necessary to realign the perimeter components using push props or tethers. Dimensional checks should be made on column verticality, position and squareness during these operations.

If the orientation of the floor slab is such that a beam, or wall, is loaded non-symmetrically, or the fixing sequence is such that this effect cannot be avoided, the rotational stability of the support member should be ensured. This may be in the form of props with cross heads, or brackets at the strategic positions such as beam–column, wall–column or wall–wall connections. The temporary supports should remain in position until such time that the forces causing the rotations are nullified by membrane action in the floor slab, as shown in Fig. 10.19(b). Similarly the connection at the end of beam or wall should be capable of resisting the torsional moments created under the temporary or permanent floor loading arrangement.

Connector levels are positioned (by some but not all designers) up to 10 mm lower than the correct level to enforce levelling shims to be used with due allowance for squashing. It is important that the shims are accurately located to prevent large eccentricities occurring at the column face. The shim should be large enough to prevent local bearing problems because it cannot be assumed that any insitu grout placed around the connector will penetrate to the bearing surfaces. The rules for using neoprene pads between concrete surfaces (e.g. column corbels) are similar, but for excessive thicknesses of packing a sandwich construction comprising a steel plate inserted between two neoprene pads is favoured. After the beam has been fixed to its supporting member pourable grout, or concrete using a small 6 to 10 mm size aggregate, of minimum grade C25 fills the space around the connection. This applies to beams connected to continuous columns as well as continuous beams fixed to column heads. An expanding agent may be added to the mix to enable a greater water content to be used. Leakage of grout (cement fines, etc.) should be prevented. Any gaps between the precast members should be filled using dry pack mortar. The strength of the dry pack should be equal to that of the (weaker) parent precast components.

If the torsional stability of the connection relies on the insitu infill, 24 hours maturity should be provided prior to fixing floor slabs on to the beam. The strength of the grout should attain 10 N/mm$^2$ before the beam is loaded in a manner that will stress the connection. Alternatively the beam may be propped, in which case floor slabs may be fixed immediately afterwards. Connections offering immediate torsional stability allow the construction to proceed without delay. There is obviously some merit in this, particularly in buildings with a small number of beams per floor level.

### 10.5.2.3 Stability walls

Construction methods differ according to the type of wall. However, in all cases the walls should be braced against lateral movement, either by using:

- immediate permanent mechanical connections, such as angle cleats bolted or welded to the precast components
- immediate temporary connectors, such as clamps or temporary angle cleats

- by propping, using either a minimum of four compression props where the wall is internal and bound on both sides by floor slabs, or a minimum of two push-pull props where the wall is at an edge.

The props should subtend an angle of approximately 45° and be located at approximately one-fifth position along the length of the wall (Fig. 10.24). Where several wall panels are used to complete a storey height wall, all individual panels should be braced to the main frame, and *not* to each other. The temporary bracing should remain in position until the insitu concrete ring beam to the floor slab above the level of the walls is completed and matured. Walls that are connected to the frame using permanent mechanical devices do not impose such restrictions on the fixing sequence.

Stability walls which rely on composite action with the framework via a cementitious interface should be fixed and concreted in position at least 24 hours before the floor slabs are positioned. This is because the floor slabs contribute a large part of the dead weight at each storey level. The narrow annulus between the wall and framework should be equal around the perimeter of the wall, and should not be less than 10 mm if grout is being used, nor 20 mm if small aggregate concrete is used. The grout or concrete is usually grade C25 minimum.

In the case where projecting reinforcement is used to tie walls and wall/columns together it may be necessary to bend the projecting reinforcement against the face of the mould, later being pulled out by hand. The reinforcement used for this purpose should be sufficiently ductile to be bent through a right angle. If radial cracks are seen around the embedded reinforcement the surface concrete around the bar should be broken out in what will probably be a conical piece about 20 to 30 mm deep. The surface should be reinstated using an epoxy mortar or similar. Minor damage to the concrete, e.g. surface spalling, will be hidden beneath the infill grout and need not cause concern.

### 10.5.2.4    *Two-dimensional units, including floor slabs*

Precast units of large plan area present few fixing problems if adequate bearing is provided and the unit can be handled near to the horizontal. A broad distinction in the fixing procedure is made between units on *continuous* bearings (e.g. hollow core flooring) and *localized* bearings (e.g. double-tee units). With continuous bearings, uneven or cambered beams (or other supporting components) may cause 'rocking' in units exceeding about 2.0 m in width which, if restrained, may lead to torsional cracking across one of the corners.

The problem is overcome by wet bedding on mortar. Minimum bearing widths are achieved because flooring units are manufactured to positive tolerances. Flooring units are not propped unless a composite construction utilizing propping forces is specified.

The sequence of fixing floor slabs should be continuous in one direction, progressing away from the frame stabilizing components. Hollow core floor units are levered into their position using the edge of an adjacent unit as the fulcrum point for a (1 m long) crow bar as shown in Fig. 10.30. The hollow cores should be closed off at a distance of not more than 100 mm from the ends of the unit, or 100 mm beyond the end of broken out cores to prevent grout loss into the core. Plastic or polystyrene plugs are used for this.

Bearing lengths at either end should be made equal except where, for reasons of restricted access at one end, the correct bearing length may be provided at one end and a greater length at the other end. The minimum permitted bearing length is equal to the nominal length minus the tolerance distance of 15 mm (see Section 7.6). If the bearing

**Fig. 10.30** Levering hollow core floor units into final position.

length is more than about 10 mm less than this value the designer should be consulted as to whether the unit should be replaced or if additional insitu concrete infill is necessary to make an extended bearing from the precast unit to the supporting member.

Double-tee units are usually more massive in size and require careful attention during fixing. Four point lifting enables the unit to be accurately located. The units should be positioned onto $150 \times 150 \times 10$ mm neoprene, or similar, bearing pads. The bearing length on to the supporting element, i.e. through the pad, should not be less than 130 mm, and the bearing width on to the pad not less than 140 mm (see also Fig. 7.34). Contact with all four pads should be made and, if not, an additional thickness of pad should be provided at the non-contact point. Welded connections between adjacent floor units and between floor units and supporting beams are made after the completion of a bay of flooring when cumulative tolerances can be checked and the units adjusted where necessary.

Welded connections at the sides and at the ends of double-tee units are formed between pairs of plates cast into the slabs and beams. In order to cater for site fixing tolerances, a short intermediate mild steel bar is placed in the gap between the plates. The diameter of the bar varies from 8 to 20 mm typically and is about 100 mm long. A range of dowels is kept on site and the selection is made to suit the circumstances. A fillet weld is made to both sides of the bar using electric arc techniques and grade 43 electrodes. Care should be taken not to generate large amounts of heat during the welding operation, particularly at low ($< 5°$C) temperatures, as large thermal changes may cause cracking of the concrete at the ends of the plate. In extreme situations the crack will cause a large triangular piece of the flange to completely break off, thus exposing the plate from below. A small crack reliever – a slot cut into the concrete or a compressible filler – will prevent cracking.

The plates used for making welded connections between double-tee slabs and supporting beams are located over the tops of the webs of the slabs. This is the only place they can be located in cases where the end flanges have been removed for service holes. Because of the need to allow for fairly large tolerances in the length of the slabs the gap between the plates in the ends of the slabs and the beam may be as great as 30 mm. In this case a specially cut plate is used as the intermediate filler. The welded connection is made at one end of the slab to allow movement to take place between the ends of the slabs. The slabs will of course be fully tied to the frame when the insitu structural screed is applied.

## 10.6   INSITU CONCRETE

### 10.6.1   General specification

This refers to insitu concrete and grout, whether site batched or delivered to site ready mixed. The following information should be given on either the contract drawings and/or in the specification for the frame:

(1)  Mix specifications:

- compressive cube strength at 3 and 28 days
- workability
- aggregate size
- additives.

(2)  Procedures for placement of wet concrete and grout.
(3)  Procedures for placement of dry-pack grout.
(4)  Methods of compaction.
(5)  Dimensional details and deviations.
(6)  Reinforcement details and deviations.
(7)  Formwork striking information.
(8)  Defect assessment and making good information.

Insitu concrete can be supported in three ways: on permanent shutters (soffit units), over beams or on formwork. In the case of permanent shutters, these should be tight-fitting one to another and their size should be such that the total area of concrete is supported. Permanent shutter units should be examined for cracking or other damage before any wet concrete is placed on them. These may need propping until the concrete is adequately cured. Where props are used these should be installed vertically (or as directed by the manufacturers), and where necessary they should be adequately braced.

The most common use of insitu concrete is as make-up to floor units and as such final dimensions may vary slightly from those shown on the contract drawing. Where major upward variances occur, e.g. in the order of +50 mm in width for 1200 mm wide floor units, or proportionately less for narrower floor units, the precast contractor should consult the designer with the situation so that any design implications can be assessed.

When formwork is used, this must be of sufficient strength to support its own weight, and that of the concrete, together with reinforcement and the weight of any operatives or plant engaged in the work. Care must be taken when positioning props to support the formwork on to precast components below. In extreme cases small components, such as 150 mm wide precast floor units, may not be strong enough to carry the reactions from the props.

When using a mobile poker, care should be taken that the power unit is stood on a firm, level base and operated in accordance with the manufacturer's instructions. Any damage to the precast units during compaction should be noted and the appropriate repair carried out.

Sample cubes should be made and tested according to BS 1881 [10.11]. Compressive cube strength results (N/mm$^2$ units), cube densities, and any unusual failure modes should be made known to both the precast contractor and designer at 24 hours and at the time of removal of the formwork. If the strength of the concrete is less than required the precast contractor should postpone removal of formwork until further cube test results give strengths in excess of the required value.

### 10.6.2   Concrete screeds and joint infill in floors

Structural topping screeds which are being used for horizontal floor diaphragm action should be placed before the construction of the second next upper storey is completed, unless temporary horizontal bracing is provided, e.g. diagonal ties. Concrete can be delivered to the point of placement by either concrete pump, crane and skips, or hoist and wheelbarrow (for small areas).

The reinforcing bars should be lapped a distance equal to the anchorage bond length, i.e. at least 40 diameters for welded fabric. The cover to the top of the screed must be maintained at all positions, including laps, which will inevitably mean that at certain places the reinforcement will be in contact with the precast slab. Cover spacers should be used elsewhere to lift the reinforcement clear of the precast slab.

The concrete grade is usually grade C25 to C30 containing a 10 mm size aggregate. The three day cube strengths of the concrete should be at least 10 N/mm$^2$. The workability should give a slump of between 50 and 100 mm. The wet concrete should be uniformly distributed over the floor area, and should not be piled in excess of about 500 mm high at any point. It should be spread evenly over the floor area as quickly as possible to within 50 to 75 mm of the specified final level. As with all large floor areas the concrete is laid in strips or patches of (typically) 6 m dimension. Mechanical vibrating beams are used to compact the concrete. The screed may be de-watered and power floated, or rough tamped in the usual manner depending on the nature of the floor finishes.

Compressive cube strength results are required for all structural concrete. Day joints should be left rough and irregular.

Insitu concrete ring beams should be prepared as much as possible as the flooring is being laid in order to minimize the time between the completion of the precast and the pouring of the insitu concrete. An area of about 1000 m$^2$ is the recommended maximum area of flooring that should be fixed without insitu tie beams being cast. Insitu concrete infill of high workability is compacted using a mechanical poker in the ends of each precast floor unit.

The longitudinal joints between precast units should be filled using concrete grade C25 to C30. The floor units should be dampened, without surface ponding, prior to placement of insitu concrete. This may sound rather crude, but the best method of placing the infill is to sweep it into the gap using a stiff broom. Mechanical compaction is not often used.

### 10.6.3    Grouting

Grout is sand–cement mix and is used in situations where course aggregate concrete cannot penetrate. Wet grout may contain an expanding agent and a latex bonding agent which may be added to the mix in accordance with the manufacturer's instructions. Alternatively bonding agents may be applied (by painting) to cold joints.

Dry-pack grout should be used in circumstances where it is not possible to place the grout on the soffit or inclined surfaces of a precast component. The function of the grout is often identified on the contract drawing as either 'structural grout' or 'non-structural infill'. As well as meaning strength, the term 'structural' also refers to the protection of precast components or connections in which the integrity of the grout is necessary to ensure correct structural behaviour. Thus, strength tests are required for structural grout. The cube strength of the grout should be equal to the lesser design strength of the precast components in contact with the grout.

Non-structural infill is often used to complete a fascia, e.g. vertical joint at a column face, but this is not serving any structural function and need not be subject to such close inspection nor the strength tests referred to above.

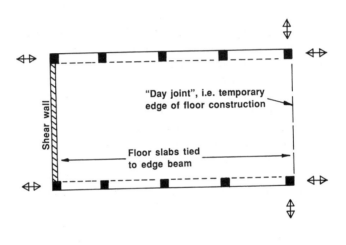

**Fig. 10.31**  Location of props to maintain temporary stability.

### 10.7    HAND-OVER

As explained in Chapter 8 phased hand-overs of precast frames take many forms. There is a great temptation with prefabricated construction to allow following trades to move in immediately after the flooring has been fixed. The recommendation is that the precaster should not work directly above other labour and that the following trades should resist doing further work until the precast at the next upper floor has been fixed.

Precast frames are readily released in self-contained and fully stable blocks. However, certain temporary props should be left in position even though the framework is stabilized at some point. Figure 10.31 shows the positions where props should be retained in partially completed frames. For this reason a phased hand-over should be carefully monitored because the dominion of the frame is still with the precaster whenever the temporary bracing is present.

Final hand-over is, of course, achieved when all props and other temporary bracing is removed, and the insitu infill has achieved its design strength. The precaster will inspect the precast frame in the company of the engineer and/or architect to agree that the precast work is complete. The 'topping out' ceremony will follow shortly.

# References

## Chapter 1

1.1 White, R. B. (1965) *Prefabrication: A History of its Development in Great Britain*. National Building Studies Special Report **36**, Ministry of Technology, Building Research Station, UK.

1.2 Morris, A. E. J. (1978) *Precast Concrete in Architecture*. George Godwin Ltd, p. 571.

1.3 Glover, C.W. (1964) *Structural Precast Concrete*. C. R. Books Ltd, London.

1.4 National Building Frame Manufacturers' Association. *The National Building Frame*, UDC 693.9.691.327. Not dated.

1.5 Diamant, R. M. E. (1964) *Industrialised Buildings*. Iliffe Books Ltd, London.

1.6 Mokk, L. ((in Hungarian) 1955, (in English) 1964) *Prefabricated Concrete for Industrial and Public Buildings*. Akademiai Kiado, Budapest.

1.7 Goodchild, G. H. & Seilen, P. (1992) *Project Profile: Merchant House and Woodchester House*. Publication 97.324, British Cement Association, Crowthorne, p. 13.

1.8 Nilson, A. H. (1987) *Design of Prestressed Concrete*, 2nd edn. John Wiley, p. 592.

1.9 Fédération Internationale de la Précontrainte (1986) *FIP Recommendations. Design of Multi-Storey Precast Concrete Structures*. Thomas Telford, London, p. 27.

1.10 Lewicki, B. (1966) *Building With Large Prefabricates*. Elsevier, London, p. 460.

1.11 Sarja, A. (1992) *Industrialised Building Technology as a Tool for the Future International Building Market*. Concrete Precasting Plant and Technology, B + FT, **58**, No. 11. Nov.

1.12 Wilson, J. G. (1962 & 1964) *Exposed Concrete Finishes*, **1** & **2**. CR Books, London.

1.13 Monks, W. (1985) *Appearance Matters*, Nos 1 to 9. Cement and Concrete Association, Wexham Springs, Slough. Publication 47–101 to 47–109.

1.14 Taylor, H. P. J. (1992) *Precast Concrete Cladding*. Edward Arnold, London, p. 307.

1.15 Precast Concrete Strategic Group (1989) *Precast Concrete Study Summary Report*. Arup Economic Consultants, London. Nov.

1.16 Van Acker, A. (1990) *General Introduction, Prefabrication of Concrete Structures*. International Seminar, Delft University of Technology, Delft University Press, Oct. pp. 7–12.

1.17 Scott, N. (1994) *Precast Prestressed Concrete Beyond Year 2000 in the United States*. Concrete Precasting Plant and Technology, B + FT, **60**, No. 7, July, pp. 46–58.

1.18 Elliott, K. S. & Filipowicz, S. (1994) High Specification Precast Concrete Structures. *Proc. XII Congress, PCI/FIP '94*, Washington DC.

1.19 Elliott, K. S. (1992) The Design of Modern Precast Concrete Multistorey Framed Structures in The United Kingdom. *PCI Journal*, **37**, No. 3, May/June, pp. 32–47.

1.20 Brigginshaw, G. (1987) A New Era for Structural Frames. *Concrete*, **21**, No. 10, Oct.

1.21  Precast Concrete Frames Association (1985) *Frames for Multi-Storey Buildings – An Economic Comparison.* 60 Charles Street, Leicester, UK, p. 16.

1.22  Bruggeling, A. S. G. & Huyghe, G. F. (1991) *Prefabrication with Concrete.* Balkema, Rotterdam, p. 380.

1.23  Elliott, K. S. & Tovey, A. (1992) *Precast Concrete Frame Buildings – A Design Guide.* British Cement Association, Wexham Springs, Slough, May, p. 88.

1.24  Precast Concrete Industry Training Association (1992) *Code of Practice for the Safe Erection of Precast Concrete.* 60 Charles Street, Leicester, UK, p. 86.

1.25  Precast Flooring Federation (1992) *Code of Practice for the Safe Erection of Precast Concrete Flooring.* 60 Charles Street, Leicester, UK, p. 84.

1.26  Tremi, S. (1994) *The New RAI Technical Telecommunications Centre, Rome.* Concrete Precasting Plant and Technology, B + FT, **60**, No. 5, May, pp. 67–82.

## Chapter 2

2.1  Fédération Internationale de la Précontrainte (1994) *FIP Recommendations. Planning and Design of Precast Concrete Structures.* FIP Commission on Prefabrication, SEKO, Institution of Structural Engineers, London, May.

2.2  Elliott, K. S. & Tovey, A. (1992) *Precast Concrete Frame Buildings – A Design Guide.* British Cement Association, Wexham Springs, Slough, May, p. 88.

2.3  British Standards Institution (1985) *The Structural Use of Concrete.* BSI, London, BS 8110.

2.4  Institution of Structural Engineers (1978) *Structural Joints in Precast Concrete.* IStructE, London, August, p. 56.

2.5  Prestressed Concrete Institute (1992) Design Handbook, 4th edn. PCI, Chicago, USA.

2.6  Fédération Internationale de la Précontrainte (1986) *FIP Recommendations. Design of Multi-Storey Precast Concrete Structures.* Thomas Telford, London, p. 27.

2.7  Fédération Internationale de la Précontrainte (1988) *FIP Recommendations. Precast Prestressed Hollow Cored Floors*, FIP Commission on Prefabrication, Thomas Telford, London, p. 31.

2.8  *Guidelines for the Use of Structural Precast Concrete in Buildings.* Study Group of the New Zealand Concrete Society, and National Society of Earthquake Engineering, Christchurch, New Zealand, 1991, p. 174.

2.9  Nilson, A. H. (1987) *Design of Prestressed Concrete.* 2nd edn. John Wiley, p. 592.

2.10  Sheppard, D. A. & Phillips W. R. (1989) *Plant-Cast Precast and Prestressed Concrete.* 3rd edn. McGraw-Hill, p. 791.

2.11  Bruggeling, A. S. G. & Huyghe, G. F. (1991) *Prefabrication with Concrete.* Balkema, Rotterdam, p. 380.

2.12  Lewicki, B. (1966) *Building with Large Prefabricates.* Elsevier, London, p. 460.

2.13  Haas, A. M. (1983) *Precast Concrete Design and Application.* Applied Science Publishers, London.

2.14  Bljuger, F. (1988) *Design of Precast Concrete Structures.* Ellis Horwood, Chichester, UK, p. 296.

2.15  Levitt, M. (1982) *Precast Concrete.* Applied Science Publishers, London.

2.16  Kromer, R. (1992) *Tammer Elementti – A Nearly Automated Precast Works.* Concrete Precasting Plant and Technology, B + FT, **58**, No. 3, March, pp. 116–180.

2.17  Richardson, J. G. (1973) *Precast Concrete Production.* Cement & Concrete Association, London.

2.18 Richardson, J. G. (1991) *Quality in Precast Concrete – Design – Production – Supervision.* Longman Group, Harlow, UK, p. 395.

2.19 Fédération Internationale de la Précontrainte Secretariat. Institution of Structural Engineers, 11 Upper Belgrave Street, London, SW1X 8BH, UK.

2.20 Prestressed Concrete Institute Secretariat. 175 West Jackson Boulevard, Chicago, Illinois 60604, USA.

2.21 Department of Environment and CIRIA (1973) *The Stability of Precast Concrete Structures.* Seminar, London, March.

2.22 *Prefabrication of Concrete Structures* (1990) International Seminar. Delft University of Technology, Delft University Press, Oct. p. 214.

2.23 Noteworthy Developments in Prestressed and Precast Concrete (1989) Singapore, November.

2.24 Institute of Engineers Malaysia (1993) *Trends, Innovation and Performance of Prestressed and Precast Concrete.* Institute of Engineers Malaysia, Kuala Lumpur, May.

## Chapter 3

3.1 Bruggeling, A. S. G. & Huyghe, G. F. (1991) *Prefabrication with Concrete.* Balkema, Rotterdam, p. 380.

3.2 Building Research Establishment (1993) Some Recent Work at the Building Research Establishment. *Concrete Plant and Production*, Jul./Aug., p. 104.

3.3 Evans, B. (1993) Underfloor Heating and Cooling. *The Architect's Journal*, Aug., pp. 40–41.

3.4 A. B. Strangbetong, PO Box 5074, S 131 05, Nacka, Sweden.

3.5 Fédération Internationale de la Précontrainte (1988) *FIP Recommendations. Precast Prestressed Hollow Cored Floors.* FIP Commission on Prefabrication, Thomas Telford, London, p. 31.

3.6 Partek Ergon, System de Construction Ergon, Marnixdreef B–2500 Lier, Belgium.

3.7 Elliott, K. S. & Tovey, A. (1992) *Precast Concrete Frame Buildings — A Design Guide.* British Cement Association, Wrexham Springs, Slough, May, p. 88.

3.8 Goodchild, C. H. (1995) *Hybrid Structures.* Reinforced Concrete Council, British Cement Association, Crowthorne, UK.

## Chapter 4

4.1 Elliott, K. S. & Tovey, A. (1992) *Precast Concrete Frame Buildings – A Design Guide.* British Cement Association, Wexham Springs, Slough, May, p. 88.

4.2 British Standards Institution (1985) *The Structural Use of Concrete.* BSI, London, BS 8110.

4.3 BRE (1988) *Design of Normal Concrete Mixes.* Building Research Establishment, Watford, UK, p. 42.

4.4 Prestressed Concrete Institute (1989) *Manual for Structural Design of Architectural Precast Concrete.* 2nd edn. PCI, Chicago, USA, p. 340.

4.5 Institution of Structural Engineers (1978) *Structural Joints in Precast Concrete.* IStructE., London, August, p. 56.

4.6 Raths, C. H. (1984) Spandrel Beam Behaviour and Design. *PCI Journal*, March–April 1984, pp. 62–131.

4.7 Mayfield, B., Davies, G. & Kong, F. K. (1970) Some Tests on the Transmission Length and

Ultimate Strength of Pretensioned Concrete Beams Incorporating Dyform Strand. *Magazine of Concrete Research*, **22**, No. 73, Dec., pp. 219–226.

4.8 British Standards Institution (1984) *Code of Practice for Dead and Imposed Loads*. Part 1. BSI, London, BS 6399.

4.9 Adlparvar, M. R. (1993) *Torsional Behaviour of Precast Concrete Beams and Connections*. M. Phil. Thesis, University of Nottingham.

4.10 Elliott, K. S., Davies, G. & Adlparvar, R. M. (1993) Torsional Behaviour of Joints and Members in Precast Concrete Structures. *Magazine of Concrete Research*, **164**, Sept., pp. 157–168.

4.11 Mitchell, D. & Marcakis, K. (1980) Precast Concrete Connections with Embedded Steel Inserts. *PCI Journal*, July, Chicago, USA.

4.12 Holmes, M. & Posner, C. D. (1971) The Connection of Precast Concrete Structural Members. *CIRIA Report 28*, Feb. London, UK.

4.13 Cook, W. D. & Mitchell, D. (1988) Studies of Disturbed Regions near Discontinuities in Reinforced Concrete Members. *ACI Structural Journal*, March.

4.14 Martin, L. D. & Korkosz, W. J. (1982) *Connections for Precast Prestressed Concrete Buildings, including Earthquake Resistance*. The Consulting Engineers Group, Inc., Illinois 60025, March, p. 285.

4.15 Prestressed Concrete Institute (1992) *Design Handbook*. 4th edn. PCI, Chicago, USA.

4.16 Prestressed Concrete Institute (1990). *Design and Typical Details of Connections for Precast and Prestressed Concrete*. 2nd edn. PCI, Chicago, USA.

4.17 Beeby, A. W. (1968) Short Term Deformations of Reinforced Concrete Members. *Technical Report 42.408*, Cement & Concrete Association, Wexham Springs.

4.18 Clark, L. A. & Thorogood, P. (1988) Serviceability Behaviour of Reinforced Concrete Half Joints. *The Structural Engineer*, **66**, No. 18, Sept., pp. 295–302.

4.19 Mattock, A. H. & Chan, T. C. (1979) Design and Behaviour of Dapped-end Beams. *PCI Journal*, **24**, No. 6, pp. 28–45.

4.20 Windisch, A. (1991) *Strut-Crack-and-Tie Model in Structural Concrete*. Structural Concrete, IABSE Colloquium, Stuttgart, pp. 539–544.

4.21 Mitchell, D. & Cook, W. D. (1991) *Design of Disturbed Regions*. Structural Concrete, IABSE Colloquium, Stuttgart, pp. 533–538.

4.22 Ife, J. S., Uzumeri, S. M. & Huggins, M. W. (1968) Behaviour of the 'Cazalay Hanger' Subjected to Vertical Loading. *PCI Journal*, **13**, No. 6, Dec., pp. 48–66.

4.23 Elliott, K. S., Davies, G. & Mahdi, A. A. (1992) Semi-rigid Joint Behaviour on Columns in Precast Concrete Buildings. *COST C1 Proceedings of the First State of the Art Workshop, Semi-rigid Behaviour of Civil Engineering Structural Connections, ENSAIS*, Strasbourg, Oct., pp. 282–295.

4.24 Mahdi, A. A. (1992) Moment Rotation Effects on the Stability of Columns in Precast Concrete Structures. PhD thesis, University of Nottingham.

4.25 Cranston, W. (1972) Analysis and Design of Reinforced Concrete Columns. *Research Report 20*, Cement & Concrete Association, Wexham Springs, Aug.

## Chapter 5

5.1 British Standards Institution (1985) *The Structural Use of Concrete*. BSI, London, BS 8110.

5.2 Institution of Structural Engineers (1978) *Structural Joints in Precast Concrete*. IStructE., London, Aug., p. 56.

5.3   Elliott, K. S. & Tovey, A. (1992) *Precast Concrete Frame Buildings – A Design Guide.* British Cement Association, Wexham Springs, Slough, May, p. 88.

5.4   Fédération Internationale de la Précontrainte. *FIP Recommendations. Precast Prestressed Hollow Cored Floors.* FIP Commission on Prefabrication, Thomas Telford, London, p. 31.

5.5   CEN TC 229 (1995) *Floors of Precast Prestressed Hollow Cored Elements.* European Committee for Standardisation, Draft document.

5.6   Prestressed Concrete Institute (1985) *Manual for the Design of Hollow-core Slabs.* PCI, Chicago, USA.

5.7   Walraven, J. C. & Mercx, W. (1983) *The Bearing Capacity of Prestressed Hollow Core Slabs.* Heron, **28** No. 3, University of Delft, Netherlands, p. 46.

5.8   Johnsson, E. (1988) *Shear Capacity of Prestressed Extruded Hollow Core Slabs.* Nordic Concrete Research, **7**, Oslo, pp. 27–31.

5.9   Pajari, M. (1988) *Load Carrying Capacity of Prestressed Hollow Core Slabs.* Nordic Concrete Research, **7**, pp. 232–249.

5.10  Pisanty, A. (1992) The shear strength of extruded hollow core slabs. *Materials and Structures*, **25**, pp. 224–230.

5.11  Juvas, K. (1987) *The Workability of No-Slump Concrete for Use in Hollow-Core Slabs.* Nordic Concrete Research, **6**. The Nordic Concrete Federation, Oslo, pp. 121–130.

5.12  Den Uijl, J. A. (1985) *Bond Properties of Strands in Connection with Transmission Zone Cracks.* Concrete Precasting Plant and Technology, B + FT, **51**, No. 1, Jan.

5.13  B. Venkateswarlu *et al.* (1982) Roof and Floor Slabs Assembled with Precast Concrete Hollow Cored Units. *ACI Journal*, Jan., pp. 50–55.

5.14  Stanton, J. (1987) Proposed Design Rules for Load Distribution in Precast Concrete Decks. *ACI Structural Journal*, Sept.–Oct., pp. 371–382.

5.15  Aswad, A. & Jacques, F. J. (1992) Behaviour of Hollow Core Slabs Subject to Edge Loads. *PCI Journal*, **37**, No. 2, March–April, pp. 72–84.

5.16  Fédération Internationale de la Précontrainte (1982) *FIP Technical Report. Design Principles for Hollow Core Slabs Regarding Shear and Transverse Load Capacity.* FIP Commission on Prefabrication, Wexham Springs, Slough.

5.17  Kong, F. K. & Evans, R. H. (1987) *Reinforced and Prestressed Concrete*, 3rd edn. Van Nostrand, p. 508.

5.18  British Standard Institution (1980) *Specification for High Tensile Steel Wire and Strand for the Prestressing of Concrete.* BSI, London, BS 5896.

5.19  Girhammer, U. A. (1992) Design Principles for Simply Supported Prestressed Hollow Core Slabs. *Structural Engineering Review*, **4**, No. 4, Oxford, UK, pp. 301–316.

5.20  Mayfield, B., Davies, G. & Kong, F. K. (1970) Some Tests on the Transmission Length and Ultimate Strength of Pretensioned Concrete Beams Incorporating Dyform Strand. *Magazine of Concrete Research*, **22**, No. 73, Dec., pp. 219–226.

5.21  Lin Yang (1994) Design of Prestressed Hollow Core Slabs with Reference to Web Shear Failure. *Journal of Structural Engineering*, ASCE, **120**, No. 9, Sept., pp. 2675–2696.

5.22  ACI Standard 318–89 (1989) *Building Code Requirements for Reinforced Concrete.* American Concrete Institute, Detroit.

5.23  Akesson, M. (1993) *Fracture Mechanics Analysis of the Transmission Zone in Prestressed Hollow Core Slabs.* Publication 93.5. Chalmers University of Technology, Göteborg, Sweden, Nov., p. 62.

5.24  Tassi, G. (1988) Bond Properties of Prestressing Strands. *Proceedings of FIP Symposium*, Jerusalem, Israel, p. 8.

5.25  Gylltoft, K. (1979) *Bond Properties of Strand in Fatigue Loading.* Research Report TULEA

1979:22, Lulea University of Technology, Lulea, Sweden, p. 73.

5.26   Barney, G. B., Corley, W. G., Hanson, J. M. & Parmelee, R. A. (1977) Behaviour and Design of Prestressed Concrete Beams with Large Web Openings. *PCI Journal*, Nov.–Dec., pp. 32–60.

5.27   Adus, S. S. & Harrop, J. (1979) Prestressed Concrete Beams with Circular Holes. *ASCE Journal of the Structural Division*, March, pp. 635–652.

5.28   Savage, J., Tadros, M., Einea, A. & Fischer, L. (1994) *Precast Prestressed Concrete Double Tees with Web Openings*. Proc. X11 Congress, PCI/FIP '94, Washington DC, pp. C48–54.

5.29   Land, H. (1994a) *Precast Concrete Floors, Special Points on Section Values Determination Design and Construction*. Part 1, Concrete Precasting Plant and Technology, B + FT, **60**, No. 5, May.

5.30   Land, H. (1994b) *Precast Concrete Floors, Special Points on Section Values Determination, Design and Construction*. Part 2, Concrete Precasting Plant and Technology, B + FT, **60**, No. 6, June.

## Chapter 6

6.1   British Standards Institution (1985) *The Structural Use of Concrete*. BSI, London, BS 8110.

6.2   Fédération Internationale de la Précontrainte (1982) *FIP Recommendations. Shear at the Interface of Precast and Insitu Concrete*. Wexham Springs, Slough, UK, Jan., p. 31.

6.3   Vesa, M. (1979) *Shear Strength at the Interface Between Topping and Concrete Element*. Helsinki Technical University, p. 96.

6.4   Gustavsson, K. (1980) *Shear at the Interface – Tests on T-slabs with Thin Top Layers*. FIP Commission on Prefabrication, Lulea, Sweden, June, p. 7.

6.5   British Standards Institution (1990) *Guide to Accuracy in Building*. BSI, Milton Keynes, BS 5606.

6.6   Evans, R. H. & Parker, A. S. (1955) Behaviour of Prestressed Concrete Composite Beams. *ACI Journal*, **51**, No. 43, May, pp. 861–881.

6.7   CEN TC 229 (1995) *Floors of Precast Prestressed Hollow Cored Elements*. European Committee for Standardisation, Draft document.

6.8   Kajfasz, S., Sommerville, G. & Rowe, R. E. (1963) *An Investigation of the Behaviour of Composite Concrete Beams*. Cement & Concrete Association Research Report No. 15, London, Nov.

## Chapter 7

7.1   Stanton, J. F., Anderson, R. G., Dolan, C. W. & McCleary, D. E. (1986 & 1987) *Moment Resistant Connections and Simple Connections*. Final Report to PCI, Specially Funded R & D Projects Nos 1 & 4, 1986, and *PCI Journal*, **32**, No. 2, March–April 1987, pp. 62–74.

7.2   Coleman, D. G. (1982) *A Study of Precast Concrete Beam to Column Joints Using Steel Connections*. Applied Science Publishers, London.

7.3   Vambersky, J. N. J. A. (1990) *General Design Philosophy, Prefabrication of Concrete Structures*. International Seminar. Delft University of Technology, Delft University Press, Oct., pp. 15–28.

7.4   Vambersky, J. N. J. A. (1990) *Mortar Joints Loaded in Compression, Prefabrication of*

*Concrete Structures*. International Seminar. Delft University of Technology, Delft University Press, Oct., pp. 167–180.

7.5 Bljuger, F. (1981) On an Estimation Methodology for Wall Bearing Capacity in Precast Concrete Construction. *Proc. 10th International Congress of the Precast Concrete Industry*, Jerusalem, Sept., pp. 259–271.

7.6 British Standards Institution (1985) *The Structural Use of Concrete*. BSI, London, BS 8110.

7.7 Bljuger, F. (1978) *Non-linear Characteristics of Joints in Compression*. RILEM-CEB-CIB, Athens, Greece, Sept., **1**, pp. 159–169.

7.8 Bljuger, F. *Some Problems of Design of Prefabricated Multistorey Buildings, Strength and Analysis of Multistorey Structures*, ed. G. Lvov & F. Bljuger, MNIITEP, Moscow, pp. 5–42.

7.9 *Guidelines for the Use of Structural Precast Concrete in Buildings* (1991) Study Group of the New Zealand Concrete Society, and National Society of Earthquake Engineering, Christchurch, New Zealand, p. 174.

7.10 Prestressed Concrete Institute (1989) *Manual for Structural Design of Architectural Precast Concrete*, 2nd edn. PCI, Chicago, USA, p. 340.

7.11 Walraven, J. C. & Mercx, W. (1983) *The Bearing Capacity of Prestressed Hollow Core Slabs*. Heron, **28**, No. 3, University of Delft, Netherlands, p. 46.

7.12 Bljuger, F. (1976) Determination of Deformability Characteristics of Vertical Shear Joints in Precast Buildings. *Building and Environment*, **11**, pp. 277–282.

7.13 Bljuger, F. (1982) Wall Clamping of Floor Slabs. *Building and Environment*, **17**, No. 1, pp. 17–21.

7.14 Clarke, J. L. & Simmonds, R. M. (1978) *Tests on Embedded Steel Billets for Precast Concrete Beam–Column Connections*. Technical Report No. 42.523, Cement & Concrete Association, Wexham Springs, Aug., p. 12.

7.15 Bruggeling, A. S. G. & Huyghe, G. F. (1991) *Prefabrication with Concrete*. Balkema, Rotterdam, p. 380.

7.16 Fawzi, H. T. (1976) *Concrete Bearing with Steel Inserts in Precast Column to Beam Connections*. M.Phil. thesis, University of Nottingham.

7.17 Institution of Structural Engineers (1978) *Structural Joints in Precast Concrete*. IStructE, London, Aug., p. 56.

7.18 Park, R. (1986) *Seismic Design Considerations for Precast Concrete Construction in Seismic Zones*. Seminar on Precast Concrete Construction in Seismic Zones, Japan Society for the Promotion of Science & US National Science Foundation, Tokyo, **1**, pp. 1–38.

7.19 Pillai, S. U. & Kirk, D. W. (1981) Ductile Beam–Column Connections in Precast Concrete. *ACI Journal*, Nov./Dec., pp. 480–487.

7.20 Mahdi, A. A. (1992) *Moment Rotation Effects on the Stability of Columns in Precast Concrete Structures*. PhD thesis, University of Nottingham.

7.21 Fédération Internationale de la Précontrainte (1988) *FIP Recommendations. Precast Prestressed Hollow Cored Floors*. FIP Commission on Prefabrication, Thomas Telford, London, p. 31.

7.22 Elliott, K. S. & Tovey, A. (1992) *Precast Concrete Frame Buildings – A Design Guide*. British Cement Association, Wexham Springs, Slough, May, p. 88.

7.23 Prestressed Concrete Institute (1992) *Design Handbook*. 4th edn. PCI, Chicago, USA.

7.24 Fogarasi, G. (1988) *Precast Concrete in Buildings: Design, Manufacture and Construction*. Precast Concrete in Buildings, Seminar, Precast Concrete Manufacturers Association of NSW and Concrete Institute of Australia, April pp. 1–45.

7.25 Regan, P. E. (1981) *Billet Insert Joints for Precast Concrete Frames–Two Beam Joints*

*Under Symmetrical Loads*. Report to Dowmac Concrete Ltd, Polytechnic of Central London, Oct.

7.26  Marcakis, K. & Mitchell, D. (1980) Precast Concrete Connections with Embedded Steel Members. *PCI Journal*, **25**, No. 4, Jul./Aug., pp. 88–115.

7.27  Mattock, A. H. & Gaafar, G. H. (1982) Strength of Embedded Steel Sections as Brackets. *ACI Journal*, Mar./Apr., pp. 83–91.

7.28  Holmes, M. & Bond, D. (1963) Tests on a Beam to Column Connection for Precast Concrete. *The Structural Engineer*, **41**, No. 9, Sept., pp. 293–297.

7.29  Holmes, M. & Posner, C. D. (1970) Factors Affecting the Strength of Steel Plate Connections Between Precast Concrete Elements. *The Structural Engineer*, **48**, No. 10, Oct., pp. 399–406.

7.30  British Standards Institution (1985) *Structural Use of Steelwork in Buildings, Part 1, Code of Practice for Design in Simple and Continuous Construction: Hot Rolled Sections*. BSI, London, BS 5950.

7.31  Holmes, M. & Posner, C. D. (1971) *The Connection of Precast Concrete Structural Members*. CIRIA Report 28, London, Feb.

7.32  Comité Euro-Internationale du Beton (1994) *Fastenings to Concrete and Masonry Structures*, CEB Report, Thomas Telford, London, 249 p.

7.33  Mohamed, S. A. M. (1992) *Behaviour of Sleeved Bolt Connections in Precast Concrete Building Frames*. PhD Thesis, University of Southampton, UK.

7.34  Bruggeling, A. S. G. (1992) *Development Towards Structural Concrete, Concrete Precasting Plant and Technology*. B + FT, 58 No. 3, March.

7.35  Bruggeling, A. S. G. & Huyghe, G. F. (1991) *Prefabrication with Concrete*. Balkema, Rotterdam, p. 380.

7.36  Kris, L. B. & Raths, C. H. (1965) Connections in Precast Concrete Structures: Strength of Corbel Brackets. *PCI Journal*, **10**, No. 1, pp. 16–61.

7.37  Somerville, G. (1972) *The Behaviour and Design of Reinforced Concrete Corbels*. Publication No. 12.069, Cement & Concrete Association, Wexham Springs, p. 60.

7.38  Cook, W. D. & Mitchell, D. (1988) Studies of Disturbed Regions near Discontinuities in Reinforced Concrete Members. *ACI Structural Journal*, Mar.

7.39  Zeller, W. (1991) *Conclusions from Tests on Corbels, Structural Concrete*. IABSE Colloquium, Stuttgart, p. 577–582.

7.40  Lindberg, R. & Keronen, A. (1992) *Semi-rigid Behaviour of a RC Portal Frame*. COST C1 Proceedings of the First State of the Art Workshop, Semi-rigid Behaviour of Civil Engineering Structural Connections, ENSAIS, Strasbourg, Oct., pp. 53–63.

7.41  Park, R. (1993) *Trends, Innovation and Performance of Precast Concrete in Buildings*. Seminar, Trends, Innovation and Performance of Prestressed and Precast Concrete, Institute of Engineers Malaysia, Kuala Lumpur, May.

7.42  Chuan-Chien Shu & Hawkins, N. M. (1992) Behaviour of Columns Continuous Through Concrete Floors. *ACI Structural Journal*, Jul./Aug.

7.43  Kuttab, A. & Dougill, J. W. (1988) Grouted and Dowelled Jointed Precast Concrete Columns: Behaviour in Combined Bending and Compression. *Magazine of Concrete Research*, **40**, No. 144, Sept.

7.44  Bergström, S. (1994) *Inspanning Av Betongelementpelare Med Grundskruv Och Stalfot*. Chalmers University of Technology, Göteborg, Report 94:1, May.

7.45  Somerville, G. (1967) *Tests on Column–Column Joints for the Ministry of Public Building and Works*. Cement & Concrete Association, Wexham Springs, Note DW/30, July.

7.46  Korolev, L. V. & Korolev, H. V. (1962) Joint Between Prefabricated Reinforced Column and Foundation. *Promyshlennoe Sroitel'stvo*, **16**, No. 9, Sept.

# Chapter 8

8.1   British Standards Institution (1972) *Wind Loads* CP3: Chapter 5: Part 2. BSI, London.

8.2   Cook, N. (1985) *The Designer's Guide to Wind Loading of Building Structures*. Butterworth.

8.3   Pearce, D. J. & Matthews, D. D. (1973) *An Appraisal of the Design of Shear Walls in Box Frame Structures*. Property Services Agency, Apr.

8.4   Institution of Civil Engineers and Institution of Structural Engineers (1985) *Manual for the Design of Reinforced Concrete Building Structures*. Joint publication, London.

8.5   Moustafa, S. E. (1981) Effectiveness of Shear-friction Reinforcement in Shear Diaphragm Capacity of Hollow-core Slabs. *Journal of the Prestressed Concrete Institute*, **26**, No. 1, Jan./Feb., pp. 118–132.

8.6   Svensson, S. (1985) *Diaphragm Action in Precast Hollow-core Floors. Description of a Pilot Test Series*. The Nordic Concrete Federation, Nordic Concrete Research, Publication No. 4.

8.7   Stroband, J. & Kolpa, J. J. (1985) The Behaviour of a Demountable Floor. *Proc. Symposium on Demountable Concrete Structures*, Delft University Press, Delft, pp. 201–216.

8.8   Sarja, A. (1978) *Analysis of the Statical In-plane Behaviour of Prefabricated Hollow Core Slabs Field*. Concrete Laboratory Report No. 51, Espoo, Finland, June, p. 36.

8.9   Svensson, S., Engstrom, B. & Cederwall, K. (1986) *Diaphragm Action of Precast Floors with Grouted Joints*. Nordisk Betong, Feb., pp. 123–128.

8.10  Stroband, J. (1990) *Diaphragm Action, Prefabrication of Concrete Structures*. International Seminar, Delft University of Technology, Delft University Press, Oct., pp. 203–214.

8.11  Menengotto, M. (1989) *Seismic Resistant Extruded Hollow Core Slabs*. International Symposium on Noteworthy Developments in Prestressed and Precast Concrete, Singapore, Nov.

8.12  Menengotto, M. (1988) *Hollow Core Floors Tests for Seismic Action*. FIP Symposium. Jerusalem, Sept., pp. 477–484.

8.13  Abdul-Wahab, H. & Sarsam, S. (1988) Strength of Vertical Plane Joints Between Large Precast Concrete Panels. *The Structural Engineer*, **66**, 14, July.

8.14  British Standards Institution (1985) *The Structural Use of Concrete*. BSI, London, BS 8110.

8.15  Mast, R. F. (1968) *Auxiliary Reinforcement in Concrete Connections*. Proc. American Society of Civil Engineers, **94**, No. ST 6, pp. 1485–1504.

8.16  Jimenez, R., Gergely, P. & White, R. N. (1978) *Shear Transfer Across Cracks in Reinforced Concrete*. New York, Cornel University, Dept of Structural Engineering, Report No. 78–4, Aug.

8.17  Tassios, P. T. & Vintzeleou, E. N. (1987) Concrete to Concrete Friction. *Journal of Structural Engineering*, Proc. ASCE, **113**, No. ST4, Apr., pp. 832–849.

8.18  Divinkar, M. P., Fafitis, A. & Shah, S. P. (1987) Constitutive Model for Shear Transfer in Cracked Concrete. *Journal of Structural Engineering*, **113**, No. 5, May, pp. 1046–1062.

8.19  Cholewicki, A. (1991) *Shear Transfer in Longitudinal Joints of Hollow Core Slabs*. Concrete Precasting Plant and Technology, B + FT, **57**, No. 4, Apr., pp. 58–67.

8.20  Konstrucksjonshandbok Betongelementer (1970), Norges Betongindustriforbund, Oslo.

8.21  Walraven, J. C. (1990) *Diaphragm Action in Floors, Prefabrication of Concrete Structures*. International Seminar, Delft University of Technology, Delft University Press, Oct., pp. 143–154.

8.22  Millard, S. G. & Johnson, R. P. (1984) Shear Transfer Across Cracks in Reinforced Concrete due to Aggregate Interlock and to Dowel Action. *Magazine of Concrete Research*, **36**, No. 126, Mar., pp. 9–21.

8.23 Walraven, J. C. & Reinhardt, H. W. (1981) *Concrete Mechanics: Part A: Theory and Experiments on the Mechanical Behaviour of Cracks in Plain and Reinforced Concrete Subjected to Shear Loading.* Heron, **26**, No. 1A, pp. 1–68.

8.24 Davies, G., Elliott, K. S. & Omar, W. (1990) Horizontal Diaphragm Action in Precast Hollow Core Floors. *The Structural Engineer*, **68**, No.2, Jan., pp. 25–33.

8.25 Elliott, K. S., Davies, G. & Omar, W. (1992) Experimental and Theoretical Investigation of Precast Hollow Core Slabs Used as Horizontal Diaphragms. *The Structural Engineer*, **70**, No. 10, May.

8.26 Omar, W. (1990) *Diaphragm Action in Precast Concrete Floor Construction.* PhD thesis, University of Nottingham.

8.27 Elliott, K. S., Davies, G. & Bensalem, K. (1993) *Precast Floor Slab Diaphragms Without Structural Screeds, Concrete 200.* Economic and Durable Construction Through Excellence, Dundee, Sept., pp. 617–632.

8.28 British Standards Institution (1972) *Methods for the Assessment of Surface Texture.* Part 1, BSI, London, BS 1134.

8.29 Private communication Dec. 1990.

8.30 de Roo, A. M. & Straman, J. P. (1991) *Krachtsverdeling in en vormverandering van een in zynvlak belaste prefab vloerconstructie.* PhD thesis, Delft Precast Concrete Institute, Delft University of Technology.

8.31 Prestressed Concrete Institute (1992) *Design Handbook.* 4th edn. PCI, Chicago, USA.

8.32 Bensalem, K. (1994) Unpublished research at University of Nottingham, March.

8.33 Institution of Structural Engineers (1988) *Stability of Buildings.* IStructE, London, Dec.

8.34 Hekenyi, M. (1946) *Beams on Elastic Foundations.* University of Michigan Press.

8.35 Stafford-Smith, B. & Carter, C. (1969) *A Method of Analysis for Infill Frames.* Paper 7218, Proc. Instn. Civ. Engrs, **44** Sept.

8.36 Mainstone, R. J. (1972) *On the Stiffnesses and Strengths of Infill Frames.* Building Research Establishment Paper CP2/72, Feb.

8.37 Wood, R. H. (1978) *Plasticity, Composite Action and Collapse Design of Unreinforced Shear Wall Panels in Frames.* Paper 8110, Proc. Instn. Civ. Engrs, **65**, June.

8.38 Kwan, K-H. & Liaum, T-C. (1982) *Non-linear Analysis of Multi-storey Infilled Frames.* Paper 8577, Proc. Instn. Civ. Engrs, 73, June.

8.39 Kwan, K-H. & Liaum, T-C. (1983) *Plastic Theory of Infilled Frames with Finite Interface Shear Strength.* Paper 8718, Proc. Instn. Civ. Engrs. **75**, Dec.

8.40 Wright, R. (1981) *Design of Wind Walls.* Internal Paper, Trent Concrete Ltd, June.

8.41 British Standards Institution (1978) *Code of Practice for Use of Masonry.* BSI, London, BS 5628.

8.42 Chakrabarti, S. C., Nayak, G. C. & Paul, D. K. (1988) Shear Characteristics of Cast-in-place Vertical Joints in Storey High Precast Wall Assembly. *ACI Structural Journal*, Jan./Feb., pp 30–45.

8.43 Chakrabarti, S. C., Bhise, N. N. & Sharma, K. N. (1981) Failure Criterion of Vertical Shear Key Joints in Prefabricated Wall Panels. *Indian Concrete Journal*, Mar., pp. 63–67.

8.44 Foerster, H. R., Rizkalla, S. H. & Scott Heuvel, J. (1989) Behaviour and Design of Shear Connections for Loadbearing Wall Panels. *PCI Journal*, **34**, No. 1, Jan./Feb., pp. 102–118.

8.45 CIRIA (1973) *The Behaviour of Large Panel Structures.* Construction Industry Research and Information Association, Report 45, London, March.

8.46 Straman, J. P. (1990) *Precast Concrete Cores and Shear Walls, Prefabrication of Concrete Structures.* International Seminar, Delft University of Technology, Delft University Press, Oct., pp. 41–54.

8.47 Bruggeling, A. S. G. & Huyghe, G. F. (1991) *Prefabrication with Concrete*. Balkema, Rotterdam, p. 380.


## Chapter 9

9.1 British Standards Institution (1985) *The Structural Use of Concrete*. BSI, London, BS 8110.

9.2 Mahdi, A. A. (1992) *Moment Rotation Effects on the Stability of Columns in Precast Concrete Structures*. PhD thesis, University of Nottingham.

9.3 *Guidelines for the Use of Structural Precast Concrete in Buildings*. (1991) Study Group of the New Zealand Concrete Society, and National Society of Earthquake Engineering, Christchurch, New Zealand. p. 174.

9.4 Engström, B. (1992) *Ductility of Tie Connections in Precast Structures*. PhD thesis, Chalmers University of Technology, Goteborg, Sweden.

9.5 COST C1 Proceedings of the First State of the Art Workshop (1992) *Semi-rigid Behaviour of Civil Engineering Structural Connections*. ENSAIS, Strasborg, Oct.

9.6 Astbury, N. F. *et al.* (1970) *Gas Explosions in Loadbearing Brick Structures*. British Ceramic Research Association, Report No. 68.

9.7 The Building Regulations 1976 (1978) No. 1676, HMSO, London.

9.8 Omar, W. (1990) *Diaphragm Action in Precast Concrete Floor Construction*. PhD thesis, University of Nottingham.

9.9 Fédération Internationale de la Précontrainte (1988) *FIP Recommendations. Precast Prestressed Hollow Core Floors*. FIP Commission on Prefabrication, Thomas Telford, London, p. 31.

9.10 Regan, P. E. (1974) *Catenary Tests on Composite Precast–Insitu Concrete Composite Floors*. Report to the Department of Environment, The Polytechnic of London.

9.11 Schultz, D., Burnett, E. & Fintel, M. (1978) *A Design Approach to General Structural Integrity – Design and Construction of Large Panel Concrete Structures*. Supplemental Report A, Portland Cement Association, Stokie, USA.

9.12 Hanson, N. & Burnett, E. (1977) *Wall Cantilever and Slab Suspension Tests, Design and Construction of Large Panel Concrete Structures*. Report 4, Portland Cement Association, Stokie, USA.

9.13 Wilford, M. J. C. & Yu, C. W. (1973) *Catenary Action in Damaged Structures*. Paper No. 6, Department of Environment and CIRIA, The Stability of Precast Concrete Structures, Seminar, London, March, pp. 1–24.

9.14 Odgard, A. (1975) *Connections Between Precast Floor Slabs*. CIB Symposium, Bearing Walls, Sept.

9.15 Engström, B. (1990) Connections Between Precast Components, Nordisk Betong. *Journal of the Nordic Concrete Federation*, No. 2–3, pp. 53–56.

9.16 Bruggeling, A. S. G. & Huyghe, G. F. (1991) *Prefabrication with Concrete*. Balkema, Rotterdam, p. 380.

9.17 Elliott, K. S. & Tovey, A. (1992) *Precast Concrete Frame Buildings – A Design Guide*. British Cement Association, Wexham Springs, Slough, May, p. 88.


## Chapter 10

10.1 British Standards Institution (1985) *The Structural Use of Concrete*. BSI, London, BS 8110.

10.2  McCaully, C. H. A. (1975) *Chain Tester's Handbook*, 2nd edn. Chain Tester's Association of Great Britain.

10.3  Adlparvar, M. R. (1993) *Torsional Behaviour of Precast Concrete Beams and Connections.* M Phil. thesis, University of Nottingham.

10.4  Kong, F. K. & Evans, R. H. (1987) *Reinforced and Prestressed Concrete*, 3rd edn. Van Nostrand, p. 508.

10.5  Precast Concrete Industry Training Association (1992) *Code of Practice for the Safe Erection of Precast Concrete.* Leicester, UK, p. 86.

10.6  Engineering Science Data Unit (1970) *Fluid Forces and Moments on Flat Plates.* ESDU Item No. 70015, London, Sept.

10.7  Engineering Science Data Unit (1971) *Fluid Forces, Pressures and Moments on Rectangular Blocks.* ESDU Item No. 71016, London, Sept.

10.8  Cook, N. (1985) *The Designer's Guide to Wind Loading of Building Structures.* Butterworth.

10.9  Institution of Structural Engineers (1978) *Structural Joints in Precast Concrete.* IStructE, London, Aug., p. 56.

10.10  Elliott, K. S. & Tovey, A. (1992) *Precast Concrete Frame Buildings – A Design Guide.* British Cement Association, Wexham Springs, Slough, May, p. 88.

10.11  British Standards Institution (1983) *Methods of Testing Concrete.* BSI, London, BS 1881.

# Index

abutment bearing, 297–8
accidental loading, 510
additional bending moments, 191
admixtures, 125–6
aggregate interlock, 448–51
alternative load path, 516–18
architectural components, 12, 19–21, 68–9, 94
array of tendons, 152

balconies, 25, 111–13, 118, 235
balcony ties, 534–5
bar bearing stress, 169
bar bending radius, 169
bar sizes, 126
base plates, 30, 409, 419–28
beam and block floor, 75
beam and plank floor, 75, 253–5
beam–beam connection, 97, 160, 387–96
beam–column connection, 24, 25, 95–97, 293–6, 336–87, 470
beam end shear design, 160–88
beams, 91–5, 130
beams on elastic foundation, 479
beam–slab connection, 23, 326–33
bearing lengths or widths, 132, 151, 170, 179, 304, 322, 326, 373, 380
bearing pad, 23, 137, 322
bearing plate, 164–5, 167–8
bearing stress, 167, 320–25, 344, 380
bearing types, 322
bent-up bars, 160
billet connector, 337, 340, 344, 346–52, 356–60
bitumen sealing strips, 128
bolts, 127, 312, 368, 401
bolts in sleeves, 368

boot, 91–2, 131, 136–41
boot reinforcement design, 138–41, 148–50, 156
Bourse, The, Leeds, 39, 430–31
boxes, 102, 497
braced structures, 23, 28, 122–3, 192, 465, 471–4,
bracing, 100–103, *see also* shear walls
brick infill walls, 482–4, 487–9
brickwork, 19, 482–9
bridging elements, 518
buckling in bars, 515
bursting force, 140, 167, 170, 348, 411
bursting force coefficients, 170–71
bursting reinforcement, 348, 350

camber, 229
Canary Wharf, London, 8
cantilevers, 84, 86, 87, 111–13, 235
cantilever shear walls, 478–9, 497, 509
  failure modes, 498
cantilever wall, 101–3
cast-in sockets, 366–7
castellated joints, 313–15, 413, 445, 499
catenary action, 523, 537–40
'Cazalay hanger', 188, 390
centroid of tendons, 153, 155–9
chamfer, 152
cladding, 8–10, 13
clear connector, 337–8, 341, 345
codes of practice, 61–3
coefficient of friction, 136, 168, 328
colouring pigments, 126
column–base connections, 409–29
column–floor splices, 408–9
column haunch, 379–84
column insert, 24, 96, 107, 109, 295, 339–69

columns, 97–100, 116–18, 188–98, 463–5, 468–9, 476–8
column splices, 24, 26, 396–409
  coupler, 397–400
  grouted sleeve, 399, 402, 403, 406–7
  steel shoe, 403–5
  welded lap, 399, 402–3
  welded plate, 398, 400–402, 405–6, 536
column ties, 523–4, 530–31
company organization, 45
component details, 56–8
composite beams, 283-92
composite construction, 258–92
composite floors, 271–83
composite floor slab,
  design, 248–53
  data, 251
  voided slab, 248, 250–51
composite plank floor, 75, 84–5, 247–53
composite beams in torsion, 143–4
compression field, 162, 172, 180
compression joint, 304–11, 396–7
concealed connection, 95–6
concept design, 33
concrete crushing, 304
concrete grade, 123, 259
concrete specification, 123, 579–81
connections, 31, 160, 293–429
  column–base, 409–29
  column haunch, 379–84
  column top, 384–7
  concealed, 95–6
  corbels, 369–78
  definitions, 301–2
  design criteria, 302–4
  development of, 293
  design brief, 296–301
  functional behaviour, 297
  maturity, 569
  moment resisting, 325
  multi-sided, 356–65
  pinned, 325
  single-sided, 344–56
  slab–beam, 23, 326–33
  temporary, 568–9
  testing, 299–300, 347, 349
  tolerances, 301
  torsion, 325, 516

  wall–base, 505–6
  wall–beam, 493
  wall–column, 495
  wall–wall, 490, 505–9
confinement links, 350–52, 360, 364
connectors
  billet, 337, 340, 344, 346–52, 356–60
  cleat, 337–8, 341, 345
  narrow plate, 167, 185–8, 365–6
  sliding plate, 338, 341
  welded plate, 338, 341
construction, 36–42, 561–82
construction traffic allowance, 273–4
contact breadth, 265
continuity ties, *see* stability ties
continuous framework, 22
continuous joints, 333
corbel, 24, 96, 294, 369–78
cost exercises, 77, 100, 282–3
coupler splice, 297–300
coupling bars, 446–8, 495
cover, 132, 145, 165, 169
crack inclination, 164
crack inspection, 556–7
crack width, 449–50, 556–60
cranage, 119
creep loss, 155, 227
creep strain, 126
cube penetration, 580
cube strengths, 123, 259

D-regions, 372
deep boot in beam, 136, 138
deep corbel, 376–8
definitions, 65–7, 301
deflection, 146
deformability of joints, 307–11, 317–18
design calculations, 48, 50
design documents, 63–4
design philosophy, 120
design stages, 48
design stresses, 123
detensioning strength, 123
development length, 232–4
diaphragm action, 25, 28, 29, 441–62
  depth of floor, 452
  details, 443–4
  finite elements, 460–62

moments, 451–4, 459
reinforcement, 442, 445, 448, 451–4, 457
shear stress, 454–60
structural models, 445–51
tests, 454–60
differential shrinkage, 265–70
dimensional deviation, 128
direct method for structural integrity, 518
displaced bearing, 515, 540, 545
distributed moments, 190, 196
distribution of horizontal loads, 431–41
double-tee slabs
bearings, 328–31
design, 241–6
floor, 76, 81–4, 240–46, 273
hoisting, 551, 554–6
longitudinal joints, 332
manufacture, 242–4
shear cage, 244
welded connections, 245
dowel action, 314–15, 448–50
dropped boot beam, 91, 92
dry pack, 322, 385
ductile frame design, 510–15
ductility factor, 510–11

eccentricity, 152, 189, 306, 348, 350, 356
edge beams, 91–5, 131
effective breadth, 283–5, 279–80
effective lengths, 189–95, 467, 471, 475–7, 484
effective span, 128
effective thickness, 465, 494
elastic shortening loss, 155
elastomeric bearings, 128, 322
epoxy fusion bonded reinforcement, 126
epoxy mortars, 128
erection methods, 37, 41, 60, 550–55, 561–82
expanding agents, 126
expansion joints, 473
extended base plate, 420–28
extended bearing, 430
extrusion, 73, 209–12

factory communication, 55
'FIELDS', 162, 372

finishes, 71, 84, 87
finite elements, 160, 164, 234, 372, 460–62
fire rating, 78, 82, 84
fixing gang, 570
fixing rates, 39
fixings, 95, 100
flexicore, 73
flexural design, 134, 146–7, 150–60, 224–9, 273–82, 287–92
flexural joint, 319–20
flexural tension, 151
floor plate action, 25, 28, 432–65
flooring options, 206–10
floors, 5, 17, 28–9, 51–3, 72–85, 206–57
beam and block, 75
beam and plank, 75, 253–5
composite design, 271–83
composite plank, 75, 84–5, 247–53
diaphragm action, 25, 28–9, 441–62
double tee, 76, 81–4, 240–46, 273, 328–31, 551–6
hollow core, 73–81, 207–41, 329–30, 462, 552–3
flush base plate, 420, 426–7
formwork, 579
foundations, 30–32, 409–29
foundations for props, 570–71
frame analysis, 20, 21
frame design exercise, 111–19
frame erection procedure, 570–82
framework, 22
framing plan, 34, 71
friction, 136, 168, 410–11
friction restraint steel, 168–9, 173, 176, 178
functional system, 9, 34

gable ties, 527–9
grade of concrete, 123
Grand Island, Manchester, 86
grouted sleeve column splice, 25, 294, 399, 402–3, 406–7
grouted sleeves, 31, 32, 312, 334, 385, 428–9
grouting, 581

H-frame, 429–30
hand over, 581–2
handling, 556–8

halving joint, 23, 162–73, 294
hanger stirrups, 140, 173, 389–91
haunch, 379–84
head plates to props, 561
'heavy' period, 2
height limitations, 36
hierarchical building, 9
helical strand, 127
hidden connections, 24, 337
Highbury College, Porstmouth, 3, 6
history, 1
hoisting, 551–6, *see also* lifting
holding down bolt, 420, 506
hollow core cantilever shear walls, 500–503
    design, 501–2
    profile, 500
hollow core infill walls, 491–7
    design method, 492–6
    edge profile, 493
hollow core floor slab, 8, 11, 73–81
    bearings, 329–30
    cantilever, 234–6
    catenary action, 538–40
    cross-section, 211–13, 215–18
    data, 239–41
    design, 214–33, 237–9, 255–7
    development length, 232–4
    edge profile, 214, 216, 443, 454–5
    lateral load distribution, 220–24
    lifting, 552–3
    load vs span data, 217
    longitudinal joints, 532
    manufacture, 208–13
    propped slab, 281–2
    reinforcement, 218–19
    shrinkage, 462
    spalling stress, 219, 235
    transmission length, 216
    wet cast, 236–7
hollow core wall, 103, 491–7
horizontal loading, 430–509
    distribution, 431–41
    parapet, 142
horizontal shear, 218, 281, 288, 446, 454–60
hybrid, 70, 103–11

indirect tie method, 517, 519
infill shear wall, 101, 475–97

brickwork wall, 482–4, 487–9
concrete walls, 481, 483–5
hollow core walls, 491–7
structural analysis, 481–5, 487–8
without framing, 489–92
initial design, 43
insert design, 339–66
inserts, 95, 100, 337–66
insitu concrete, 579–81
integrity, 26
interface shear, 263–5, 281, 288
inverted tee beams, 91, 150–60, 205
isolated joints, 333
isolated members, 128

joints, 31
    compression, 304–11, 396–7
    definitions, 301
    deformability, 307–11
    design criteria, 302–4
    flexural, 319–20
    shear, 313–19
    stiffness, 319, 512–13
    tensile, 311–13, 397–405
    tests, 309

key elements, 516–18
keyed joints, 313–15, 413, 445

L-beams, 91–5, *see also* edge beams
landing joints, 201–3
landings, 85–90
lapped joint, 311, 314, 320–21, 334
layout drawings, 48, 53
ledger beams, 131–50
lever arm, 134, 139, 410, 445, 451, 468
lifting design, 197, 544–56
lifting devices, 542, 546, 548–9, 553, 555
lifting points, 542, 546–56
lifting strengths, 123, 544–5
lifting weights, 119
'lighter' period, 2
links, 135–40, 148, 152, 156, 348, 350–52, 360, 364
literature, 65
load vs span data, 133, 217
London Docklands, 3, 8
losses in prestressed concrete, 151, 227

market data, 15, 18
materials specification, 122
Merchant House, London, 3, 7
mesh reinforcement, 29, 126, 272
MGM, Las Vegas, USA, 3, 8
minimum section sizes, 128, 132, 151
mix design, 124
mix proportions, 125
mixed construction, *see* hybrid
modular coordination, 7, 9
modular grid, 7
modularization, 7
modular ratio, 268, 274, 276, 277
modulus of rupture, 162
moment connection, 297, 325–6
moment of resistance, 134–5, 155–7, 159,
    228, 274–6, 278–82, 287, 414, 421–
    3
moment resisting pockets, 410–19
mortar joints, 24, 305–6, 322

narrow plate, 167, 185–8, 365–6
National Building Frame, 2, 4–6
neoprene, 127, 139, 322, 328, 379
neutral axis, 134, 152
nib, 96
non-cementitious materials, 127
non-composite beam design, 131–60
non-isolated members, 128
non-symmetrical loading, 543, 559
normal links, 147
Northgate House, Halifax, 69

office practice, 46
openings in beams, 150, 152
openings in slabs, 5, 78–81, 84
openings in walls, 499
organization, 45
Orpington, Kent, 99

parapet loads, 142
partially braced structure, 28, 122, 193, 465,
    474–6
patch loading, 356
patents, 293
pigments, 126
pinned base plate, 424
pinned connection, 325

pinned joint, 21
pitching design, 197, 549–51
pitching points, 550–51
pocket depth, 412–13
pocketed beam end, 160–61
pockets, 30, 409–19
polymers, 127
polystyrene, 128
polysulphide sealants, 128
portal frame, 12, 14
pozzolana, 125, 126
precamber, 229
precast company organization, 45
precast-insitu comparison, 37
precast pockets, 415–16
prefabricated shear box, 160, 179–88, 497
prestressed design
    beams, 150–60
    slabs, 224–34
    stairflight, 199–200
    walls, 496
prestressing force, 152–3
prestressing losses, 151, 155
prestressing steel, 126
primary columns, 71
procurement, 44–6
programming, 38
progressive collapse, 515-17
propped cantilever, 193
propping, 276, 281–2, 288, 561–8, 571–
    5, 577, 581
protected members, 516–18
Public Building Frame, 2
punching shear, 139
push-pull props, 575

quotations, 48

RAE, Rome, 40, 298
re-alignment methods, 574
rebar bursting stress, 169
rebar bending stress, 169
recessed beam end, 160–79
retarding agents, 126
reinforcement specification, 126
reinforcement volumes, 129–31
relaxation loss, 127, 155, 227
reverse bending, 300

robustness, 26, 510–40
rolled steel sections, 127, 179, 340, 346–52, 356, 360–65
Ronan Point, 515–17
roughness factor, 454–6

scarf joint, 201–5
schedules, 48, 49, 52, 56, 58
scope of project, 44
second-order deflections, 191, 477
secondary columns, 71
section modulus, 153, 274, 279, 291
semi-rigid connections, 22
sensitivity exercise, 129, 228–9
service holes, 5, 78–81, 84, 150, 152, 499
serviceability moment of resistance, 154–5, 225–8, 273–5
shallow beam root, 136
shallow corbel, 369–76
shear boxes, 160, 179–88, 497
shear cage, 165–79
shear centre, 143, 432–41
shear cores, 465
shear deformability, 317–18
shear design, 147, 156, 229–33, 278, 280–81, 390
shear friction, 313–15, 444, 448
shear joints, 313–19, 480
shear keys, 313–15, 413, 445
shear span ratio, 371–3
shear stiffness, 316–19, 448–50, 456, 511
shear stirrups, 135–6
shear stresses, 136, 372–3, 377, 440–41, 454, 456, 480, 485, 488, 495
shear tests, 164, 166, 232, 318, 347, 454–60
shear transfer mechanism, 449
shear walls, 100–103, 432, 465, 472, 475–6, 478–509
shelf angle, 202, 204
shrinkage
  cracks, 449
  induced deflection, 269
  prestressing loss, 155, 227
  strain, 126, 265–70, 445, 462–5
simple supports, 129
site communication, 55
site practice, 36–42, 541
site preparation, 570

skeletal structure, 12, 14, 70
slab–beam connection, 23, 326–33
slab bearing, 23, 137, 143, 209, 230–31, 322–5, 328–36
slabs, 72–85, *see also* floors
  basic properties, 76
  economic comparison, 77
slab–wall connections, 333–6
slender walls, 480, 484–7
slenderness, 141, 145, 191
sliding plate connector, 338, 341
slip forming, 73, 209–12
slump, 208
solid cantilever-shear walls, 503–9
  design method, 504
  connections, 504, 508–9
spalling, 323–4, 346, 382, 514, 545
span vs load data, 123, 217
spandrel beams, 91–3, 467, 470–71
splice, 26, 294, 396–409, 505–9
spreader lifting beams, 547
stability, 26, 27, 430–509
stability ties, 24, 27, 28, 119, 516, 520–37
stacking, 556–8
stainless steel, 126
staircase joints, 201–3
staircases, 85–90, 198–205
  lifting, 555
  stringer, 200
standard deviation, 122
standardization, 9
starter bars, 421, 500
steel bearing plate, 167, 170, 175, 177–8, 295, 323, 377, 380, 383
steel shoe column splice, 403–5
steel specification, 127
steelwork to hollow core connection, 110
stiffness
  flexural, 319, 469, 511
  joint, 319, 512–13
  shear, 316–19, 448–50, 456, 511
stiffness parameter, 482
strand, 127, 218–19, 446, 522
strength of concrete, 123
stringer staircase design, 200
structural integrity, 510
structural screeds, 272–3, 283, 442, 460–62, 526, 580

structural steelwork, 104–10, 127, 179
structural zone, 25
strut-and-tie, 137–8, 162, 168–73, 338,
    372–5, 389
substructures, 121
super strand, 127
surface roughness, 261–3, 454–6
surface textures, 259–63
Surrey Docks, London, 3, 7

technical systems, 9
temporary stability, 27, 541, 561–79
    beams, 575–6
    columns, 571–5
    connections, 568–70
    floor diaphragm, 564
    shear walls, 565–7
    slabs, 577–9
    sway modes, 563
    temporary loads, 567–8
    walls, 577–9
tender drawings, 49
tendons, 150
tensile joints, 311–13
tensile stress, 151, 225–6
terminology, 127
Termodeck, 77
thermal movement, 462–5
thin plate connector, 339
tie forces, 526–37
tie steel, 24, 29, 444, *see also* stability ties
timber roof, 70, 108
tolerances, 165, 301, 323, 356, 544
torsion, 142–3, 221
Training Associations, 37
transfer stress, 153–5
transfer strength, 151
transformed sections, 274, 276
transportation, 59
transmission length, 233–4
trimmer angles, 79, 82
twin facade, 90

ultimate moment of resistance, 134–5, 156,
    228, 275–6
unbraced structure, 22, 28, 122, 192, 465–71

unidirectionally braced structures, 474
upstand, 91–2, 131, 141–4

Vauxhall Bridge, London, 9
vehicle impact, 142
vertical load transfer, 465–78
vertical stability ties, 520, 523, 534–7
Virendeel action, 245, 445
visual concrete, 68–9
voided composite slab, 248, 250–51
voids in components
    beams, 150, 152
    slabs, 5, 78–81, 84
    walls, 499
volume of concrete, 13
volumetric changes, 462–5

wall connections, 333–6, 505, 508–9
wall design exercise, 112–13
wall frame, 11, 13
walls, 100–103, 432, 465, 472, 475–6, 478–
    509
wall shoe connector, 505
wall splice plate, 509
Weaver's Mill, 1
wedging action, 448
welded inserts, 360–61, 365
welded joints, 312–14, 334, 401, 470
welded lap column splice, 399, 402–3
welded plate column splice, 398, 400–402,
    405–6, 534
welded plate connector, 338, 341
welded reinforcement, 300, 352–4
Western House, Swindon, 3, 6
wet cast hollow core slab, 234–6
wind posts, 123
wind loads, 25, 113, 430, 440
wire, 126

Young's modulus of brickwork, 488
Young's modulus of concrete, 124, 259
    effective, 307, 310
Young's modulus of infill, 482
Young's modulus of steel, 127

zero slump, 208